# Nitrate Handbook

# ADVANCES IN TRACE ELEMENTS IN THE ENVIRONMENT

**Series Editor: H. Magdi Selim**
Louisiana State University, Baton Rouge, USA

**Trace Elements in Waterlogged Soils and Sediments**
edited by Jörg Rinklebe, Anna Sophia Knox, Michael Paller

**Phosphate in Soils: Interaction with Micronutrients, Radionuclides and Heavy Metals**
edited by H. Magdi Selim

**Permeable Reactive Barrier: Sustainable Groundwater Remediation**
edited by Ravi Naidu and Volker Birke

**Nitrate Handbook Environmental, Agricultural, and Health Effects**
edited by Christos Tsadilas

# Nitrate Handbook

## Environmental, Agricultural, and Health Effects

Edited by
Christos Tsadilas

CRC Press
Taylor & Francis Group
Boca Raton London New York

CRC Press is an imprint of the
Taylor & Francis Group, an **informa** business

First edition published 2022
by CRC Press
6000 Broken Sound Parkway NW, Suite 300, Boca Raton, FL 33487-2742

and by CRC Press
2 Park Square, Milton Park, Abingdon, Oxon OX14 4RN

© 2022 Taylor & Francis Group, LLC

CRC Press is an imprint of Taylor & Francis Group, LLC

*Library of Congress Cataloging-in-Publication Data*
Names: Tsadilas, Christos, editor.
Title: Nitrate handbook : environmental, agricultural, and
health effects / edited by Christos Tsadilas.
Description: First edition. | Boca Raton : CRC Press, 2022. |
Series: Advances in trace elements in the environment, 2475-6253 |
Includes bibliographical references and index.
Identifiers: LCCN 2021020231 (print) | LCCN 2021020232 (ebook) |
ISBN 9780367338220 (hardback) | ISBN 9780429326806 (ebook)
Subjects: LCSH: Nitrates. | Nitrates–Environmental aspects. |
Nitrates–Health aspects. | Plants–Effect of nitrates on.
Classification: LCC QD181.N1 N55 2022 (print) |
LCC QD181.N1 (ebook) | DDC 546/.71124–dc23
LC record available at https://lccn.loc.gov/2021020231
LC ebook record available at https://lccn.loc.gov/2021020232

ISBN: 978-0-367-33822-0 (hbk)
ISBN: 978-1-032-11805-5 (pbk)
ISBN: 978-0-429-32680-6 (ebk)

DOI: 10.1201/9780429326806

Typeset in Times
by Newgen Publishing UK

The Open Access version of chapter 15 was funded by Sarah Fossen Johnson.

*This book is dedicated to the memory of the recently deceased Professor Magdi H. Selim, Editor of the series* "Advances in Soil Science" *by CRC Press, Taylor & Francis Group, an important scientist and a kind man who taught modesty and solidarity.*

# Contents

Preface......................................................................................................xi
Editor......................................................................................................xiii
List of Contributors..............................................................................xv
List of Abbreviations..........................................................................xix

## PART 1    Nitrate in Soils

**Chapter 1**    Biogeochemistry of Nitrate in Soils.....................................................3

                  *Roberto García-Ruiz, Milagros Torrús-Castillo,*
                  *Pablo Domouso de Agar, Roberto García-Perán,*
                  *Gustavo Ruiz-Cátedra, and Julio Calero*

**Chapter 2**    Nitrate in Agricultural Soils ...............................................................25

                  *Mélida Gutiérrez, Maria Teresa Alarcón-Herrera,*
                  *Esperanza Y. Calleros-Rincón, and Melissa Bledsoe*

**Chapter 3**    Nitrate Transport in Agricultural Systems..........................................45

                  *Sara Vero and Matthew Ascott*

## PART 2    Nitrate in Plants

**Chapter 4**    Nitrate in Plant Physiology.................................................................73

                  *Jitu Chauhan, Simran Chachad, Zoya Shaikh,*
                  *Johra Khan, and Ahmad Ali*

**Chapter 5**    Nitrate in Plant Nutrition....................................................................91

                  *Ya-Yun Wang*

**Chapter 6**    Nitrogen Fertilizers and the Environment .......................................103

                  *Eric Walling and Celine Vaneeckhaute*

**Chapter 7**    Nitrate Management by Using Innovative Techniques....................137

                  *Eleftherios Evangelou and Christos Tsadilas*

**Chapter 8**    Estimation of Nitrogen Status in Plants ........................................... 163

*Miguel Garcia-Servin, Luis Miguel Contreras-Medina,*
*Irineo Torres-Pacheco, and Ramón Gerardo Guevara-González*


## *PART 3    Nitrate in Water*

**Chapter 9**    Nitrate in Fresh Waters ..................................................................... 185

*Daniel F. Gomez Isaza and Essie M. Rodgers*


**Chapter 10**   Inorganic Nitrogen Ions of Surface and Ground Waters
of Sterea Hellas, Central Greece: A Case Study ............................. 209

*Christos Tsadilas, Miltiadis Tziouvalekas,*
*Eleftherios Evangelou, Alexandros Tsitouras,*
*Christos Petsoulas, and Antonios Peppas*


**Chapter 11**   Determination of Nitrate in Waters ................................................. 235

*Md Eshrat E. Alahi, Fahmida Wazed Tina, and*
*Subhas Mukhopadhyay*


## *PART 4    Nitrate in the Food Chain*

**Chapter 12**   Nitrate in Plant and Animal Foods .................................................. 265

*Małgorzata Karwowska*


**Chapter 13**   Nitrate as Food Additives: Reactivity, Occurrence,
and Regulation .................................................................................. 281

*Teresa D'Amore, Aurelia Di Taranto, Giovanna Berardi,*
*Valeria Vita, and Marco Iammarino*


## *PART 5    Nitrate in Human Body*

**Chapter 14**   Nitrate and Human Health: An Overview ........................................ 303

*Keith R. Martin and Richard J. Bloomer*

**Chapter 15** Nitrates and Methemoglobinemia .....................................................347

*Sarah Fossen Johnson*

**Chapter 16** Inorganic Nitrate and Nitrite: Dietary Nutrients or Poisons?...........357

*Nathan S. Bryan*

**Chapter 17** Nitrate and Cardiovascular Systems.................................................375

*Andrew Xanthopoulos, Apostolos Dimos, Alexandros Zagouras, Nikolaos Iakovis, John Skoularigis, and Filippos Triposkiadis*

## PART 6   *Regulations*

**Chapter 18** Regulations on Nitrate Use and Management ..................................405

*Fayçal Bouraoui, Panos Panagos, Anna Malagó, Alberto Pistocchi, and Christophe Didion*

**Index**.................................................................................................................425

# Preface

Nitrate, the most important soil nitrogen form for plants, has attracted increased attention in recent decades due to its significance both for crop production and the environmental impact. Furthermore, through the food chain, nitrate may enter the human body where, after peculiar transformations, it may affect human health. However, a debate exists on the possible influences of nitrate – that is, whether they are adverse or beneficial to human health. The dominant aspect is that nitrate is responsible for a number of serious diseases, such as methemoglobinemia in infants and specific cancers. However, at the same time, several scientists claim that nitrate is not responsible for these diseases but, in contrast, that nitrate favors the formation of NO in the human body, which has beneficial effects on blood pressure and on cardiovascular disease factors. We believe that this topic is well illuminated in the respective chapter of this book. Special regulations on the nitrate content of drinking water were adopted by WHO and EU to protect human health.

Since N is the one of the most significant essential nutrients for plant growth, seriously affecting global plant production – and the amount of nitrate applied every year to the soils are huge at a global scale – clarification of the issues related to the role of nitrate in the environment, agriculture, and human health is important. The present book aspires to offer a comprehensive update on all aspects related to nitrate paths from atmosphere, soil, water, and plants to the food chain and, through them, to the human body with possible consequences.

Considering the continuous increase of the human population and the subsequent increase of food demands, the expected amounts of N needed for meeting the crop requirements will substantially increase. N management must be a high priority, while any practice or innovative technique developed so far must be used to increase nitrogen use efficiency and reduce nitrate losses in the environment.

The book is divided into six parts. Part 1, containing three chapters, provides an in-depth update of the basic biological, chemical, and physical conditions controlling the behavior of nitrate, what are nitrate, where they exist, and how they behave (Chapter 1). Further, a focus is made on the existence of nitrate and behavior, especially in agricultural soils, where nitrate abundance and activity is very high (Chapter 2). Finally, the mobility of nitrate, which is crucial for its easy transport to the environment, is an issue that must be seriously taken into account to avoid environmental pollution.

Part 2, which contains five chapters, is devoted to nitrate involvement in plant physiology (Chapter 4), the estimation of plant nutrition, with nitrate's special role (Chapter 5), as well as the role of nitrogen fertilizers in plant production and their effects on the environment (Chapter 6). A special mention is made on the new innovative technologies developed in nitrate management, especially through the variable rate of its application (Chapter 7).

In Part 3, a general description of nitrate's existence in fresh waters from the various sources – rivers, lakes, and groundwaters – is presented (Chapter 9). The main results of a case study from an intensively cultivated area of a Mediterranean country focused on nitrate and other inorganic nitrogen ions, such as nitrite and ammonium, are presented, explaining the reasons for water contamination with inorganic nitrogen ions (Chapter 10). A detailed description of the methods used for determination of nitrate in waters is also given (Chapter 11).

Part 4 presents the nitrate content in animal and plant foods, showing the main ways nitrate follow in their transfer into human body (Chapter 12) as well as the use of nitrate as food protective agents is clearly shown, revealing an interesting use of this substance (Chapter 13).

Part 5 focuses on clearly medical issues with respect to nitrate impacts in human health. In Chapter 14 a general overview of the nitrate relationship with human health is given. Chapter 15 analyses in detail the possible relationship of nitrate and the methemoglobinemia in babies fed with water rich in nitrate. Chapter 16 tries to answer the question of whether nitrate and nitrite are dietary nutrients or poisons, explaining very well how exactly these two ions are involved in human health. Chapter 17 focuses especially on the nitrate influence on the human cardiovascular system mechanisms.

Part 6 (Chapter 18) gives a detailed description of the regulations and legislation established mainly in the European Union for nitrate use and management.

It is believed that this handbook adds important new information from recent research on a very important issue that affects the environment, agricultural production and human health in a variety of ways. From this position, I would like to express my sincere thanks to the many authors for their very valuable contributions, which have made this book possible, and which, I hope will provide a good assistance to researchers, teachers, and students. Finally, warm acknowledgments for the editorial staff and, especially, to Mrs Irma Britton and Rebecca Pringle and Kawiya Bakthavatchalam for their irreplaceable help without of which publication of this book would not have been possible.

# Editor

**Christos Tsadilas, PhD,** graduated from the Agricultural University of Athens in 1975 and earned his master's degree and PhD from the same university. From 1977 to 1984, he worked at the Institute of Soil Mapping and Classification of the Ministry of Rural Development and Foods as a soil surveyor and has been a senior scientist of the National Project of Soil Survey Mapping in Greece. Later, he worked as a researcher at the National Agricultural Research Foundation in Greece, which merged with the current Hellenic Agricultural Organization, DEMETER. He has taken post-doc studies at the University of Reading (United Kingdom), University of Kentucky, and University of Lincoln (Nebraska), sponsored by the Royal Society of England, OECD, and the Fulbright Foundation, respectively. His main research interests include soil survey and classification, soil chemistry and fertility, soil pollution and remediation, and waste management. He also has work experience on reclamation of disturbed lands and the new technologies used in precision farming systems.

Dr. Tsadilas served as director of the Institute of Soil Mapping and Classification (later Institute of Industrial and Forage Crops) which, along with the Agricultural University of Athens, was appointed by the government to make the Soil Map of Greece. He also served as president of the Hellenic Soil Science Society for several years. Dr. Tsadilas has published more than three hundred papers in more than ninety peer-reviewed journals, more than 15 chapters in books, and papers in international conference proceedings. Particular reference is made in the books *The Soils of Greece* (Yassoglou, N., C. Tsadilas, and C. Kosmas, 2017, Springer) and *Nickel in Soils and Plants* (C.D. Tsadilas, J. Rinklebe, and H. Magdi Selim[+] by CRC Press, Taylor & Francis Group). He has served as an associate editor for several journals, and as a reviewer for decades of international journals.

# Contributors

**Md Eshrat E. Alahi**
The School of Engineering, Macquarie University, Sydney, Australia

**Maria Teresa Alarcón-Herrera**
Department of Geography, Geology and Planning, Missouri State University, Springfield, Missouri, USA

**Ahmad Ali**
Department of Life Sciences, University of Mumbai, Mumbai, Maharashtra, India

**Matthew Ascott**
British Geological Survey, Crowmarsh Gifford, Wallingford, Oxfordshire, UK

**Giovanna Berardi**
Instituto Zooprofilattico Sperimentale della Puglia e della Basilicata, Foggia, Italy

**Melissa Bledsoe**
Department of Geography, Geology and Planning, Missouri State University, Springfield, Missouri, USA

**Richard J. Bloomer**
Center for Nutraceutical and Dietary Supplement Research, College of Health Sciences University of Memphis, Memphis, Tennessee, USA

**Fayçal Bouraoui**
European Commission, Joint Research Centre, Ispra, Italy

**Nathan S. Bryan**
Baylor College of Medicine, Department of Molecular and Human Genetics, Houston, Texas, USA

**Julio Calero**
Department of Geology, University of Jaén, Spain

**Esperanza Y. Calleros-Rincón**
Department of Geography, Geology and Planning, Missouri State University, Springfield, Missouri, USA

**Simran Chachad**
Department of Life Sciences, University of Mumbai, Mumbai, Maharashtra, India

**Jitu Chauhan**
Department of Life Sciences, University of Mumbai, Mumbai, Maharashtra, India

**Luis Miguel Contreras-Medina**
Autonomous University of Queretaro, Queretaro, QUE, Mexico

**Teresa D'Amore**
Istituto Zooprofilattico Sperimentale della Puglia e della Basilicata, Foggia, Italy

**Pablo de Agar**
Domouso Department of Plant and Animal Biology and Ecology, University of Jaén, Jaén, Spain

**Christophe Didion**
European Commission, DG Environment, Brussels, Belgium

**Apostolos Dimos**
Department of Cardiology, Larissa University General Hospital, Larissa, Greece

**Aurelia Di Taranto**
Instituto Zooprofilattico Sperimentale della Puglia e della Basilicata, Foggia, Italy

**Eleftherios Evangelou**
Hellenic Agricultural Organization
DEMETER, Institute of Industrial
and Forage Crops, Larissa, Greece

**Sarah Fossen Johnson**
Minnesota Department of Health,
St. Paul, Minnesota, USA

**Roberto García-Perán**
Hamburg University of Technology,
Germany

**Roberto García-Ruiz**
University Institute of Research in Olive
Groves and Olive Oils, University of
Jaén, Jaén, Spain

**Miguel Garcia-Servin**
Autonomous University of Queretaro,
QUE, Mexico

**Ramón Gerardo Guevara-González**
Faculty of Engineering, Autonomous
University of Queretaro, QUE, Mexico

**Mélida Gutiérrez**
Department of Geography,
Geology and Planning, Missouri
State University, Springfield,
Missouri, USA

**Nikolaos Iakovis**
Department of Cardiology, Larissa
University General Hospital,
Larissa, Greece

**Marco Iammarino**
Instituto Zooprofilattico Sperimentale
della Puglia e della Basilicata,
Foggia, Italy

**Daniel F. Gomez Isaza**
University of Queensland, Brisbane,
QLD, Australia

**Małgorzata Karwowska**
Department of Meat Technology
and Food Quality, University of Life
Sciences, Lublin, Poland

**Johra Khan**
Department of Medical Laboratory
Sciences, College of Applied Medical
Sciences, Majmaah University,
Majmaah, Saudi Arabia

**Anna Malagó**
European Commission, Joint Research
Centre, Ispra, Italy

**Keith R. Martin**
Center for Nutraceutical and Dietary
Supplement Research College of Health
Sciences, University of Memphis,
Memphis, Tennessee, USA

**Subhas Mukhopadhyay**
Department of Engineering, Macquarie
University, Sydney, Australia

**Panos Panagos**
European Commission, Joint Research
Centre, Ispra, Italy

**Antonios Peppas**
ETME Peppas and Associates,
Athens, Greece

**Christos Petsoulas**
Hellenic Agricultural Organization
DEMETER, Institute of Industrial and
Forage Crops, Larissa, Greece

**Alberto Pistocchi**
European Commission, Joint Research
Centre, Ispra, Italy

**Essie M. Rodgers**
School of Biological Sciences,
University of Canterbury, New Zealand

**Gustavo Ruiz-Cátedra**
Department of Plant and Animal
Biology and Ecology, University of
Jaén, Jaén, Spain

**Zoya Shaikh**
Department of Life Sciences, University
of Mumbai, Vidyanagari, Mumbai,
Maharashtra, India

**John Skoularigis**
Department of Cardiology, Larissa
University General Hospital,
Larissa, Greece

**Fahmida Wazed Tina**
Nakhon Si Thammarat Rajabhat
University, Tha Ngio, Thailand

**Miltiadis Tziouvalekas**
Hellenic Agricultural Organization
DEMETER, Institute of Industrial and
Forage Crops, Larissa, Greece

**Irineo Torres-Pacheco**
Autonomous University of Queretaro,
QUE, Mexico

**Milagros Torrús-Castillo**
Department of Plant and Animal
Biology and Ecology, University of
Jaén, Jaén, Spain

**Filippos Triposkiadis**
Department of Cardiology, Larissa
University General Hospital,
Larissa, Greece

**Christos Tsadilas**
Hellenic Agricultural Organization
DEMETER, Institute of Industrial and
Forage Crops, Larissa, Greece

**Alexandros Tsitouras**
Hellenic Agricultural Organization
DEMETER, Institute of Industrial and
Forage Crops, Larissa, Greece

**Celine Vaneeckhaute**
Centre de recherche sur l'eau, Université
Laval, Québec, Canada

**Sara Vero**
Agricultural Catchments Programme,
Teagasc, Environmental Research
Centre, Johnstown Castle, Co.,
Wexford, Ireland

**Valeria Vita**
Instituto Zooprofilattico Sperimentale
della Puglia e della Basilicata,
Foggia, Italy

**Andrew Xanthopoulos**
Department of Cardiology, Larissa
University General Hospital,
Larissa, Greece

**Eric Walling**
Research team on green process
engineering and biorefineries,
Chemical Engineering Department,
Université Laval, Québec, Canada

**Ya-Yun Wang**
National Taiwan University, Institute
of Plant Biology, Department of Life
Science, Taiwan

**Tina Fahmida Wazed**
Faculty of Science and Technology,
Nakhon Si Thammarat Rajabhat
University,
Tha Ngio, Thailand

**Alexandros Zagouras**
Department of Cardiology, Larissa
University General Hospital,
Larissa, Greece

# Abbreviations

| | |
|---|---|
| **ADI** | Acceptable Daily Intake |
| **AI** | Adequate Intake |
| **ANN** | Artificial Neutral Network |
| **ANS** | Additives and Nutrient Sources |
| **BOD** | Biological Oxygen Demand |
| **CAC** | Codex Alimentarius Commission |
| **CAP** | Common Agricultural Policy |
| **CFR** | Code of Federal Regulations |
| **CIRE** | Chlorophyl Index Red Edge |
| **CLCs** | Chlorate Channels |
| **COPD** | Chronic Obstructive Pulmonary Disease |
| **DAP** | Days After Planting |
| **DASH** | Dietary Approach to Stop Hypertension |
| **DNRA** | Dissimilatory Nitrate Reduction to Ammonium |
| **DRI** | Dietary Reference Intake |
| **EAR** | Estimated Average Intake |
| **EC** | Electrical Conductivity, European Commission |
| **EFSA** | European Food Safety Authority |
| **EIS** | Electrochemical Impedance Spectroscopy |
| **EPA** | Environmental Protection Agency |
| **EU** | European Union |
| **FAD** | Flavin Adenine Dinucleotide |
| **FA** | Food and Agriculture Organization |
| **FATIMA** | Farming Tools for External Nutrient Inputs and Water Management |
| **FDA** | Food and Drug Administration |
| **FMD** | Flow-mediated Dilation |
| **FNT** | Formate Nitrite Transporters |
| **FSANZ** | Food Standards, Australia and New Zealand |
| **HATS** | High Affinity Transport System |
| **HBM** | Hemoglobin M Disease |
| **GC** | Gas Chromatography |
| **GCI** | Green Chlorophyl Index |
| **GDH** | Glutamate Dehydrogenase |
| **GSFA** | General Standard for Food Additives |
| **GS-GOGAT** | Glutamine Synthesase-Glutamine-2-Oxoglutarate Amino Transferase |
| **IARC** | International Agency for Research on Cancer |
| **IFA** | International Fertilizer Association |
| **IPCC** | International Panel on Climate Change |
| **ISE** | Ion Selective Electrodes |
| **JECFA** | Joint Expert Committee on Food Additives |
| **LATS** | Low Affinity Transport System |
| **LD$_{50}$** | Lethal Dose |

| | |
|---|---|
| **LHT** | Lysine/Histamine/Transporters |
| **LOT** | Limit of Detection |
| **MPL** | Maximum Permitted Levels |
| **ND** | Nitrates Directive |
| **NDRE** | Normalized Difference Red Edge |
| **NDVI** | Normalized Difference Vegetation Index |
| **NO** | Nitrogen Oxide |
| **NOS** | Nitrogen Oxide Synthase |
| **NOAEL** | No Observed Adverse Effect Level |
| **NOCl** | Nitrosyl Chloride |
| **NOEC** | No Observed Effect Concentrations |
| **NO-MC** | Nitroso-Myochromogen |
| **NPF** | Nitrate Transport Family |
| **NR** | Nitrate Reductase |
| **NTC** | Nutrient Transport Continuum |
| **NUE** | Nitrogen Use Efficiency |
| **NVZ** | Nitrate Vulnerable Zones |
| **PESA** | Planar Electronic Sensor Array |
| **PVC** | Polyvinyl Chloride |
| **RDA** | Recommended Dietary Allowance |
| **RFD** | Reference Dose |
| **SAVI** | Soil Adjusted Vegetation Index |
| **SCF** | Scientific Committee for Food |
| **SINAR** | Stressed Induced Nitrate Allocation in Root |
| **SLAC** | Slow Anion Channel |
| **SWOT** | Strengths, Weaknesses, Opportunities and Threats analysis |
| **THTDPCI** | Trihexyltetradecylphosphonium Chloride |
| **TUL** | Tolerable Upper Level Intake |
| **VRA** | Variable Rate Application |
| **WFD** | Water Framework Directive |
| **WHO** | World Health Organization |
| **WOF** | Warmed Over Flavor |

# Part 1

---

## Nitrate in Soils

# 1 Biogeochemistry of Nitrate in Soils

*Roberto García-Ruiz, Milagros Torrús-Castillo,
Pablo Domouso de Agar, Roberto García-Perán,
Gustavo Ruiz-Cátedra, and Julio Calero*

## CONTENTS

1.1 Overviewing the Biogeochemistry of Nitrate in Soils ...................................4
1.2 Main Properties of the Nitrate Molecule .........................................................5
1.3 Nitrate Can Move Fast and Far .........................................................................9
1.4 Wet and Dry Deposition, and Biological N Fixation; Main Inputs of
New Nitrogen in the Soil .................................................................................10
    1.4.1 Wet and Dry Deposition Is a Natural Abiotic Source of
        Nitrate in the Soil .................................................................................11
    1.4.2 Biological N Fixation Fuels the Soil's N Cycling .............................11
1.5 Nitrate Transformations Are Biologically Controlled .................................12
    1.5.1 Nitrogen Decomposition and Mineralization Are the
        Main N Recirculation Pathways .........................................................12
    1.5.2 Nitrification: The Only Biological Process that
        Renders Nitrate .....................................................................................13
1.6 Biological Processes Which Transform Nitrate into Other
N-compounds .....................................................................................................14
    1.6.1 Assimilatory Nitrate Reduction: A Process Which Promotes the
        Recirculation of N ................................................................................14
    1.6.2 Plant Assimilatory Nitrate Reduction ...............................................14
    1.6.3 Assimilatory Microbial Nitrate Reduction, or Biological Nitrate
        Immobilization .....................................................................................16
    1.6.4 Dissimilatory Nitrate Reduction: Gaseous Losses of N ...................18
        1.6.4.1 Dissimilatory Nitrate Reduction to
            N gases: Denitrification ......................................................18
        1.6.4.2 Dissimilatory Nitrate Reduction to Ammonium ..................20
References ..................................................................................................................21

DOI: 10.1201/9780429326806-2

**3**

## 1.1  OVERVIEWING THE BIOGEOCHEMISTRY OF NITRATE IN SOILS

The biogeochemistry of nitrate includes the transformation of other compounds containing nitrogen (e.g., ammonium) to nitrate and its transformation to other N-containing compounds (e.g., nitrous oxide) (Figure 1.1). Due to the oxidized state of the nitrate and the lack of hydrogen in the molecule, the main mechanism governing these transformations are redox reactions, mainly mediated by organisms in their search for: (i) sources of energy coming from exergonic reaction when transforming a more reduced N-containing molecule into a more oxide N-containing compound (e.g., the transformation of ammonium to nitrate); (ii) sources of compounds that can accept electrons liberated when using sources of organic carbon as fuel and building biomass (e.g., dissimilatory nitrate reduction); or (iii) sources of nitrogen to produce amino acids or other N-containing compounds (e.g., plant nitrate assimilation) (Figure 1.1).

In many cases, these transformations result in a change of the state of the compounds: from being dissolved in the liquid phase of the soil solution (e.g., nitrate in the water soil) to a gaseous (e.g., nitrous oxide or atmospheric nitrogen) or a solid state (e.g., amino acids or other N-containing compounds in plant and organism

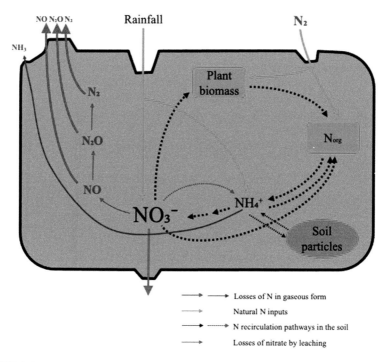

**FIGURE 1.1**  An overview of the N cycle in the soil. Red arrows stand for processes under anaerobic or very low $O_2$ conditions. Dashed lines stand for processes in which nitrogen is recirculating among compartments in the soil.

biomass or adsorbed into the anion exchangeable capacity of low or very low pH soils) (Figure 1.1). Other processes of the N cycling also involve a change from gaseous (atmospheric nitrogen) to a solid state (e.g., amino acids or other N-containing compounds in plant and organism biomass) by biological N fixation, or from solid (e.g., amino acids or other N-containing compounds in plant or organisms biomass) to a liquid (e.g., nitrate in the soil solutions) state. As already mentioned, these transformations are mainly mediated by microorganisms, but other changes of state within the N cycle are possible without the direct implication of organisms as it is, for instance, the transformation of ammonium (in the liquid state of the soil solution) to ammonia gas. The change of state after transformation is not exclusive to N-cycling while during the cycling of other nutrients such changes of states are frequent. Some transformations in the cycles of carbon and oxygen involve changes between liquid and gas states, whereas only solid or liquid states are possible during the cycling of other nutrients (e.g., phosphorus, potassium, and other micronutrients).

Nitrate is relatively easily and quickly transported within and outside the ecosystem due to its high solubility in water (Figure 1.1). Molecular diffusion and convection and advection are the main processes governing the movement of nitrate, but the former is negligible compared to the other two. The rate at which nitrate moves depends on the physical and chemical properties of the soil, plant activity, and other environmental conditions. An understanding of the basic principles influencing the distribution and movement of water in the soil is essential for minimum cost/benefit ratios, especially in regions where agriculture uses irrigation and/or fertilization for optimum production.

Some of the negative environmental impacts of nitrate stem from the ease with which nitrate moves between ecosystems. In addition, the ability of the nitrate to move and the existence of gaseous N-containing compounds within the N cycle which originate from nitrate, causes the nitrogen cycle in ecosystems to have many loopholes through which nitrogen can be lost.

## 1.2   MAIN PROPERTIES OF THE NITRATE MOLECULE

There is only one molecular structure of nitrate; where an atom of nitrogen with a positive charge and three atoms of oxygen with two negative charges are shaped following a symmetrical trigonal geometric figure (Figure 1.2). These two negative charges are shared amongst the three oxygen atoms resulting in three electronic conformations, with the same properties. Indeed, independently of the electronic conformation of the molecule and the origin of the nitrogen and oxygen of the molecule and processes which have been produced, the structure of the nitrate molecule is the same. Thus, the molecular structure of the nitrate which comes from an organic source of nitrogen or from an N-containing fertilizer that is produced synthetically is indistinguishable and shows the same properties.

Due to the oxygen atoms and the structural arrangement, nitrate has the ability to accept the electrons of other substances which are relatively reduced, and both, the oxidant and the reductant, are the components of an oxidation-reduction reaction; oxidation and reduction occur in tandem. Nitrate is the most fully oxidized compound

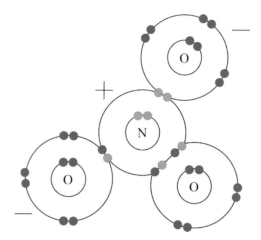

**FIGURE 1.2**  Nitrate molecule.

of the nitrogen cycle (Figure 1.3), and it is stable in oxidation. Nitrate is a strong oxidizing agent. Most of the oxidation-reduction reactions are exergonic; the transfer
of electrons between molecules is important because most of the energy stored in
atoms is in the form of high-energy electrons. Therefore, nitrate is the main acceptor
of electrons produced during the decomposition process of organic matter for some
microorganism (e.g., denitrifiers) under low, or in the absence of, oxygen (e.g., under
anaerobic conditions), or nitrate is formed by the oxidation of ammonium by nitrifying bacteria (Figure 1.3). The nitrate´s ability to act as a strong oxidant was the
reason why potassium nitrate, or saltpeter, together with other fuels such as sulfur
and charcoal, have been widely used as a propellant in firearms, artillery, rockets,
and fireworks. The importance of potassium nitrate as a source of strong oxidant for
gunpowder production in the past is evident in Güldner and colleagues (2016), where
they demonstrated that nitrate extracted by saltpeter production in the rural settlement
of Pamhagen (southern Vienna) in 1778 was equivalent to 23 percent of the total
available N in manure.

   Despite the fact that the molecular structure and properties of the nitrate are indistinguishable, there may be different nitrogen atoms in the nitrate molecule according
to the atomic mass. Measurements of the atomic weight of N, which according to the
atomic weight should be 14.00, give a value of 14.008, implying that nitrogen has
a heavier isotope. In fact, N has both a heavier ([15]N) and a lighter isotope ([13]N), but
the latter is unstable and radioactive, although has been used in short-term studies of
N uptake by plants (Von Wirèn et al., 1997). Natural nitrogen consists of two stable
isotopes: a nitrogen with an atomic mass of 14 and a nitrogen with an atomic mass of
15. Both isotopes have the same arrangement of electrons in orbits around the nucleus
and therefore the same chemical properties. However, the vast majority (about 99.6%)
of the naturally occurring nitrogen is [14]N. Nonetheless, soil biotic processes, such as
N mineralization, nitrification, denitrification, N assimilation by plants, and some
abiotic process such as $NH_3$ volatilization, discriminate against [15]N and lead to a soil

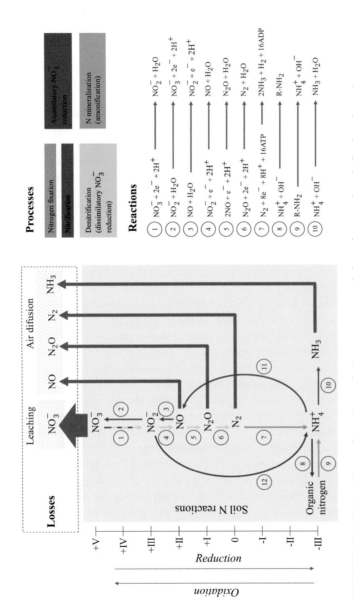

**FIGURE 1.3** Biological transformations of nitrogen compounds by redox reactions indicating the main processes involved. The interconversion of ammonium and organic nitrogen and ammonium and ammonia do not involve a change in the redox state of the nitrogen atom.

N pool with a different $^{15}N/^{14}N$ ratio ($^{15}N$ signature or $\delta^{15}N$) (Mariotti et al., 1981; Handley and Raven, 1992; Nadelhoffer and Fry, 1994; Piccolo et al., 1994; Robinson, 2001). High soil N mineralization rates are, sometimes, correlated with denitrification or leaching of nitrate (Goodale and Aber, 2001). Later, both processes lead to losses of $^{15}N$-depleted $N_2O$ and $N_2$ or nitrate, respectively, and leave the remaining inorganic N pool relatively enriched in $^{15}N$. The $\delta^{15}N$ signatures of different soil N pools are further imprinted in the $\delta^{15}N$ values of plants that utilize these soil N pools for their N nutrition (Brackin et al., 2015). Therefore, the natural abundance of stable $^{15}N$ isotopes in soils or plants has been suggested as an efficient and simple non-invasive tool to assess and monitor N dynamics in ecosystems (Frank et al., 2000). In addition, by analyzing the $\delta^{15}N$ in various nitrogen pools in the soil after adding $^{15}N$ enriched ammonium or nitrate, the rate of processes has been successfully quantified (Brakin et al., 2015).

Nitrate is an anion, and since this anion bears a relatively low and decentralized negative charge, nitrate is not strongly attracted to their cations, which makes them more easily separate. Thus, when salts (potassium nitrate, calcium nitrate, ammonium nitrate, etc.) of a cation and nitrate dissolve in the water of the soil solution, the salt dissociates to a cation and nitrate, and both acquire an electric charge; the potassium, calcium, and ammonium become $K^+$, $Ca^{2+}$ and $NH_4^+$ with a positive charge, and the nitrate acquires a negative single charge to become $NO_3^-$.

The single negative charge of $NO_3^-$, together with the relatively high solubility of nitrate, is the main reason why nitrate is hardly retained in the soil. The exchange of ions, a reversible process by which a cation or anion adsorbed on the surface is exchanged with another cation or anion in the liquid phase, occurs on the surface, mainly on soil clay minerals, but also on organic matter and secondary oxides of iron and aluminum. Although clay minerals might exhibit a negative (cation exchange capacity) and positive (anion exchange capacity) surface charge, the former is more important than the latter in a range of normal soil pH (6.0–8.5), and for most of the common soil, clay minerals have negative permanent charges independent of the pH (i.e. smectite, vermiculite, chlorite, mica). In addition, as soil pH increases, some of the $H^+$ ions are neutralized, and the negative edge charge increases, resulting in an increase in the negative charge or in the cation exchange capacity. Anions, like nitrate, are not only not adsorbed by the clay minerals but actually are repelled by it (Thomas and Swoboda, 1970), lowering the retention capacity of nitrate in the soil. Cations, such as ammonium, potassium, and calcium, are attracted to the surface of the clay.

In soils with low pH ($< 5.0$) with relatively high contents of oxides of iron and aluminum, which have a pH-dependent charge, the overall charge of the clay is positive, and soil nitrate is electrostatically attracted to the clay and/or oxide surfaces. The sorption of nitrate in these soils, which sometimes refers to abiotic nitrate immobilization (Torres-Cañabate et al., 2008), holds it back from being washed out of the soil (Duwig et al., 2003).

The electrostatic attraction of nitrate to surfaces charged positively is the basis for the use of anion exchange membranes which, when located in different positions and soil depth, can lead to the assessment of nitrate availability and some N cycling processes involving the measurement of nitrate over short periods (weeks and months), such as net nitrification (Subler et al., 1995).

## 1.3   NITRATE CAN MOVE FAST AND FAR

As mentioned above, nitrate is extremely soluble in water; about 0.5 kg of nitrate can be dissolved in one liter of water.[1] Due to the high electronegatively of oxygen, the electron density of the nitrate ions is relatively decentralized. The hydrogen atoms of a nearby water molecule are strongly attracted to the electron-rich oxygen atoms of the nitrate anion, which results into the water molecules being more strongly attracted to the nitrate anion than they are to each other. Three corollaries can be drawn from this fact: (i) the majority of the nitrate that is measured in the soil is dissolved in the soil solution; (ii) solubility is not a limiting factor of the nitrate concentration in natural water courses; and (iii) wherever the water in the soil goes, nitrate goes with it. Therefore, by understanding the movement of water throughout the soil profile, the fate of the nitrate can be delineated.

The main mechanism of movement of nitrate, which is dissolved in the soil solution, is by mass flow of the soil water. Infiltration, which describes the process by which water enters the soil pore spaces and becomes soil water, is a transitional phenomenon that takes place at the soil surface. Nitrate in the soil water solution is distributed into the infiltrated water and moves with it downwards or sidewards, depending on the difference between the energy level of water, or soil moisture potential, from the site with high potential (e.g., wetter soil, since under this condition, water molecules are not very close to the soil and thus are not held very tightly by the soil matrix) and the site of lower potential (e.g., drier soil, since the small amount of water which remains in the soil is located in small pores and thin water films, and is therefore held tightly by the soil solids having little freedom of movement). Percolation, the downward movement of the water out of the soil profile, with the nitrate dissolved in it, depends on the soil's hydraulic conductivity, which is affected by any factor affecting the size and configuration of soil pores, as the total flow rate of water in soil pores is proportional to the fourth power of the radius. Thus, when water flows through a pore with a radius of 1 mm, it is equivalent to 10,000 pores with a radius of 0.1 mm. Thus, macropores account for most water movement in saturated soils, and the presence of biopores has a marked influence on the saturated hydraulic conductivity and thus on the movement of nitrate downwards. The texture and structure of a soil horizon is also an important determinant in the saturated hydraulic conductivity. The nitrate in sandy soils and soils with a stable granular structure percolates, or escapes, down the soil easier. Therefore, nitrate leaching losses are usually less from fine-textured soils than from coarse-textured soils. In a water-unsaturated soil, the movements of water and nitrate are more complex because most of the macropores are filled with air, leaving only the fine pores to accommodate water movement. The driving forces in the water movement under unsaturated conditions are inversely related both with the moisture content and soil matrix suction, so water flux in drier soil is progressively approaching zero. When the suction potential exceeds the gravitational drainage (suctions greater than 0.1 bar), an upward movement of water occurs in the soil that can even increase the nitrate concentrations in the topsoil.

High nitrate leaching out of the plant root zone is expected when N fertilizers are applied in excess of crop needs for optimum growth/yield on highly porous coarse-textured soils with very rapid leaching potential. Under these conditions, relatively shallow groundwater may show a high level of nitrate within a few years. Indeed, increasing diffuse nitrate loading of surface waters and groundwater has emerged as a major problem in many agricultural areas of the world, resulting in contamination of drinking water resources in aquifers as well as eutrophication of freshwaters and coastal marine ecosystems (Wang and Li, 2019). The nitrates directive in the EU, and other similar legislation in other non-EU countries, include the designation of nitrate-vulnerable zones to prevent surface and groundwater contamination with nitrate. Under this area – which accounted for 61 percent of EU agricultural land by 2015 – measures have been designed to synchronize crop nitrogen demand with available nitrate.

Nitrate is transported to the root by mass flow as a result of transpirational water uptake by the plants. The quantity of nitrate reaching plant roots by mass flow is determined by the rate of water flow or the water consumption by plants and the average nitrate concentration in the soil water. As soil moisture is reduced, so does the water transport to the root surface and nitrate as well.

Loss of nitrate by the surface runoff can be significant, especially in agricultural systems. As an example, Elrashidi and colleagues (2005), found losses of N as dissolved nitrate of between 0.8 to 2.6 kg nitrate-N ha$^{-1}$ y$^{-1}$ for cropland of 12 major soils in the Wagon Train Watershed (Lancaster County, Nebraska).

## 1.4 WET AND DRY DEPOSITION, AND BIOLOGICAL N FIXATION; MAIN INPUTS OF NEW NITROGEN IN THE SOIL

Historical and contemporary views assume that nitrogen enters the earth´s land-surface ecosystems from the atmosphere. However, there is a body of evidence that bedrock is a nitrogen source that may rival atmospheric nitrogen inputs across the biomass (Houlton et al., 2018; Dynarski et al., 2019). Over billions of years, N has accumulated in rocks, largely as a product of N fixation by aquatic and terrestrial organisms that become trapped in sedimentary basins; this N has been traced back to ancient biogeochemical processes, as opposed to contemporary N fixation by free-living microbes and root-associated symbionts. The amount of N varies widely among general rock types; sedimentary and metasedimentary lithologies occupying ~75 percent of the Earth's surface have concentrations of ~500 to 600 mg N kg$^{-1}$ rock, whereas more spatially restricted igneous rocks often have much lower values (<100 mg N kg$^{-1}$ rock) (Johnson and Goldblatt, 2015; Holloway and Dahlgren, 2002). Although N-rich sediments are globally widespread, such rock N concentrations do not translate directly to N inputs. Rather, rock N availability to terrestrial soils and vegetation is determined by denudation (physical plus chemical weathering), which varies as a general function of geochemistry, relief, tectonic uplift, climate, and biology (Jenny, 1941). However, the intimate mechanisms and the magnitude and extension of this N entry on a local scale are still to be elucidated.

The main widespread recognized natural entries of N into ecosystems are atmospheric deposition and biological and anthropogenic N fixation. The two former processes, together with N fertilization, are the sources of all reactive N$_2$ in terrestrial ecosystems.

Anthropogenic N fixation is the N fertilizer production via the industrial Haber-Bosch process by which an $N_2$ gas is hydrogenated to produce biologically available ammonia ($NH_3$). The amount of reactive N introduced into the biosphere by these means is already impressive, amounting to 120 Tg $y^{-1}$ or about twice as much in comparison to what is fixed by all natural terrestrial processes combined (Fowler et al., 2013). Unfortunately, modern agricultural systems are also highly inefficient in their N use, typically losing 50–70 percent of the applied reactive to the environment (Cassman et al., 2002).

## 1.4.1 WET AND DRY DEPOSITION IS A NATURAL ABIOTIC SOURCE OF NITRATE IN THE SOIL

Atmospheric nitrate deposition is an annual entry of nitrate in terrestrial ecosystems, which can be an important input of N in some ecosystems. In many forests, a large fraction of the annual uptake and circulation of nitrogen in vegetation may be derived from the atmosphere (Kennedy et al., 2002). N compounds in the atmosphere are deposited from rain and snow as $NH_4^+$, $NO_3^-$, $NO_2^-$ and organic N. Although rather variable, about 10 to 20 percent of the $NO_3^-$ is formed during atmospheric electrical discharges, with the remainder from industrial waste gases or denitrification from soils, whereas $NH_4^+$ comes largely from industrial sites where $NH_3$ is used or manufactured, or from nearby animal husbandry. Typically, the entry of $NH_4^+$ is higher than that of $NO_3^-$, especially around areas of intense industrial activity and intense animal husbandry and, as a rule, is greater in tropical than in polar or temperate areas.

In Europe, deposition ranges from less than 5 kg N $ha^{-1}$ $y^{-1}$ in Northern Europe to greater than 60 kg N $ha^{-1}$ $y^{-1}$ in Central and Western Europe (Dise et al., 2009). At remote sites in the southern hemisphere, wet deposition is less than 1 kg N $h^{-1}$ $y^{-1}$ (Vitousek et al., 1997); hence, total pre-industrial inputs were probably below 2 kg N $ha^{-1}$ $y^{-1}$, assuming a small contribution from dry deposition.

## 1.4.2 BIOLOGICAL N FIXATION FUELS THE SOIL'S N CYCLING

Nitrogen gas is chemically inactive. This is the reason why this gas makes up 78 percent of the atmosphere, and life, as we know, would not be possible if we were surrounded by a more reactive gas. This inactivity means that nitrogen gas is not directly accessible to plants, depriving them of an abundant supply, unless some transformation boosts the reactivity of this N.

Before industrialization, biological $N_2$ fixation was the major pathway of reactive N creation in terrestrial ecosystems, and this may still be the case in more unpolluted regions (Cleveland et al., 1999). Even though atmospheric $N_2$ is the major component of the earth's atmosphere, and nitrogen is essential for all forms of life, $N_2$ cannot be used directly by biological systems to synthesize the chemicals required for growth, maintenance, and reproduction. The general chemical reaction for biological N fixation is similar to that of chemical N fixation (Haber–Bosch). Living organisms use energy derived from the oxidation of carbohydrates to reduce molecular nitrogen ($N_2$) to ammonia and ammonium.

The capacity of biological N fixers to transform $N_2$ to organic N is more than enough to maintain N pools in natural ecosystems and to replenish N losses by mainly denitrification, lixiviation, and erosion.

Significant biological $N_2$ fixation on an ecosystem scale is most often associated with symbiotic $N_2$ fixation, with the classic example of the bacterium *Rhizobium* infecting the roots of leguminous plants (e.g., peas and beans, soya bean, clover, peanut). Potential leguminous symbiotic $N_2$ annual fixation can be up to approximately 30 kg N $ha^{-1}$ $y^{-1}$ in natural ecosystems. Some reported values range from 1.5–2.0 kg N $ha^{-1}$ $y^{-1}$, 6.5–26.6 kg N $ha^{-1}$ $y^{-1}$, 2.3–3.1 kg N $ha^{-1}$ $y^{-1}$ and 1.5–3.5 kg N $ha^{-1}$ $y^{-1}$ in boreal forest and woodland, temperate forests and forested floodplains, natural grasslands, and Mediterranean shrublands, respectively (Cleveland et al., 1999). In agroecosystems, values can be as high 150 kg N $ha^{-1}$ $y^{-1}$ (Smil, 1999). Therefore, it is not surprising that legumes are often used in agricultural systems, representing a major direct source of food and forage for livestock. In low-input and organic farming systems, leguminous crops are included in rotational designs in order to provide N for other crops within the rotation (Ball et al., 2005).

Another major pathway for biological $N_2$ fixation is by heterotrophic bacteria in soils. Heterotrophic fixation during the decomposition of plant litter is thought to be important in terrestrial ecosystems and may account for 1–5 kg N $ha^{-1}$ $y^{-1}$. The net contribution of heterotrophic N fixation to ecosystem N budgets may be greater in wetland soils (Cassman and Harwood, 1995), and heterotrophic fixers contribute a substantial proportion of this total. Similarly, high rates of heterotrophic fixation may support plant production in some natural wetlands.

## 1.5 NITRATE TRANSFORMATIONS ARE BIOLOGICALLY CONTROLLED

### 1.5.1 NITROGEN DECOMPOSITION AND MINERALIZATION ARE THE MAIN N RECIRCULATION PATHWAYS

Plant litter production and decomposition are the major drivers of terrestrial N turnover. Nitrogen enters the soil organic matter pool after internal N cycling via plants (Figure 1.1).

The N mineralization is commonly known as the process by which carbon dioxide, ammonium, and later nitrate – embedded into the recognizable organic matter and soil organic matter – are released as a result of the activity of the soil microbial biomass and the rest of the soil population. For nitrogen, the process involves first the depolymerization of organic matter, which is carried out by extracellular enzymes of fungi and bacteria, and ammonification, which renders ammonium, and nitrification, which produces nitrate.

Heterotrophic bacteria dominate the breakdown of proteins in neutral and alkaline environments, with some involvement of fungi, while fungi usually dominates in acid soils. The end products of the activities of one group furnish the substrate for the next and so on until the material is decomposed.

Depolymerization of N-containing compounds, or aminization, converts proteins in residues to amino acids, amines, and urea. Ammonification is the conversion of these more readily decomposable organic nitrogen compounds into ammonium and can be affected by a wide variety of bacteria (aerobic and anaerobic) and fungi, and it is influenced by temperature, moisture, and other factors that affect such organisms. A hydroxyl ion is formed, so the process makes the soil slightly more alkaline:

$$\text{Organic N} + H_2O \rightarrow NH_4^+ + OH^- + energy$$

The $NH_4^+$ produced through ammonification can be: (i) converted to $NO_3^-$ by nitrification; (ii) absorbed by plants (plant assimilation); (iii) utilized by heterotrophic bacteria to decompose residues (assimilatory microbial nitrate reduction or biological nitrate immobilization); (iv) fixed temporarily as unavailable N in the lattice of certain clay minerals and onto organic matter ($NH_4^+$ adsorbed in the cation exchange surfaces); or (v) converted to $NH_3$, and slowly released back to the atmosphere ($NH_4^+$ volatilization) (Figure 1.1).

As a general rule, the higher the soil organic N, the higher the annual amount of organic N mineralized to $NH_4^+$. Therefore, soil and crop management strategies that conserve or increase soil organic matter will result in a greater contribution of mineralizable N to N availability ($NH_4^+$ plus $NO_3^-$) in the soil.

### 1.5.2  NITRIFICATION: THE ONLY BIOLOGICAL PROCESS THAT RENDERS NITRATE

Nitrification, probably the least understood process of the N cycle, is the aerobic oxidation of $NH_4^+$ to $NO_2^-$ and of $NO_2^-$ to $NO_3^-$. Classically, these two reactions are regarded as being catalyzed by chemolithoautotrophs exemplified by *Nitrosomonas* and *Nitrobacter* species. There is more recent evidence that these nitrification reactions can, to some extent, also be catalyzed by heterotrophs in what is referred to as heterotrophic nitrification. The energy released by first oxidizing $NH_4^+$ to $NO_2^-$ and later $NO_2^-$ to $NO_3^-$ is exploited by nitrifying microorganisms for cell growth. These bacteria are autotrophs and therefore the source of carbon to build their C-biomass is inorganic carbon; they do not need organic carbon and, therefore, it is not surprising that indicators of nitrification activity can be high in relatively degraded soil. As a matter of fact, in arable soils with relatively low organic matter content, the rate of nitrification can be as high as in high organic matter soils (Garcia-Ruiz et al., 2008). Oxygen is the acceptor of electrons, and thus nitrification is very low or lacking under low $O_2$ or anaerobic environments, which explains the relatively high ammonium concentration in watered soils. It is generally believed that oxidation of $NH_4^+$ to $NO_2^-$ (Equation 1.1) is carried out by bacteria of the genera *Nitrosomonas*, although some studies have suggested that nitrospira-type bacteria are responsible for most of the oxidation of $NH_4^+$ in cropland, whereas *Nitrosomonas* is involved in ammonia-rich environments.

$$2\,NH_4^+ + 3O_2 \rightarrow 2NO_2^- + 4H^+ + 2H_2O + energy \qquad \text{(Equation 1.1)}$$

This first stage has an acidifying effect because hydrogen ions are formed (1 mol of $NH_4^+$ produces 2 moles of $H^+$). Therefore, increasing soil acidity with nitrification is a natural process, although cropland soil acidification is accelerated with continued application of N fertilizers based on ammonia, such as ammonium sulfate, or other fertilizer $NH_4^+$ – forming fertilizers, such as urea (Latifah et al., 2011). Bacteria responsible for this first step of nitrification are very sensitive to low pH, and therefore in acid soil, organic N ($N_{org}$) mineralization is decoupled with nitrification resulting in temporary ammonium accumulation.

The second state, for which *Nitrobacter* genera is believed to be responsible, is

$$2NO_2^- + O_2 \rightarrow 2NO_3^- + energy$$

$NO_2^-$ is toxic to most soil organisms including plant roots, however rarely accumulated in the soil, because the rate of this second step is much higher than that of the first (Equation 1). When relatively high levels of $NO_2^-$ are detected in the soil, this might indicate dysfunction of the whole nitrification process as it is the case of some over-enriched soils with ammonium, likely because *Nitrobacter* activities are partially inhibited under large concentration of ammonium (Smith et al., 1997; Vadively et al., 2007).

Because the supply of $NH_4^+$ is the first requirement for nitrification, if conditions do not favor N mineralization or if $NH_4^+$-containing or $NH_4^+$-forming fertilizers are not added to the soils, nitrification is reduced. Other factors that influence nitrification in soils are i) Soil pH. Although nitrification occurs over a wide range of pH (5.0-10.0) the optimum is around 8.5, especially when there is an adequate supply of $Ca^{2+}$, ii) Soil aeration. Because nitrification is an aerobic process, the optimal soil nitrification rate takes place under atmospheric $O_2$ concentration. Thus, soil conditions (e.g., coarse-textured soil) or management (e.g., incorporation of crop residues and other organic amendments) which facilitate rapid gas exchange and good soil structure will ensure an adequate supply of $O_2$ for nitrifying bacteria, and iii) Soil moisture. Nitrification rate is optimal at soil field capacity and is lowered when soil moisture exceeds or is far below field capacity.

## 1.6 BIOLOGICAL PROCESSES WHICH TRANSFORM NITRATE INTO OTHER N-COMPOUNDS

### 1.6.1 Assimilatory Nitrate Reduction: A Process Which Promotes the Recirculation of N

Assimilatory nitrate reduction is the transformation of nitrate into N-containing compounds which are part of the biomass of plants, fungi, algae, and bacteria. This process begins with the assimilation of nitrate, which is later reduced to form essential N-containing compounds, such as proteins, nucleic acids, coenzymes, and numerous plant secondary products. This process implies the transference of N-nitrate from the pool of N-dissolved in the soil solution to the solid pool of N contained in biomass, and thus it is a process that retains N within the ecosystems (Figure 1.1).

### 1.6.2 Plant Assimilatory Nitrate Reduction

The inorganic N forms utilized by plants are $NO_3^-$ and $NH_4^+$. Nitrite uptake by plant roots is generally not considered to be of consequence, as a result of the low levels of nitrite in the soil and the reported toxicity of this anion. Although $NH_4^+$ is more reduced than nitrate, in most agricultural soils the roots of plants largely take up N as $NO_3^-$. This is because $NO_3^-$ generally occurs in higher concentrations than either nitrite or $NH_4^+$, and is free to move within the root solution due to the tendency for soils to possess an overall negative charge (Reisenauer, 1978).

Plants only rely extensively on "root interception" for the uptake of sparingly soluble nutrients such as phosphorus (P). In contrast, nitrate is mostly delivered to roots through a combination of mass flow and diffusion (De Willigen, 1986). Root interception is thought to account for ca. 1 percent of nitrate taken up (Barber, 1984; Marschner, 1995). Mass flow relies on transpiration to draw water to the roots. If the rate of N delivery in the transpiration water stream is lower than the root demand for N, then diffusion also plays a role in uptake. Diffusion depends on the concentration gradient and the diffusion coefficient for the particular form of N. Although the diffusion coefficients for $NO_3^-$ and $NH_4^+$ in water are similar ($1.9$–$2.0$ $10^{-9}$ $m^2$ $s^{-1}$), the diffusion coefficients in the soil are additionally determined by ion size and charge, viscosity of water, temperature, soil moisture, tortuosity, and the soil buffer capacity. For $NO_3^-$ the diffusion coefficient is ca. $1$ $10^{-10}$ $m^2$ $s^{-1}$ (Barber, 1984), while that of $NH_4^+$ is ca. 10-fold to 100-fold less (Owen and Jones, 2001). The corollary of this is that $NH_4^+$ is also less available in the soil for roots to take up, although when roots have access to $NH_4^+$ they take it up more readily than $NO_3^-$ (Lee and Rudge 1986; Colmer and Bloom 1998).

Nitrate is actively transported across the plasma membranes of epidermal and cortical cells of roots, but net uptake is the balance between active influx and passive efflux. This transport requires energy input from the cell over almost the whole range of concentrations encountered in the soil (Glass et al., 1992; Miller and Smith, 1996; Zhen et al., 1991). It is generally accepted that the uptake of $NO_3^-$, mainly in the root cortex of plants, is coupled with the movement of two protons down an electrochemical potential gradient and is therefore dependent on ATP supply to the $H^+$-ATPase that maintains the $H^+$ gradient across the plasma membrane (McClure et al., 1990; Meharg and Blatt, 1995; Miller and Smith, 1996).

Efflux systems have been studied less than influx systems; however, it is known that efflux is protein-mediated, passive, saturable, and selective for $NO_3^-$ (Aslam et al., 1992; Grouzis et al., 1997). Anion channels seem the most obvious route for $NO_3^-$ efflux, because the transport is thermodynamically downhill and genome analysis has identified several gene families that may fulfill this function. The $NO_3^-$-efflux system is under a degree of regulation, induced by $NO_3^-$ (Aslam et al., 1992), and it is also proportional to whole-tissue $NO_3^-$ concentrations (Teyker et al., 1988). It can therefore predict that the anion channel(s) responsible for $NO_3^-$ efflux must be $NO_3^-$-inducible. Net $NO_3^-$ uptake is regulated by whole plant demand via shoot-derived signals transported in the phloem to the roots (Imsande and Touraine, 1994; Vidmar et al., 2000). The nature of these feedback signals seems to be amino acid concentrations in the phloem, specifically glutamine (Pal'ove-Balang and Mistrik, 2002; Tillard et al., 1998). Interestingly, efflux of $NO_3^-$ has been found to be associated with slow growth rates (Nagel and Lambers, 2002). This efflux is, however, a consequence rather than a cause of slow growth. Slow-growing plants from nutrient-poor habitats may simply not be able to exploit high concentrations of $NO_3^-$, which is then effluxed.

Before $NO_3^-$ can be used in the plant, it must be reduced to a $NH_4^+$ or $NH_3$. Once the nitrate is incorporated into cells it is reduced by a series of assimilatory enzymes (Crawford et al., 1995). Nitrate ions are initially reduced to $NO_2^-$ via the enzyme nitrate reductase and further reduction to $NH_4^+$ occurs via the nitrite reductase enzyme. Ammonium is then added to C skeletons to produce a variety of amino acids via the glutamine synthetase and glutamate synthase or glutamate-2-oxoglutarate enzymes amino-transferase cycle. The activity of these N-assimilatory enzymes, like the transporters, can be regulated at a number of different levels – by the synthesis of both mRNA (transcription) and protein (translation) and the activity of the enzyme (post-translation).

### 1.6.3 ASSIMILATORY MICROBIAL NITRATE REDUCTION, OR BIOLOGICAL NITRATE IMMOBILIZATION

The precise mechanism of assimilatory microbial nitrate reduction is less known than the equivalent process described by plants. Nevertheless, the physiological and biochemical studies on $NO_3^-$ assimilation have been performed in several bacterial species (e.g., *Klebsiella oxytoca*, *Bacillus subtilis*, and *Azotobacter vinelandii*, among others), revealing significant differences in the bacterial $NO_3^-$-assimilatory systems, but it is believed that $NO_3^-$ is incorporated into the bacteria by high-affinity transport systems which are usually regulated by $NO_3^-$ induction and $NH_4^+$ repression. Once in the cell, $NO_3^-$ is reduced to $NO_2^-$, which is further reduced to $NH_4^+$ which is incorporated into carbon skeletons mainly by the glutamine synthetase/glutamate synthase pathway, similar to what happens in plant cells.

The assimilatory reduction of $NO_3^-$ and $NH_4^+$ is also referred to as N immobilization, which is basically the conversion of $NO_3^-$ and $NH_4^+$ to microbial N and is the reverse of N mineralization.

Among other factors, the magnitude of the microbial $NO_3^-$ immobilization largely depends on the C-to-N ratio of the decomposing organic matter (Figure 1.4). Due to the high activity and very low duplication time, the protein content of the microbial community is relatively high and, therefore, to keep their activities at a homeostatic steady state, high available N is demanded. During microbial decomposition, a relative high fraction (about two-thirds) of the organic carbon ($C_{org}$) of the organic matter is used in respiration, which releases $CO_2$ and provides energy, and another fraction is used to build their C-biomass, which requires available N; about one-eighth of the $C_{org}$-biomass. Thus, if decomposing organic matter contains low N and high $C_{org}$, the microorganism will assimilate $NO_3^-$ (and/or $NH_4^+$) from the soil solution, and it will be utilized by the rapidly growing microorganism, lowering the $NO_3^-$ (and/or $NH_4^+$) concentration in the soil solution (Figure 1.4). The microbial immobilization might compete effectively with plants for nitrate. $NO_3^-$ immobilization involves the transference of N from the soil solution to the biomass compartment, and this constitutes an N retention and recirculation mechanism within the ecosystems.

The magnitude and the extent of the microbial N immobilization depend on C-to-N ratio of the source of $C_{org}$ that microorganisms use as a source of energy and C for building their biomass. If the C-to-N ratio of the organic matter is low and contains high N (e.g., chicken manure), immobilization will not proceed because the

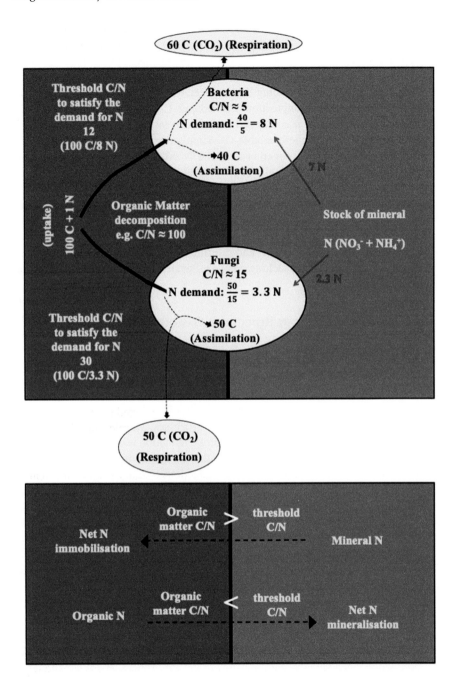

**FIGURE 1.4** Conceptual framework, and a numerical example, of N immobilization. When the C-to-N ratio of a source of organic matter (e.g., 100) exceed that of bacteria (e.g., 5) or fungi (e.g., 15), during the decomposition of that source of organic matter the soil available N (e.g., nitrate) is assimilated (assimilatory nitrate reduction) by bacteria and fungi to satisfy their N demand.

decomposing organic matter contains sufficient N to meet the microbial demand, and soil $NH_4^+$, the product of $N_{org}$ mineralization, and later $NO_3^-$, once the $NH_4^+$ is nitrified, increase in the soil.

Early studies (e.g., Iritani and Arnold, 1960) indicated that a C-to-N ratio of approximately 20:1 was the dividing line between immobilization and N mineralization. As organic matter decomposition proceeds, its C-to-N ratio decreases due to a decrease in $C_{org}$, as most of it is respired, and an increase of $N_{org}$ ($NO_3^-$ immobilized from soil solution). When the available $C_{org}$ supply decreases, microorganisms die, releasing the immobilized N together with the N taken up from the decomposing organic matter. Once this microbial-N is mineralized (remineralization) and nitrified, an increase in the pool of $NO_3^-$ in the soil solution can be observed. Thus, the assimilatory microbial $NO_3^-$ reduction is a transitional process (weeks to months). The application to soils of sources of organic matter with a high C-to-N ratio – such as wheat straw residues – promotes temporary immobilization of the $NO_3^-$ in the soil solution, increasing soil N retention and offering the potential of lowering losses of N through denitrification and leaching when soil-available N is not required by crops (Garcia-Ruiz et al., 2007).

### 1.6.4   DISSIMILATORY NITRATE REDUCTION: GASEOUS LOSSES OF N

Nitrate can act as an electron acceptor for the cell´s metabolism in low or $O_2$ lacking environments. As a consequence, most of the nitrate is reduced and not assimilated. Assimilatory nitrate reduction could also occur, but most anaerobic environments have large concentrations of ammonium and organic nitrogen, which repress this assimilatory process or make it quantitatively insignificant.

The nitrate reduction is dominated by two processes: Denitrification and nitrate reduction to ammonium. The dissimilatory nitrate reduction is distinguished from the assimilatory nitrate reduction by the fact that the nitrogen reduced is not used by the cell, and that is why one of these two nitrate reductions is called *dissimilatory* nitrate reduction. The dissimilatory nitrate reduction to dinitrogen gases, NO, $N_2O$, and $N_2$, or denitrification, is one of the major paths of losses of N from the soil, whereas in the nitrate reduction to ammonium, nitrate is reduced to ammonium, which is less mobile and thus may conserve N within the ecosystems. In both nitrate-reducing processes, the reduction equivalents (or electron donors) usually originate from the oxidation of organic compounds, but both reactions can also be driven by inorganic oxidation processes such as hydrogen and sulfur oxidation.

#### 1.6.4.1   Dissimilatory Nitrate Reduction to N gases: Denitrification

Denitrification or the dissimilatory nitrate reduction to N gases is an important path of loss of N in the ecosystem. This loss of nitrate might have negative consequences for crop production since valuable nitrate is lost to the atmosphere, and when the whole denitrification process is incomplete, nitrous oxide ($N_2O$) is produced. However, what is detrimental to terrestrial ecosystems might be positive for wastewater treatment works, as this process is desirable in lowering nitrate concentration in water, and considerable effort by engineers has been devoted to improving designs for the efficient and economical removal of nitrate from wastewater by this process.

Denitrification is also the process that returns $N_2$ to the atmosphere, taken up by biological $N_2$ fixation and by the fertilizer industry, playing an essential role in keeping the atmospheric N in balance, since oceans are virtually $NO_3^-$ free. N transference from terrestrial ecosystems to the atmosphere by denitrification roughly accounts for one-third of the total agriculture and natural N inputs from N fertilizers, animal manure, biological N fixation, and atmospheric N deposition (Bouwman et al., 2013). The importance of this process also lies in the fact that, when it is incomplete, $N_2O$ is produced. $N_2O$ is one of the main greenhouse gases (1 kg of $N_2O$ is equivalent to 298 kg of $CO_2$) and contributes to the destruction of the protective ozone layer. Denitrification is both a source and a sink of $N_2O$.

The basic reduction-oxidation steps are as follows:

$$NO_3^- \rightarrow NO_2^- \rightarrow N_2O \rightarrow N_2 \qquad\qquad \text{(Equation 1.2)}$$

Large populations of denitrifying microorganisms exist – the most common being the *Pseudomonas*, *Bacillus*, and *Paracoccus* bacteria – as do several autotrophs such as *Thiobacillus denitrificans* or *Thiobacillys thioparus*, and some of them are facultative and others strict anaerobes, the former being more abundant in agricultural soils. Most denitrifiers freshly isolated from soil have the entire pathway, $NO_3^- \rightarrow N_2$, but strains that cannot reduce $NO_3^- \rightarrow NO_2^-$ and $N_2O \rightarrow N_2$ are also common. Although the energy sources of denitrifiers include light and organic and inorganic compounds, organic sources of energy and carbon (e.g., organic matter) are the most common, both in number of genera as well as in the population dominance in soils. When using organic carbon as the source of energy, organic carbon, and electrons, the general reaction for microbial reduction of $NO_3^-$ to $N_2O$ or $N_2$ are as follow:

$$4(CH_2O) + 4NO_3^- + 4H^+ \rightarrow 4CO_2 + 2N_2O + 6H_2O$$

$$5(CH_2O) + 4NO_3^- + 4H^+ \rightarrow 5CO_2 + 2N_2 + 7H_2O$$

The level of decomposable organic carbon – such as crop residues and manure, enough soil water content that impedes the diffusion of $O_2$ through soil-enhancing anaerobic conditions, a fine soil texture soil, which promotes low $O_2$ diffusion in it while promoting micro anaerobic hot spots, relatively high temperatures, and available $NO_3^-$ – together these all promote nitrate-N losses throughout denitrification, mainly in the form of $N_2$, but also some significant $N_2O$. Therefore, in agricultural clayey or silty soils that are under intensive irrigation and subjected to high doses of organic fertilization (source of organic carbon, energy, and nitrate throughout mineralization), losses of nitrate in form of gases throughout denitrification can be rather significant. For instance, in the review of Hofstra and Bouwman (2005), it was found that annual losses of N by denitrification ranged 9 to 51 kg N ha$^{-1}$ with the highest being under a combination of animal manure and synthetic fertilizers, and with a tendency of increasing the N loss in agricultural fields with high annual N application rates and poor soil drainage. Although nitrate-N loss estimates vary widely in agricultural soils, figures are usually < 15 percent of N applied to crops.

$N_2$ losses relative to other intermediates predominate, often accounting for about 90 percent of the total N gases. NO and $N_2O$ are produced at intermediate steps during the denitrification process and may escape to the atmosphere before being reduced to $N_2$. Due to the environmental concerns related to $N_2O$, there has been a large volume of studies reporting the accumulation of intermediates, such as $N_2O$ in the denitrification pathway. Despite the lack of a general understanding of the process and its control mechanisms, the ratio electron donor (i.e., decomposable organic carbon) to-electron acceptor (i.e. nitrate), the soil pH, and the redox potential of the soil are the primary regulators of the ratio $N_2/(N_2+N_2O)$ (i.e. the relative emission of $N_2O$) (Wlodarczyk et al., 2003). In general, the lower the content of decomposable organic carbon, the soil pH, and the temperature – especially under a negative redox potential – the lower the N gases losses, but the higher the proportion of $N_2O$ relative to $N_2$ (Tiedje, 1988).

### 1.6.4.2  Dissimilatory Nitrate Reduction to Ammonium

Like assimilatory nitrate reduction, the dissimilatory nitrate reduction to ammonium (DNRA), also called fermentative ammonification or nitrate ammonification or fermentative nitrate reduction, is a two-step process involving $NO_3^-$ reduction to $NO_2^-$ followed by a further reduction to $NH_4^+$. This process has also been called "a short circuit in the biological N cycle" (Cole and Brown, 1980), as the direct transfer of $NO_3^-$ and $NO_2^-$ to $NH_4^+$ bypasses denitrification and biological $N_2$ fixation. However, it rarely occurs in cropland soils, as it is strictly an anaerobic process that is only found in reductant-rich environments. In these environments, electron acceptors are scarce, so optimal use must be made of any available oxidant to regenerate the electron donor-acceptor systems, such as $NAD^+$-NADH and, hence, sustain substrate oxidation and growth. The potential free energy of total denitrification is higher than that of DNRA, and thus the former should be favored over the latter. However, under $NO_3^-$, limiting and strongly reducing conditions which characterize highly reductant environments, DNRA has the advantage over denitrification since more electrons can be transferred per mode of $NO_3^-$. Additionally, the potential free energy calculated per mole of $NO_3^-$ is higher for DNRA than denitrification (Strohm et al., 2007), and therefore this nitrate-reduction process is one of the most favorable to anaerobes faced with a shortage of electron acceptors. The benefits of the microbial cells of DNRA are thought to be (i) detoxification of accumulated nitrite; (ii) electron sinks which allow the re-oxidation of NAHD which seems to be the most common; and (iii) the production of energy through electron transport phosphorylation, like what occurs for the denitrifying nitrite reduction.

DNRA has been found in bacteria that have fermentative rather than oxidative metabolism, which includes obligate anaerobes, such as several species of *Clostridium*, *Veillonella alkalescent*, and several species of *Desulfovibrio*, some facultative anaerobes such as *Escherichia coli*, *Salmonella typhimurium*, and some species of *Klebsiella* and *Vibrio*, and some aerobes such as several strains of *Pseudomonas* and *Bacillus* (Tiedje, 1988). DNRA activity seems to be greater in carbon-rich, electron-acceptor-poor environments and, under these conditions DNRA organisms seem to be selected. The ratio of available electron donor (e.g., labile organic carbon) and electron acceptor (e.g., nitrate) controls the extend to which nitrate is dissimilatory reduced to ammonium (DNRA) or to N-gas (denitrification). When the ratio is low, the greatest advantage is to the organism that gains the most energy per nitrate, and this is denitrification. Thus, habitats that are predominantly aerobic, such as most cropland soils, would be expected to select populations that are the most efficient and competitive in using the available carbon under respiratory conditions.

## NOTES

1 The solubility of nitric salts is variable but, in general, very high, ranging between 209 g/L $(20^{0}C)$ for the $KNO_3$ and 1900 g/L $(20^{o}C)$ for the $NH_4NO_3$.
2 Nitrogen bonded to carbon (organic N), hydrogen ($NH_3$, $NH_4^+$) or oxygen ($NO_x$ $NO_3^-$ $N_2O$) is termed reactive N because it is available for biological assimilation or transformation.

## REFERENCES

Aslam, M., Travis, R.L., Huffaker, R.C. 1992. Comparative kinetics and reciprocal inhibition of nitrate and nitrite uptake in roots of uninduced and induced barley (*Hordeum vulgare* L.) seedlings. *Plant Physiol.* 99: 1124–1133.

Ball, C.D., Bingham, I., Rees, R.M., Watson, C.A., Litterick, A. 2005. The role of crop rotations in determining soil structure and crop growth conditions. *Can. J. Soil Sci.* 85: 557–577.

Barber, S.A. 1984. *Soil Nutrient Bioavailability. A Mechanistic Approach.* John Wiley and Sons, New York.

Bouwman, A.F., Beusen, A.H.W., Griffioen, J., Van Groenigen, J.W., Hefting, M.M., Oenema, O., Van Puijenbroek, P.J.T.M., Seitzinger, S., Slomp, C.P., Stehfest, E. 2013. Global trends and uncertainties in terrestrial denitrification and $N_2O$ emissions. *Philos. Trans. R. Soc. Lond. B. Biol. Sci.* 368: 20130112.

Brackin, R., Näsholm, N.R., Guillou, S., Vinall, K., Laskhmanan, P., Schmidt, S., Inselbacher, E. 2015. Nitrogen fluxes at the root-soil interface show a mismatch of nitrogen fertilizer supply and sugarcane root uptake capacity. *Nature Scientific Reports* 5: 15727.

Cassman, K.G., Dobermann, A., Walters, D.T. 2002. Agroecosystems, nitrogen use efficiency and nitrogen management. *Ambio* 31: 132–140.

Cassman, K.G., Harwood, R.R. 1995. The nature of agricultural systems – food security and environmental balance. *Food Policy* 20: 439–454.

Cleveland, C.C., Townsend, A.R., Schimel, D.S., Fisher, H., Howarth, R.W., Heding, L.O., Perakis, S.S., Latty, E.F., Von Fischer, J.C., Elseroad, A., Wasson, M.F. 1999. Global patterns of terrestrial biological nitrogen ($N_2$) fixation in natural ecosystems. *Global Biogeochemical Cycles* 13: 623–645.

Cole, J.A., Brown, C.M. 1980. Nitrite reduction to ammonia by fermentative bacteria: a short circuit in the biological nitrogen cycle. *FEMs Microbiology Letters* 7: 65–72.

Colmer, T.D., Bloom, A.J. 1998. A comparison of $NH_4^+$ and $NO_3^-$ net fluxes along roots of rice and maize. *Plant Cell Environ.* 21: 240–246.

Crawford, N.M. 1995. Nitrate: Nutrient and signal for plant growth. *Plant Cell* 7: 859–868.

De Willigen, P. 1986. Supply of soil nitrogen to the plant during the growing season. In *Fundamental, Ecological and Agricultural Aspects of Nitrogen Metabolism in Higher Plants,* ed. H. Lambers, J. J. Neeteson and I. Stulen. pp. 417–432. Dordrecht, Boston, Lancaster: Martinus Nijhoff.

Dise, N.B., Rothwell, J.J., Gauci, V., van der Salm, C., De Vries, W. 2009. Predicting dissolved inorganic nitrogen leaching in European forests using two independent databases. *Science of the Total Environment* 407: 1798–1808.

Duwig, C., Normand, B., Vauclin, M., Vachaud, G., Green, S.R., Becquer, T. 2003. Evaluation of the WAVE Model for Predicting Nitrate Leaching for Two Contrasted Soil and Climate Conditions. *Vadose Zone Journal* 2: 76–89.

Dynarski, K., Morford, S. L., Mitchell, S.A., Houlton, B.Z. 2019. Bedrock nitrogen weathering stimulates biological nitrogen fixation. *Ecology* 100: 2–11.

Elrashidi, M.A., Mays, M.D., Fares, A., Seybold, C.A., Harder, J.L. Peaslee, S.D., VanNeste, P. 2005. Loss of nitrate-nitrogen by runoff and leaching for agricultural watersheds. *Soil Science* 170: 969–984.

Fowler, D., Coyle, M., Skiba, U., Sutton, M.A., Cape, J.N., Reis, S., Sheppard, L.J., Jenkins, A., Grizzetti, B., Galloway, J.N., Vitousek, P., Leach, A., Bouwman, A.F., Butterbach-Bahl, K., Dentener, F., Stevenson, D., Amann, M., Voss, M. 2013. The global nitrogen cycle in the twenty-first century. *Phil. Trans. Roy. Soc. B.* 368: 20130164.

Frank, D.A., Groffman, P.M., Evaus, R.D., Tracy, B. F. 2000. Ungulate stimulation of nitrogen cycling and retention in Yellowstone Parl grasslands. *Oecologia* 123, 116–121.

Garcia-Ruiz, R., Baggs, E.M. 2007. $N_2O$ emissions from soils following combined application of fertilizer-N and ground weed residues. *Plant and Soil* 299: 263–274.

Garcia-Ruiz, R., Ochoa, V., Hinojosa, M.B., Carreira, J.A. 2008. Suitability of enzyme activities for the monitoring of soil quality improvement in organic agricultural systems. *Soil Bio. Biochem.* 40: 2137–2145.

Glass, A.D.M., Shaff, J., Kochian, L. 1992 Studies of the uptake of nitrate in barley. IV. Electrophysiology. *Plant Physiol.* 99: 456–463.

Goodale, C.L, Aber, J.D. 2001. The long-term effects of land-use history on nitrogen cycling in northern hardwood forest. *Ecological Applications* 11: 253–267.

Grouzis, J-P., Pouliquin, P., Rigand, J., Grignon, C., Gibrat, R. 1997. In vitro study of passive nitrate transport by native and reconstituted corn root cells. *Biochim. Biophys. Acta.* 1325: 329–342.

Güldner, D., Krausmann, F., Winiwarter, V., 2016. From farm to gun and no way back: Habsburg gunpowder production in the eighteenth century and its impact on agriculture and soil fertility. *Reg. Environ. Change* 16: 151–162.

Handley, L.L., Raven, J.A. 1992.The use of natural abundance of nitrogen isotopes in plant physiology and ecology. *Plan, Cell and Environment* 15: 965–985.

Hofstra, N., Bouwman, A.F. 2005. Denitrification in agricultural soils: Summarizing published data and estimating global annual rates. *Nutrient Cycling in Agroecosystems* 72: 267–278.

Holloway, J.M., Dahlgren, R.Z. 2002. Nitrogen in rock: Occurrences and biogeochemical implications. *Global Biogeochem Cycles* 16: 1118, doi 10.1029/2002GB001862.

Houlton, B.Z., Morford, S.L., Dahlgreen, R.A. 2018. Convergent evidence for widespread rock nitrogen sources in Earth´s surface environment. *Science* 360: 58–62.

Imsande, J., Touraine, B. 1994. N demand and the regulation of nitrate uptake. *Plant Physiol.* 105: 3–7.

Iritani, W.M., Arnold, C.Y. 1960. Nitrogen release of vegetable crop residues during incubation as related to their chemical composition. *Soil Sci.* 9: 74–82.

Jenny, H. 1941. *Factors of Soil Formation: A System of Quantitative Pedology.* McGraw-Hill.

Johnson B., Goldblatt, C. 2015. The nitrogen budget of Earth. *Earth Sci, Rev.* 148: 150–173.

Kennedy, M.J., Hedin, L.O., Derry, L.A. 2002. Decoupling of unpolluted temperate forests from rock nutrient sources revealed by natural 87Sr/86Sr and 84Sr tracer addition. *Proceedings of the National Academy of Science* 99: 9639–9644.

Latifah, O., Ahmed, O.H., Muhamad, A.M. 2011. Reducing ammonia loss from urea and improving soil exchangeable ammonium and available nitrate in non waterlogged soils through mixing zeolite and sago (metroxylon sagu) waste water. *International Journal of Physical Sciences* 6: 866–870.

Lee, R.B., Rudge, K.A. 1986. Effects of nitrogen deficiency on the absorption of nitrate and ammonium by barley plants. *Ann. Bot.* 57: 471–486.

Mariotti, J.C., Germon, P., Hubert, P., Kaiser, R., Létolle, A., Tardieux, P. 1982. Experimental determination of nitrogen kinetic isotope fractionation: some principles; illustration for the denitrification and nitrification processes. *Plant and Soil* 62: 413–430.

Marschner, H. 1995. *Mineral Nutrition of Higher Plants*. London: Academic Press.

McClure, P.R., Kochian, L.V., Spanswick, R.M., Shaff, J.E. 1990. Evidence for cotransport of nitrate and protons in maize roots. I. Effects of nitrate on the membrane potential. *Plant Physiol.* 93: 281–289.

Meharg, A., Blat, M. 1995. $NO_3^-$ transport across the plasma membrane of *Arabidopsis thaliana* root hairs: Kinetic control by pH and membrane voltage. *J. Membrane Biol.* 145: 49–66.

Miller, A.J., Smith, S.J. 1996 Nitrate transport and compartmentation. *J. Exp. Bot.* 47: 843–854.

Nadelhoffer, K.J., Fry, B. 1994. Nitrogen isotope studies in forest ecosystems. In: *Stable Isotopes in Ecology*, ed. K. Lajtha, and R. Michener, 22–44. Oxford: Blackwell.

Nagel, O.W., Lambers, H. 2002 Changes in the acquisition and partitioning of carbon and nitrogen in the gibberellin-deficient mutants A70 and W335 of tomato (*Solanum lycopersicum* L.). *Plant Cell Environ.* 25: 883–891.

Owen, A.G., Jones, D.L. 2001. Competition for amino acids between wheat roots and rhizosphere microorganisms and the role of amino acids in plant N acquisition. *Soil Biol. Biochem.* 33: 651–657.

Pal'ove-Balang, P., Mistrik, I. 2002 Control of nitrate uptake by phloem-translocated glutamine in *Zea mays* L. seedlings. *Plant Biol.* 4: 440–445.

Piccolo, M.C., Neill, C., Cerri, C.C. 1994. [15]N Natural abundance in soils along forest-to-pasture chronosequences in the western Brazilian Amazon Basin. *Oecologia* 99: 112–117.

Reisenauer, H.M. 1978. Absorption and utilization of ammonium nitrogen by plants. In *Nitrogen in the environment*, ed D.R. Nielsen and J.G. McDonald, vol II. pp. 157–170. London and New York: Academic Press.

Robinson, D. 2001. N as an integrator of the nitrogen cycle. *Trends in Ecology and Evolution* 16: 153–162.

Smil, V. 1999. Nitrogen in crop production: an account of global flows. *Global Biogeochemical Cycles* 13: 647–662.

Smith, R.V., Burns, L.C., Doyle, R.M., Lennox, S.D., Kelson, B.H.L., Foy B., Stevens, R.J.. 1997. Free ammonia inhibition of nitrification in river sediments leading to nitrite accumulation. *Journal of Environmental Quality* 26: 1049–1055.

Strohm, T.O., Griffin, B., Zumft, W.G., Schink, B. 2007. Growth yields in bacterial denitrification and nitrate ammonification. *Appl Environ Microbiol* 73: 14201424.

Subler, S., Parmelee, R.W., Allen, M.F. 1995. Comparison of buried bag and PVC core methods for in situ measurement of nitrogen mineralization rates in an agricultural soil 1. *Communications in Soil Science and Plant Analysis* 26: 2369–2381.

Teyker, R.H., Jackson, W.A., Volk, R.J., Moll, R.H. 1988. Exogenous [15]$NO_3^-$ influx and endogenous [14]$NO_3^-$ efflux by two maize (*Zea mays* L.) in breds during nitrogen deprivation. *Plant Physiol.* 86, 778–781.

Thomas, G.W., Swoboda A.R. 1970. Anion exclusion effects on chloride movement in sols. *Soil Sci.* 110: 163–166.

Tiedje, J.M. 1988. Ecology of denitrification and dissimilatory nitrate reduction to ammonium. In *Environmental Microbiology of Anaerobes*, ed. A.J.B. Zehnder. New York; John Wiley.

Tillard, P., Passama, L., Gojon, A. 1998 Are phloem amino, acids involved in the shoot to root control of $NO_3^-$ uptake in Ricinus communis plants? *J. Exp. Bot.* 49: 1371–1379.

Torres-Cañabate, P., Davidson E.A., Bulygina, E., Garcia-Ruiz, R., Carreira, J.A. 2008. Abiotic immobilization of nitrate in two soils of relic Abies pinsapo-fir forests under Mediterranean climate. *Biogeochemistry* 91: 1–11.

Vadivelu, V.M., Keller, J., Yuan, Z. 2007. Free apmonioa and free nitrous acid inhibition on the anabolic and catabolic process of *Nitrosomonas and Nitrobacter. Water Sci. Technol.* 56: 89–97.

Vidmar, J-, Zhuo, D., Siddiqi, M., Schjoerring, J., Touraine, B., Glass, A. 2000 Regulation of high-affinity nitrate transporter genes and high-affinity nitrate influx by nitrogen pools in roots of barley. *Plant Physiol.* 123: 307–318.

Vitousek, P.M., Aber, J.D., Howarth, R.W., Likens, G.E., Matson, P.A., Schindler, D.W., Schlesinger, W.H., Tilman, D.G. 1997. Human alteration of the global nitrogen cycle: Sources and consequences. *Ecological Applications* 7: 737–750.

Von Wirèn, N., Gazzarrini, S., Frommer, W.B. 1997. Regulation of nitrogen uptake in plants. *Plant and Soils* 196: 191–199.

Wang, Z.H., Li, S.L. 2019. Nitrate N loss by leaching and surface runoff in agricultural land: a global issue (a review). *Advances in Agronomy* 156: 159–217.

Wlodarczyk, T., Stepniewska, Z., Brzezinska, M. 2003. Denitrification, organic matter and redox potential transformation in Cambisols. *Int. Agrophysics* 17, 219–227.

Zhen, R.G., Koyro, H.W., Leigh, R.A., Tomos, A.D., Miller, A.J. 1991 Compartmental nitrate concentrations in barley root-cells measured with nitrate-selective microelectrodes and by single cell sap sampling. *Planta* 185: 356–361.

# 2 Nitrate in Agricultural Soils

*Mélida Gutiérrez, Maria Teresa Alarcón-Herrera,
Esperanza Y. Calleros-Rincón, and
Melissa Bledsoe*

## CONTENTS

2.1   Introduction .................................................................................................26
2.2   The N-Cycle in Soils ...................................................................................27
    2.2.1   Nitrification .....................................................................................27
    2.2.2   Denitrification..................................................................................28
    2.2.3   N Mineralization..............................................................................28
    2.2.4   Role of Microbial Communities......................................................29
2.3   N Fertilizers.................................................................................................29
    2.3.1   N Fertilizers: Urea...........................................................................29
    2.3.2   Other Mineral Fertilizers ................................................................30
    2.3.3   Livestock Manure Application .........................................................30
2.4   N Budget ......................................................................................................31
    2.4.1   N Content in Soil and N Fluxes.......................................................31
    2.4.2   N Budget Calculations .....................................................................32
2.5   Nitrate Management Optimization...............................................................34
    2.5.1   Timing and Split Application ...........................................................34
    2.5.2   Cover Crops and Tillage...................................................................34
    2.5.3   N Stabilizers and Inhibitors.............................................................35
    2.5.4   Field Sensors ....................................................................................35
    2.5.5   Optimization Modeling ....................................................................35
2.6   Sustainable Soil Management ......................................................................36
    2.6.1   Soil Degradation...............................................................................36
    2.6.2   Water Pollution.................................................................................36
    2.6.3   Air Pollution.....................................................................................37
2.7   Case Study: Comarca Lagunera of Mexico..................................................37
    2.7.1   Description of the Area.....................................................................37
    2.7.2   Manure and Fertilizers.....................................................................38
    2.7.3   Impact on Human Health .................................................................38
2.8   Conclusions ..................................................................................................38
References.................................................................................................................39

DOI: 10.1201/9780429326806-3

## 2.1 INTRODUCTION

Soil is a non-renewable, natural resource that holds the enormous challenge of producing food for eight billion people, 821 million of whom are undernourished (Food Agricultural Organization, FAO, 2019). As an integral part of agro-ecosystems, soil provides a matrix, water, and the nutrients needed for plants to grow. Soil is therefore vital to the people who depend upon it for food and non-food products. Additionally, soil plays a primary role in the biogeochemical cycling of elements vital to life, such as carbon (C), N, phosphorous (P), and potassium (K) (Michalski et al. 2004, Robertson and Vitousek 2009, Espie and Ridgway 2020).

Nitrogen is a primary component of nucleotides and proteins, an essential nutrient for carbohydrate metabolism and therefore for plant growth (Robertson and Vitousek 2009). Although other elements are also required by plants, N is generally the limiting nutrient for production in agricultural soils and, thus, it is applied in larger amounts than any other nutrient, including P or K. In the last few decades, technological improvements in farming and fertilization have continued to increase crop yield, but have also exposed soil to one or more types of degradation by erosion, salinization, compaction, acidification, and/or chemical contamination (White et al. 2012, FAO 2018). It is currently estimated that one-third of soils worldwide have been degraded (FAO 2018).

From 1950 to the 1970s, large amounts of synthetic fertilizers were first applied to soils, causing crop yields to increase significantly (Galloway et al. 2008). Some countries lagged behind others in adopting these fertilization practices but, eventually, these applications became common worldwide (Lu and Tian 2017). The precise impact of fertilizers to crop yields is hard to calculate since other technological advances in farming and machinery occurred during the same period (Erisman et al. 2008, Shi et al. 2020). However, the role of synthetic N-fertilizers is undeniable and has been estimated to be responsible for about half of crop production worldwide (Erisman et al. 2008, Hansen et al. 2017).

In the decades that followed the application of synthetic fertilizers, farmers focused on the maximization of crop yield, and the detrimental environmental consequences of its overuse were largely overlooked. These non-sustainable practices in fertilizer use and crop management resulted in the degradation of soil, water, and air (Sebilo et al. 2013, Congreves et al. 2016, Robertson and Vitousek 2009, Shi et al. 2020). The resulting N losses from cropping systems also resulted in significant profit losses for farmers (Norton and Ouyang 2019, Shi et al. 2020). Thus, minimizing N losses while maintaining soil health and food production is the ultimate goal for farmers and soil scientists alike, especially since the survival of soil as a resource is at stake. As viable solutions to maintain soil health are being sought, most stakeholders agree that farming should be shifted to more efficient and sustainable practices to minimize soil and environmental degradation (Wise 2019, Veldhuizen et al. 2020, Xia et al. 2020).

Plants generally acquire N from soil-absorbing mineral forms such as nitrate and ammonium. Some plants – legumes (alfalfa, bean, pea) – are also able to utilize N from $N_2$ in the atmosphere by forming nodules holding symbiotic $N_2$-fixing bacteria, e.g., rhizobia. With the atmosphere as an N source, farmers generally do not fertilize leguminous crops with additional mineral N sources. Plants without these symbiotic relationships, such as maize, wheat, and rice, rely on increased N supply in the soil and respond to N fertilization with significant yield increases (Guo et al. 2020, Shi et al. 2020). To increase the yield of non $N_2$-fixing crops, N is added to the soil either in the form of a mineral (synthetic) fertilizer such as urea or in an organic form such as manure (FAO 2018). Nowadays, more soil and plant scientists look for ways to optimize the application of fertilizer using a variety of approaches, from crop rotation and field sensors that quantify N losses to complex mathematical models (John et al. 2017, Espie and Ridgway 2020, Khoshnevisan et al. 2020, Veldhuizen et al. 2020).

In spite of advances in maximizing efficiency and minimizing N losses, the N-fertilizer consumption worldwide continues on an increasing trend at a compound annual growth rate of 1.5 percent. Global consumption of N fertilizers is expected to reach 119 million tons in 2020 (FAO 2017a).

In most soils, N in fertilizer is quickly transformed to nitrate, a soluble form of N that is available for plant uptake, a process known as nitrification. The rate of the chemical reactions is a function of temperature, soil pH, substrate availability, and soil biota (Norton and Ouyang 2019). Because of the solubility of nitrate, any excess of nitrate in soil solution readily leaches into streams or groundwater with negative environmental consequences and loss of profit to the farmer (John et al. 2017). A small amount of nitrate is also present in the air after oxidation of NOx and from aerosols (Michalski et al. 2004). The causes and management of soil nitrate and nitrification are discussed in more detail in the sections below.

## 2.2   THE N-CYCLE IN SOILS

N occurs in nature in various forms and valence states. Major inorganic forms in soils include nitrate $NO_3^-$, nitrite $NO_2^-$, nitrous oxide $N_2O$, dinitrogen gas $N_2$, ammonium $NH_4^+$, and ammonia $NH_3$, with valences +5, +3, +1, 0, −3, and −3, respectively (Robertson and Vitousek 2009). Organic N substances are present in soil mainly as amino acids and amino sugars. Such organic forms of N in the soil are generally unavailable to plants (Stevenson 1982, Castaldelli et al. 2020).

The N cycle is complex, yet very important in agricultural soils, where N transformations include $N_2$-fixation, assimilation, nitrification, ammonification, and denitrification. These processes are facilitated by bacteria and may require a catalyst to proceed, for example, the natural enzyme urease in ammonification.

### 2.2.1   NITRIFICATION

Nitrification in soil commonly takes place in two steps. In the first step, ammonia oxidizing bacteria such as *Nitrosomonas*, *Nitrosospira*, and *Nitrososphaera* oxidize $NH_4^+$ to $NO_2^-$, releasing $H^+$ in the process. The second step consists of further oxidation of $NO_2^-$ to $NO_3^-$ by nitrite oxidizing bacteria *Nitrobacter* or *Nitrospira*. Some *Nitrospira* bacteria will transform $NH_4^+$ to $NO_3^-$ in one step only, in a process known as commamox (complete ammonia oxidation) (Daims et al. 2015, Beekman et al. 2018).

Nitrification occurs rapidly in most soils, almost as rapidly as $NH_4^+$ forms (Schmidt 1982). Variables also speed up the conversion of ammonia to nitrate by ammonia-oxidizing microbial population, such as an increase in temperature, moisture, and long-term fertilization, thus favoring nitrification and leaching of N (Stevenson 1982). $NO_3^-$ is a soluble form of N that is stable under oxic conditions and readily leaches from soil to surface water and groundwater (Hansen et al. 2017, Gutiérrez et al. 2018). The two nitrification steps are shown in Eqs. 2.1 and 2.2, respectively:

$$NH_4^+ + 1.5O_2 \rightarrow NO_2^- + H_2O + 2H^+ \tag{2.1}$$

$$NO_2^- + 0.5O_2 \rightarrow NO_3^- \tag{2.2}$$

## 2.2.2 Denitrification

Denitrification removes N from the system as gaseous forms in several steps, where the intermediate forms $NO_2^-$, NO, and $N_2O$ are generated before the final form of $N_2$ (dinitrogen gas). The combined reaction is shown in Eq. 2.3. This reaction takes place helped by denitrifying bacteria under anoxic and oxic-anoxic interface conditions present in wet and water-logged soils and in soils with a large amount of organic matter. Other paths of denitrification have been identified, including anammox (anaerobic ammonium oxidation) and aerobic denitrification (Seitzinger et al. 2006). Denitrifying bacteria are a diverse and ubiquitous group of bacteria, including *Pseudomonas* and *Bacillus*. This high diversity facilitates denitrifying bacteria to cover a wide range of environmental conditions, for example, high salinity and high-temperature environments (Seitzinger et al. 2006).

Agricultural soils leach nitrate into surrounding canals or pools that form in flat areas. Flat topography and poor drainage promote denitrification and therefore a subsequent loss of N to the atmosphere. N loss by denitrification is hard to quantify in agricultural soils due to the fluxes of N in the soil with rapidly changing water conditions throughout the growing season (Castaldelli et al. 2020). Equation 2.3 shows the denitrification process.

$$2NO_3^- + 12H^+ + 10e^- \rightarrow N_2 + 6H_2O \tag{2.3}$$

## 2.2.3 N Mineralization

Mineralization converts organic forms such as protein and other nitrogenous organic substances (generally unavailable forms for plant uptake) to mineral or inorganic forms. Mineralization processes include proteolysis and aminization (Eq. 2.4) and ammonification (Eq. 2.5)

$$Organic \cdot N \rightarrow \cdot R - NH_2 \tag{2.4}$$

$$R - NH_2 \rightarrow NH_3 ; NH_3 + H_2O \rightarrow NH_4^+ + OH^- \tag{2.5}$$

## 2.2.4 Role of Microbial Communities

Soil microorganisms enable the N transformations shown in Eqs. 2.1 to 2.5. $NH_4^+$ to $NO_2^-$ oxidizers present in soil include *Nitrosomonas, Nitrosospira,* and *Nitrosococcus*. Bacteria oxidizing $NO_2^-$ to $NO_3^-$ belong to the genus *Nitrobacter*, and some *Nitrospira* will transform $NH_4^+$ to $NO_3^-$ directly (Daims et al. 2015). $NH_4^+$ to $NO_2^-$ can also be carried out by Nitrososphaera, an ammonia oxidizing archaea (AOA) bacteria (Beekman et al. 2018, Norton and Ouyang 2019). Also, in fertilized soils with high $NH_4^+$ concentrations, methane-oxidizing bacteria may transform $NH_4^+$ to $NO_2^-$ (Schmidt 1982, Seitzinger et al. 2006). Other factors affecting the work of bacteria are the presence of fungi, root exudates produced by plants, and the presence of nutrients such as organic C (Norton and Ouyang 2019, Espie and Ridgway 2020).

Soils that have received a long-term application of fertilizers will have well-adapted nitrifiers, and nitrification processes may occur more rapidly (Quan et al. 2016). Besides the abundance and diversity of these specialized bacteria, the rate and extent of nitrification depend on the availability of substrate ($NH_4^+$, $NH_3$, $NO_2^-$, $O_2$), environmental conditions, and the presence or absence of nitrification inhibitors (Norton and Ouyang 2019).

## 2.3 N FERTILIZERS

Global fertilizer consumption varies greatly from region to region and continues an increasing trend (Lu and Tian 2017, FAO 2018), except for Europe. Europe has reduced their N-fertilizer applications since mid-1980s and has maintained low rates following, likely a result of awareness of the environmental impacts and the European Union Nitrate Directive policy of 1991 (Sommers et al. 2004, FAO 2018).

### 2.3.1 N Fertilizers: Urea

There are many different forms of commercially available synthetic (mineral) N fertilizers. The most common in dry granular form is urea, accounting for about 57 percent of N fertilizer use worldwide (IFA 2016). Large-scale production of urea fertilizer became possible after the Haber-Bosch process was discovered, a process by which $N_2$ from the air is transformed into anhydrous ammonia. Urea, also called carbamide, is a solid that is highly soluble in water and has a high N-content (46.7% weight). Because of the high N-content, ease of application and low transportation costs, carbamide is used widely as an N-fertilizer worldwide (Sommers et al. 2004). Urea decomposes in a reaction known as ammonification. Urea is converted to ammonia or to ammonium in the soil in the presence of water and urease – an enzyme naturally present in soils (Eq. 2.6). This reaction is temperature-dependent; from 7–10 days at a temperature of 10 °C to about two days at 30 °C. Ammonia and ammonium maintain an equilibrium depending on pH value (Eq. 2.7).

$$\left(NH_2\right)_2 CO + H_2O \xrightarrow{\text{urease}} 2NH_{3(gas)} + CO_{2(gas)} \tag{2.6}$$

$$NH_{3(gas)} + H_2O \rightleftharpoons NH_4^+ + OH^- \qquad (2.7)$$

$NH_3$ reacts with water to form $NH_4^+$, a form of N that is available to plants. Alkaline soil conditions shift Eq. 2.7 to the left, favoring the formation of $NH_{3(gas)}$ that is lost to the atmosphere. Increased temperature also favors volatilization. An average of about 20 percent and upwards to 50 percent of N from a broadcast application of urea may be lost to volatilization (DelMoro et al. 2017). Prevention of such $NH_3$ losses is of high priority to users and is aided by one or more of the following practices, (a) incorporating the fertilizer concurrently with moisture (e.g., irrigation); (b) regulating the effect of urease to slow down the reaction shown in Eq. 2.6: and (c) adding a time-release mechanism to the fertilizer (Robertson and Vitousek 2009).

Coating urea granules with a substance or membrane to reduce solubility slow-releases urea and generally results in an increase of fertilizer efficiency and a reduction of gaseous N loss. The first coating materials were sulfur-based, sealed with a layer of polyethylene or wax, but now a wide variety of coatings are available, among them humate, lime, and guano (Robertson and Vitousek 2009, Espy and Ridgway 2020). Release rates delivered by specific coatings can be adjusted to particular conditions of soil temperature and moisture. Increases in N efficiency have been reported in the order of 10 percent in the grass (Espy and Ridgway 2020) and 30 percent in maize (Zheng et al. 2020) using coated urea.

### 2.3.2 OTHER MINERAL FERTILIZERS

Other mineral fertilizers include a variety of commercial soluble forms of N, usually combined with nutrients such as P and K. N commercial fertilizers include ammonium bicarbonate, ammonium sulfate, ammonium chloride, hydrous and anhydrous ammonia, and nitrate nitrogen. Some plants (e.g., sugar beets, radishes) show a preference for nitrate nitrogen fertilizers. Nitrate nitrogen includes sodium nitrate, calcium nitrate, and ammonium nitrate (Robertson and Vitousek 2009). Several of these may be corrosive or combustive, requiring special storage and care.

### 2.3.3 LIVESTOCK MANURE APPLICATION

Livestock manure is widely used as a fertilizer because of its high content of N and organic carbon ($C_{org}$), and other minor nutrients such as zinc and sulfur (FAO 2018, Cui et al. 2020, Guo et al. 2020). Organic matter applied to soils is also known to improve soil structure and moisture-holding capacity, facilitate water infiltration and mitigate N losses and greenhouse gas emissions (Galloway et al. 2008).

In the 53 years from 1961 to 2014, the global input of N to agricultural land through N fertilizer, both in manure and synthetic forms, showed a small increase of manure (18 to 28 million tons of N) compared to the 12 to 102 million tons increase of N synthetic fertilizers (FAO 2018). The livestock species contributing the largest amount of manure-N to agricultural land are cattle, swine, and poultry (FAO 2018).

In agricultural systems, a mixture of manure and fertilizer is commonly utilized to meet nutrient requirements, with producers seeking a balance between crop yields and possible environmental impacts (Cui et al. 2020, Guo et al, 2020). For instance, nitrate leaching has been found to be reduced in systems using manure applications but, under certain circumstances, a large amount of P may be released as well (Cui et al. 2020).

## 2.4 N BUDGET

The N budget of the fluxes in agricultural soils is key to determine where and how to reduce N losses, and is vital information to either determine best management practices of a particular agricultural system or to assess their effectiveness. Figure 2.1 depicts the main variables involved in the gains and losses of an N budget.

N budgets are reported on a yearly basis for an entire crop cycle, usually involving more than one crop. Measurements needed to calculate this budget are difficult to quantify due to the dispersed nature of both gaseous N losses and leachates (Galloway et al. 2008). In addition, these measurements require considerable expense and effort (Oertel et al. 2016, Robertson and Vitousek 2009). Therefore, few studies include a complete N budget. Another hindrance is that most of the available N balance studies are conducted in temperate regions and less is known about N fluxes in tropical areas (Galloway et al. 2008, Robertson and Vitousek 2009).

### 2.4.1 N CONTENT IN SOIL AND N FLUXES

Inorganic N concentrations in soils vary widely on soil type and management (Table 2.1). Even within one location, such N pools vary by type of crop, timing in the crop cycle, and crop-management techniques.

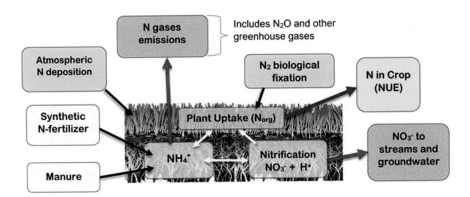

**FIGURE 2.1**  Schematic representation of the main nitrogen fluxes operating in agricultural soils. N losses include gaseous emissions to the atmosphere and nitrate leaching to the surface and groundwater. These losses, which may account for up to 50 percent of the applied N, represent a loss of profit to the farmer and may be deleterious to the environment.

**TABLE 2.1**
**Content of Total Nitrogen (TN) Nitrate and Ammonium in Some**
**Representative Agricultural Soils**

| Crop, Region | Depth cm | TN g kg$^{-1}$ | NO$_3$-N mg kg$^{-1}$ | NH$_4^+$-N mg kg$^{-1}$ | Soil Texture | Reference |
|---|---|---|---|---|---|---|
| **Rice** Shangai China | 0–15 | 2.3 | 2.4 | 14.9 | silty clay loam | Cui et al. (2020) |
| **Maize** N. China Plain | 0–20 | 1.00 | 25.0 | 5.1 | silt loam | Guo et al. (2020) |
| **Maize** Zimbabwe | 0–50 | 0.30 | n.r. | n.r. | sand | Masvaya et al. (2017) |
| **Chile, Onion** New Mexico | 0–60 | n.r. | 14-16 | n.r. | clay | Sharma et al. (2012) |
| **Wheat-Maize** N. China Plain | 0–40 40–100 | 0.68 0.40 | 9.1 6.3 | 5.9 4.9 | sandy loam silt loam | Zheng et al. (2020) |

Note: n.r. = not reported

The N budget in soils is conventionally reported in g N ha$^{-1}$, except for crop yield, which is reported as nitrogen use efficiency (NUE) in %. NUE is a way to measure the effectiveness of a particular change to farming with respect to a control plot. NUE can be measured in four different ways: the apparent recovery efficiency of applied N index (RE$_N$) kg taken up kg$^{-1}$ N applied; partial factor productivity (PFP$_N$), physiologic efficiency (PE$_N$), and agronomic efficiency (AE$_N$). The most commonly used is RE$_N$, which is calculated as follows:

$$RE_N = \frac{\left(U_N - U_0\right)}{F_N} \tag{2.8}$$

U$_N$ is the plant N uptake (kg ha$^{-1}$) measured in aboveground biomass at physiological maturity in a plot that received N at a rate of F$_N$ (kg ha$^{-1}$) and U$_0$ is the N uptake measured in aboveground biomass in a plot without the addition of N.

With over 50 percent loss of the total applied N for major crops in the world, the practices that lead to an efficient uptake of N need to be identified and implemented (Congraves et al. 2016, John et al. 2017, Cui et al. 2020, Guo et al. 2020). In research field trials, a variety of factors have been found to increase NUE up to 80–90 percent in irrigated maize, rice and wheat (Balasubramanian et al. 2004). However, NUE in most research rain-fed plots is about 40 percent, lower than that obtained in irrigated plots. In contrast, too little N not only affects the crop yield but may lead to soil degradation (Rocha et al. 2020).

## 2.4.2   N BUDGET CALCULATIONS

The N budget can be calculated by nitrogen inputs (fertilizer, fixation, mineralization and atmospheric deposition), N outputs (volatilization, denitrification, uptake by plants, and runoff), and change in N stored in soil, per year, according to the following equation (Castaldelli et al. 2020):

$$\Delta N_{Soil} = \text{inputs } (N_{fert} + N_{fix} + N_{min} + N_{dep}) - \text{outputs}$$
$$(N_{vol} + N_{denit} + N_{uptake} + N_{runoff}) \quad\quad (2.9)$$

N leaching can be measured directly with lysimeters and other field procedures, but for long-term measurements, the amount of $NO_3^-$ leached can also be obtained from the equation below (Meisinger and Randall 1991, John et al. 2017):

$$NO_{3\,leached}^- = \text{inputs } (N_{fert} + N_{fix} + N_{min} + N_{dep}) - \text{outputs}$$
$$(N_{vol} + N_{denit} + N_{uptake}) - (\text{soil } NO_{3\,final}^- - \text{soil } NO_{3\,initial}^-) \quad (2.10)$$

Simplifications are generally made for some of these parameters. For example, $N_{dep}$ data may be approximated from reports of a nearby air quality monitoring station, and volatilization amounts for the particular fertilizer may be assumed to be similar to the value for that area reported in the literature (John et al. 2017). Some primary N budget parameters have been reported for agricultural soils under a variety of conditions, as listed in Table 2.2.

With the variability in N budgets, determinations from a particular site or region cannot be applied to a different region or to a different cropping system within the

**TABLE 2.2**
**NUE and N Losses by $NO_3^-$-N Leaching and by N Volatilization ($NH_4^+$-N) Reported by Different Cropping Systems of Selected Studies**

| Crop (cycle) | NUE % | $NO_3^-$-N kg ha$^{-1}$ | $NH_4^+$-N kg ha$^{-1}$ | Reference Region |
|---|---|---|---|---|
| Maize-soybean-wheat | 70 | 204 | 51–10 | Congraves et al. (2016) Ontario, Canada |
| Best management of above | 84 | 105 | 9–7 | Congraves et al. (2016) Ontario, Canada |
| Maize (50% manure) | 56 | 30 | 0.05 | Guo et al. (2020) North China Plain |
| Grass-maize | 85 | 0.5 | 12–1 | Rocha et al. (2020) Sao Paulo, Brazil |
| Winter wheat | 63 | 12–15 | n.r. | Sebilo et al. (2013) France |
| Onion | 31 | 144.6 | 1.4* | Sharma et al. (2012) New Mexico, USA |
| Chile | 51 | 130.3 | 3.4* | New Mexico, USA |
| Wheat-Maize | 45–67 | 119 | 613* | Zheng et al. (2020) North China Plain |
| Rice | 32 | 23–25 | 78 | Shi et al. (2020) Hubei, China |

* Undifferentiated gaseous emissions including $NH_3$ and denitrification.

same region (Robertson and Vitousek 2009). Additionally, while the N recovery is determined from one growing season, real-world N budgets would have additional factors from cycling with previous and subsequent crops.

## 2.5  NITRATE MANAGEMENT OPTIMIZATION

Besides the major options of commercial fertilizers and using a fertilizer-manure combination, other management practices are implemented to further increase effectiveness and reduce N losses. Some of the most commonly used practices are discussed below.

### 2.5.1  Timing and Split Application

One critical management technique to reduce N losses is to schedule fertilizer applications to coincide nutrient availability in the soil with times of rapid nutrient uptake and utilization by the plants (Cui et al. 2020). A technique to ensure N availability throughout the growing season is split applications, or to divide the total rate of fertilizer into two or more applications, for example, split into basal, first topdressing, and second topdressing. Split applications have been found to increase yield, improve grain quality (Grahmann et al. 2014, Norton and Ouyang 2019), and reduce N losses (Cui et al. 2020). The quantification of losses after each application can be used to identify which application leads to greater losses and to identify best management practices (Cui et al. 2020).

### 2.5.2  Cover Crops and Tillage

Cover crops (also known as catch crops) are planted amongst or following primary crops to cover the soil surface and avoid bare ground. Some common cover crops include legumes (alfalfa, clover, pea), non-legumes (spinach, flax) grasses (ryegrass, barley), and brassicas (radishes, turnip) (John et al. 2017, Abdalla et al. 2019). Depending on the soil and climate conditions, these crops can perform one or more of the following functions: increase N in the soil (legumes), entrap nitrate (non-legumes) that otherwise would leach out, increase soil organic carbon, water capacity, and soil porosity, and reduce soil erosion. The use of cover crops retains additional N in organic matter but NUE needs to be carefully monitored to avoid deleterious effects of the primary crop (Norton and Ouyang 2019). Novel practices combining cover crop and ecological engineering have also been explored to reduce both nitrate leaching and soil erosion, for example, in citrus orchards (Li et al. 2020).

Tillage is another practice that has been shown to directly affect the leaching of nitrate from soils. Nitrate leaching is decreased when plant residues are left on the soil surface (Grahmann et al. 2014, Grahman et al. 2018, Hess et al. 2020). Studies report the effects to crop yield and nitrate leaching after varying tillage practices, for example, no-tillage, tillage-straw partially retained, tillage-burned straw retained (Grahmann et al. 2014), and also with respect to other variables such as split application and storm events (Hess et al. 2020).

### 2.5.3   N Stabilizers and Inhibitors

N stabilizers and inhibitors are added to fertilizers to either promote denitrification to reduce N leaching (Wen et al. 2020) or to delay transformations of the fertilizer that would result in N losses such as leaching of $NO_3^-$, $NH_3$ volatilization. and $N_2O$ emissions (Wang et al. 2020a). One such practice involves the coating of the fertilizer with a stable form of organic carbon such as humate or a combination of humate with guano or lime (Espie & Ridgway 2020, Holub et al. 2020).

Nitrification inhibitors (NI) reduce nitrification reactions and gaseous N emissions from fertilizers (Wang et al. 2020a). Many different NI products are commercially available (Beeckman et al. 2018), including some biological chemicals (e.g., root exudates). NI products should be tested prior to their application to the field, as some inhibitors that reduce nitrification also increase $NH_3$ emissions (Wang et al. 2020a).

### 2.5.4   Field Sensors

Field sensors help collect the information needed to monitor crops and fertilizer efficiency, and to calculate the N balance. Sensors may focus on estimating NUE, nitrate leaching, and/or gaseous emissions to estimate N balances. Because of the complex and diffuse nature of these fluxes, a particular sensor may have an optimal range of performance, which may be a limitation in some situations. Additionally, sensors are being adapted to be more user-friendly and affordable (Oertel et al. 2016, Insausti et al. 2020).

Sensors are often used to continuously measure nitrate in agricultural soil systems. The conventional method to measure nitrate leaching from an agricultural system is to analyze a representative sample of leachate water. This can be done using an automatic sensor at a drainage discharge location (Jones et al. 2018), by analyzing water samples collected by a lysimeter, canal, or runoff pool (Cui et al. 2020, Holub et al. 2020, Shi et al. 2020), or by resin samplers installed in the soil (Grahmann et al. 2018).

Gaseous N emissions from soils can be measured in the field or in the laboratory, or indirectly by airborne and mathematical models (Oertel et al. 2016). A common field method is a chamber-based analysis that consists of a box or cylinder placed onto the soil surface directing gases to accumulate in its chamber headspace (Shi et al. 2020). The collected gases can be analyzed by chromatography, chemiluminescence, or photoacoustics. Another commonly utilized method is the micrometeorological method composed of a 3-D ultrasonic anemometer and a gas analyzer attached to a tower or mast (Wagner-Riddle et al. 2007, Oertel et al. 2016).

Other technologies that are rapidly advancing in agriculture are remote sensing methods, including special satellite imagery and unmanned aerial vehicle (UAV) technologies (Maes and Steppe 2018, Holub et al. 2020). UAVs can carry sensors to help measure N fluxes needed to complete an N budget and to test the effectiveness of best practices (Oertel et al. 2016, Wang et al. 2020b).

### 2.5.5   Optimization Modeling

Meteorological and soil parameters (e.g., texture, organic matter content, temperature) can serve as input data to mathematical models that include general physical and chemical processes. Such process-oriented models can model daily decomposition, nitrification, ammonia volatilization, and $CO_2$ production (Oertel et al. 2016). Once the model is validated, it can obtain results from a variety of management practices and scenarios. For example, models have been used to optimize N consumption rates in temperate zones (Khoshnevisan et al. 2020) and rice yields in humid areas (Shi et al. 2020).

## 2.6  SUSTAINABLE SOIL MANAGEMENT

The sustainable management of the soil resource is key to its existence, and it is also needed to address vital services to mankind such as supporting agriculture, food security, and fighting poverty (White et al. 2012, FAO 2018, Coomes et al. 2019). The environmental consequences of excess N fertilization affect the agricultural systems at the local and regional scale through the contamination of groundwater, acidification of soil, and eutrophication of surface water. On a global scale, these N imbalances are contributors to acid rain, climate change, ozone-layer depletion, and coastal dead zones (Hansen et al. 2017, Jones et al. 2018).

### 2.6.1  SOIL DEGRADATION

Chemical and structural degradations of soil often result following intensive agriculture practices with high fertilizer inputs. While higher crop yields are obtained from this practice, organic carbon is depleted from the soil, causing a decrease in the soil biological functionality. Soil chemical properties such as pH, cation exchange capacity, organic matter, and macro- and micronutrient availability also change as a result of N transformations. Under long-term N fertilization, non-buffered soil becomes acidic (Huang et al. 2019). Acidification results from atmospheric deposition of $NO_x$ and from fertilization through the oxidation of ammonia from either synthetic or manure fertilizers (see Eq. 2.1). Soil physical properties are also degraded, including damage to soil structure leading to compaction, root growth restrictions, and water-logging (White et al. 2012, FAO 2018). This in turn increases crop susceptibility to other stresses, including limited water and nutrient access, and diseases and insect pressures. All of these stresses directly affect land productivity (White et al. 2012, FAO 2018).

### 2.6.2  WATER POLLUTION

In agricultural systems, water pollution from nutrients occurs when fertilizers are applied at a greater rate than they are fixed by soil particles or taken up by plants. With only 25 to 85 percent of applied N estimated to be taken up by plants, excess N remains in the soil. Excess N forms soluble nitrate and leaches into groundwater or moves via surface runoff into waterways (FAO 2017b, Xia et al. 2020). In many agricultural areas, underlying aquifers are used as a drinking water source (Mateo-Sagasta et al. 2017). High nitrate concentration in drinking water can have negative effects on human health, such as methemoglobinemia in infants, cancer, and thyroid affectations (Ward et al. 2005, Hansen et al. 2017).

The N-load in surface water and groundwater has also caused eutrophication of freshwater and marine ecosystems, leading to the formation of localized hypoxia or "dead zones" in coastal marine ecosystems (Boesch 2019, Hess et al. 2020). In marine ecosystems, eutrophication results in diminished water clarity, the massive growth of toxic algae in phytoplanktonic communities, and altered macrovegetation in nearshore environments (Boesch 2019). The decomposition of the dead organisms can produce toxic compounds and vectors for infections (Galloway et al. 2008, Boesch 2019). Practices to entrap N carried by runoff from agricultural areas include ecological ditches and constructed wetlands. These may remove up to 5.2 kg N $km^{-1}$ $day^{-1}$ depending on the type of vegetation and harvesting times (Xia et al. 2020).

### 2.6.3 AIR POLLUTION

Atmospheric N and consequently $NO_x$ deposition has increased significantly (2- to 4-fold) since the onset application of synthetic fertilizers (Michalski et al. 2004, Galloway et al. 2008). Global problems associated with N compounds in the atmosphere, especially the powerful greenhouse gas $N_2O$, include global climate change and destruction of the ozone layer. An increase in atmospheric N oxides also results in a decrease in agricultural productivity and, from a human health standpoint, respiratory and cardiovascular problems (Galloway et al. 2008).

## 2.7   CASE STUDY: COMARCA LAGUNERA OF MEXICO

### 2.7.1 DESCRIPTION OF THE AREA

The Comarca Lagunera is a region in northern Mexico within the Chihuahuan Desert. This area is located within 102°03'09" and 104°46'12" longitude and 24°22'21" and 26°52'54" latitude, has an arid climate and average annual precipitation of 273 mm. Topographically, the area is flat and slightly depressed, with a mean elevation of 1130 masl. In the past, about a dozen endorheic shallow lagoons existed in the area, most of which are now dry (Descroix 2004). Their presence indicates that runoff will converge into the area where playa lakes were once located, from which part will evaporate and part will infiltrate into the subsurface.

The Comarca Lagunera encompasses over 4 million ha from which over 90 percent are desert, fallow land, and 130,000 ha (1,300 $km^2$) are irrigated agricultural land (Fortis-Hernandez et al. 2010). About 89,500 ha are used to grow alfalfa, oats, and maize as fodder for cattle and dairy cows, but other crops including sorghum, wheat, and cotton are also grown. The agricultural portion of this region is renowned for its dairy farms and dairy industry. Land modifications to increase and enhance agriculture and dairy production in the past hundred years or so have changed the drainage patterns and the recharge of the playa lakes. With little to no surface water, groundwater has been withdrawn from aquifers faster than they are being recharged, with a reported hydrostatic level drop of 1.4 m yr.[1] These changes in water input have caused some irrigated areas to return to desert conditions (Descroix 2004).

### 2.7.2 MANURE AND FERTILIZERS

The Comarca Lagunera contains the largest number of dairy cattle in Mexico, with an average of 500,000 cows, accounting for 20 percent of the total number of cows in the country (Fortis-Hernandez et al. 2010). The dairy cow's metabolism is characterized by low efficiency in the use of nutrients, especially of N. Cows assimilate a mere 30 percent of the N they ingest and excrete the rest (Van Horn et al. 1996).

The manure production of the region is 619,000 tons year$^{-1}$ (Fortis-Hernandez et al. 2010) and, when spread onto land, amounts to an average of 130 tons manure ha$^{-1}$ year$^{-1}$. These large application rates, along with poor manure-management practices, generate a significant contamination risk to groundwater despite the deep hydrostatic levels of the underlying aquifers (Figueroa et al. 2009). Concentrations up to 45 mg L$^{-1}$ of NO$_3$-N have been measured in the deep aquifer used as a drinking-water source, whereas shallow and surface water used for irrigation had 83 to 124 mg L$^{-1}$ NO$_3$-N (Calleros-Rincón, 2012a). Both the shallow and deep-water samples greatly exceed the Mexican guideline of 10 mg L$^{-1}$ NO$_3$-N.

Human activities have profoundly disturbed the N cycle in the form of synthetic fertilizers and manure added to agricultural soils, inflows from sewage and septic tanks, and waste flows from dairy farms. To a lesser extent, additional sources of N include NOx gases released by industry, motor vehicle exhaust, and fossil-fuel combustion, as well as other gases such as ammonia produced by dairy farms and cattle-confined operations.

### 2.7.3 IMPACT ON HUMAN HEALTH

In the Comarca Lagunera, the application of N fertilizers and manure to soils supporting the intensive agriculture and dairy farms have contaminated aquifers with NO$_3^-$, with levels varying between 3 and 45 mg L$^{-1}$ NO$_3$-N (Calleros-Rincón et al. 2012a). Health problems may arise due to ingestion of drinking water of high NO$_3^-$ content, among them methemoglobinemia (Ward et al. 2005). Although acute levels of methemoglobinemia have not been detected in the region, the blood methemoglobin limit of 1.5 percent has been surpassed in some instances (Calleros-Rincón et al. 2012b). Other reported health impacts relate to hyperthyroidism and human reproduction, the latter causing longer-term pregnancies (Calleros-Rincón et al. 2018). The high concentration of NO$_3^-$ in drinking water may have also contributed to the cases of hyperthyroidism that are common in the region, but more research is needed to verify this association.

## 2.8 CONCLUSIONS

The role of N in soils and plants of agricultural systems is complex, and it has become harder to estimate due to major changes in N applications in farming systems. N losses through nitrification and gas emissions amount for about half of the applied N. This is a considerable loss in profit to the farmer and a cause of environmental degradation. Best agricultural practices are sought to reduce N losses while increasing crop yield, thus aiming towards soil sustainability.

One major difficulty in N management is that a best practice found for one cropping situation may not be valid for another. Measuring N fluxes in agricultural soils remains a challenge, resulting in incomplete N budgets. Promising results to reach a sustainable N balance in agricultural soils include advances in technology (e.g., more precise sensors, nitrification inhibitors, and time-release fertilizers) as well as advances in mathematical models that can accommodate a large spectrum of variables, including losses, fertilizer prices, and farmers' cultural values and viewpoints. As such, a case study in a prominent agricultural area in northern Mexico sheds some light on the complexities that can be encountered at some of the world's agricultural regions.

## REFERENCES

Abdalla, M., A. Hastings, K. Cheng, et al. 2019. A critical review of the impacts of cover crops on nitrogen leaching, net greenhouse gas balance and crop productivity. *Glob Chang Biol* 25, 2530–2543.

Balasubramanian, V., B. Alves, M. Aulakh, et al. 2004. Crop, environmental, and management factors affecting nitrogen use efficiency. In *Agriculture and the Nitrogen Cycle*, ed. A.R. Mosier, J.K. Syers and J.R. Freney, 3–18, SCOPE (Scientific Committee on Problems of the Environment) Report 65, Paris.

Beekman, F., H. Motte, T. Beeckman. 2018. Nitrification in agricultural soils: impact, actors and mitigation. *Curr Opin Biotechnol* 50: 166–173.

Boesch, D.F. 2019. Barriers and bridges in abating coastal eutrophication. *Front Mar Sci* 6: 123. doi.org/10.3389/fmars.2019.00123.

Calleros-Rincón, E.Y., M.T. Alarcón-Herrera, R. Pérez, et al. 2012a. Evaluación de riesgo sistémico y niveles de metahemoglobina en niños que consumen agua contaminada por nitratos, *Ingeniería* 16: 183–194.

Calleros-Rincón, E.Y., M.T. Alarcón-Herrera, J. Morán-Martínez, et al. 2012b. Caracterización de una zona contaminada por nitratos y su impacto en la salud humana. In *Género Ambiente y Contaminación por Sustancias Químicas*, ed. L.A. Cedillo and F.K. Cano Robles, Secretaría de Medio Ambiente y Recursos Naturales, Mexico.

Calleros-Rincón, E.Y., R. Pérez Morales, A. González Zamora, et al. 2018. Metahemoglobina y cuerpos de Heinz como biomarcador de exposición a nitratos en niños. Abstract XVII Congreso Internacional XXIII Congreso Nacional de Ciencias Ambientales, Mexico 9(21): 15–29.

Castaldelli, G., F. Vincenzi, E.A. Fano, et al. 2020. In search for the missing nitrogen: Closing the budget to assess the role of denitrification in agricultural watersheds. *Appl Sci* 10: 2136. doi:10.3390/app10062136.

Congreves K.A., B. Dutta, B.B. Grant, et al. 2016. How does climate variability influence nitrogen loss in temperate agroecosystems under contrasting management systems? *Agric Ecosys Environ* 227: 33–41. doi.org/10.1016/j.agee.2016.04.025.

Coomes, O.T, B.L. Barham. G.K. MacDonald et al. 2019. Leveraging total factor productivity growth for sustainable and resilient farming. *Nat Sustain* 2: 22–28.

Cui, N., M. Cai, X. Zhang et al. 2020. Runoff loss of nitrogen and phosphorous from a rice paddy field in the east of China: Effects of long-term chemical N fertilizer and organic manure applications. *Global Ecol Conserv* 22: e01011. doi.org/10.1016/j.gecco.2020. e01011.

Daims, H., E.V. Lebedeva, P. Pjevac, et al. 2015. Complete nitrification by *Nitrospira* bacteria. *Nature* 528: 504–509 doi.org/10.1038/nature14461.

DelMoro S.K., D.M. Sullivan, D.A. Horneck. 2017. Ammonia volatilization from broad-cast urea and alternative dry nitrogen fertilizers. Soil Sci. *Soc. Am. J.* 81: 1629–1639 doi:10.2136/sssaj2017.06.0181.

Descroix. L. 2004. Hidrografía de las lagunas de Mayrán y de Viesca: endorreísmo y antropismo, In *Las playas del desierto chihuahuense (parte mexicana),* ed. O. Gnienberger, V.M. Reyes-Gómez, and J. Janeau. Instituto de Ecologia, Mexico.

Erisman J.W., M.A. Sutton, J. Galloway, et al. 2008. How a century of ammonia synthesis changed the world. *Nature Geoscience.* doi.org/10.1038/ngeo325.

Espie, P., H. Ridgway. 2020. Bioactive carbon improves nitrogen fertilizer efficiency and eco-logical sustainability. *Sci Reports,* 10: 3227. doi.org/10.1038/s41598-020-60024-3.

FAO. 2017a. *World Fertilizer Trends and Outlook to 2020, Summary Report.* Food and Agriculture Organization of the United Nations, Rome.

FAO. 2017b. *Voluntary Guidelines for Sustainable Soil Management. Food and Agriculture Organization of the United Nations.* Rome. 26 pp.

FAO. 2018. Nitrogen inputs to agricultural soils from livestock manure: New statistics. Food and Agriculture Organization of the United Nations, Rome. www.fao.org/3/I8153EN/i8153en.pdf.

FAO. 2019. *The State of Food Security and Nutrition in the World: Safeguarding against Economic Slowdowns and Downturns.* Food and Agriculture Organization of the United Nations, Rome. 190 pp.

Figueroa V.U., G.G. Nuñez, J.A. Delgado, et al. 2009. Estimación de la producción de estiércol y de la excreción de nitrógeno, fósforo y potasio por bovino lechero en la Comarca Lagunera, In *Agricultura Orgánica, Segunda Parte,* ed. C.I. Orona, S.E. Salazar, H.M. Fortis. Sociedad Mexicana de Ciencia del Suelo, Universidad Juarez del Estado de Durango. Durango. Mexico, pp. 128–151.

Fortis-Hernandez, M., R.J.A. Leos, C.I. Orona, et al. 2010. Uso de estiércol bovino en la Comarca Lagunera, In: *Agricultura Orgánica, Segunda Parte,* ed. C.I. Orona, S.E. Salazar, H.M. Fortis. Sociedad Mexicana de Ciencia del Suelo, Universidad Juarez del Estado de Durango. Durango. Mexico, pp. 104–127.

Galloway, J.N., A.R. Townsend, J.W. Erisman, et al. 2008. Transformation of the nitrogen cycle: recent trends, questions, and potential solutions. *Science,* 320: 889–892.

Grahmann K., N. Verhulst, L.R.J. Peña, A, Buerkert. 2014. Durum wheat (*Triticum durum* L.) quality and yield as affected by tillage-straw management and nitrogen fertiliza-tion practice under furrow-irrigated conditions. *Field Crops Res* 164: 166–177. doi.org:/10.1016/j.fcr.2014.05.002.

Grahmann K., N. Verhulst, L. Mora Palomino, et al. 2018. Ion exchange resin samplers to estimate nitrate leaching from a furrow irrigated wheat-maize cropping system under different tillage-straw systems. *Soil Tillage Res* 175: 91–100. doi.org:/10.1016/j.still.2017.08.013.

Guo S., J. Pan, L. Zhan, et al. 2020. The reactive nitrogen loss and GHG emissions from a maize system after a long-term livestock manure incorporation in the North China Plain. *Sci Tot Environ* 720: 137558. doi.org/10.1016/j.scitotenv.2020.137558.

Gutiérrez, M., R.N. Biagioni, M.T Alarcón-Herrera et al. 2018. An overview of nitrate sources and operating processes in arid and semiarid aquifer systems, *Sci Tot Environ* 624: 1513–1522.

Hansen, B., L. Thorling, J. Shullehner, et al. (2017) Groundwater nitrate response to sustain-able nitrogen management, *Sci Rep* 7: 1–12. doi.org/10.1038/s41598-017-07147-2.

Hess, L.J.T., E.S. Hinckley, G.P, Robertson, et al. 2020. Rainfall intensification increases nitrate leaching from tilled but not no-till cropping systems in the U.S. Midwest. *Agric Ecosys Environ* 290: 106747. doi.org/10.1016/j.agee.2019.106747.

Holub, P., K. Klem, I. Tuma, et al. 2020. Application of organic carbón affects mineral nitrogen uptake by winter wheat and leaching in subsoil: Proximal sensing as a tool for agronomic practice. *Sci Total Environ* 717: 137058. doi.org/10.1016/j.scitotenv.2020.137058.

Huang, L., C.W. Riggins, S. Rodriguez-Zas, et al. 2019. Long-term N fertilization imbalances potential N acquisition and transformations by soil microbes. *Sci Tot Environ* 601:562–571. doi.org/10.1016/j.scitotenv.2019.07.154.

IFA, 2016. International Fertilizer Industry Association, http://ifadata.fertilizer.org/, accessed June 9, 2020.

Insausti, M., R. Timmis, R. Kinnersley, et al. 2020. Advances in sensing ammonia from agricultural sources. *Sci Tot Environ* 706:135124. doi.org/10.1016/j.scitotenv.2019.135124.

John, A.A., C.A. Jones, S.A. Ewing, et al. 2017. Fallow replacement and alternative nitrogen management for reducing nitrate leaching in a semiarid region. *Nutr Cycl Agroecosyst* 108: 279–296. doi.org/10.1007/s10705-017-9855-9.

Jones, C., S. Kim, T.F. Wilton, et al. 2018. Nitrate uptake in an agricultural stream estimated from high-frequency, in-situ sensors. *Environ Monit Assess* 190: 226

Khoshnevisan B., S. Rafiee, J. Pan, Y. Zhang, H. Liu. 2020. A multi-criteria evolutionary-based algorithm as a regional scale decision support system to optimize nitrogen consumption rate; A case study in North China Plain. *J Cleaner Prod* 256: 120213. doi.org/10.1016/j.jclepro.2020.120213.

Li, H., N. Zhu, S., Wang, et al. 2020. Dual benefits of long-term ecological agricultural engineering: Mitigation of nutrient losses and improvement of soil quality. *Sci Tot Environ* 721: 137848. doi.org/10.1016/j.scitotenv.2020.137848.

Lu, C., H. Tian. 2017. Global nitrogen and phosphorous fertilizer use for agriculture production in the past half century: Shifted spots and nutrient imbalance. *Earth Syst Sci Data* 9:181–192. doi.org/10.5194/essd-9-181-2017.

Maes W.H., and K. Steppe 2018. Perspectives for remote sensing with unmanned aerial vehicles in precision agriculture. *Trends Plant Sci* 24: 152–164. doi.org/10.1016/j.tplants.2018.11.007.

Masvaya, E.N., J. Nyamangara, K. Descheemaeker, K.R. Giller. 2017. Tillage, mulch and fertilizer impacts on soil nitrogen availability and maize production in semi-arid Zimbabwe. *Soil Tillage Res*, 168: 125–132. doi.org/10.1016/j.still.2016.12.007.

Mateo-Sagasta, J., S. Marjani Zadeh, H. Turral. 2017. Water pollution from agriculture: a global review FAO, Rome, 29 pp. (www.fao.org/land-water/overview/global- framework/global-framework).

Meisinger J.J. and G.W. Randall. 1991. *Estimating nitrogen budgets for soil-crop systems, in Managing nitrogen for groundwater quality and farm profitability*, R.F. Follet, D.R. Keeney, R.M. Cruse (eds.). Soil Science Society of America, Madison WI, 85–124.

Michalski, G., T, Meixner, M. Fenn et al. 2004, Tracing atmospheric nitrate deposition in a complex semiarid ecosystem using $\Delta^{17}O$, *Environ Sci Technol*, 38: 2175–2181 doi.org/10.1021/es034980+.

Norton, J., Y. Ouyang. 2019. Controls and adaptive management of nitrification in agricultural soils. *Front Microbiol* 10:1931. doi.org/10.3389/fmicb.2019.01931.

Oertel, C., J. Matschullat, K. Zurba, et al. 2016. Greenhouse gas emissions from soils – A review. 2016. *Chem Erde* 76: 327–352. doi.org/10.1016/j.chemer.2016.04.002.

Quan, Z., B. Huang, C. Lu, et al. 2016. The fate of fertilizer nitrogen in a high nitrate accumulated agricultural soil. *Sci Rep* 6: 21539. doi.org/10.1038/srep21539.

Robertson G.P., P.M. Vitousek. 2009. Nitrogen in agriculture: Balancing the cost of an essential resource, *Annu Rev Environ Resour* 34: 97–125. doi.org/10.1146/annurev.environ.032108.105046.

Rocha, K.F., M. de Souza, D.S. Almeida, et al. 2020. Cover crops affect the partial nitrogen balance in a maize-forage cropping system. *Geoderma*, 360: 114000. doi.org/10.1016/j.geoderma.2019.114000.

Schmidt E.L. 1982. Nitrification in Soil, 253–288. In: *Nitrogen in Agricultural Soils*, ed. F.J. Stevenson. Agronomy Monograph no. 22, ASA-CSSA-SSSA, Madison, WI.

Sebilo, M., B. Mayer, B. Nicolardot, et al. 2013. Long-term fate of nitrate fertilizer in agricultural soils, *PNAS*, 45: 18185–89. doi.org/10.1073/pnas.1305372110.

Seitzinger, S., J.A. Harrison, J.K. Böhlke, A., et al. 2006. Denitrification across landscapes and waterscapes: a synthesis. *Ecol Appl*, 16:2064–2090.

Sharma, P., M.J. Shukla, T.W. Sammis, et al. 2012. Nitrate-nitrogen leaching from three specialty crops of New Mexico under furrow irrigation system. *Agr Water Manag* 109: 71–80. doi.org/10.1016/jagwat.2012.02.008.

Shi, X., K. Hu, W.D. Batchelor, et al. 2020. Exploring optimal management strategies to mitigate nitrogen losses from paddy soil in the middle reaches of the Yangtze River. *Agr Water Manag*, 228: 105877. doi.org/10.1016/j.agwat.2019.105877.

Sommer, S.G., K. Schjørring, O.T. Denmead. 2004. Ammonia emission from mineral fertilizers and fertilized crops. *Adv Agron*. 82: 557–622. doi:10.1016/S0065-2113(03)82008-4.

Stevenson F.J. 1982. Organic Forms of Soil Nitrogen, 67–122 In: *Nitrogen in Agricultural Soils*, ed. F.J. Stevenson, Agronomy Monograph no. 22, ASA-CSSA-SSSA, Madison, WI.

Van Horn H.H., G.L. Newton, W.E. Kunkle. 1996. Ruminant nutrition from an environmental perspective: factors affecting whole farm nutrient balance. *J Anim Sci* 3082–3102.

Veldhuizen L.J.L., K.E. Giller, P. Oosterveer, et al. 2020. The missing middle: Connected action on agriculture and nutrition across global, national and local levels to achieve Sustainable Development Goal 2. *Global Food Secur* 24: 100336. doi.org/10.1016/j.gfs.2019.100336.

Wagner-Riddle C., A. Furon, N.L. McLaughlin, et al. 2007. Intensive measurement of nitrous oxide emissions from a corn-soybean-wheat rotation under two contrasting management systems over 5 years. *Global Change Biol*, 13, 1722–1736, doi.org/10.1111/j.1365-2486.2007.01388.x.

Wang H., S. Kobke, K. Dittert. 2020a. Use of urease and nitrification inhibitors to reduce gaseous nitrogen emissions from fertilizers containing ammonium nitrate and urea. *Global Ecol Conserv* 22: e00933. doi.org/ 10.1016/gecco.2020.e00933.

Wang H., A.K. Mortensen, P. Mao, et al. 2020b. Estimating the nitrogen nutrition index in grass seed crops using a UAV-mounted multispectralcamera, *Int. J. Remote Sensing* 40: 2467–2482. doi.org/10.1080/01431161.2019.1569783.

Ward, M.H., T.M. deKok, P. Levallois et al. 2005. Workgroup report: Drinking water nitrate and health-recent findings and research needs. *Environ Health Perspect* 113: 1607–1614.

Wen, Y., B. Freeman, Q. Ma, et al. 2020. Raising the groundwater table in non-growing season can reduce greenhouse gas emission and maintain crop productivity in cultivated fen peats. *J Cleaner Prod*. 262: 121179. doi.org/10.1016/j.jclepro.2020.121179.

White, P.J., J.W. Crawford, M.C. Díaz Álvarez, et al. 2012. Soil management for sustainable agriculture. *Appl Environ Soil Sci* 2012:850739. doi.org/10.1155/2012/850739.

Wise, T.A. 2019. *Eating Tomorrow: Agribusiness, Family Farmers, and the Battle for the Future of Food*, The New Press, New York, 313 pp.

Xia, Y., M. Zhang, D.C.W. Tsang, et al. 2020. Recent advances in control technologies for non-point source pollution with nitrogen and phosphorous from agricultural runoff: current practices and future prospects. *Appl Biol Chem* 63: 8. doi.org/10.1186/s13765-020-0493-6.

Zheng, W., Y. Wan, Y. Li, et al. 2020. Developing water and nitrogen budgets of a wheat-maize rotation system using auto-weighing lysimeters: Effects of blended application of controlled-release and un-coated urea. Environ Pollut, *Environ Pollut* 263: 114383. doi.org/10.1016/j.envpol.2020.11483.

# 3 Nitrate Transport in Agricultural Systems

*Sara Vero and Matthew Ascott*

## CONTENTS

3.1 Introduction ........................................................................................................45
3.2 The Source-Pathway-Receptor Model ..............................................................47
3.3 The Overland Pathway ......................................................................................49
3.4 The Unsaturated Subsurface Pathways .............................................................50
    3.4.1 Leaching and Vertical Transport in the Unsaturated Zone ................50
    3.4.2 Lateral Flow in the Unsaturated Zone ...............................................55
    3.4.3 Timing of Unsaturated Transport ......................................................56
3.5 The Groundwater Pathway ................................................................................56
    3.5.1 Fundamentals of Groundwater Flow and Solute Transport ...............56
    3.5.2 Nitrate Transport Processes in the Groundwater Pathway ................58
        3.5.2.1 Nitrate Attenuation Through Denitrification .....................58
        3.5.2.2 Mixing of Groundwater of Different Ages ..........................58
3.6 Time Lag ............................................................................................................59
3.7 Managing Nitrate Transport ..............................................................................62
3.8 Summary ............................................................................................................63
Acknowledgments .....................................................................................................63
References ..................................................................................................................63

## 3.1 INTRODUCTION

Intensification of agricultural production has increased demand for chemical N fertilizers (world production increasing from 12 Tg N yr$^{-1}$ in 1960 to 110 Tg N yr$^{-1}$ by 2013 (Battye et al., 2017), concurrent with greater excretion from livestock systems (Liu et al., 2017). Consequently, increased loads of reactive N, including nitrate ($NO_3^-$) are made available for either utilization or loss due to intentional application to the soil surface as chemical or organic fertilizer ($N_{fert}$), bacterial fixation ($N_{fix}$), deposition by grazing livestock ($N_{man}$), and by atmospheric deposition ($N_{dep}$). In a perfectly balanced system, this N would be taken up by the growing crop and taken

DOI: 10.1201/9780429326806-4

**45**

off as a farm output ($N_{withdrawn}$) or recycled back through the system as feed or bedding (Figure 3.1). This balance is expressed in Eq. 3.1:

$$N_{budget} = N_{fix} + N_{dep} + N_{fert} + N_{man} - N_{withdrawn} \qquad (3.1)$$

Unlike phosphorus, nitrogen cycles rapidly and dynamically within the soil system, shifting from occluded forms in soil organic matter, gaseous forms (ammonia ($NH_3$), dinitrogen ($N_2$), nitrous oxide ($N_2O$)), and soluble forms (ammonium ($NH_4^+$), $NO_3^-$, and nitrite ($NO_2^-$)). These dynamics are controlled by microbial activity, inputs to the system, soil saturation, temperature, and other temporally transient and spatially variable factors. The complexity of nitrogen within the agricultural system has resulted in "black box" approaches to quantification and modeling becoming commonplace. Consequently, it is challenging to quantify soil N storage and to disentangle the origin of N loads in watercourses. It has been estimated that peak storage of $NO_3^-$ in the unsaturated zone ranges between 605-1814 Tg, with the greatest accrual in North America, China, and Europe (Ascott et al., 2017). Understanding the transport of these $NO_3^-$ loads in agricultural watersheds is crucial for preventing or managing losses and for predicting water-quality impacts.

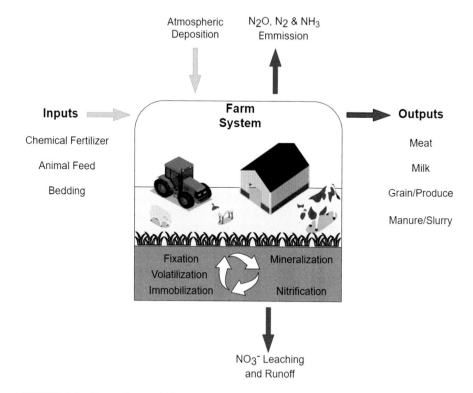

**FIGURE 3.1**   Farm nitrogen balance.

In this chapter the principles of $NO_3^-$ transport are described. Section 3.2 discusses the overarching concept of the source-pathway-receptor model, before considering the individual transport pathways in Sections 3.3–3.5 The issue of time lag is addressed in Section 3.6, followed by a discussion of management strategies for $NO_3^-$ transport in Section 3.7.

## 3.2 THE SOURCE-PATHWAY-RECEPTOR MODEL

Transport of $NO_3^-$ away from the rhizosphere presents a financial cost as nutrients that are not taken up by the crop are essentially a failed investment. Furthermore, they present a significant environmental risk to water quality (see Part 3, Chapter 9). It is estimated that within the EU c. 39% of anthropogenic $NO3^-$ delivered to surface and groundwaters originates from livestock production systems (Uwizeye et al., 2020). Potability standards are related to risks of methemoglobinemia in humans (a disorder which impairs the capacity of the hemoglobin molecule to transport oxygen through the bloodstream), and threshold values for drinking water are generally in the vicinity of 50 mg $l^{-1}$ as $NO_3^-$ or 11.3 mg $l^{-1}$ as $NO_3^-$-N, as recommended by the World Health Organisation (WHO, 2007).

In 2015 the Court of Justice of the EU ruled that Greece had failed to adequately protect its waters from nitrate pollution by agricultural sources. This resulted in the imposition of financial sanctions until the establishment of Nitrate Vulnerable Zones and appropriate $NO_3^-$ control programs is achieved, or until a further court judgment (European Commission, 2019). Control of $NO_3^-$ is therefore not solely an issue for those directly involved in agriculture or dependent upon water sources in rural areas, but is also of national concern. Diffuse nutrient transfer within an agricultural landscape can be conceptually organized according to the Source-Pathway-Receptor model (SPR) (Wall et al., 2011). The SPR model consists of the following stages:

**Sources**: N applied to the soil surface as organic or inorganic fertilizers, as atmospheric deposition or fixed by legumes and symbiotic bacteria represents a nutrient source which may be taken up by growing crops, cycled through the soil and atmosphere or transported via water. Detailed discussion on N in soils is provided in Part 1, Chapter 2, and so will not be discussed here. However, it should be recognized that other sources of N exist in the agricultural landscape. These largely fall into the "point source" category – that is, nutrient sources that are discharged at a single identifiable location. Within the agricultural landscape, these point sources include farmyard and industrial discharges, wastewater treatment facilities, septic tanks, and livestock drinking points in streams. While not all of these are directly related to agricultural production (e.g., septic tanks), they frequently represent significant contributions to nutrient loads arriving to waterbodies.

**Pathways**: This refers to the routes and mechanisms by which N is physically conveyed from a source to a receptor. For nutrients (such as phosphorus) that are prone to sorption, this stage can be subdivided as per the Nutrient Transport Continuum (NTC) model described by Haygarth and colleagues (2005) to include both mobilization and transport phases. Briefly, mobilization refers to the initiation of nutrient movement either by solubilization or detachment of soil particles. The latter is particularly

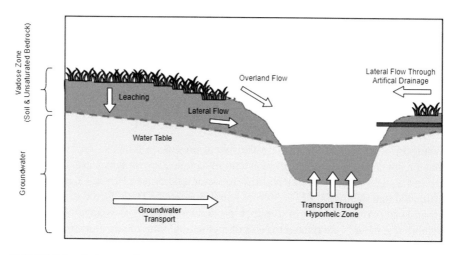

**FIGURE 3.2**   Pathways of nitrate transport in the agricultural landscape.

relevant in the case of phosphorus, which is frequently highly sorbed to particles and most heavily concentrated near the surface of the soil profile. As $NO_3^-$ does not readily adsorp to soil particles, it is vulnerable to transport in solution by water flowing over or percolating through the soil matrix. This difference in propensity from attenuation to mobilization is therefore a controlling factor on which pathway is dominant. Transport refers to the movement of nutrients through the landscape via hydrologic pathways (Figure 3.2). Hydrologic pathways are conceptually divided:

- The overland pathway
- The soil pathways
  - The vertical soil pathway
  - The lateral soil pathway
- The groundwater pathway

These individual pathways will be discussed in detail in sections 3.3–3.5. Determining the flow pathway by which $NO_3^-$ (or any contaminant) reaches the waterbody is crucial for designing effective best-management practices (Rittenburg et al., 2015).

**Receptor:** Receptors refer to any waterbody to which the nutrient is delivered (see Part III, Chapter 9). They include groundwater, lakes, rivers and streams, and estuaries. The impact of $NO_3^-$ is quantified by the change in water chemistry and/or the influence of these changes on the ecology of the waterbody. Chemistry can be scored relative to environmental or drinking water thresholds, such as those set under the European Union Water Framework Directive (WFD). Ecological impacts are typically assessed based on the presence or absence of indicator species.

The individual pathways involved in transport will be discussed in the following sections.

## 3.3   THE OVERLAND PATHWAY

Overland flow refers to the transport of water over the surface of the terrain, whether that is soil or an impervious surface such as concrete or tarmac. As this chapter focuses on agricultural landscapes, the primary focus will be on soil. Overland flow can be divided into sheet flow, in which water flows in a broad, diffuse layer across a relatively homogeneous surface, and rill or channel flow, in which it is focused into distinct rivulets (Rittenburg et al., 2015). Overland flow may be triggered by two mechanisms: infiltration excess flow and saturation excess flow. Infiltration excess flow occurs when precipitation is applied at a rate that exceeds the infiltration capacity of the soil surface (within the agricultural context). Infiltration capacity is defined by the maximal rate at which water may flow vertically through the soil profile. It is a function of soil structure and explicitly related to the saturated hydraulic conductivity ($k_{sat}$ or $k_s$). High soil conductivity is related to low bulk density, high porosity, and typically, low clay content. Conversely, saturation excess overland flow occurs when soil pores are filled with water and do not have the capacity to receive any further liquid, irrespective of the rate of application. Observation of nutrient concentrations in overland flow confirms that while it is not the dominant pathway for $NO_3^-$ loss, dilution is the key control on quantities lost during rainfall events (Kleinman et al., 2006).

Once generated, the overland flow will proceed from areas of high energy to low energy – in other words, will follow the slope until it reaches a point where it may infiltrate, pool, or discharge into a watercourse. When the latter occurs, it may be referred to as "runoff," although the term is often casually used as analogous to overland flow. In summary, the partitioning of water into overland flow or infiltration, therefore, depends upon the nature of precipitation (intensity, form) and the characteristics of the soil.

Overland flow is not considered to be the primary pathway for nitrate transport in agriculture in most scenarios (Vadas et al., 2015). Loads of $NO_3^-$ arriving at watercourses are frequently dominated by baseflow contributions. For example, Schilling (2002) attributed >61% of N load arriving at watercourses in four sub-catchments to baseflow contributions. The proportion of $NO_3^-$ loads arriving via baseflow has been found to be linearly related to the baseflow index (ratio of mean baseflow relative to total streamflow) (Tesoriero et al., 2013). Unlike phosphorus, $NO_3^-$ typically does not accumulate near the top of the soil surface. However, if rainfall occurs shortly after the application of fertilizer there may be some losses via the overland pathway (Outram et al., 2016). Fertilizer type and application methods also influence losses via the overland pathway. A rainfall-simulation trial on the application of poultry manure, dairy manure, ammonium sulfate, and urea-based fertilizers (plus inhibitors in the latter case) indicated the greatest cumulative $NO_3^-$ losses via runoff from surface-applied poultry manure followed by ammonium sulfate, urea, and dairy manure, with incorporated poultry manure demonstrating the least losses (Kibet et al., 2016). The proportion of N lost in runoff as urea, $NO_3^-$, or $NH_4^+$ shifted over the four-week monitoring period, with $NO_3^-$ becoming dominant over time. This reflects the conversion of $NH_4^+$ to $NO_3^-$, which may provide a signal of storm-event transfer in catchment studies (Shore et al., 2016). Considering the burgeoning evidence of overland flow as a potentially important pathway for $NO_3^-$ loss, further research is needed into its impacts and consequences for fertilizer management.

## 3.4  THE UNSATURATED SUBSURFACE PATHWAYS

The unsaturated or vadose zone consists of unsaturated bedrock and soil above the water table. Soil pores fluctuate in saturation over time. The degree of saturation dictates how connected pores are. This in turn controls transport through those pores. When pores are unsaturated there can be no transport of $NO_3^-$. It should be noted that when rainfall occurs after periods of prolonged drought, microbial respiration may spike causing mineralization of N stored in soil organic matter. This can cause spikes of $NO_3^-$ in receiving waterbodies (see Section 4.1) but should not be confused with fertilizer N already in transit through the system.

### 3.4.1  LEACHING AND VERTICAL TRANSPORT IN THE UNSATURATED ZONE

Leaching is the vertical movement of $NO_3^-$ through the soil profile. As the soil consists of both macro-and mesopores, and a matrix of micro-pores between individual particles, transport will occur at different rates across relatively small areas. Rapid flow through cracks and biopores is known as preferential flow, while slower, more uniform flow through mesopores (spaces between individual peds) is known as matrix flow. When the soil approaches the permanent wilting point flow ceases as water is tightly held in films on the surface of soil particles (Figure 3.3). Although larger pores empty most quickly as moisture content decreases, some preferential flow still occurs under unsaturated conditions (Nimmo, 2012). In such instances, the preferential flow may proceed due to connectivity of the flow path through different parts of the pore, although it is not full. Consequently, preferential leaching of nitrate does not occur exclusively under saturated conditions, although it is quickest at these times. Loads of leached $NO_3^-$ are correlated with the volume of drained water (Rakotovolona et al., 2019), hence, management of soil water is crucial to mitigating potential losses.

The propensity for leaching of $NO_3^-$ is influenced by soil texture, structure, and depth to groundwater. Ranking of the leaching potential of soil series (along with climate, geologic, and land-use parameters) is employed to generate groundwater vulnerability maps. Generally, more freely drained soils exhibit a greater likelihood of $NO_3^-$ leaching as water drains if the nutrient is not aligned with crop requirements. Sandy soils have low bulk density and high porosity, facilitating rapid leaching. Conversely, clay soils have low permeability, which promotes slow percolation of water through the profile. However, texture alone as a descriptor of soil-water relations is frequently over-simplistic and insufficient to conceptualize reality (Vero et al., 2014). The order, thickness, and characteristics of individual soil layers influence water dynamics within the profile and movement of $NO_3^-$ in the unsaturated zone (He et al., 2013). A study in wheat-maize cropping systems indicated that the depth and thickness of a clay-rich horizon was crucial in determining the losses of $NO_3^-$, both as a result of its control on water infiltration and as it influenced the ability of the plant to utilize soil N stores (He et al., 2013). Depth of the water table, stoniness of

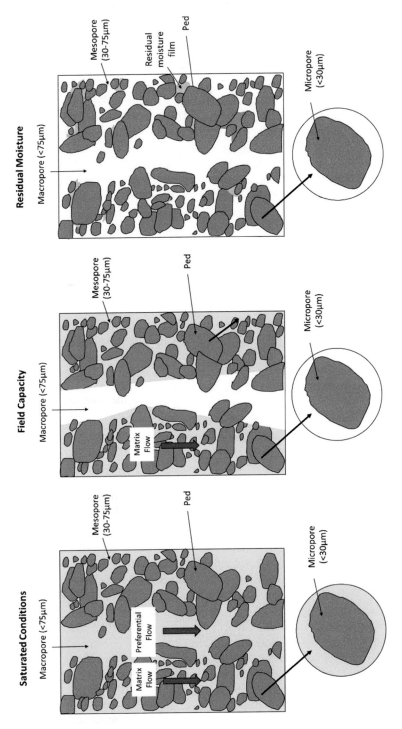

**FIGURE 3.3** Leaching under various soil moisture contents.

the profile, organic matter content, and depth of the upper two horizons were found in an Australian study to drive variability in leaching rates between soil series (Lilburne and Webb, 2002).

Slow infiltration in clay-rich soils delays physical transport of $NO_3^-$ and simultaneously facilitates the reduction of $NO_3^-$ to $N_2O$, and finally to $N_2$ via denitrification. Multiple studies have observed that soil processes transition from primary nitrification to denitrification at c. 60–70% water-filled pore space (WFPS) (Davidson et al., 1991; Bateman et al., 2004; Scheer et al., 2009; Cardenas et al., 2017). However, peak denitrification rates occur at greater WFPS as clay content increases (Cardenas et al., 2017). At low suction values (i.e., closer to saturation) sandy soils exhibit greater hydraulic conductivity than clay-rich soils (Hillel, 2004). This causes them to drain more quickly. However, as clay soils retain their water content for longer at the equivalent suction their WFPS is greater for a more prolonged period compared to coarser soils. This facilitates greater denitrification (as demonstrated by McTaggart et al., 2002; Jahangir et al., 2012; and others), and consequently, less $NO_3^-$ becomes available for leaching.

Leaching of $NO_3^-$ has become a prominent issue in intensive row crop production under irrigation in many regions, including the Midwestern United States (Gehl et al., 2005), China (Lu et al., 2019), and throughout Europe. As such, these are worst-case scenarios in which high rates of chemical fertilization are employed to meet the needs of demanding crops concurrent with artificially increased soil moisture amplifying the opportunity for leaching. Closely matching both nutrient and water demands of crops in freely drained soils is, therefore, crucial to mitigate losses. N loss to watercourses is not restricted to these scenarios, however, and is a notable concern in temperate grassland systems (Ryan et al., 2006), glasshouse vegetable production (Dahan et al., 2014), concentrated animal feeding operations (CAFOs) (Weldon and Hornbuckle, 2006), and a wide variety of other systems. In the face of this evidence, it cannot be considered as characteristic of any one production style, but rather a ubiquitous feature of modern intensive systems.

While it is beyond the scope of this chapter to discuss $NO_3^-$ transport in all soil types individually, it is worth mentioning the case of vertisols. These are soils exhibiting ≥30% clay with shrink–swell properties (such as montmorillonite or kaolin) to a depth of 50 cm (Brady and Weil, 2017). Due to shrink–swell behavior, vertisols exhibit cracks through which preferential flow can occur. The orientation of these cracks is spatially variable, site-specific, and will influence the direction and pattern of flow (Dinka and Lascano, 2012). Further, shrink–swelling behavior is hysteretic, meaning that modeling of water and solute transport of these soils is particularly challenging. Vertisols represent 2.3% of land area worldwide, most extensively in Australia and India, although also in many other countries including the United States, China, Mexico, Spain, and Israel (Kurtzmann et al., 2016). While the presence of deep cracks presents a pathway for preferential transport of contaminants, groundwater $NO_3^-$ contamination beneath these soils is typically low (Kurtzmann et al., 2016), in contrast to other agricultural soils worldwide and particularly sandy-loam soil types. This is primarily attributed to high denitrification rates (Sigunga et al., 2002) although it is speculated that adsorption onto the surfaces of kaolinite clays may play a role in retarding the transport of $NO_3^-$, while not manifestly reducing the total mass within the system (Kurtzmann et al., 2016). The behavior of $NO_3^-$ in verti-sols thus demonstrates the interdependent aspects of soil chemistry and hydrology controlling transport.

Transport through karst systems is also notable. These landscapes consist of typically thin soils overlying porous limestone aquifers characterized by matrix, fracture, and conduit flow pathways (Yue et al., 2019. Transport through these indi-vidual pathways and rates of retention and denitrification means that the delivery of N to groundwater is asynchronous. Integrated assessment of $NO_3^-$ loss from an intensive agricultural platform on a karst landscape in southeastern Ireland showed the highest concentrations (mg l$^{-1}$) of $NO_3^-$ delivery arriving during storm events through the quick-flow conduits. Nevertheless, the greatest loads (kg ha$^{-1}$) were delivered via matrix pathways subject to significant time lags measuring several years. While the latter enables natural denitrification subject to soil microbial activity, it is difficult to influence anthropogenically and likely to constrain the achievement of water-quality goals beyond short-term timelines. Conversely, the authors of that study suggested that implementation of ecologically engineered so-lutions such as bioreactors at easily accessible pathways such as spring discharges would provide a rapid and financially quantifiable solution to peak concentrations (Fenton et al., 2017).

It should be noted that the vadose zone may also include deep unsaturated bed-rock below the base of the soil profile. Modeling indicates that shallow groundwater influences 22% to 32% of land area (Fan et al., 2013); in which case, vadose trans-port scenarios such as those described above prevail. However, many landscapes (e.g., Loess Plateau in China (Turkeltaub et al., 2018), parts of the United States (Pratt et al., 1972), England (Wang et al., 2012), and France (Chen et al., 2019) exhibit unsaturated zones tens of meters thick above deep water tables. Infiltration and lateral flow also occur through these unsaturated regions (Rempe and Dietrich, 2018). Assessment of $NO_3^-$ transport in these environments has been made through repeated porewater profiles (Foster et al., 1982), analysis of deep soil cores (Harter et al., 2005), and detailed local and catchment scale modeling (Mathias et al., 2007). More recently, the significance of deep unsaturated zone on $NO_3^-$ transport has been assessed at regional (Turkeltaub et al., 2019), national (Ascott et al., 2016; Wang et al., 2016), and global (Ascott et al., 2017) scales. The low availability of soluble carbon substrate and lack of reducing conditions in deep vadose regions does not support microbial activity and, hence, denitrification may be limited (Harter et al., 2005; Rivett et al., 2008). Furthermore, the distribution of $NO_3^-$ in this region is highly heterogeneous due to preferential transport through rock fractures (Harter et al., 2005; Onsoy et al., 2005).

Evidence of leaching is perhaps most simply ascertained by trends in ground-water concentrations (Stuart et al., 2007). These have been correlated with agricul-tural intensification over the past century (as demonstrated in Denmark: Hansen et al., 2017). Although recent best-management practices have been reflected in curtailment of groundwater N in some instances, levels may remain elevated as a result of legacy N (discussed later). Consequently, groundwater may not be the most reliable means by which current agricultural practices may be assessed and assessment of $NO_3^-$ trans-port pathways within the soil profile allows early detection of trends.

Dye tracing is used to highlight flow pathways that are visual indicators of the propensity for leaching. This consists of the application of staining or fluorescent dyes (e.g., brilliant blue FCF or uranane, respectively (Gerke et al., 2013)) to the soil surface or to specific hydrologic pathways. These tracers may then be assessed visually (e.g., in excavated soil profiles or lysimeters (Kramers et al., 2012)) or detected using fluorometers or fluorescent spectrometers. This approach is perhaps most commonly applied in karstic settings, where preferential pathways are pronounced.

An alternative to dye tracing is the application of a chemical tracer that exhibits similar transport characteristics to the contaminant being mimicked. In the case of $NO_3^-$, key characteristics are conservative (unreactive) behavior, low environmental toxicity, not prevalent in the environment, and detectability through either laboratory analysis or using in-situ devices. Bromide and chloride are common proxies for $NO_3^-$ in unsaturated and saturated transport studies (Kramers et al., 2012; Perkins et al., 2011; Owens et al., 1985). These tracers are detected in laboratory analysis of water samples taken over the study period and plotted over time to create a breakthrough curve. Pre-application concentrations should be obtained to provide a baseline. The breakthrough curve can be disaggregated into different phases, including initial breakthrough (when the tracer or nutrient first becomes apparent), peak (indicating the greatest concentration), the center of mass (indicating the bulk of the tracer has reached the measurement point), and the tail (in which concentrations decline over time) (Vero et al., 2014). Finally, exit occurs at the point when concentrations return to baseline, although this may not always be discernible. Isotopic tracers such as $\delta^{15}N$ allow identification of specific N sources and differentiation from background N (Kellman and Hillaire-Marcel, 2003) via mass spectrometry and hydrograph separation (Gilbert et al., 2019; Jung et al., 2020). Tracers may be extracted from soil pore water, groundwater, or ultimately, surface water receptors.

Due to the prolonged lag of $NO_3^-$ transport through the unsaturated zone, detection at the water table may exceed study durations or be difficult to disentangle from $NO_3^-$ already resident within the system. Extraction of water from within pores in the soil and unsaturated zones allows earlier detection of nitrate movement through the profile, both vertically and laterally (Perkins et al., 2011). This may be achieved through porous cups, lysimetry or, where water tables are deep, cored boreholes (Foster et al., 1982). There are several different approaches from excavation and casing of intact soil monoliths and collection of leachate from the base under gravity or applied suction, or less invasive methods involving the insertion of a porous cup into the soil profile and extraction of water using suction (Singh et al., 2017; Di Bonito et al., 2008).

It is common for field measurements of leaching to be accompanied by numerical modeling to account for variability in water flow. This may take the form of pore- or groundwater data coupled with a hydrologic model – such as Hydrus, MODFLOW, and others (e.g., Vero et al., 2017a; Wei et al., 2019). Alternatively, measures of soil water and mineral N in soil cores may be coupled with an elution model, such as LIXIM) (e.g., Rakotovolona et al., 2019). Rakotovolona and colleagues (2019) suggested that the former approach is environmentally oriented to determine the movement of N to watercourses, while the latter is agronomically oriented, indicating drivers of N utilization and loss within the soil-crop-water interface. Where measured field data are used as inputs to the model, this inversely computes antecedent processes during the observation period. Conversely, models may be used in a predictive fashion to anticipate future leaching patterns and test various scenarios based on initial soil conditions, boundary parameters, and weather data.

### 3.4.2 LATERAL FLOW IN THE UNSATURATED ZONE

Lateral flow is the horizontal transport of water and dissolved solutes and may provide a pathway for transport to a surface water receptor without interaction with groundwater. Such transport may occur in porous and fractured soils but is perhaps most well documented in land that has been drained for agricultural purposes. The primary objective of drainage is to remove excess water from the soil, providing more optimal conditions for crop growth. Although artificial drainage of agricultural land is not new (Valipour et al., 2020), intensification of production in recent decades has increased the extent of land subject to this practice. Lack of documentation, the difference in design, and challenges in assessment make identification of the precise extent of artificial drainage challenging. However, it has been estimated that the worldwide extent of artificially drained land amounts to >166 million ha$^{-1}$ (Feick et al., 2005). The extent of drained land varies greatly between individual countries and regions. Estimates based on analysis of land cover, soil type, and administrative data in GIS suggest that, in the leading "corn belt" states of the United States, the percentage of cropland with artificial subsurface drainage is 48.3% – Ohio, 47.8% – Illinois, 42.2% – Indiana, 32.4% – Iowa, and 28.7% – Michigan (Sugg, 2007). Considering the extent of cropland in these states, this represents a major pathway for potential nutrient transport. In freely drained soils the direction of water flow is primarily vertical. Consequently, groundwater becomes the initial receptor for contaminants transported via this pathway. Conversely, tile drainage systems route soil water laterally across the profile, providing a rapid conduit to surface receptors (Figure 3.2) (Dinnes et al., 2002). The reduction in infiltration lowers the water table and decreases the extent and duration of soil saturation. Water moving laterally across the field through drains is discharged into a main receiving drain, which may be open (i.e., a French drain) or piped. This provides an accelerated pathway to a receiving watercourse, which bypasses the typically slower groundwater component and reduces the opportunity for denitrification.

In the absence of artificial drainage, lateral interflow can still occur when water percolating through the soil profile meets a layer of lower conductivity and is routed horizontally (Zhang et al., 2011). This may result from changes in texture (namely, increases in clay content – Zhu and Lin, 2009) or the presence of compacted layers (Etana et al., 2013). As a consequence of lateral interflow, $NO_3^-$ may be subject to short residence times allowing lesser denitrification than under more prolonged leaching and groundwater transport (Hesser et al., 2010). Modeling of interflow can improve estimates of $NO_3^-$ transport at catchment scale; however, it incurs site-specific data demands, including soil morphology, digital elevation mapping, and electromagnetic induction survey (Zhu and Lin, 2009).

### 3.4.3 TIMING OF UNSATURATED TRANSPORT

Transport of $NO_3^-$ through the unsaturated zone is strongly influenced by temporal factors that reflect (a) the timing of agricultural operations and (b) meteorological patterns (via their influence on soil moisture). Generally speaking, the greatest $NO_3^-$ losses will occur when periods or areas of high $NO_3^-$ availability coincide with large drainage volumes (Di and Cameron, 2002). Agricultural practices influencing $NO_3^-$ transport include those that add exogenous N to the system, for example, repeated application of fertilizer N, or which cause it to be mineralized from soil stores. Mineralization on endogenous N in 30 grassland systems worldwide was found to be driven by temperature during the wettest quarter, microbial biomass, clay content, and bulk density (Risch et al., 2019), though incubation studies indicate an interaction between moisture and temperature (Guntinas et al., 2012). However, the transport of that $NO_3^-$ once it has become available is strongly driven by precipitation and drainage (Morecroft et al., 2000). Soil $NO_3^-$ tends to be concentrated in the upper soil horizons during periods of fertilization and harvest, but migrates through the profile during the recharge period, typically over winter (van Es et al., 2006; Rakotovolona et al., 2019; Zhang et al., 2019). As with all contaminant transport, losses may be considered in terms of loads and concentrations. While large amounts of $NO_3^-$ may be transported during high flow periods (total loads), concentrations may be highest during or after drought periods in which high rates of mineralization occur with minimal opportunity for dilution, either in soil water or in receiving waterbodies (Morecroft et al., 2000).

The timing of precipitation relative to fertilizer application is the primary driver of $NO_3^-$ losses (van Es et al., 2006; Kibet et al., 2016). Ideally, application of fertilizers is avoided prior to heavy precipitation; however, this may not always be reliably predicted because frequency of rainfall in some parts of the world constrains viable application days, and even when relatively dry conditions are assured, trade-offs will occur with gaseous losses due to volatilization. Furthermore, greater durations between the application of N fertilizers and rainfall events provide an opportunity for N to be utilized by the growing plant, provided that water or other nutrients are not limiting. Aligning N application with plant requirements may be challenging, particularly in scenarios where the storage capacity of organic manures is limited, forcing application at sub-optimal times.

## 3.5 THE GROUNDWATER PATHWAY

Nitrate that has reached the water table will undergo transport and, potentially, attenuation in the saturated zone. In this section, the fundamentals of solute (such as $NO_3^-$) transport in saturated groundwater systems from recharge to discharge are described, before providing a brief overview of more specific nitrate transport processes occurring in the groundwater pathway.

### 3.5.1 FUNDAMENTALS OF GROUNDWATER FLOW AND SOLUTE TRANSPORT

Nitrate transport in groundwater is principally controlled by the nature of groundwater flow in the saturated zone. Groundwater flow occurs through saturated pores

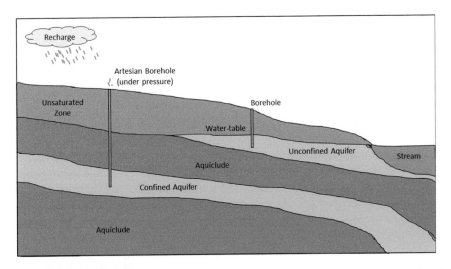

**FIGURE 3.4** Conceptual model of groundwater flow systems illustrating confined and unconfined aquifer systems.

and voids and from areas of high to low hydraulic heads typically from recharge areas to discharge areas (groundwater-fed streams, springs, and wetlands, pumping boreholes, the coast). The geometry of aquifers results in different groundwater conditions, as illustrated in Figure 3.4. Where a water table separates the saturated zone from the unsaturated zone this is known as an "unconfined aquifer." A "confined aquifer" is an aquifer that is contained between two low-permeability geological formations (known as "aquitards"). In confined aquifers, water is held under pressure, which results in the groundwater level in boreholes penetrating the aquifer to rise to the potentiometric surface. Unconfined and confined aquifers often have different hydrogeochemical conditions that can affect the fate of nitrate in these different settings.

Individual aquifers have different groundwater flow and transport characteristics, which are principally controlled by the nature of their hydraulic conductivity K $(L\,T^{-1})$ – the ease of flow of water through a porous media – and porosity $(n)$, dimensionless, the fraction of a given volume of porous media, which is void space. In some aquifers, groundwater flow occurs through intergranular pore space (e.g., sandstone and alluvial aquifers), whereas in others flow occurs through fractures and fissures (e.g., carbonate and crystalline hard rock aquifers). Groundwater flow is controlled by the saturated hydraulic conductivity of the aquifer material and the slope of the water table (known as the hydraulic gradient, dh/dx). In an idealized, homogeneous, and isotropic saturated porous intergranular media, the rate of groundwater flow, Q $[L^3\,T^{-1}]$, is described by Darcy's Law:

$$Q = -KA\frac{dh}{dx}$$

Where A is cross-sectional area of flow [$L^2$]. The average linear groundwater velocity, $v$ [$L\,T^{-1}$], is calculated as

$$v = \frac{Q}{An_e}$$

where it is the effective porosity, the fraction of void space in which groundwater flows. Nitrate transport is directly linked to groundwater flow rates and velocities through the processes of advection and hydrodynamic dispersion. Advective transport of $NO_3^-$ is directly controlled by groundwater velocities, whereas hydrodynamic dispersion is the mechanical dispersion of solutes due to variations in groundwater flow velocity associated with the tortuosity of groundwater flow paths in pore spaces. Hydrodynamic dispersion acts to spread and lower concentrations of a contaminant plume.

### 3.5.2 NITRATE TRANSPORT PROCESSES IN THE GROUNDWATER PATHWAY

The mechanisms of groundwater flow and solute transport detailed in Section 3.5.1 are simplified representations of the hydrogeological processes occurring in the saturated zone. For a given aquifer, significant additional complexity exists associated with spatial heterogeneity of hydrogeological properties and structure, and variability in transport processes across scales from the individual pore and fracture scale to catchment and aquifer scale. It is beyond the scope of this book to provide an exhaustive account of hydrogeological properties and processes affecting solute transport in groundwater, and the reader is referred to specific text on hydrogeology – for example, Fetter (2000) – for further details. There are, however, a number of specific processes that are pertinent to $NO_3^-$ transport in the groundwater pathway and which are applicable across a range of aquifer settings, which are discussed herein.

#### 3.5.2.1 Nitrate Attenuation Through Denitrification

Under suitable hydrogeochemical conditions, $NO_3^-$ can be subject to attenuation. This principally occurs by heterotrophic denitrification – the reduction of nitrate under anaerobic conditions to nitrogen gas, via $NO_2^-$ and $N_2O$, due to the removal of oxygen by consumption of organic carbon by heterotrophic microorganisms. Denitrification in groundwater has been subject to reviews by Rivett and colleagues (2008) and Stuart et al. (2018) and is generally only considered to occur in specific settings; microzones in the zone of water-table fluctuation, the hyporheic zone of groundwater–surface water interaction, confined aquifer systems.

#### 3.5.2.2 Mixing of Groundwater of Different Ages

When evaluating $NO_3^-$ concentrations at groundwater discharge points (e.g., pumping boreholes, streams), an important consideration is the "age" of groundwater at these discharge points. In comparison to $NO_3^-$ transport via surface runoff and in rivers, groundwater travel times are long. The age of groundwater may be in the order of decades up to centuries and millennia, with the age typically increasing according to

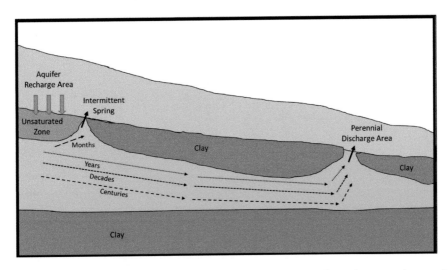

**FIGURE 3.5** Conceptual model of groundwater flow systems illustrating groundwater of different ages.

the depth of the flow system (Figure 3.5). Groundwater discharge to long, screened boreholes and streams is a mixture of groundwater of different ages, with older low $NO_3^-$ groundwater likely to dilute younger high $NO_3^-$ groundwater. Older groundwater is typically less likely to have higher $NO_3^-$ concentrations, as it may have been recharged during a period when leaching through the soil profile was limited, or when attenuation has occurred along the flow path. This combination of mixing of different aged groundwater and variable $NO_3^-$ attenuation can interpret concentrations at groundwater discharge points. The use of indicator species for redox conditions conducive to denitrification (Stuart et al., 2018), nitrate isotopes, and groundwater flow modeling (Rivett et al., 2007) can support this.

## 3.6 TIME LAG

Time lag, sometimes referred to as legacy effect, retardation factor, residence time or memory effect (Cook et al., 2003; Bechman et al., 2008; Tesoriero et al., 2013), is the inherent delay in the response of waterbodies to measures imposed to improve water quality. The time lag may impede the ability of waterbodies to attain qualitative thresholds set by legislation if those timescales are not catchment-specific and reflective of delays intrinsic to those settings. Meals and colleagues (2010) categorized elements of time lag as follows (Figure 3.6):

- *Project management component*: This element relates to delays in the design and implementation of measures.
- *Effect component*: This is the time it takes for the measure, once implemented, to produce an effect. For example, a reduction in fertilizer application will take a certain amount of time to affect a reduction in nutrient loss from the soil

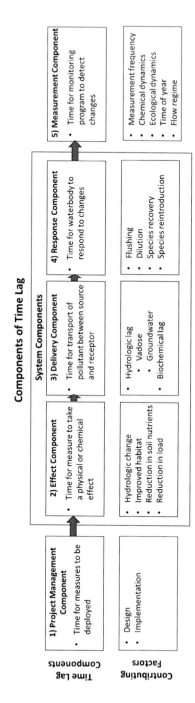

**FIGURE 3.6**  Components of time lag in response to water quality mitigation measures. (Adapted from Meals et al., 2010.)

as residual stores become depleted over years, whereas exclusion of livestock from watercourses will have a rapid response. Typically, mitigation of diffuse losses is associated with longer effect components than infrastructural solutions to point discharges.

- *Delivery component*: This element relates to the time required for the contaminant to be physically transmitted from the source to the receptor. This is the biophysical time lag through the landscape. Biophysical lag is influenced by (a) the route (direct discharge to the watercourse, overland flow, or subsurface transport through vadose and saturated pathways), (b) distance from the receptor/length of the pathway, and (c) travel rate (Meals et al., 2010). Travel rate depends on both rainfall patterns, the physical characteristics of the pathway, and chemical parameters influencing attenuation, mobilization, and transformation of the pollutant.

- *Response component*: This is the time it takes for the waterbody itself to respond to the measure, after delivery (that is, to reductions in nutrient loads delivered). This may depend upon dilution, on flushing of legacy nutrients from stream sediment, or for recovery of the indicator species. This latter factor presents as particularly challenging considering the interactive effects of multiple stressors (Davis et al., 2019) and the requirement for the reintroduction of species where they have been extirpated. While surface water bodies are often considered the final receptor of transported $NO_3^-$, groundwater is also a receptor as well as a pathway. This is particularly crucial where groundwater is extracted for human consumption.

- *Measurement component*: The temporal resolution and success of a water-quality monitoring regime will influence its ability to detect response to imposed measures. Regimes with low observation frequency create a "statistical lag" delaying confirmation of a response (Meals et al., 2010).

Neither the project management nor measurement components are innate parts of the system, but they may impair accurate interpretation of water-quality parameters and, hence, the perceived efficacy of imposed measures.

The delivery component refers to the time lag associated with the transport phase of the SPR continuum. Total time lag ($t_T$) within the subsurface can be divided into the unsaturated component ($t_U$) and the saturated component ($t_S$) (Vero et al., 2014). Sousa et al. (2013) stated that the relative proportion ($t_r$) spent in either component can be expressed as

$$t_r = \frac{t_U}{t_T} \, or \, \frac{t_S}{t_T}$$

Under the EU Water Framework Directive (WFD), goals for water quality are set based on six-year reporting periods. The objective of member states is to attain nutrient concentrations in waterbodies that do not exceed fixed threshold environmental quality standards expressed in mg $L^{-1}$. Taking Ireland as an example, modeling of hydrologic time lags (Fenton et al., 2011; Vero et al., 2017a) indicates that

$t_T$ (and in many catchments $t_U$ alone) will exceed reporting periods, and more closely align with decadal time frames. This has been similarly reported in the UK (Wang et al., 2013; Stuart et al., 2016), throughout Europe, Canada, and the United States (Vero et al., 2017b), China (Zhang et al., 2019), Australia (Lamontagne, 2003), and elsewhere. It is crucial that water-quality goals be correctly framed within the geographic setting and hydrogeologic realities if realistic, achievable objectives and effective programs of measures are to be devised. Considering the well-documented example of the Gulf of Mexico hypoxic zone, modeling of $t_T$ estimates that >50% of nitrate exported from the catchment exceeds 30 years' residence (Van Meter et al., 2018). This legacy of historic land management is the main driver for the current excessive loading, and scenario testing indicates that it would be sufficient to maintain N concentrations above targets on a decadal timeframe. The issue of time lag must be acknowledged in policy and included when designing water-quality targets, both quantitatively and temporally. Failure to do so means that goals are unrealistic and may result in perceived failure of environmental measures before they have been given due time to come to fruition.

## 3.7   MANAGING NITRATE TRANSPORT

Details of $NO_3^-$ management are detailed in Chapter 7, however a brief discussion of strategies for managing the transport of $NO_3^-$ is provided here. Once the vertical movement of $NO_3^-$ through the soil profile has been initiated, there are limited opportunities to mitigate its delivery to groundwater. Substantial nitrate reservoirs exist within the unsaturated zone as a legacy of historical N application. This has been colorfully referred to as the "nitrate time bomb" (Wang et al., 2013), reflecting the uncertainties and potential impacts of its delivery to watercourses. Strategies for the management of $NO_3^-$ transport through unsaturated pathways include minimizing surplus N by aligning available N with crop requirements. This is made challenging by difficulties in devising an effective N soil test and chemical indices (McDonald et al., 2014). Advisory efforts have been employed that attune fertilizer recommendations to growth stages and estimates of crop requirements modeled using meteorological data (Sela et al., 2018). Legislative approaches to minimize nutrient loss are also common. These prescribe seasonal "open" and "closed" periods, maximum application rates, and designated buffer zones near watercourses or vulnerable pathways such as karst features. Nevertheless, implementation of these measures on a "one-size-fits-all" basis is unlikely to achieve the desired improvements and remains a contentious issue.

The sequence of crop rotation has been found to influence $NO_3^-$ losses, though optimal rotations vary depending on geographic location. The use of legumes in grassland swards has been proposed as a means to minimize fertilizer application and as a cover crop to mitigate losses over winter (De Notaris et al., 2018; Askegaard et al., 2006; and many others). The method of fertilizer application also offers potential protection against surface losses where incorporation of manures limits overland flow concentrations versus broadcast application (Kibet et al., 2016). Practices that facilitate denitrification (such as minimum tillage) show potential for reducing $NO_3^-$ leaching but may increase losses via other pathways such as $N_2O$ emissions (Wang and Zou, 2020). Pollution trade-offs must be considered.

Engineered solutions have the potential to mitigate losses within the agricultural landscape. Bioreactors that provide a carbon substrate (such as woodchips) that provide an energy source for denitrifying bacteria can be used to process contaminated water (Schipper et al., 2010). Routing of effluent from farmyards or diffuse losses to constructed wetlands can facilitate denitrification and uptake of available N by plants (Matheson and Sukias, 2010; Lee et al., 2009). The efficacy of N removal is influenced by residence time; hence, this is a crucial element in the siting, design, and scale of constructed wetlands. This technology is well established in the United States, but restrictions on permitting have delayed implementation in parts of the European Union (Healy and Cawley, 2002).

## 3.8 SUMMARY

This chapter has provided an overview of $NO_3^-$ transport in agricultural systems, building on the source-pathway-receptor concept. Nitrate transport can occur through overland flow, vertical and lateral transport through the unsaturated zone, and transport in saturated groundwater systems. The relative significance of these different pathways varies depending on soil, hydrogeological, and landscape types. Which pathway dominates $NO_3^-$ transport has significant implications for understanding responses in receptors to changes in $NO_3^-$ losses in the soil zone. Transport via overland flow is likely to be relatively rapid in comparison to, for example, transport via the unsaturated zone and saturated zone. In the latter case, delays in the impact on receptors of measures put in place to reduce nitrate leaching may occur. Attenuation via denitrification can reduce $NO_3^-$ loadings at receptors; however, this is only likely to occur in the soil zone, the hyporheic zone, and in areas of the saturated zone where redox conditions are favorable. Consideration of the different nitrate transport pathways in a given agricultural system and the implications of the pathways on time lags and attenuation can help improve management of diffuse pollution and support sustainable agriculture.

## ACKNOWLEDGMENTS

Sara Vero would like to acknowledge the funding of the Interreg VA CatchmentCARE project. Matthew Ascott is funded by the British Geological Survey's Environmental Change, Adaptation and Resilience Challenge National Capability program (UK Research and Innovation), and publishes with permission of the Director, British Geological Survey.

## REFERENCES

Ascott, M.J., Gooddy, D.C., Wang, L., Stuart, M.E., Lewis, M.A., Ward, R.S. and Binley, A.M. 2017. Global patterns of nitrate storage in the vadose zone. *Nature Communications*. 8(1): (1416)1416. DOI:10.1038/s41467-017-01321-w.

Ascott, M.J., Wang, L., Stuart, M.E., Ward, R.S. and Hart, A. 2016. Quantification of nitrate storage in the vadose (unsaturated) zone: A missing component of terrestrial N budgets. *Hydrological Processes*. 30: 1903–1915 DOI:10.1002/hyp.10748.

Askegaard, M., Oleson, J.E. and Kristensen, K. 2006. Nitrate leaching from organic arable crop rotations: effects of location, manure and catch crop. *Soil Use and Management*. 21(2): 181–188.

Bateman, E., Cadisch, G. and Baggs, E. 2004. Soil water content as a factor that controls N2O production by denitrification and autotrophic and heterotrophic nitrification. *Controlling Nitrogen Flows and Losses*, 12th Nitrogen Workshop, University of Exeter, UK, 21–24 September 2003, 290–292, 2004.

Battye, W., Aneja, V.P. and Schlesinger, W.H. 2017. Is nitrogen the next carbon? *Earth's Future*. 5: 894–904.

Bechmann, M., Deelstra, J., Stålnacke, P., Eggestad, H.O., Øygarden, L. and Pengerud, A. 2008. Monitoring catchment scale agricultural pollution in Norway: policy instruments, implementation of mitigation methods and trends in nutrient and sediment losses. *Environmental Science and Policy*. 11: 102–114.

Brady, N.C. and Weil, R.R. 2017. *The Nature and Properties of Soils*. 15th Edition. Columbus: Pearson Education.

Cardenas, L.M., Bol, R., Lewicka-Szcezebak, D., Gregory, A.S., Matthews, G.P., Whalley, W.R., Misselbrook, T.H., Scholefield, D. and Well, R. 2017. Effect of soil saturation on denitrification in a grassland soil. *Biogeosciences*. 14: 4691–4710.

Chen, N., Valdes, D., Marlin, C., Blanchoud, H., Guerin, R., Rouelle, M. and Ribstein, P., 2019. Water, nitrate and atrazine transfer through the unsaturated zone of the Chalk aquifer in northern France. *Sci. Total Environ.*, 652: 927–938. DOI:https://doi.org/10.1016/j.scitotenv.2018.10.286.

Cook, P.G., Jolly, I.D., Walker, G.R. and Robinson, N.I. 2003. From drainage to recharge to discharge: Some timelags in subsurface hydrology. In: Alsharhan, A.S., Wood, W.W. (Eds), *Water Resources Perspectives: Evaluation, Management and Policy*. Elsevier Science, Amsterdam, pp. 319–326.

Dahan, O., Babad, A., Larazovitch, N., Russak, E.E. and Kurtzmann, D. 2014. Nitrate leaching from intensive organic farms to groundwater. *Hydrology and Earth System Sciences*. 18: 333–341.

Davidson, E.A., Hart, S.C., Shanks, C.A., and Firestone, M.K. 1991. Measuring gross nitrogen mineralization, immobilization, and nitrification by N-15 isotopic pool dilution in intact soil cores. *Journal of Soil Science*. 42: 335–349.

Davis, S.J., O'hUallachain, D., Mellander, P-E., Kelly, A-M., Matthaei, C.D., Piggott, J.J. and Kelly-Quinn, M. 2018. Multiple-stressor effects of sediment, phosphorus and nitrogen on stream macroinvertebrate communities. *Science of the Total Environment*. 637–638: 577–587.

De Notaris, C., Rasmussen, J., Sorensen, P. and Oleson, J.E. 2018. Nitrogen leaching: A crop rotation perspective on the effect of N surplus, field management and use of catch crops. *Agriculture, Ecosystems and Environment*. 255: 1–11.

Di, H.J. and Cameron, K.C. 2002. Nitrate leaching in temperate agroecosystems: sources, factors and mitigating strategies. *Nutrient Cycling in Agroecosystems*. 46: 237–256.

Di Bonito, M., Breward, N., Crout, N, Smith, B. and Young, S. 2008. Overview of selected soil pore water extraction methods for the determination of potentially toxic elements in contaminated soils: Operational and technical aspects. In: Environmental Geochemistry.

Dinka, T.M. and Lascano, R.J. 2012. A technical note: Orientation of cracks and hydrology in a shrink-swell soil. *Open Journal of Soil Science*. 2: 91–94.

Dinnes, D.L., Karlen, D.L., Jaynes, D.B., Kaspar, T.C., Hatfield, J.L., Colvin, T.S. and Cambardella, C.A. 2002. Nitrogen management strategies to reduce nitrate leaching in tile-drained Midwestern soils. *Agronomy Journal*. 94: 153–171.

Etana, A., Larsbo, M., Keller, T., Arvidsson, J., Schjonning, P., Forkman, J. and Jarvis, N. 2013. Persistent subsoil compaction and its effects on preferential flow patterns in a loamy till soil. *Geoderma*. 192: 430–436.

European Commission. 2019. Press Release: Nitrates: Commission decides to refer Greece to the Court of Justice and asks for financial sanctions. Brussels. 7th March 2019.

Fan, Y., Li, H. and Miguez-Macho, G. 2013. Global patterns of groundwater table depth. *Science*. 339.

Feick, S., Siebert, S. and Döll, P. 2005. A digital global map of artificially drained agricultural areas. Frankfurt Hydrology Paper.

Fenton, O., Mellander, P-E., Daly, K., Wall, D.P., Jahangir, M.M.R., Jordan, P., Hennessey, D., Huebsch, M., Blum, P., Vero, S.E. and Richards, K.G. 2017. Integrated assessment of agricultural nutrient pressures and legacies in karst landscapes. *Agriculture, Ecosystems and Environment*. 239: 246–256.

Fenton, O., Schulte, R.P.O., Jordan, P., Lalor, S.T.J. and Richards, K.G. 2011. Time lag: A methodology for the estimation of vertical and horizontal travel and flushing timescales to nitrate threshold concentrations in Irish aquifers. *Environmental Science and Policy*. 14(4): 419–431.

Fetter, C.W. 2000. Applied Hydrogeology 4th Edition. Long Grove, IL: Waveland Press.

Foster, S.S.D., Cripps, A.C., Smith-Carington, A., 1982. Nitrate Leaching to Groundwater. *Philosophical Transactions of the Royal Society*. 296(1082): 477–489. DOI:10.1098/rstb.1982.0021.

Gehl, R.J., Schmidt, J.P., Stone, L.R., Schlegel, A.J. and Clark, G.A. 2005. In situ measurements of nitrate leaching implicate poor nitrogen and irrigation management on sandy soils. *Journal of Environmental Quality*. 34: 2243–2254.

Gerke, K.M., Sidle, R.C. and Mallants, D. 2013. Criteria for selecting fluorescent dye tracers for soil hydrological applications using Uranine as an example. *Journal of Hydrology and Hydromechanics*. 61(4): 313–325.

Gilbert, P.M., Middelburg, J.J., McClelland, J.W. and Vander Zanden, M.J. 2019. Stable isotope tracers: Enriching our perspectives and questions on sources, fates, rates, and pathways of major elements in aquatic systems. *Limnography and Oceanography*. 64: 950–981.

Guntinas, M.E., Leirós, M.C., Trasar-Cepeda, C. and Gil-Sotres, F. 2012 Effects of moisture and temperature on net soil nitrogen mineralization: A laboratory study. *European Journal of Soil Biology*. 48: 73–80.

Hansen, B., Thorling, L., Schullehner, J., Termansen, M. and Dalgaard, T. 2017. Groundwater nitrate response to sustainable nitrogen management. *Scientific Reports*. 7: 8566 DOI:10.1038/s41598-017-07147-2

Harter, T., Onsoy, Y., Heeren, K., Denton, M., Weissman, G., Hopmans, J.W. and Horwath, W.R. 2005. Deep vadose zone hydrology demonstrates fate of nitrate in eastern San Joaquin Valley. *California Agriculture*. 59(2): 124–132.

Haygarth, P.M., Condron, L.M., Heathwaite, A.L., Turner, B.L. and Harris, G.P. 2005. The phosphorus transfer continuum: Linking source to impact with an interdisciplinary and multi-scaled approach. *Science of the Total Environment*. 344(1–3): 5–14.

He, Y., Hu, K.L., Wang, H., Huang, Y.F., Chen, D.L., Li, B.G. and Li, Y. 2013. Modelling of water and nitrogen utilization of layered soil profiles under a wheat-maize cropping system. *Mathematical and Computer Modelling*. 58(3–4): 596–605.

Healy, M. and Cawley, A.M. 2002. Nutrient processing capacity of a constructed wetland in Western Ireland. *Journal of Environmental Quality*. 31(5): 1739–1747.

Hesser, F.B., Franko, U. and Rode, M. 2010. Spatially distributed lateral nitrate transport at the catchment scale. *Journal of Environmental Quality*. 39: 193–203.

Hillel, D. 2004. *Introduction to Environmental Soil Physics.* Oxford: Elsevier.

Jahangir, M.M.R., Khalil, M.I., Johnson, P., Cardenas, L.M., Hatch, D.J., Butler, M., Barrett, M., O'Flaherty, V. and Richards, K.G. 2012. Denitrification potential in subsoils: A mechanism to reduce nitrate leaching to groundwater. *Agriculture, Ecosystems and Environment.* 14: 13–23.

Jung, H., Koh, D-C., Kim, Y.S., Jeen, S-W. and Lee, J. 2020. Stable isotopes of water and nitrate for the identification of groundwater flowpaths: A review. *Water.* 12(138). DOI:10.3390/w12010138.

Kellman, L.M. and Hillaire-Marcel, C. 2003. Evaluation of nitrogen isotopes as indicators of nitrate contamination sources in an agricultural watershed. *Agriculture, Ecosystems and Environment.* 95: 87–102.

Kibet, L.C., Bryant, R.B., Buda, A.R., Kleinman, P.J.A., Saporito, L.S., Allen, A.L., Hashem, F.M. and May, E.B. 2016. Persistence and surface transport of urea-nitrogen: A rainfall simulation study. *Journal of Environmental Quality.* 45: 1062–1070.

Kleinman, P.J.A., Srinivasan, M.S., Dell, C.J., Schmidt, J.P., Sharpley, A.N. and Bryant, R.N. 2006. Role of rainfall intensity and hydrology in nutrient transport via surface runoff. *Journal of Environmental Quality.* 35: 1248–1259.

Kramers, G. Holden, N.M., Brennan, F., Green, S. and Richards, K.G. 2012. Water content and soil type effects on accelerated leaching after slurry application. *Vadose Zone Journal.* 11(1). DOI:10.2136/vzj2011.0059.

Kurtzman, D., Baram, S. and Dahan, O. 2016. Soil–aquifer phenomena affecting groundwater under vertisols: A review. *Hydrology of Earth System Sciences.* 20: 1–12.

Lamontagen, S. 2003. Groundwater delivery rate of nitrate and predicted change in nitrate concentration in Blue Lake, South Australia. *Marine and Freshwater Research.* 53(7): 1129–1142.

Lee, C., Fletcher, T.D. and Sun, G. 2009. Nitrogen removal in constructed wetlands. *Engineering Life Sciences.* 9(1): 11–22.

Lilburne, L.R. and Webb, T.H. 2002. Effect of soil variability, within and between soil taxonomic units, on simulated nitrate leaching under arable farming, New Zealand. *Australian Journal of Soil Research.* 40(7): 1187–1199.

Liu, Q., Wang, J., Bai, Z., Ma, L. and Oenema, O. 2017. Global animal production and nitrogen and phosphorus flows. *Soil Research.* 55: 451–462.

Lu, J., Bai, Z., Velthof, G.L., Wu, Z., Chadwick, D. and Ma, L. 2019. Accumulation and leaching of nitrate in soils in wheat-maize production in China. *Agricultural Water Management.* 212: 407–415.

Matheson, F.E. and Sukias, J.P. 2010. Nitrate removal processes in a constructed wetland treating drainage from dairy pasture. *Ecological Engineering.* 36(10): 1260–1265.

Mathias, S.A., Butler, A.P., Ireson, A.M., Jackson, B.M., McIntyre, N. and Wheater, H.S., 2007. Recent advances in modelling nitrate transport in the Chalk unsaturated zone. *Quarterly Journal of Engineering Geology and Hydrogeology.* 40(4): 353–359. DOI:10.1144/1470-9236/07-022.

McDonald, N.M.T., Watson, C.J., Lalor, S.T.J. Laughlin, R.J. and Wall, D.P. 2014. Evaluation of soil tests for predicting nitrogen mineralization in temperate grass-land soils. *Soil Science Society of America Journal.* 78: 1051–1064. doi:10.2136/sssaj2013.09.0411.

McTaggart, I.P., Akiyama, H., Tsuruta, H. and Ball, B.C. 2002. Influence of soil physical properties, fertilizer type and moisture tension on $N_2O$ and NO emissions from nearly saturated Japanese soil. *Nutrient Cycling in Agroecosystems.* 63: 207–217.

Meals, D.W., Dressing, S.A. and Davenport, T.E. 2010. Lag time in water quality response to best management practices: A review. *Journal of Environmental Quality.* 39: 89–96.

Morecroft, M.D., Burt, T.P., Taylor, M.E. and Rowland, A.P. 2000. Effects of the 1995–1997 drought on nitrate leaching in lowland England. *Soil Use and Management.* 16: 117–123.

Nimmo, J. 2012. Preferential flow occurs in unsaturated conditions. *Hydrological Processes.* 26: 786–789.

Onsoy, Y.S., Harter, T., Ginn, T.R. and Horwath, W.R. 2005. Spatial variability and transport of nitrate in a deep alluvial vadose zone. *Vadose Zone Journal.* 4: 41–54.

Outram, F.N., Cooper, R.J., Sünnenberg, G., Hiscock, K.M. and Lovett, A.A. 2016. Antecedent conditions, hydrological connectivity and anthropogenic inputs: Factors affecting nitrate and phosphorus transfers to agricultural headwater streams. *Science of the Total Environment.* 545–546(1): 184–199.

Owens, L.B., Van Keuren, R.W. and Edwards, W.M. 1985. Groundwater quality changes resulting from a surface bromide application to a pasture. *Journal of Environmental Quality.* 14: 543–548.

Perkins, K.S., Nimmo, J.R., Rose, C.E. and Coupe, R.H. 2011. Field tracer investigation of unsaturated zone flow paths and mechanisms in agricultural soils of northwestern Mississippi, USA. *Journal of Hydrology.* 396: 1–11.

Pratt, P.F., Jones, W.W. and Hunsaker, V.E. 1972. Nitrate in deep soil profiles in relation to fertiliser rates and leaching volume. *Journal of Environmental Quality.* 1(1): 97–101.

Rakotovolona, L., Beaudoin, N., Ronceux, A., Venet, E. and Mary, B. 2019. Driving factors of nitrate leaching in arable organic cropping systems in Northern France. *Agriculture, Ecosystems and Environment.* 272: 38–51.

Rempe, D.M. and Dietrich, W.E. 2018. Direct observations of rock moisture, a hidden component of the hydrologic cycle. *Proceedings of the National Academy of Science.* 115(11): 2664–2669.

Risch, A.C., Zimmerman, S., Ochoa-Hueso, R., Schütz, M., Frey, B., Firn, J.L., Fay, P.A., Hagedorn, F., Borer, E.T., Seabloom, E.W., Harpole, W.S., Knops, J.M.H., McCulley, R.L., Broadbent, A.A.D., Stevens, C.J., Silveira, M.L., Adler, P.B., Báez, S., Biederman, L.A., Blair, J.M., Brown, C.S., Caldeira, M.C., Collins, S.L., Daleo, P., di Virgilio, A., Ebeling, A., Eisenhauer, N., Esch, E., Eskelinen, A., Hagenah, N., Hautier, Y., Kirkman, K.P., MacDougall, A.S., Moore, J.L., Power, S.A., Prober, S.M., Roscher, C., Sankaran, M., Siebert, J., Speziale, K.L., Tognetti, P.M., Virtanen, R., Yahdjian, L. and Moser, B. 2019. Soil net nitrogen mineralisation across global grasslands. *Nature Communications.* 10:4981. https://doi.org/10.1038/s41467-019-12948-2.

Rittenburg, R.A., Squires, A.L., Boll, J., Brooks, E.S., Easton, Z.M. and Steenhuis, T.S. 2015. Agricultural BMP effectiveness and dominant hydrological flow paths: Concepts and review. *Journal of the American Water Resources Association.* 51(2): 305–329.

Rivett, M.O., Buss, S.R., Morgan, P., Smith, J.W. and Bemment, C.D., 2008. Nitrate attenuation in groundwater: a review of biogeochemical controlling processes. *Water Research.* 42(16): 4215–4232.

Rivett, M.O., Smith, J.W.N., Buss, S.R., Morgan, P. 2007. Nitrate occurrence and attenuation in the major aquifers of England and Wales. *Quarterly Journal of Engineering Geology & Hydrogeology.* 40(4): 335–352. DOI:10.1144/1470-9236/07-032.

Ryan, M., Brophy, C., Connolly, J., McNamara, K. and Carton, O.T. 2006. Monitoring of nitrate leaching on a dairy farm during four drainage seasons. *Irish Journal of Agricultural and Food Research.* 45: 115–134.

Scheer, C., Wassman, R., Butterbach-Bahl, K., Lamers, J.P.A. and Martius, C. 2009. The relationship between $N_2O$, NO, and $N_2$ fluxes from fertilised and irrigated dryland soils of the Aral Sea basin, Uzbekistan. *Plant and Soil.* 314(1–2): 273–283.

Schilling, K.E. 2002. Chemical transport from paired agricultural and restored prairie watersheds. *Journal of Environmental Quality.* 31: 1184–1193.

Schipper, L.A., Robertson, W.D., Gold, A.J., Jaynes, D.B. and Cameron, S.C. 2010. Denitrifying bioreactors – An approach for reducing nitrate loads to receiving waters. *Ecological Engineering*. 36: 1532–1543.

Sela, S., van Es, H.M., Moebius-Clune, B.N., Marjerison, R. and Kneubuhner, G. 2018. Dynamic model-based recommendations increase the precision and sustainability of N fertilization in Midwestern US maize production. *Computers and Electronics in Agriculture*. 153: 256–265.

Shore, M., Jordan, P., Melland, A.R., Mellander, P-E., McDonald, N. and Shortle, G. 2016. Incidental nutrient transfers: Assessing critical times in agricultural catchments using high-resolution data. *Science of the Total Environment*. 553: 404–415.

Sigunga, D.O., Janssen, B.H. and Oenema, O. 2002. Denitrification risks in relation to fertilizer nitrogen losses from vertisols and phaoezems. *Communications in Soil Science and Plant Analysis*. 33: 3–4.

Singh, G., Kaur, G., Williard, K., Schoonover, J. and Kang, J. 2017. Monitoring of water and solute transport in the Vadose Zone: A review. *Vadose Zone Journal*. 17: 1–23. doi:10.2136/vzj2016.07.0058.

Sousa, M.R., Jones, J.P., Frind, E.O. and Rudolph, D.L. 2013. A simple method to assess unsaturated zone time lag in the travel time from ground surface to receptor. *Journal of Contaminant Hydrology*. 144: 138–151.

Stuart, M.E., Ascott, M.J., Talbot, J.C., Newell, A.J., 2018. *Mapping Groundwater Denitrification Potential: Methodology Report*, British Geological Survey, Keyworth, UK.

Stuart, M.E., Chilton, P.J., Kinniburgh, D.G., and Cooper, D.M. 2007. Screening for long-term trends in groundwater nitrate monitoring data. *Quarterly Journal of Engineering Geology and Hydrogeology*. 40: 361–376.

Stuart, M.E., Gooddy, D.C., Bloomfield, J.P. and Williams, A.T. 2018. *Review of Denitrification Potential in Groundwater of England*, British Geological Survey, Keyworth, UK.

Stuart, M.E., Wang, L., Ascott, M.J., Ward, R.S., Lewis, M.A., 2015. *Modelling the Nitrate Legacy*, British Geological Survey, Keyworth, UK.

Sugg, Z. 2007. *Assessing US Farm Drainage: Can GIS Lead to Better Estimates of Subsurface Drainage Extent?* World Resources Institute.

Tesoriero, A.J., Duff, J.H., Saad, D.A., Spahr, N.E., and Wolock, D.M., 2013. Vulnerability of streams to legacy nitrate sources. *Environmental Science and Technology*. 47. 3623–3629.

Turkeltaub, T., Ascott, M.J., Gooddy, D.C., Xiaoxu, J., Shao, M. and Binley, A., 2019. On the application of global scale models for prediction of regional scale groundwater recharge and nitrate storage in the vadose (unsaturated) zone. *Hydrological Processes* (submitted).

Turkeltaub, T., Jia, X., Zhu, Y., Shao, M-A. and Binley, A. 2018. Recharge and nitrate transport through the deep vadose zone of the Loess Plateau: A regional-scale model investigation. *Water Resources Research*. 54(7): 4332–4346.

Uwizeye, A., de Boer, I.J.M., Opio, C.I., Schulte, R.P.O., Falcucci, A., Tempio, G., Teillard, F., Casu, F., Rulli, M., Galloway, J.N., Leip, A., Erisman, J.W., Robinson, T.P., Steinfield, H. and Gerber, P.J. 2020. Nitrogen emissions among global livestock supply chains. *Nature Food*. 1: 437–446.

Vadas, P.A., Busch, D.L., Powell, J.M. and Brink, G.E. 2015. Monitoring runoff from cattle-grazed pastures for a phosphorus loss quantification tool. *Agriculture, Ecosystems and Environment*. 199: 124–131.

Valipour, M., Krasilnikof, J., Yannopoulos, S., Kumar, R., Deng, J., Roccaro, P., Mays, L., Grismer, M.E. and Angelakis, A.N. 2020. The evolution of agricultural drainage from the earliest times to the present. *Sustainability*. 12(1): 416. DOI:10.3390/su12010416.

van Es, H.M., Sogbedji, J.M. and Schindelbeck, R.R. 2006. Effect of manure application timing, crop, and soil type on nitrate leaching. *Journal of Environmental Quality.* 35(2): 670–679.

van Meter, K.J., van Capellen, P. and Basu, N.B 2018. Legacy nitrogen may prevent achievement of water quality goals in the Gulf of Mexico. *Science.* 360: 427–430.

Vero, S.E., Healy, M.G., Henry, T., Creamer, R.E., Ibrahim, T.G., Richards, K.G., Mellander, P-E., McDonald, N.T. and Fenton, O. 2017a. A framework for determining unsaturated zone water quality time lags at catchment scale. *Agriculture, Ecosystems and Environment.* 236: 234–242.

Vero, S.E., Basu, N.B., Van Meter, K., Richards, K.G., Mellander, P-E., Healy, M.G. and Fenton, O. 2017b. Review: The environmental status and implications of the nitrate time lag in Europe and North America. *Hydrogeology Journal.* DOI 10.1007/s10040-017-1650-9.

Vero, S.E., Ibrahim, T.G., Creamer, R.E., Grant, J., Healy, M.G., Henry, T., Kramers, G., Richards, K.G. and Fenton, O. 2014. Consequences of varied soil hydraulic and meteorological complexity on unsaturated zone time lag estimates. *Journal of Contaminant Hydrology.* 170: 53–67.

Wall, D.P., Jordan, P., Melland, A.R., Buckley, C., Reaney, S.M. and Shortle, G. 2011. Using the nutrient transfer continuum concept to evaluate the European Union Nitrates Directive National Action Programme. *Environmental Science and Policy.* 14(6): 664–674.

Wang, J. and Zou, J. 2020. No-till increases soil denitrification via its positive effects on the activity and abundance of the denitrifying community. *Soil Biology and Biochemistry.* 142: 107706.

Wang, L., Butcher, A., Stuart, M., Gooddy, D., Bloomfield, J., 2013. The nitrate time bomb: a numerical way to investigate nitrate storage and lag time in the unsaturated zone. *Environ. Geochem. Health*, 35(5): 667–681.

Wang, L., Stuart, M.E., Bloomfield, J.P., Butcher, A.S., Gooddy, D.C., McKenzie, A.A., Lewis, M.A. and Williams, A.T., 2012. Prediction of the arrival of peak nitrate concentrations at the water table at the regional scale in Great Britain. *Hydrol. Processes*, 26(2): 226–239. DOI:10.1002/hyp.8164.

Wang, L., Stuart, M.E., Lewis, M.A., Ward, R.S., Skirvin, D., Naden, P.S., Collins, A.L. and Ascott, M.J., 2016. The changing trend in nitrate concentrations in major aquifers due to historical nitrate loading from agricultural land across England and Wales from 1925 to 2150. *Sci. Total Environ.*, 542: 694–705. DOI:10.1016/j.scitotenv.2015.10.127.

Wei, X., Bailey, R.T., Records, R.M., Wible, T.C. and Arabi, M. 2019. Comprehensive simulation of nitrate transport in coupled surface-subsurface hydrologic systems using the linked SWAT-MODFLOW-RT3D model. *Environmental Modelling and Software.* 122. 104242.

Weldon, M.B. and Hornbuckle, K.C. 2006. Concentrated animal feeding operations, row crops and their relationship to nitrate in Eastern Iowa rivers. *Environmental Science and Technology.* 40(10): 3168–3173.

WHO. 2007. *Nitrate and nitrite in drinking-water. Background document for development of WHO Guidelines for drinking-water quality.* World Health Organization, Geneva (WHO/SDE/WSH/07.01/16).

Yue, F-J., Waldron, S., Li, S-L., Wang, Z-J., Zeng, J., Xu, S., Zhang, Z-C. and Oliver, O. 2019. Land use interacts with changes in catchment hydrology to generate chronic nitrate pollution in karst waters and strong seasonality in excess nitrate export. *Science of the Total Environment.* 696. https:doi.org/10.1016/j.scitotenv.2019.134062.

Zhang, B., Tang, J.L., Gao, Ch. and Zepp, H. 2011. Subsurface lateral flow from hillslope and its contribution to nitrate loading in streams through an agricultural catchment during subtropical rainstorm events. *Hydrology and Earth Systems Sciences.* 15: 3153–3170.

Zhang, H., Yang, R., Wang, Y. and Ye, R. 2019. The evaluation and prediction of agriculture-related nitrate contamination in groundwater in Chengdu Plain, southwestern China. *Hydrogeology Journal*. 27: 785–799.

Zhu, Q. and Lin, H.S. 2009. Simulation and validation of concentrated subsurface lateral flow paths in an agricultural landscape. *Hydrology and Earth Systems Sciences*. 13: 1503–1518.

# Part 2

Nitrate in Plants

# 4 Nitrate in Plant Physiology

*Jitu Chauhan, Simran Chachad, Zoya Shaikh,*
*Johra Khan, and Ahmad Ali*

## CONTENTS

4.1 Introduction .................................................................................................74
4.2 Importance of Nitrate .................................................................................74
4.3 Uptake Mechanisms for Nitrate ................................................................75
    4.3.1 A Symbiotic Relationship .............................................................76
    4.3.2 Nitrogen Gain from Its Surroundings..........................................76
    4.3.3 Nitrate Uptake and Assimilation Reactions.................................77
4.4 Assimilation of Nitrate (Enzymes and Genes) and Regulation.....................79
    4.4.1 Nitrogen Assimilation ...................................................................79
    4.4.2 Nitrogen Fixation ..........................................................................80
    4.4.3 Root-to-Shoot Nitrogen Transport ...............................................80
    4.4.4 Nitrogen Transported in Mesophyll Cells ...................................81
    4.4.5 Nitrate Transporters......................................................................81
        4.4.5.1 NRT1 Transporters......................................................81
        4.4.5.2 NRT2 Transporters......................................................81
        4.4.5.3 NAR1 Transporters .....................................................81
        4.4.5.4 Nitrate Reductase........................................................82
        4.4.5.5 Nitrite Reductase.........................................................82
        4.4.5.6 Nitrogen Storage .........................................................82
        4.4.5.7 Regulation of Nitrogen Transport ..............................82
4.5 Nitrate and Nitric Oxide Link ...................................................................83
4.6 Interactions between Carbon and Nitrogen................................................84
4.7 Nitrogen-Use Efficiency .............................................................................84
4.8 Nitrate Pollution ........................................................................................85
4.9 Conclusion ..................................................................................................86
References.............................................................................................................87

DOI: 10.1201/9780429326806-6

## 4.1   INTRODUCTION

Plants are very versatile as far as their life cycles and growth are concerned. They can use simple raw materials to synthesize their building blocks and provide food for the organism. Nitrogen (N) is a microelement found in the environment in inorganic form. It is then converted into organic components like amino acids by the nitrogen assimilation process. Nitrogen is an important component of most of the biomolecules like nucleic acids, proteins, and vitamins (Reichardt and Timm 2020). Nitrogen assimilation is an indispensable process controlling plant development and growth. The most frequent inorganic nitrogen sources used by plants for these purposes are $NH_4^+$ (ammonia) and $NO_3^-$ (nitrate). Assimilation of inorganic nitrogen occurs, leading to the production of amino acids, glutamate, glutamine, aspartate, and asparagine. These molecules are the source of nitrogen carriers in plants (Chaffey, Ohyama 2010).

Biosynthesis of amino acids that carry nitrogen takes place with the help of enzymes like asparagine synthetase (AS), aspartate aminotransferase (AT), glutamate dehydrogenase (GDH), glutamate synthase (GOGAT), and glutamine synthetase (GS) (Krapp 2015). Nitrate reduction is conveyed by two steps. In the first step, nitrate is converted into nitrite in the cytosol by using the enzyme nitrate reductase and with the help of NADH or NADPH. Next, nitrite is reduced to ammonia in chloroplast with the help of ferredoxin-dependent nitrite reductase. In the chloroplast, glutamate is utilized as substrate, and glutamate synthase integrates ammonia as the amide group of glutamine. Amide group is transferred onto the 2-oxoglutarate molecule by glutamate synthase, this produces glutamate. Next, transamination occurs so that formation of other amino acids such as glutamine and asparagine can take place (Masclaux-Daubresse et al. 2010, Ali 2020).

## 4.2   IMPORTANCE OF NITRATE

Soil, as we know, is considered to be the main resource of nutrition for plant development. There are three main plant nutrients, nitrogen (N), phosphorus (P), and potassium (K). The trio is commonly recognized as NPK. Some of the supplementary essential nutrients required by plants are calcium, magnesium, and sulfur (Chaffey). Nitrate remains a major source of nitrogen for all plants because of their versatile functions in both plant nutrition and physiological regulation. Nitrate ($NO_3^-$) is composed of oxygen and nitrogen. Inorganic nitrate is formed by the aerobic bacteria in the biogeochemical cycle of nitrogen, depicted in Figure 4.1, and are an essential component of agricultural soil (Robertson 1997). Bacteria use nitrate for different purposes: assimilatory and dissimilatory nitrate assimilation processes. Organic nitrate is a manufactured compound and, since it generates oxygen when heated, it is used in explosives such as nitrocellulose and nitroglycerin (Crawford and Glass 1998, Jones 1997).

Nitrogen is an essential component of chlorophyll and plant proteins. Nitrogen is added to the soil by nitrate. Plants use nitrate as a supply of nitrogen, which is needed to make amino acids for protein synthesis for a nutritious development (Jones 1997). Plant roots help in the uptake of nitrate through water (Morgan and Connolly 2013). If there is a lack of nitrate, the amount of chlorophyll in leaves is decreased. Hence, the process of photosynthesis is reduced in the plant, leading to improper growth.

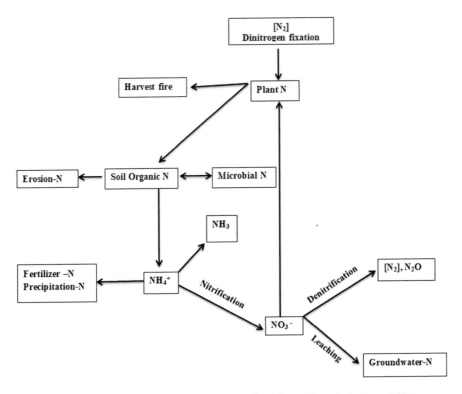

**FIGURE 4.1**    A simple illustration of nitrogen cycle. Adapted from Robertson (1997).

This in turn reduces crop yield. Plants suffer from a variety of deficiency diseases when they lack essential vitamins or mineral ions needed for their further growth and development. For example, in the absence of nitrate plants are unable to grow and produce seeds and flowers. Fertilizers contain a very high level of nitrate. Thus, chemical or natural fertilizers are added to increase the amount of nitrate in the soil to support plant growth and development (Näsholm, Kielland, and Ganeteg 2009, Ohyama 2010, Bijay, Yadvinder, and Sekhon 1995).

Soil nitrate levels increase through rapid nitrification in well-aerated soils with a pH of 6–8 and temperatures that are greater than 50 degrees Fahrenheit. As we know, wet, cold, or acidic soils do not contain high levels of nitrate. At the same time, the microbial processes of denitrification and dissimilatory nitrate reduction to ammonium (DNRA) are two significant nitrate-reducing mechanisms in soil, and which are accountable for the loss of nitrate as well as the production of the potent greenhouse gas, nitrous oxide ($N_2O$). Also, nitrate above 150 parts per million (ppm) in the environment can cause nutrient burn in the roots (Benincasa, Guiducci, and Tei 2011).

## 4.3   UPTAKE MECHANISMS FOR NITRATE

Nitrogen is a colorless and odorless gas comprising about 78 percent of the atmospheric volume. It helps plants in their growth, reproduction, and the process of

photosynthesis. Nitrogen is an important component of DNA, RNA, and protein –
acknowledged as the building blocks of life. But the nitrogen found in the atmosphere is the form of dinitrogen ($N_2$) (Ohyama 2010). Plants are not able to convert
dinitrogen found in the air into a biologically useful form. This is where the process of nitrogen uptake and assimilation comes in. The atmospheric nitrogen either
combines with hydrogen to form ammonia ($NH_3$) or with oxygen to form nitrate
($NO_3$). Nitrogen (N) accessibility is one of the main factors that influence a plant's
growth and its yield, also acting as a signal for the commencement of a variety of
processes (Ohyama 2010). Internationally, high amounts of nitrogenous fertilizer
are used to ensure the utmost crop productivity. In plants, nitrogen is obtained
from the soil in the form of nitrate and ammonium (Morgan and Connolly 2013).
Nitrate is mostly found in aerobic soils, whereas ammonia is common in anaerobic
and acidic soil environments. According to the availability of nitrogen resources,
plants have developed efficient signaling and transport mechanisms. Nitrate itself
serves as a systemic and local signal in regulating floral induction, seed dormancy,
root morphology, leaf expansion, and genome-wide gene expression (Asim et al.
2020, Imsande and Touraine 1994).

### 4.3.1  A SYMBIOTIC RELATIONSHIP

Many times, the levels of nitrogen in the soil are found to be low. This is when a
symbiotic relationship in some plants is established. For instance, the relationship between legumes and rhizobium/bradyrhizobium/azorhizobium, or between
parasponia and its root nodules with rhizobia helping them fix atmospheric $N_2$. The
inorganic nitrogen acquired by the plant is reduced to amino acids in roots, nodules,
or leaves, depending on the species of the plant and its environment (Chaffey 2014;
Jones Jr 1997). This takes place through the following steps:

* Bacterial infection of the root;
* Rhizobia-host specificity;
* Bacterium penetrating the host cell wall to enter the space between the cell wall
  and the plasma membrane;
* Release of bacteria into the host cell;
* Budding;
* Division and differentiation of cells into nitrogen-fixing cells known as
  bacteroids;
* Vascular connection between them;
* Vascular connections help import photosynthetic carbon to the nodules and ex-
  ports fixed nitrogen from the nodules to the plant.

### 4.3.2  NITROGEN GAIN FROM ITS SURROUNDINGS

Dinitrogen ($N_2$) has to be further processed to convert it into a biologically func-
tional variety. For each $N_2$ molecule reduced, one molecule of hydrogen is generated.
Nitrogen is gained by the plant through nitrate or ammonium (Keys et al. 1978).
The uptake of nitrate is swift due to high particle mobility. Nitrate is preferred over

ammonium in most plants. Hence, the uptake of ammonium is less in comparison to nitrate. Roots take up the ammonium that is attached to the clay particles in the soil. Most of the ammonium is nitrified beforehand. The soil bacteria convert ammonium into nitrate (Nitrification) (Näsholm, Kielland, and Ganeteg 2009, Tischner 2000). During this process, nitrous oxide and nitric oxide are released into the atmosphere. Denitrification is favored due to a shortage of oxygen (e.g. due to waterlogging). The soil bacteria convert nitrate to gaseous nitrous oxide, nitric oxide, and nitrogen, and release it into the atmosphere. Immobilized nitrogen cannot instantly be available for plant consumption because it needs to be mineralized. Mineralization transforms mineral nitrogen into biomass or humus. Ammonium is released into the soil via the mineralization of soil organic matter. Ammonium is then converted to ammonia through volatilization and given up into the atmosphere; this is supported by elevation in soil pH and temperature (McDermitt and Loomis 1981, Krapp 2015, Reichardt and Timm 2020).

### 4.3.3 Nitrate Uptake and Assimilation Reactions

After the absorption of nitrate from the soil, it is either reduced in the roots or goes through root-to-shoot transportation with the help of xylem. It is then reduced in the leaves. The nitrate assimilation pathway initiates with the uptake of nitrate followed by nitrate reduction that results in ammonium. It is then fixed into the amino acids, glutamine, and glutamate in most plants. Glutamine is the main N-compound that is taken up to the shoot in many ammonium-grown plants (Garma and Bloom 2014, Masclaux-Daubresse et al. 2010). Nitrate reductase (NR) performs the first step of nitrate reduction in the cytosol, which leads to the production of nitrite (Tischner 2000).

$$NO_3^- + NADH + H^+ + 2e^- \xrightarrow{\textit{Nitrate-reductase}} NO_2 + NAD^+ + H_2O \quad (4.1)$$

Nitrite then enters the plastid in the roots (chloroplast in the shoot), followed by its reduction to ammonium by nitrite reductase (NIR). Nitrate reductases of higher plants are homo-dimers made of two identical subunits each, with a molecular mass of a hundred kDa. Each subunit has three prosthetic groups: (i) FAD (flavin adenine dinucleotide); (ii) the heme group; (iii) molybdenum complex. Nitrate reductase converts nitrate into nitrite. Nitrite, being a very unstable compound, is immediately transferred from the site of its formation, that is, the cytosol to the chloroplast in leaves, plastids, and roots. Further, with the help of nitrite reductase, nitrite is converted into ammonia. Ammonium is fixed into amino acids (glutamine/glutamate) with the help of the GS/GOGAT pathway. These amino acids act as substrates for transamination reactions for the production of other proteins and amino acids (Tischner 2000, Ali et al. 2008).

$$NO_2^- + 6Fd_{red} + 8H^+ + 6e^- \xrightarrow{\text{Nitrite Reductase}} NH_4^+ + 6Fd_{ox} + 2H_2O \quad (4.2)$$

Nitrite reductase is encrypted in the nucleus, synthesized as a precursor carrying the nitrogen terminus as well as the transit peptide. Nitrate can be simulated in both the root and shoot of the plant depending upon the level of nitrate in the plant and the plant species. While some plants only assimilate nitrite in the roots, others only assimilate nitrite in the shoots, and some in both. To avoid ammonium toxicity, plants rapidly convert ammonia into amino acids with the help of glutamine synthetase or glutamine dehydrogenase (Ali, Sivakami, and Raghuram 2007).

$$\text{Glutamate} + NH_4 + ATP \xrightarrow{\text{Glutamine Synthetase}} \text{Glutamine} + ADP + Pi \qquad (4.3)$$

Or

$$2 - \text{oxoglutarate} + NH_4 + NADH \xrightarrow{\text{Glutamine dehydrogenase}} \text{Glutamine} + H_2O + NAD^+$$
$$(4.4)$$

During the vegetative phase, leaves and budding roots are the major nitrogen sinks. At the reproductive stage, seeds, flowers, and fruits are the main N assimilate-importing sinks. The root to shoot movement of nitrogen occurs in xylem and partitioning of nitrogen from leaf base to sinks occurs at phloem. Sink organs, which are highly active metabolically, and rapidly growing tissues like meristems and immature leaves, typically exhibit slight xylem intake as an outcome of their low rates of transpiration (Crawford and Glass 1998, Tegeder and Masclaux-Daubresse 2018).

Phloem loading takes place in the collection phloem of the leaf minor vein networks, and phloem unloading occurs at free phloem of sink organs. The interconnection of collection phloem to transporting phloem assists the loading, unloading, and exchange of N compounds. It also helps in the linkage of phloem and xylem throughout the pathway of transport in the stem, roots, pod wall, and major leaf veins. The plasma membrane-localized transport proteins play an important role in the source to sink transfer of N from the soil. The regulation of N root uptake, seed loading, leaf-to-sink, and root-to-shoot transport, N accessibility, and consumption for vegetative growth largely depends on the uptake of nitrogen from the soil, its incorporation, and its transfer to sinks. The occurrence of these extremely varied uptake mechanisms allows the plant root to alter according to the various changes that take place in the soil, to N composition, and concentration throughout the growth period, even under nutrient stress conditions (Crawford and Glass 1998) (Tegeder and Masclaux-Daubresse 2018).

Low- and high-affinity nitrate transporters mediate the uptake of nitrate from the soil. They are transporters of the Nitrate Transporter1 (NRT1) family, also known as the Nitrate Transporter1/Peptide Transporter Family (NPF) (Léran et al. 2014). These are mostly low-affinity systems except in Arabidopsis and rice (Oryza sativa). NRT1.1B, possesses both low and high affinity for nitrate. In Arabidopsis, 16 out of 53 NPF/NRT1 proteins have been described to date. Seven high-affinity nitrate transporters belonging to NRT2 family have been studied so far. NPF/NRT1 and NRT2 transporters are proton-coupled importers, except NPF7.3/NRT1.5 transporters, which are bidirectional, and NPF2.7/Nitrate Excretion Transporter1 (NAXT1), which mediates nitrate efflux. Some of the nitrate transporters are present in the Chloride Channel (CLC) family, consisting of anion channels or anion proton exchangers. Among Arabidopsis, up to six transporters are responsible for nitrate uptake via roots. NPF6.3/NRT1.1 and NPF4.6/NRT1.2 function under high nitrate supply, while NRT2.1, NRT2.2, NRT2.4 and NRT2.5 operate in low nitrate concentrations (Sanz-Luque et al. 2015).

The uptake and assimilation of ammonium are highly regulated in plants because excess ammonium can be hazardous for plant life. In plants, ammonium transport and homeostasis are managed by high-affinity and low-affinity uptake systems. Ammonium Transporters (AMTs) are the high affinity, and aquaporins, or cation channels, are the low-affinity uptake systems that handle ammonium transport in plants (González-Ballester, Camargo, and Fernández 2004). Previous studies have been conducted to analyze AMT genes. Six AMT genes were studied in Arabidopsis, 14 in poplar (*Populus trichocarpa*), and 10 in rice. Three AMT genes have also been discovered in pine (*Pinus pinaster*). In Arabidopsis, four AMTs are involved in ammonium root attainment, of which AMT1;1, AMT1;3 and AMT1;5 take part in direct soil uptake via the epidermis. AMT1;2 helps in apoplastic absorption of ammonium (Miller and Chapman 2011).

The majority of amino acids are localized in the plasma membrane. They function in the cellular import of a variety of amino acids in co-transport with protons. Root uptake systems are linked to three families of the APC group: Proline, Glycine and Betaine Transporters (ProTs), Lysine/Histidine-like Transporters (LHTs), and Amino Acid Permeases (AAPs). Usually, substrate affinity and specificity may differ among different transporter families (Tegeder and Masclaux-Daubresse 2018). AAPs possess moderate affinity and broad substrate specificity. In Arabidopsis, AAP1 helps in root uptake of glutamate and neutral amino acids (Lee et al. 2007), while AAP5 plays a role in acquiring basic amino acids (Svennerstam, Ganeteg, and Näsholm 2008). LHTs are high-affinity transport systems. They import neutral and acidic amino acids into the root (Chen and Bush 1997).

## 4.4 ASSIMILATION OF NITRATE (ENZYMES AND GENES) AND REGULATION

### 4.4.1 NITROGEN ASSIMILATION

Nitrite reductase and nitrate catalyze the reduction of nitrate to ammonium in the cytosol and plastids respectively. Nitrite Transporter NITR2; 1 carries nitrite into plastids 1 (Maeda et al. 2014). Assimilation of ammonium into amino acids is carried out by Asparagine Synthetases (ASs), Glutamine Oxoglutarate Aminotransferase (GOGATs), and Glutamine Synthetases (GSs) (Masclaux-Daubresse et al. 2010). The GS2 present in chloroplast helps in primary N assimilation and re-assimilation of photorespiratory ammonium of mesophyll. GS1 enzymes of the cytosol function to contribute to primary N assimilation in roots under high nitrate conditions. GS1 isoforms are found in the companion cells of mature leaves; this promotes phloem loading for N translocation to sinks. The activity of GS1 is induced in mesophyll cells (Tegeder and Masclaux-Daubresse 2018).

Ferredoxin-dependent GOGAT isoforms are located in the mesophyll chloroplasts, and NADH-GOGAT is found in plastids of the leaf and root companion cells. It is involved in N assimilation and partitioning (Suzuki and Knaff 2005). The cytosolic

AS is involved in ammonium assimilation (Carvalho et al. 2003). Biosynthesis of proteinogenic amino acids and membrane proteins occurs in chloroplasts and cytosol (Tegeder et al. 2006). To date, two chloroplast amino acid transporters have been identified. They are glutamate/malate exchanger dicarboxylate transporter (DiT2) in Arabidopsis (Renné et al. 2003) and the Petunia hybrida Cationic Amino Acid Transporter (PhpCAT) ,which functions to export aromatic amino acids (Widhalm et al. 2015).

### 4.4.2 Nitrogen Fixation

$N_2$ fixation in legumes takes place in root nodules by bacteroids. This produces ammonia, which is transferred via ammonia channel Nod 26 (Nodulin 26) to nodule host cells. It is reduced further to glutamine and asparagine (Fortin, Morrison, and Verma 1987).

### 4.4.3 Root-to-Shoot Nitrogen Transport

The root-to-shoot flow of nitrate is carried out through elements of xylem. A hydrostatic pressure gradient is created in xylem due to transpiration at the leaf surface; this determines the movement to the above-ground tissue (Tyree 2003). Nitrate, amino acids, or ureides are transported through xylem to the shoot according to the location of N assimilation, which can be nodule, root, or source leaf. Xylem loading is supported by delivering N compounds to the apoplast from the nodule or root endodermis, pericycle, or vascular parenchyma. This is done through passive transportation (Tegeder and Masclaux-Daubresse 2018). Arabidopsis roots pericycle achieve nitrate extraction by targeting proton-N-transporting mechanism with NPF7.3 / NRT1.5 (Lin et al. 2008). Hence, the root-to-shoot nitrate transport, xylem pH, and transporter activity are linked. NPF2.3 in Arabidopsis is expressed in the pericycle. It exports nitrate under salt stress. Root-to-shoot nitrate translocation is regulated by NPF7.2 / NRT1.8 and NPF2.9 / NRT1.9. The plasma membrane of root parenchyma cells possesses NPF7.2/ NRT1.8, while NPF2.9/NRT1.9 is present in phloem companion cells (Tegeder and Masclaux-Daubresse 2018).

Transpiration helps in loading N compounds into xylem. These N compounds are then transported to the leaves, although, during vegetative growth, some amount of N is taken out from xylem for metabolism or N storage (Bailey and Leegood 2016). Besides, N can be directly transferred to phloem after its extraction from xylem, so that it can be immediately delivered to rapidly growing sinks (Pate, Sharkey, and Lewis 1975). The transmission that occurs between xylem and phloem needs N compounds to return from the pathway of xylem parenchyma cells and transfer through proton support, along with N movement and apoplastic loading action in the phloem. The xylem parenchyma contains NPF7.2 / NRT1.8 and AAP6. An increase in nitrate content of npf7.2 / nrt1.8 xylem sap and reduction in the concentration of amino acid in AAP6 phloem indicates that NPF7.2 / NRT1.8 and AAP6 act upon N, and nitrate regenerates xylem (Li et al. 2010).

#### 4.4.4 NITROGEN TRANSPORTED IN MESOPHYLL CELLS

Limited information is available regarding transporters involved in N xylem unloading and mesophyll import. Many studies predict that they may strongly influence leaf apoplastic N concentrations. This can further impact shoot regulation of N uptake from the soil (Tegeder 2012). Arabidopsis NPF6.2/NRT1.4 may retrieve nitrate for storage in petiole or import it into mesophyll cells with the help of leaf NPFs/NRTs. Ammonium ingestion can be handled by AtAMT1; 1 and AtAMT2; 1. After reaching the mesophyll cells, ammonium and nitrate are either reduced to amino acids or stored in the vacuole (Tegeder and Masclaux-Daubresse 2018).

LHT1 contributes in importing amino acid into leaf cells. It was found that knockdown of LHT1 decreased the import of amino acids into mesophyll and amino acid accumulation in leaf apoplast. This in turn affects plant growth (Hirner et al. 2006). UPS1 plasma membrane transporters are involved in the import of ureide into leaf cells. Incorporated amino acids and ureides are transported to transitory storage pools located in the cytosol or vacuole for storage, redistribution, or metabolism (Tegeder and Masclaux-Daubresse 2018).

#### 4.4.5 NITRATE TRANSPORTERS

#### 4.4.5.1 NRT1 Transporters

The NRT1 transporters belong to PTR transporters (Peptide Transporters), also known as NPF for NRT1/PTR Family (Léran et al. 2014). Most of the plant NRT1 genes encode low-affinity nitrate transporters. In Arabidopsis, CHL1 (AtNRT1.1 or AtNPF2.6) function as a basic protein. AtNRT1.1 can be called a transceptor with sensor and carrier functions (Tsay et al. 1993). AtNRT1.1 has dual affinity; it can switch from high to low affinity for nitrate according to the status of nitrate, and phosphorylation/dephosphorylation of threonine T101. CIPK23 protein kinase phosphorylates AtNRT1.1 when the level of nitrate is low; hence tNRT1.1 functions as high-affinity transporter. On the other hand, when the nitrate level is high, AtNRT1.1 gets dephosphorylated and acts as low-affinity transporter (Tegeder and Masclaux-Daubresse 2018).

#### 4.4.5.2 NRT2 Transporters

NRT2 proteins belong to the super family MFS. Genomes of organisms that can assimilate nitrate (plants, algae, fungi, yeast, bacteria) possess NRT2 genes. NRT2 proteins can transport nitrate or nitrite. Some of these proteins are two-component systems that require a second protein (NAR2) to be operational. NAR2 is rooted in the membrane through a transmembrane domain. NRT2.1/NAR2 in Chlamydomonas is bispecific high-affinity nitrate/nitrite transporter, while its NRT2.2/NAR2 is a high-affinity nitrate transporter (Tegeder and Masclaux-Daubresse 2018).

#### 4.4.5.3 NAR1 Transporters

The NAR1 (Nitrate Assimilation-Related component 1) family is produced by FNT (Formate Nitrite Transporters). NAR1.2 (LCIA) is a chloroplast envelope

HCO$_{3-}$ transporter. It is also part of a $CO_2$-concentration mechanism (CCM) that operates at very low $CO_2$ (Sanz-Luque et al. 2015).

#### 4.4.5.4  Nitrate Reductase

The Nitrate Reductase Enzyme reduces nitrate to nitrite by utilizing electrons obtained from NAD(P)H. Among eukaryotes, NR is a homodimeric protein of about 100–120 kDa. It contains one of three prosthetic groups each: FAD, b557 heme, and Molybdenum cofactor (Moco) (Ali 2020). Firstly, NAD(P)H-dehydrogenase or diaphorase, catalyzes electron transport from pyridine nucleotides to artificial electron acceptors such as dichlorophenol indophenol, ferricyanide, or cytochrome c. Moco domain is not involved in this activity. Secondly, the terminal activity catalyzes the transfer of electrons from artificial electron donors like bromophenol blue, flavins, or viologens to nitrate. Moco domain is needed in this step (Campbell 1999).

#### 4.4.5.5  Nitrite Reductase

Nitrite reductase (NiR) enzyme catalyzes the conversion of nitrite to ammonium in the stroma of chloroplasts. In this process, the photosynthetic electron transfer produces reduced ferredoxin (Fdred), which acts as an electron donor for nitrite reduction. Nitrite reduction in the dark can occur from NADPH, which is produced during the oxidative pentose phosphate cycle, through Fd-NADP+ oxido-reductase intervention (Canvin and Atkins 1974).

#### 4.4.5.6  Nitrogen Storage

Comparatively large quantities of N might build up in short- or long-term storage pools before or later than the transportation. The form, as well as quantity, of N compounds present in a variety of resource organs, sub cellular compartments such as vacuole, plastids, or chloroplasts along with tissues ought to have significant effects on N transport and partitioning to sinks. When cytosolic levels of nitrate elevate, it is stored in the vacuole. It is then sent back to the cytosol when N is found in limited quantity for assimilation. As a result, in roots as well as in leaves, elevated levels of nitrate are observed in the vacuole against the cytosol. AtCLCa brings about the transfer of nitrate into the vacuole through the proton anti-port (Rossato, Lainé, and Ourry 2001). AtNRT2.7 is a tonoplast transporter, and its location is constrained to seeds. In rice, OsNPF7.2 is found in tonoplast for both huge and tiny root spaces. Erosion of OsNPF7.2 results in a stunt growth of nitrate, which enhances its role in disrupting cell N during the temporary addition of nitrate to a vacuole. To avoid cell toxicity, excessive amounts of ammonium molecules are trapped in the vacuole. Their concentration is found up to 1 mM to uphold cytosolic ammonium concentrations < 15 mM. Normally, the ammonium levels are elevated in old/senescing and young leaves as opposed to in mature leaves, as a consequence of amino acid catabolism and photorespiratory recycling, respectively (Tegeder and Masclaux-Daubresse 2018).

#### 4.4.5.7  Regulation of Nitrogen Transport

Nitrate regulates transcription in many nitrate carriers, including NPF7.3/NRT1.5, NPF6.3/NRT1.1, NPF7.2/NRT1.8, NRT2.1, and NRT2.2. In Arabidopsis, nitrate recognition consists of transceptor (transporter/receptor) NPF6.3/NRT1.1 (Krouk et al. 2010). It is involved in many physiological and morphological actions related to nitrate, such as lateral root formation and seed dormancy. Calmodulin interacting protein kinases CIPK8, CIPK23, and the transcription factors SPL9 (Squamosa Promoter Binding Protein-Like 9), TGA1/4 (BZIP proteins), and NLP7 (Nodule Inception-Like Protein 7) are involved in nitrate signaling pathways (Liu et al. 2017, Hu, Wang, and Tsay 2009). It is known those nitrate and nitrite reductases are brought about by nitrate and sugars, which are subdued by glutamine, ammonium, and darkness at transcriptional and post-transcriptional stages. In addition, nitrate uptake and reduction are restricted by light. It is also associated with phytochrome and sugar-signaling pathways. Furthermore, nitrate, nitrite reductase, and GS and GOGAT activities are supported by sugar-signaling pathways via transcriptional and post-transcriptional regulation. Phytochrome signaling triggers the expression of assimilation genes with the help of transcription factors HY5/HYH (Elongated Hypocotyl 5/HY5 homolog). It also leads to the suppression of AtAMT1:2 and NPF6.3/NRT1.1 (Tegeder and Masclaux-Daubresse 2018).

## 4.5 NITRATE AND NITRIC OXIDE LINK

Nitric oxide (NO) is a very important signal and a second messenger in plants. It is an inorganic free radical. Several studies have given insights of NO production pathways. Different functions carried out by NO in plants have also been revealed. Major functions like root and shoot development, cell differentiation and lignification, programmed cell death, plant-pathogen interactions, stomatal movement, senescence and maturation, and flowering are involved, controlled through NO (Planchet and Kaiser 2006).

Nitrate reductase enzyme (NR) is a source of NO in plants (Dean and Harper 1988). It is reported that the NO production capacity of NR is 1 percent of its usual nitrate reduction capacity (Rockel et al. 2002). A study that involved *nia* mutants and plants made NR-free by growth conditions has established that NR is involved in NO production. While NR reduces nitrate (+5) to nitrite (+3) by utilizing NAD(P)H, it also catalyzes a 1-electron transfer from NAD(P)H to nitrite, leading to the formation of NO (+2). The NO production in plants is modulated by variation of NR activity through reversible serine phosphorylation. Diurnal NO emission pattern, as opposed to the wild type (low in light, high in the dark), was observed in plants that constitutively expressed NR upon exchange of the regulatory serine with aspartic acid. Hence, NR plays an important role in NO production (Planchet and Kaiser 2006).

A recent study showed that in vitro and in vivo post-translational modulation of NR also modulated the rate of NO production. It was also indicated that NO production has relatively high $K_m$ for nitrite (100 μM). Also, it is inhibited competitively in low nitrate concentrations ($K_i$=50 μM) (Rockel et al. 2002). Although studies reveal that NO plays a significant role in plants, the conditions under which it is produced through nitrate reductase and regulated is still not clear.

## 4.6   INTERACTIONS BETWEEN CARBON AND NITROGEN

Over the span of the past twenty years a number of studies have reached the same conclusions regarding nitrogen and carbon atoms associating. The main metabolism of carbon is noticed when it is dependent on nitrogen to associate, since the maximum amount of nitrogen inside the plants is infused in creating proteins. On the contrary, an association of nitrogen amounts to a continuous energy supply and carbon atoms. The resultant photosynthesis needs to be segregated into carbohydrate and amino acids synthesis (Kumar, Polisetty, and Abrol 1993).

Photorespiratory nitrogen metabolism is an important part of carbon and nitrogen interactions. Carbon cycles are provided by a Tricarboxylic acid (TCA) cycle for the biosynthesis of amino acid. It functions in light, along with the carbon, which flows from photosynthesis into the TCA cycle. It is regulated by ammonium. Also, nitrate and nitrite assimilation are involved in stimulating the cycling activity. An association appears between the nitrogen-use efficiency (NUE) of plants and carbon assimilation pathways. Plants that have higher NUE of C4 require lower nitrogen investment in photosynthetic carboxylation enzymes. Also, they have efficient distribution and redistribution of nitrogen (Kumar, Polisetty, and Abrol 1993).

Carbon (C) and nitrogen (N) are vital for the fundamental cellular activities of plants. Compounds of C include carbohydrates (glucose and sucrose), which provide energy, and C-skeletons for the assimilation of ammonium during amino acid biosynthesis. Inorganic N compounds like nitrate and ammonium and organic amino acids are produced by incorporation of ammonium into the C-skeletons, as shown in Figure 4.2. Thus, resultant proteins and amino acids act as building blocks of the cell. An adequate supply of C and N nutrients is essential for cellular functions and plant growth. During photosynthesis, assimilation of $CO_2$ occurs and Suc and Glc are produced. The TCA cycle and glycolysis convert this Suc and Glc to 2-oxoglutarate (2OG) or α-ketoglutarate. On the other hand, nitrate ($NO_3^-$) is reduced by nitrate reductase to nitrite ($NO_2^-$) and then by nitrite reductase to ammonium ($NH_4^+$). Glutamate (Glu) is synthesized using 2OG as C skeleton and incorporation of photorespiratory $NH_4^+$. Next, glutamine (Gln) is produced by incorporating $NH_4^+$ obtained from primary N assimilation into Glu. Later, $NH_4^+$ is given out by Glu and Gln, which is utilized for amino acid synthesis (Zheng 2009).

## 4.7   NITROGEN-USE EFFICIENCY

Nitrogen is the most important factor limiting crop productivity worldwide. The capability of plans to acquire nitrogen from fertilizers is one of the most critical steps limiting the efficient use of nitrogen. Nitrogen fertilizers have been used worldwide for centuries, approximately a hundred billion kilograms per annum. Loss of nitrogen from the soil to groundwater, rivers, and oceans is directly dependent upon the physicochemical properties of soil. Nitrate efflux from roots results in a reduction of net nitrogen uptake and, as a result, external nitrogen fertilizers are added to the soil to increase the nitrogen concentration. Proficient use of nitrogen in crops and plants can be explicated as per unit generation of nitrogen in accordance with its availability (Benincasa, Guiducci, and Tei 2011). The processes are divided as follows:

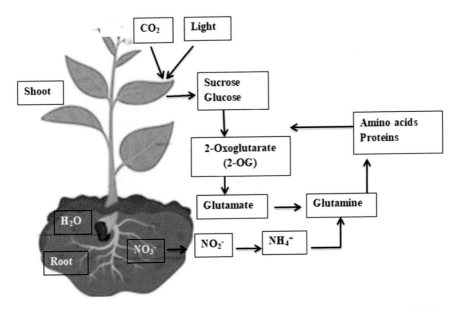

**FIGURE 4.2**   Coordination of C and N metabolism in plants. Adapted from Zheng (2009).

- Ingestion Capability: The absolute potential of plants and crops to captivate ammonium and nitrate ions from nitrate inside the soil.
- Utilizing efficiency: The ability of the plant to transfer the nitrogen to the grain.
- Enhancing nitrogen use efficiency: Plant and crop response to applied nitrogen fertilizers and use efficiency are important criteria for evaluating crops' nitrogen requirement for high production.

The recovery of nitrogen in plants is usually less than 50 percent. The loan recovery is related to leaching surface, denitrification, and volatilization. Low recovery of nitrogen causes higher cost for the product and also affects the environment. Hence, nitrogen use efficiency is desirable to improve production, minimize the cost, and maintain environmental quality. Integrated nitrogen management strategies help improve fertilizers along with production and the soil. Nitrogen supply with crop demand is necessary to ensure quantity, uptake, utilization, and maximum production (Bijay, Yadvinder, and Sekhon 1995). Leaching of nitrate takes place chiefly through the winter and follows the period when rainfall sweeps away the residue and mineralized nitrate beneath the root zone. Precise fertilization increases NUE and reduces the danger of bleeding throughout growth periods (Kraft and Stites 2003).

## 4.8   NITRATE POLLUTION

Nitrogen fertilizers have been used for centuries, and their use has been rising over the past fifty years. Approximately 76 percent of the world population lives in developed countries, where more nitrogen fertilizers are currently used. The main source

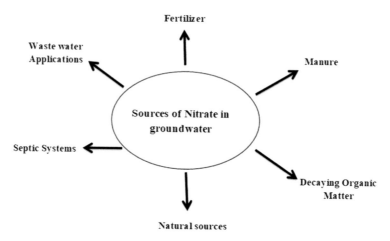

**FIGURE 4.3**   Sources of nitrate in groundwater. Adapted from Menció et al. (2016).

of nitrogen pollution is agriculture. Only 33 percent of the applied nitrogen fertilizer is taken up by the plant in the form of $NO_3$ (nitrate). This uptake is also known as nitrogen-use efficiency (NUE). The remaining 67 percent remains in the soil and seeps into the surrounding environment and enters into water bodies, which increases potential groundwater nitrogen pollution (Shen et al. 2011).

Nitrate from biochemical activities of microorganism or synthesized chemically, enters into the water bodies, depicted in Figure 4.3.

It is highly soluble in water and its low retention by soil particles makes it a major component of groundwater (Menció et al. 2016). Nitrogen enters the human body by groundwater and causes a number of health disorders such as methemoglobinaemia (hemoglobin molecule is functionally altered and prevented from carrying oxygen), gastric cancer, goiter, and hypertension (Gupta et al. 2010). Nitrogen and phosphorus are nutrients that are a natural part of the aquatic ecosystem. They help in the growth of algae. Algae provide food for fishes and small organisms that live in water. Too much nitrogen in water causes faster growth of water algae. A significant increase in algae harms water quality. Algal bloom decreases the oxygen in water, leading to illness and death for most aquatic life. Some algal bloom is harmful to humans because it produces toxins that can cause sickness (Benincasa, Guiducci, and Tei 2011, Camargo and Alonso 2006).

## 4.9   CONCLUSION

Plants require nitrogen compounds (nitrate and ammonium) in large amounts for proper growth and development. The amino acid, glutamate, glutamine, aspartate, and asparagine are produced in plants by assimilation of inorganic nitrogen. Asparagine synthetase (AS), aspartate aminotransferase (AT), glutamate dehydrogenase (GDH), glutamate synthase (GOGAT), and glutamine synthetase (GS) are the enzymes involved in this process. Various studies have revealed complex mechanisms involved in the nitrate assimilation pathway along with its regulation. Different families of nitrate transporters have also been discovered. However, detailed investigations are required to know more about NO production via nitrate reductase. Also, nitrogen-use efficiency should be widely applied to improve crop yield and reduce nitrate pollution.

## REFERENCES

Ali, Ahmad. 2020. "Nitrate assimilation pathway in higher plants: Critical role in nitrogen signalling and utilization." *Plant Science Today* 7 (2): 182–192.

Ali, Ahmad, Pamela Jha, Kuljeet Singh Sandhu, and Nandula Raghuram. 2008. "Spirulina nitrate-assimilating enzymes (NR, NiR, GS) have higher specific activities and are more stable than those of rice." *Physiology and Molecular Biology of Plants* 14 (3): 179–182.

Ali, Ahmad, S. Sivakami, and Nandula Raghuram. 2007. "Effect of nitrate, nitrite, ammonium, glutamate, glutamine and 2-oxoglutarate on the RNA levels and enzyme activities of nitrate reductase and nitrite reductase in rice." *Physiology and Molecular Biology of Plants* 13 (1): 17.

Asim, Muhammad, Zia Ullah, Fangzheng Xu, Lulu An, Oluwaseun Olayemi Aluko, Qian Wang, and Haobao Liu. 2020. "Nitrate signaling, functions, and regulation of root system architecture: Insights from Arabidopsis thaliana." *Genes* 11 (6): 633.

Bailey, Karen J., and Richard C. Leegood. 2016. "Nitrogen recycling from the xylem in rice leaves: dependence upon metabolism and associated changes in xylem hydraulics." *Journal of Experimental Botany* 67 (9): 2901–2911.

Benincasa, Paolo, Marcello Guiducci, and Francesco Tei. 2011. "The nitrogen use efficiency: Meaning and sources of variation – Case studies on three vegetable crops in central Italy." *HortTechnology hortte* 21 (3): 266. doi: 10.21273/horttech.21.3.266.

Bijay, Singh, Singh Yadvinder, and G.S. Sekhon. 1995. "Fertilizer-N use efficiency and nitrate pollution of groundwater in developing countries." *Journal of Contaminant Hydrology* 20 (3): 167–184. doi: https://doi.org/10.1016/0169-7722(95)00067-4.

Camargo, Julio A., and Álvaro Alonso. 2006. "Ecological and toxicological effects of inorganic nitrogen pollution in aquatic ecosystems: A global assessment." *Environment International* 32 (6): 831–849.

Campbell, Wilbur H. 1999. "Nitrate reductase structure, function and regulation: Bridging the gap between biochemistry and physiology." *Annual Review of Plant Biology* 50 (1): 277–303.

Canvin, D.T., and C.A. Atkins. 1974. "Nitrate, nitrite and ammonia assimilation by leaves: Effect of light, carbon dioxide and oxygen." *Planta* 116 (3): 207–224.

Carvalho, Helena G., Inês A. Lopes-Cardoso, Ligia M. Lima, Paula M. Melo, and Julie V. Cullimore. 2003. "Nodule-specific modulation of glutamine synthetase in transgenic Medicago truncatula leads to inverse alterations in asparagine synthetase expression." *Plant Physiology* 133 (1): 243–252.

Chaffey, N. *Raven biology of plants, 8th edn*: Ann Bot. 2014 Jun;113(7): vii. doi: 10.1093/aob/mcu090. Epub 2014 May 8.

Chen, Lishan, and Daniel R. Bush. 1997. "LHT1, a lysine-and histidine-specific amino acid transporter in Arabidopsis." *Plant Physiology* 115 (3): 1127–1134.

Crawford, Nigel M., and Anthony D.M. Glass. 1998. "Molecular and physiological aspects of nitrate uptake in plants." *Trends in Plant Science* 3 (10): 389–395.

Dean, John V., and James E. Harper. 1988. "The conversion of nitrite to nitrogen oxide (s) by the constitutive NAD (P) H-nitrate reductase enzyme from soybean." *Plant Physiology* 88 (2): 389–395.

Fortin, Marc G., Nigel A. Morrison, and Desh Pal S. Verma. 1987. "Nodulin-26, a peribacteroid membrane nodulin is expressed independently of the development of the peribacteroid compartment." *Nucleic Acids Research* 15 (2): 813–824.

Garma, Martin Burger, and Arnold J. Bloom. 2014. "Root strategies for nitrate assimilation." *Root Engineering: Basic and Applied Concepts* 40: 251.

González-Ballester, David, Antonio Camargo, and Emilio Fernández. 2004. "Ammonium transporter genes in Chlamydomonas: the nitrate-specific regulatory gene Nit2 is involved in Amt1; 1 expression." *Plant molecular biology* 56 (6): 863–878.

Gupta, Sunil, R.C. Gupta, A.B. Gupta, Sevgi Eskiocak, Evs Prakasa Rao, K. Puttanna, and Aditi Singhvi. 2010. "Pathophysiology of nitrate toxicity in humans and its mitigation measures." *ING Bulletins on Regional Assessment of Reactive Nitrogen, Bulletin* (20).

Hirner, Axel, Friederike Ladwig, Harald Stransky, Sakiko Okumoto, Melanie Keinath, Agnes Harms, Wolf B. Frommer, and Wolfgang Koch. 2006. "Arabidopsis LHT1 is a high-affinity transporter for cellular amino acid uptake in both root epidermis and leaf mesophyll." *The Plant Cell* 18 (8): 1931–1946.

Hu, Heng-Cheng, Ya-Yun Wang, and Yi-Fang Tsay. 2009. "AtCIPK8, a CBL-interacting protein kinase, regulates the low-affinity phase of the primary nitrate response." *The Plant Journal* 57 (2): 264–278.

Imsande, John, and Bruno Touraine. 1994. "N demand and the regulation of nitrate uptake." *Plant Physiology* 105 (1): 3.

Jones Jr, J Benton. 1997. *Plant Nutrition Manual*: CRC Press.

Keys, A.J., I.F. Bird, M.J. Cornelius, P.J. Lea, R.M. Wallsgrove, and B.J. Miflin. 1978. "Photorespiratory nitrogen cycle." *Nature* 275 (5682): 741–743.

Kraft, George J., and Will Stites. 2003. "Nitrate impacts on groundwater from irrigated-vegetable systems in a humid north-central US sand plain." *Agriculture, Ecosystems & Environment* 100 (1): 63–74. doi: https://doi.org/10.1016/S0167-8809(03)00172-5.

Krapp, Anne. 2015. "Plant nitrogen assimilation and its regulation: a complex puzzle with missing pieces." *Current Opinion in Plant Biology* 25: 115–122.

Krouk, Gabriel, Benoît Lacombe, Agnieszka Bielach, Francine Perrine-Walker, Katerina Malinska, Emmanuelle Mounier, Klara Hoyerova, Pascal Tillard, Sarah Leon, and Karin Ljung. 2010. "Nitrate-regulated auxin transport by NRT1. 1 defines a mechanism for nutrient sensing in plants." *Developmental cell* 18 (6): 927–937.

Kumar, P.A., Raghuveer Polisetty, and Y.P. Abrol. 1993. "Interaction between carbon and nitrogen metabolism." In: *Photosynthesis: Photoreactions to Plant Productivity*, edited by Y.P. Abrol, P. Mohanty, and Govindjee, 339–350. Dordrecht: Springer Netherlands.

Lee, Yong-Hwa, Justin Foster, Janet Chen, Lars M. Voll, Andreas P.M. Weber, and Mechthild Tegeder. 2007. "AAP1 transports uncharged amino acids into roots of Arabidopsis." *The Plant Journal* 50 (2): 305–319.

Léran, Sophie, Kranthi Varala, Jean-Christophe Boyer, Maurizio Chiurazzi, Nigel Crawford, Françoise Daniel-Vedele, Laure David, Rebecca Dickstein, Emilio Fernandez, and Brian Forde. 2014. "A unified nomenclature of nitrate transporter 1/peptide transporter family members in plants." *Trends in Plant Science* 19 (1): 5–9.

Li, Jian-Yong, Yan-Lei Fu, Sharon M Pike, Juan Bao, Wang Tian, Yu Zhang, Chun-Zhu Chen, Yi Zhang, Hong-Mei Li, and Jing Huang. 2010. "The Arabidopsis nitrate transporter NRT1. 8 functions in nitrate removal from the xylem sap and mediates cadmium tolerance." *The Plant Cell* 22 (5): 1633–1646.

Lin, Shan-Hua, Hui-Fen Kuo, Geneviève Canivenc, Choun-Sea Lin, Marc Lepetit, Po-Kai Hsu, Pascal Tillard, Huey-Ling Lin, Ya-Yun Wang, and Chyn-Bey Tsai. 2008. "Mutation of the Arabidopsis NRT1. 5 nitrate transporter causes defective root-to-shoot nitrate transport." *The Plant Cell* 20 (9): 2514–2528.

Liu, Kun-hsiang, Yajie Niu, Mineko Konishi, Yue Wu, Hao Du, Hoo Sun Chung, Lei Li, Marie Boudsocq, Matthew McCormack, and Shugo Maekawa. 2017. "Discovery of nitrate–CPK–NLP signalling in central nutrient–growth networks." *Nature* 545 (7654): 311–316.

Maeda, Shin-ichi, Mineko Konishi, Shuichi Yanagisawa, and Tatsuo Omata. 2014. "Nitrite transport activity of a novel HPP family protein conserved in cyanobacteria and chloroplasts." *Plant and Cell Physiology* 55 (7): 1311–1324.

Masclaux-Daubresse, C., F. Daniel-Vedele, J. Dechorgnat, F. Chardon, L. Gaufichon, and A. Suzuki. 2010. "Nitrogen uptake, assimilation and remobilization in plants: Challenges for sustainable and productive agriculture." *Ann Bot* 105 (7): 1141–1157. doi: 10.1093/aob/mcq028.

McDermitt, D.K., and R.S. Loomis. 1981. "Elemental composition of biomass and its relation to energy content, growth efficiency, and growth yield." *Annals of Botany* 48 (3): 275–290.

Menció, Anna, Josep Mas-Pla, Neus Otero, Oriol Regàs, Mercè Boy-Roura, Roger Puig, Joan Bach, Cristina Domènech, Manel Zamorano, and David Brusi. 2016. "Nitrate pollution of groundwater; all right …, but nothing else?" *Science of the Total Environment* 539: 241–251.

Miller, Anthony J., and Nick Chapman. 2011. "Transporters involved in nitrogen uptake and movement." In: *The Molecular and Physiological Basis of Nutrient Use Efficiency in Crops*, 193–210. Wiley Online Library.

Morgan, JB, and EL Connolly. 2013. "Plant-soil interactions: nutrient uptake." *Nature Education Knowledge* 4 (8): 2.

Näsholm, Torgny, Knut Kielland, and Ulrika Ganeteg. 2009. "Uptake of organic nitrogen by plants." *New Phytologist* 182 (1): 31–48.

Ohyama, Takuji. 2010. "Nitrogen as a major essential element of plants." *Nitrogen Assim. Plants* 37: 1–17.

Pate, J.S., P.J. Sharkey, and O.A.M. Lewis. 1975. "Xylem to phloem transfer of solutes in fruiting shoots of legumes, studied by a phloem bleeding technique." *Planta* 122 (1): 11–26.

Planchet, Elisabeth, and Werner M. Kaiser. 2006. "Nitric oxide production in plants: Facts and fictions." *Plant Signaling & Behavior* 1 (2): 46–51. doi: 10.4161/psb.1.2.2435.

Reichardt, Klaus, and Luís Carlos Timm. 2020. "How plants absorb nutrients from the soil." In: *Soil, Plant and Atmosphere: Concepts, Processes and Applications*, 313–330. Cham: Springer International Publishing.

Robertson, G. Philip. 1997. "Nitrogen use efficiency in row-crop agriculture: Crop nitrogen use and soil nitrogen loss." *Ecology in Agriculture* 347–365.

Rockel, Peter, Frank Strube, Andra Rockel, Juergen Wildt, and Werner M. Kaiser. 2002. "Regulation of nitric oxide (NO) production by plant nitrate reductase in vivo and in vitro." *Journal of Experimental Botany* 53 (366): 103–110.

Rossato, L, Philippe Lainé, and A Ourry. 2001. "Nitrogen storage and remobilization in Brassica napus L. during the growth cycle: Nitrogen fluxes within the plant and changes in soluble protein patterns." *Journal of Experimental Botany* 52 (361): 1655–1663.

Sanz-Luque, Emanuel, Alejandro Chamizo-Ampudia, Angel Llamas, Aurora Galvan, and Emilio Fernandez. 2015. "Understanding nitrate assimilation and its regulation in microalgae." *Frontiers in Plant Science* 6 (899). doi: 10.3389/fpls.2015.00899.

Shen, Y. Shen, Yan-jun, Lei, H. Lei, Hui-min, Yang, Dw, Dawen, Kanae, S. Kanae, and ShinjisM. 2011. "Effects of agricultural activities on nitrate contamination of groundwater in a Yellow River irrigated region."

Suzuki, Akira, and David B. Knaff. 2005. "Glutamate synthase: Structural, mechanistic and regulatory properties, and role in the amino acid metabolism." *Photosynthesis Research* 83 (2): 191–217.

Svennerstam, Henrik, Ulrika Ganeteg, and Torgny Näsholm. 2008. "Root uptake of cationic amino acids by Arabidopsis depends on functional expression of amino acid permease 5." *New Phytologist* 180 (3): 620–630.

Tegeder, Mechthild. 2012. "Transporters for amino acids in plant cells: Some functions and many unknowns." *Current Opinion in Plant Biology* 15 (3): 315–321.

Tegeder, Mechthild, and Céline Masclaux-Daubresse. 2018. "Source and sink mechanisms of nitrogen transport and use." *New Phytologist* 217 (1): 35–53.

Tegeder, M., A. Weber, W.C. Plaxton, and M.T. McManus. 2006. *Control of Primary Metabolism in Plants*. Oxford: Wiley.

Tischner, R. 2000. "Nitrate uptake and reduction in higher and lower plants." *Plant, Cell & Environment* 23 (10): 1005–1024.

Tsay, Yi-Fang, Julian I. Schroeder, Kenneth A. Feldmann, and Nigel M. Crawford. 1993. "The herbicide sensitivity gene CHL1 of Arabidopsis encodes a nitrate-inducible nitrate transporter." *Cell* 72 (5): 705–713.

Tyree, Melvin T. 2003. "Plant hydraulics: the ascent of water." *Nature* 423 (6943): 923.

Widhalm, Joshua R., Michael Gutensohn, Heejin Yoo, Funmilayo Adebesin, Yichun Qian, Longyun Guo, Rohit Jaini, Joseph H. Lynch, Rachel M. McCoy, and Jacob T. Shreve. 2015. "Identification of a plastidial phenylalanine exporter that influences flux distribution through the phenylalanine biosynthetic network." *Nature Communications* 6 (1): 1–11.

Zheng, Zhi-Liang. 2009. "Carbon and nitrogen nutrient balance signaling in plants." *Plant Signaling & Behavior* 4 (7): 584–591. doi: 10.4161/psb.4.7.8540.

# 5 Nitrate in Plant Nutrition

*Ya-Yun Wang*

## CONTENTS

5.1 Introduction ..................................................................................................91
5.2 Nitrate in the Soil .........................................................................................92
5.3 Nitrate Transport in Plants...........................................................................92
    5.3.1 Nitrate Uptake by Roots....................................................................93
    5.3.2 Long-distance Nitrate Transport from Roots to Shoots ....................95
    5.3.3 Nitrate Storage and Allocation in Leaves and Reproductive
        Organs ................................................................................................96
5.4 Nitrate Movement under Stress Conditions .................................................96
    5.4.1 Stress-induced Nitrate Allocation in Roots (SINAR).......................97
    5.4.2 Nitrate Remobilization under Nitrogen Starvation ..........................97
5.5 Nitrate Transporters and Nitrogen Use Efficiency ......................................97
    5.5.1 Manipulating Remobilization of Nitrate to Developing
        Organs ................................................................................................97
5.6 Conclusions and Perspectives......................................................................98
References............................................................................................................98

## 5.1 INTRODUCTION

Nitrogen (N) is vital for all living organisms, and is one of the building blocks for amino acids, nucleotides (genetic materials), and proteins. Plants, which are autotrophs, require N to grow and reproduce. We human beings, are heterotrophs, and must ingest proteins to support our lives. These proteins are made directly or indirectly by plants or other autotrophs. Therefore, understanding how plants acquire and metabolize nitrogen sources is very important.

Plants can uptake two forms of nitrogen sources, organic and inorganic. In boreal forests, organic forms of nitrogen sources, such as amino acids, are the major N source, and transporters mediating amino acid uptake have been found in plant roots (Näsholm et al., 2009). In most cropland or well-aerated soil, inorganic N, such as nitrate and ammonium, are the major forms. Both are often provided in fertilizers. Due to multiple factors, such as climate change, reduction of arable land, and increased world population, more and more fertilizers are used to increase crop production. However, a large proportion of N-containing fertilizers are lost by leaching (Hirel et al., 2011). Plants can only use 50 to 75 percent of N input from fertilizers (Gutiérrez, 2012). The leached N then accumulates in the lakes or coastal regions, resulting in eutrophication. Excess application of N fertilizer also increases the

DOI: 10.1201/9780429326806-7

**91**

production of nitrous oxide, a key greenhouse gas contributing to global warming (Vidal et al., 2020). Since nitrate is a major N source for most plants and is more easily leached by rainfall, a better understanding of nitrate nutrition will help us to maintain crop production while reducing the excess N input into the environment.

## 5.2   NITRATE IN THE SOIL

Although we are surrounded by 80 percent $N_2$ gas, neither plants nor humans can use it directly. $N_2$ gas has to first convert into ammonium ($NH_4^+$). This process is called nitrogen fixation. Later, in the presence of oxygen ($O_2$), the ammonium is further converted into nitrate ($NO_3^-$). This process is called nitrification. Then plants can absorb ammonium and/or nitrate as the N sources. Naturally, both nitrogen fixation and nitrification are carried out by prokaryotic microorganisms in the soil. No eukaryotic organisms can fix $N_2$ into ammonium so far. In addition to $N_2$ fixation, plant or animal waste (dead matter or feces) can also be digested into ammonium via ammonification by soil microorganisms. However, in cropland, biological fixation via soil microorganisms is not fast enough to support crop production, so the industrial fixation via the Haber-Bosch process for producing N fertilizer is crucial for agricultural practices.

Soil is a rather complex environment, containing biotic (biological waste) and abiotic materials (soil particles, mostly made of silicates). Most of time, the soil is negatively charged so that cations, such as potassium ($K^+$) and calcium ($Ca^{2+}$) can be held by the soil surface, while the anions, such as nitrate ($NO_3^-$) or phosphate ($PO_4^{2-}$) are readily dissolved in the soil solution. Due to this difference, anions are more easily taken up by plant roots than cations. On the other hand, with heavy rainfall or irrigation, anions are easily leached to underground water, rivers, lakes, and oceans. Accumulation of that nutrition in those open water areas enhances the growth of algae or microorganisms, which consume the $O_2$ dissolved in the waters, thereby releasing toxic compounds. Therefore, other aquatic animals cannot survive resulting in eutrophication. Therefore, enhancing nitrate uptake or nitrogen use efficiency (NUE) by plants is a way to reduce pollution and sustain crop production.

## 5.3   NITRATE TRANSPORT IN PLANTS

To improve nitrate uptake or nitrogen use efficiency (NUE), we first need to understand nitrate movement in plants. The biological membranes are made of two layers of phospholipids. Due to the hydrophobicity of the lipids, the biological membrane is selectively permeable to ions and organic molecules. In other words, ions and organic molecules cannot move across the membrane freely and directly. They require the help of membrane transport proteins. These membrane transport proteins create a passageway for these substrates and also play a regulating role in the substrate movement and accumulation. Being a negatively charged ion, nitrate requires membrane transport proteins to mediate its movement or translocation within plants. Four nitrate transporter families have been identified in *Arabidopsis thaliana*, a model plant. They are Nitrate Transporter 1/Peptide Transporter Family (NPF or NRT1, 53 members), Nitrate Transporter 2 (NRT2, 7 members), Chlorate Channels (CLCs, 7 members), and Slow Anion Channel –Associated 1 Homologues (SLAC/SLAH, 5 members) (Wang et al., 2012; Krapp et al., 2014; Wang et al., 2018). Since most of our understanding of the functions of these nitrate transporters are from the model plant, I will summarize the current knowledge about nitrate transport in Arabidopsis. However, interesting findings in other plant species will also be included. Several review articles about nitrate transport can provide additional information (Wang et al., 2012; Krapp et al., 2014; Wang et al., 2018; Vidal et al., 2020).

### 5.3.1 Nitrate Uptake by Roots

As mentioned previously, nitrate is easily absorbed by plants and lost by rainfall or irrigation. Therefore, the nitrate concentration in the soil can vary a lot, about a hundred-fold (Lark et al., 2004). In order to optimize nitrate acquisition, plants have evolved two different nitrate uptake systems to cope with the changing environment: a high-affinity transport system (HATS) for low nitrate condition and a low-affinity transport system (LATS) for high nitrate conditions (Crawford and Glass, 1998; Forde, 2000). Each transport system is mediated by different transporter families. In Arabidopsis, HATS is mainly supported by the NRT2 family (NRT2.1, NRT2.2, NRT2.4, NRT2.5), and LATS is supported by NPF/NRT1 family (NPF4.6/NRT1.2 and NPF6.3/NRT1.1/CHL1), with one exception, NPF6.3 (Figure 5.1) (Tsay et al., 1993; Huang et al., 1999; Cerezo et al., 2001; Filleur et al., 2001; Kiba et al., 2012; Lezhneva et al., 2014). Once nitrate is taken up into plant cells, it will be reduced into nitrite by nitrate reductase (NR) in the cytosol. Nitrite is later translocated into plastids and then further reduced into ammonium by nitrite reductase (NiR). At this moment, glutamine and glutamate are synthesized using ammonium as a reactant through the GS-GOGAT cycle. (GS: glutamine synthetase; GOGAT: glutamine-2-oxoglutarate aminotransferase or glutamate synthase). So, the inorganic form of nitrogen, nitrate, is now incorporated into the organic form of nitrogen, amino acids.

NPF6.3/CHL1 in Arabidopsis was the first identified nitrate transporter in plants (Tsay et al., 1993). Although it belongs to the NPF/NRT1 family, NPF6.3 is a dual-affinity nitrate transporter, and the switch between two affinities is regulated by phosphorylation at the 101 Threonine residue (Liu et al., 1999; Liu and Tsay, 2003). The phosphorylation is affected by external nitrate concentration. Under low nitrate conditions, the Thr101 is *hyper*-phosphorylated, whereas Thr101 is *hypo*-phosphorylated under high nitrate conditions (Liu and Tsay, 2003). In addition to nitrate transport, NPF6.3 is also a nitrate sensor that induces the primary nitrate responses (PNRs) in response to various external nitrate concentrations (Ho et al., 2009).

Rice can also use both nitrate and ammonium as the N sources. It has been shown that as much as 40 percent of total N taken up in irrigated rice is nitrate, due to the nitrification in the rhizosphere (Kronzucker et al., 2000; Kirk and Kronzucker, 2005). Therefore, understanding the nitrate uptake and movement in rice is also beneficial to enhancing rice production. The closest orthologue of Arabidopsis NPF6.3 in rice (*Oryza sariva*) is OsNPF6.5/NRT1.1B (Hu et al., 2015). OsNPF6.5 is also a dual-affinity nitrate transporter. Interestingly, a single-nucleotide polymorphism (SNP) resulting in Thr327Met substitution of OsNPF6.5 in the rice *indica* cultivar leads to better nitrate uptake, root-to-shoot transport, nitrate assimilation, and nitrogen use efficiency

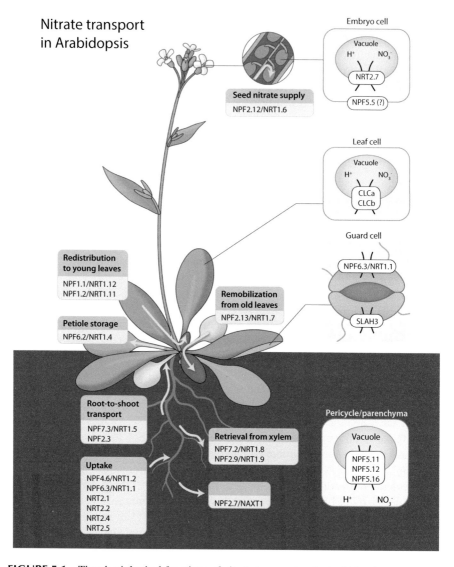

**FIGURE 5.1**  The physiological functions of nitrate transporters in Arabidopsis.

Four nitrate transporter families have been characterized: NPF (Nitrate Transporter 1/Peptide Transporter Family), NRT2 (Nitrate Transporter 2), CLC (Chloride Channel), and SLAC/SLAH (Slow Anion Channel – Associated Homologues).Source: Figure by Ying Wang (ywangstudio).

(NUE) than the *japonica* cultivar (Hu et al., 2015). This finding gives a molecular explanation for a long-known phenomenon, and also provides a target for molecular breeding. Furthermore, more genera of microbiota in the rhizosphere of *indica* cultivar have nitrogen metabolism functions than those of *japonica* cultivar (Zhang et al.,

2019). These data suggest that OsNPF6.5/OsNRT1.1B-*indica* not only enhances nitrate uptake but also influences the microbiota in the soils. Besides OsNPF6.5, OsNPF2.4 also participates in low-affinity nitrate acquisition (Xia et al., 2015).

Several members of the NRT2 functioning in HATS, are also involved in nitrate acquisition. *NRT2.1* and *NRT2.2* genes in Arabidopsis are linked, and they have been demonstrated to mediate the high-affinity nitrate uptake (Cerezo et al. 2001; Filleur et al. 2001). However, under extreme low nitrate conditions or nitrogen starvation conditions, AtNRT2.4 and AtNRT2.5 mediate the extraction of scarce nitrate from the environment (Kiba et al., 2012; Lezhneva et al., 2014). Different from NPF transporters, the nitrate transport activity of some NRT2 proteins requires interaction with NAR2/NRT3 (Okamoto et al., 2006; Orsel et al., 2006). Nitrate uptake activity is highly enhanced when NAR2.1 is coexpressed with AtNRT2.1, AtNRT2.2, and AtNRT2.5, in Xenopus oocytes (Feng et al., 2011; Kotur et al., 2012), but NAR2 is not required for the nitrate transport activity of AtNRT2.4 and AtNRT2.7 (Chopin et al., 2007; Kiba et al., 2012).

In rice, OsNAR2.1, OsNRT2.1, and OsNRT2.2 are also important for high-affinity nitrate uptake (Cai et al., 2008; Feng et al., 2011; Yan et al., 2011). Overexpression of OsNRT2.1 increases the nitrate influx rate and crop yield (Luo et al., 2018).

### 5.3.2 Long-distance Nitrate Transport from Roots to Shoots

After nitrate is taken up by root cells, it can be assimilated or stored in roots, or it can be transported to the shoots via root-to-shoot long-distance transport. Where to assimilate nitrate is species-dependent and also regulated by environmental conditions (Smirnoff and Stewart, 1985). Several NPF nitrate transporters in Arabidopsis and rice have been shown to mediate this process. Defects in this long-distance transport may reduce plant growth and yield.

In Arabidopsis, several NPF transporters are involved in regulating root-to-shoot nitrate transport under normal and stressed conditions. NPF7.3/NRT1.5 is expressed in the xylem-pole pericycle cells, mediating the xylem loading of nitrate (Lin et al., 2008). NPF7.2/NRT1.8, expressed in the root xylem parenchyma cells, mediates the retrieval of nitrate from the xylem sap (Li et al., 2010). Both transporters are involved in SINAR (stressed-initiated nitrate allocation to roots) under stresses which will be explained more in section 5.4. NPF2.9/NRT1.9 is a root phloem-localized nitrate transporter, mediating the downward movement of nitrate (Wang and Tsay, 2011). This fine-tuned regulation in nitrate transport seems to relate to plant growth regulation. Three tonoplast-localized NPF proteins, NPF5.11, NPF5.12, and NPF5.16, mediate the nitrate efflux from root vacuoles and also influence the nitrate partition between roots and shoots (He et al., 2017).

In rice, in addition to nitrate uptake, OsNPF6.5/NRT1.1B and OsNPF2.4 also participate in root-to-shoot nitrate transport (Hu et al., 2015; Xia et al., 2015). On the other hand, with the expression in the xylem parenchyma cells in both roots and shoots, OsNPF2.2 mediates the nitrate loading into root xylem but regulates the nitrate unloading from shoot xylem (Li et al., 2015). The vascular development of leaves and branches of rice is also altered in *osnpf2.2* mutants. Defects in *OsNPF6.5*, *OsNPF2.4*, or *OsNPF2.2* reduce plant growth dramatically.

In addition to NPF transporter, *OsNRT2.3a*, expressed in the xylem parenchyma cells, mediates nitrate loading into the xylem sap (Tang et al., 2012). Loss-of-function of *OsNRT2.3a* shows a reduced growth phenotype in a nitrate-dependent manner.

### 5.3.3 NITRATE STORAGE AND ALLOCATION IN LEAVES AND REPRODUCTIVE ORGANS

Once loaded into the root xylem, nitrate enters the xylem, a superhighway for water and mineral nutrients. The tension created by transpiration drives the xylem flow to the largest leaves. The petioles seem to be the primary nitrate storage, and NPF6.2/NRT1.4 in Arabidopsis mediates the nitrate storage in the petioles (Chiu et al., 2004). Less nitrate is accumulated in the petioles when *NPF6.2* is mutated. Besides, the width: length ratio of the third leaves in an *npf6.2* mutant is larger than the wild type. The molecular mechanisms of leaf development mediated by *NPF6.2* require further investigation.

Since the xylem flow is mainly driven by transpiration, the young or developing leaves, or "sinks," with lower transpiration require different strategies to obtain nitrate. *NPF1.1/NRT1.12* and *NPF1.2/NRT1.11* in Arabidopsis are expressed in the companion cells of the major vein in the expanded leaves (Hsu and Tsay, 2013). In *npf1.1npf1.2* double mutant, more nitrate is accumulated in the mature and large leaves, suggesting that *NPF1.1* and *NPF1.2* mediate the redistribution of nitrate from nitrate source (leaves with stronger transpiration) to nitrate sink (leaves with less transpiration). This redistribution is necessary for high-nitrate-enhanced leaf growth (Hsu and Tsay, 2013).

In addition to the young leaves, reproductive organs are also nitrate sinks. In previous studies, *NPF2.12/NRT1.6* in Arabidopsis, highly expressed in the vascular tissues and the funiculus in the silique, mediates the nitrate supply to the Arabidopsis developing seeds (Almagro et al., 2008). Without *NPF2.12*, the early development of embryos is affected and the seed abortion rate is increased, suggesting that delivering nitrate to the seeds is crucial for the normal embryo development in Arabidopsis. Another NPF nitrate transporter, NPF5.5, also contributes to the nitrogen accumulation in the Arabidopsis embryos (Léran et al., 2015). One of NRT2 nitrate transporter, *NPF2.7*, is highly expressed in the mature embryos of Arabidopsis (Chopin et al., 2007). The *npf2.7* mutants show lower seed nitrate content and germination rate. Unlike other NRT2 genes in Arabidopsis, *NPF2.7* is localized to the tonoplast, and its nitrate transport activity does not require NAR2 (Chopin et al., 2007).

## 5.4 NITRATE MOVEMENT UNDER STRESS CONDITIONS

Recent studies have shown that several nitrate transporters regulate nitrate distribution and also participate in stress responses. These results give molecular links or hubs between nutrition/growth and acclimation strategies. These hubs may serve as manipulating targets for developing stress-tolerant plants.

### 5.4.1 STRESS-INDUCED NITRATE ALLOCATION IN ROOTS (SINAR)

Arabidopsis NPF7.2/NRT1.8 and NPF7.3/NRT1.5 regulate the root-to-shoot nitrate transport in opposite ways. Interestingly, these two genes are differently regulated by abiotic stresses. In roots, *NPF7.2*, unloading nitrate from the xylem sap, is induced by salinity, whereas *NPF7.3*, loading nitrate into the xylem, is repressed by salinity. The opposite regulations result in the accumulation of nitrate in the roots under abiotic stress conditions, which leads to better tolerance to the stresses (Li et al., 2010; Chen et al., 2012). This phenomenon is called SINAR. The expression regulations of *NPF7.2* and *NPF7.3* are mediated by ethylene/jasmonic acid signaling (Zhang et al., 2014). Recently, it has been shown that *NPF6.3/NRT1.1* is also involved in the SINAR by regulating the expression of *NPF7.2* and *NPF7.3* and participates in cadmium tolerance (Jian et al., 2019). Due to the SINAR, less nitrate will be transported to the shoots under stress. *NPF2.3* in Arabidopsis, constitutively expressed in the root pericycle cells, support the minimum nitrate supply for shoots under stress conditions (Taochy et al., 2015).

### 5.4.2 NITRATE REMOBILIZATION UNDER NITROGEN STARVATION

Nitrate is a mobile nutrient, which means that it can be remobilized from the old organs to the developing ones through the phloem to support growth under nitrate starvation conditions. Arabidopsis NPF2.13/NRT1.7 is responsible for this process (Fan et al., 2009). *NPF2.13* is expressed in the phloem of the minor veins in the older leaves. In *npf2.13* mutants, more nitrate is accumulated in the old leaves. The growth of *npf2.13* mutants is reduced under nitrogen starvation conditions. These results suggest that nitrate remobilization from old to young leaves mediated by NPF2.13 is important to sustain plant growth under nitrogen starvation.

## 5.5 NITRATE TRANSPORTERS AND NITROGEN USE EFFICIENCY

Many studies have shown that manipulating nitrate transporters will enhance nitrogen use efficiency (NUE) summarized by these reviews (Fan et al., 2017; Li et al., 2017; Wang et al., 2018). Here are the latest studies. Co-expression of *OsNRT2.3a* and *OsNAR2.1* exhibited increased grain yields by approximately 24.6 percent compared with the wild-type rice (Chen et al., 2020). Similar to OsNRT1.1B/OsNPF6.5, OsNRT1.1A/OsNPF6.3 is also a dual-affinity nitrate transporter, but it is localized to the tonoplast (Wang et al., 2018). The *osnpf6.3* mutants exhibit late flowering, reduced nitrogen utilization and grain yield. On the other hand, overexpression of *OsNPF6.3* reduces the maturation time and enhances grain yield (Wang et al., 2018).

### 5.5.1 MANIPULATING REMOBILIZATION OF NITRATE TO DEVELOPING ORGANS

Tissue-specific expression of an "engineered" nitrate transporter will also increase nitrogen use efficiency. In order to understand which part of Arabidopsis NPF6.3/NRT1.1 protein is responsible for the high-affinity nitrate uptake, domain swabbing was performed between NPF6.3 and NPF4.6/NRT1.2. When substituted the

2nd to 4th transmembrane regions of NPF4.6 with NPF6.3, the chimera protein, named NC4N, shows dual-affinity nitrate uptake with an enhanced low-affinity nitrate uptake activity (Chen et al., 2020). When expressed NC4N in Arabidopsis, tobacco, and rice under the *NPF2.13/NRT1.7* promoter, the transgenic plants display higher source-to-sink translocation of nitrate and increased productivity under normal or nitrogen starvation conditions (Chen et al., 2020). This result indicates that manipulating nitrate transporters is a feasible way to improve nitrogen use efficiency.

## 5.6  CONCLUSIONS AND PERSPECTIVES

From soil to the needed plant cells, nitrate may cross multiple layers of cell membrane. When, where, and how much to transport nitrate across the membrane is determined by those nitrate transporters. Therefore, understanding the physiological functions and regulations of nitrate transporters is important for us to improve nitrogen use efficiency (NUE) and reduce the nitrogen input to the environment. So far, about half of the NPF members in Arabidopsis have been studied, but only 10 out of 93 NPF members in rice have been characterized (Fan et al., 2017; Li et al., 2017; Wang et al., 2018). More and more studies have shown that one NPF transporter may have more than one substrate (Corratge-Faillie and Lacombe, 2017). For example, NPF6.3/NRT1.1 can transport nitrate as well as auxin, and the auxin transport activity is inhibited in the presence of nitrate for lateral root initiation (Krouk et al., 2010). Low-affinity nitrate transporter NPF4.6/NRT1.2 also mediates the movement of abscisic acid (ABA) (Kanno et al., 2012; Kanno et al., 2013; Chiba et al., 2015; Zhang et al., 2021). Recently, a study showed that NPF7.3/NRT1.5 is an indole-3-butric acid (IBA) transporter involved in root gravitropism (Watanabe et al., 2020). The diverse substrates in NPF transporters give plants more flexibility when dealing with a harsh environment.

## REFERENCES

Almagro, A., Lin, S.H., and Tsay, Y.F. (2008) Characterization of the Arabidopsis nitrate transporter NRT1.6 reveals a role of nitrate in early embryo development. *Plant Cell* 20: 3289–3299.

Cai, C., Wang, J.Y., Zhu, Y.G., Shen, Q.R., Li, B., Tong, Y.P., and Li, Z.S. (2008) Gene structure and expression of the high-affinity nitrate transport system in rice roots. *J Integr Plant Biol* 50: 443–451.

Cerezo, M., Tillard, P., Filleur, S., Munos, S., Daniel-Vedele, F., and Gojon, A. (2001) Major alterations of the regulation of root NO(3)(-) uptake are associated with the mutation of Nrt2.1 and Nrt2.2 genes in Arabidopsis. *Plant Physiol* 127: 262–271.

Chen, C.Z., Lv, X.F., Li, J.Y., Yi, H.Y., and Gong, J.M. (2012) Arabidopsis NRT1.5 is another essential component in the regulation of nitrate reallocation and stress tolerance. *Plant Physiol* 159: 1582–1590.

Chen, J., Liu, X., Liu, S., Fan, X., Zhao, L., Song, M., Fan, X., and Xu, G. (2020) Co-overexpression of OsNAR2.1 and OsNRT2.3a increased agronomic nitrogen use efficiency in *Transgenic Rice Plants*. 11:1245. doi: 10.3389/fpls.2020.01245.

Chen, K.E., Chen, H.Y., Tseng, C.S., and Tsay, Y.F. (2020) Improving nitrogen use efficiency by manipulating nitrate remobilization in plants. *Nat Plants* 6: 1126–1135.

Chiba, Y., Shimizu, T., Miyakawa, S., Kanno, Y., Koshiba, T., Kamiya, Y., and Seo, M. (2015) Identification of Arabidopsis thaliana NRT1/PTR FAMILY (NPF) proteins capable of transporting plant hormones. *J Plant Res* 128: 679–686.

Chiu, C.C., Lin, C.S., Hsia, A.P., Su, R.C., Lin, H.L., and Tsay, Y.F. (2004) Mutation of a nitrate transporter, AtNRT1:4, results in a reduced petiole nitrate content and altered leaf development. *Plant Cell Physiol* 45: 1139–1148.

Chopin, F., Orsel, M., Dorbe, M.F., Chardon, F., Truong, H.N., Miller, A.J., Krapp, A., and Daniel-Vedele, F. (2007) The Arabidopsis ATNRT2.7 nitrate transporter controls nitrate content in seeds. *Plant Cell* 19: 1590–1602.

Corratge-Faillie, C., and Lacombe, B. (2017) Substrate (un)specificity of Arabidopsis NRT1/PTR Family (NPF) proteins. *J Exp Bot* 68: 3107–3113.

Crawford, N.M., and Glass, A.D.M. (1998) Molecular and physiological aspects of nitrate uptake in plants. *Trends in Plant Science* 3: 389–395.

Fan, S.C., Lin, C.S., Hsu, P.K., Lin, S.H., and Tsay, Y.F. (2009) The Arabidopsis nitrate transporter NRT1.7, expressed in phloem, is responsible for source-to-sink remobilization of nitrate. *Plant Cell* 21: 2750–2761.

Fan, X., Naz, M., Fan, X., Xuan, W., Miller, A.J., and Xu, G. (2017) Plant nitrate transporters: from gene function to application. *Journal of Experimental Botany* 68: 2463–2475.

Feng, H., Yan, M., Fan, X., Li, B., Shen, Q., Miller, A.J., and Xu, G. (2011) Spatial expression and regulation of rice high-affinity nitrate transporters by nitrogen and carbon status. *J Exp Bot* 62: 2319–2332.

Filleur, S., Dorbe, M.F., Cerezo, M., Orsel, M., Granier, F., Gojon, A., and Daniel-Vedele, F. (2001) An Arabidopsis T-DNA mutant affected in Nrt2 genes is impaired in nitrate uptake. *FEBS Lett* 489: 220–224.

Forde, B.G. (2000) Nitrate transporters in plants: structure, function and regulation. *Biochim Biophys Acta* 1465: 219–235.

Gutiérrez, R.A. (2012) Systems biology for enhanced plant nitrogen nutrition. *Science* 336(6089): 1673–1675.

He, Y.N., Peng, J.S., Cai, Y., Liu, D.F., Guan, Y., Yi, H.Y., and Gong, J.M. (2017) Tonoplast-localized nitrate uptake transporters involved in vacuolar nitrate efflux and reallocation in Arabidopsis. *Sci Rep* 7: 6417.

Hirel, B., Tétu, T., Lea, P.J., and Dubois, F. (2011) Improving nitrogen use efficiency in crops for sustainable agriculture. *Sustainability* 3.

Ho, C.H., Lin, S.H., Hu, H.C., and Tsay, Y.F. (2009) CHL1 functions as a nitrate sensor in plants. *Cell* 138: 1184–1194.

Hsu, P.K., and Tsay, Y.F. (2013) Two phloem nitrate transporters, NRT1.11 and NRT1.12, are important for redistributing xylem-borne nitrate to enhance plant growth. *Plant Physiol* 163: 844–856.

Hu, B., Wang, W., Ou, S., Tang, J., Li, H., Che, R., Zhang, Z., Chai, X., Wang, H., Wang, Y., Liang, C., Liu, L., Piao, Z., Deng, Q., Deng, K., Xu, C., Liang, Y., Zhang, L., Li, L., and Chu, C. (2015) Variation in NRT1.1B contributes to nitrate-use divergence between rice subspecies. *Nat Genet* 47: 834–838.

Huang, N.C., Liu, K.H., Lo, H.J., and Tsay, Y.F. (1999) Cloning and functional characterization of an Arabidopsis nitrate transporter gene that encodes a constitutive component of low-affinity uptake. *Plant Cell* 11: 1381–1392.

Jian, S., Luo, J., Liao, Q., Liu, Q., Guan, C., and Zhang, Z. (2019) NRT1.1 regulates nitrate allocation and cadmium tolerance in Arabidopsis. *Frontiers in Plant Science* 10: 384.

Kanno, Y., Hanada, A., Chiba, Y., Ichikawa, T., Nakazawa, M., Matsui, M., Koshiba, T., Kamiya, Y., and Seo, M. (2012) Identification of an abscisic acid transporter by functional screening using the receptor complex as a sensor. *Proc Natl Acad Sci U S A* 109: 9653–9658.

Kanno, Y., Kamiya, Y., and Seo, M. (2013) Nitrate does not compete with abscisic acid as a substrate of AtNPF4.6/NRT1.2/AIT1 in Arabidopsis. *Plant Signal Behav* 8: e26624.

Kiba, T., Feria-Bourrellier, A.B., Lafouge, F., Lezhneva, L., Boutet-Mercey, S., Orsel, M., Brehaut, V., Miller, A., Daniel-Vedele, F., Sakakibara, H., and Krapp, A. (2012) The Arabidopsis nitrate transporter NRT2.4 plays a double role in roots and shoots of nitrogen-starved plants. *Plant Cell* 24: 245–258.

Kirk, G.J., and Kronzucker, H.J. (2005) The potential for nitrification and nitrate uptake in the rhizosphere of wetland plants: a modelling study. *Ann Bot* 96: 639–646.

Kotur, Z., Mackenzie, N., Ramesh, S., Tyerman, S.D., Kaiser, B.N., and Glass, A.D. (2012) Nitrate transport capacity of the Arabidopsis thaliana NRT2 family members and their interactions with AtNAR2.1. *New Phytol* 194: 724–731.

Krapp, A., David, L.C., Chardin, C., Girin, T., Marmagne, A., Leprince, A.-S., Chaillou, S., Ferrario-Méry, S., Meyer, C., and Daniel-Vedele, F. (2014) Nitrate transport and signalling in Arabidopsis. *Journal of Experimental Botany* 65: 789–798.

Kronzucker, H.J., Glass, A.D.M., Siddiqi, M.Y., and Kirk, G.J.D. (2000) Comparative kinetic analysis of ammonium and nitrate acquisition by tropical lowland rice: implications for rice cultivation and yield potential. *New Phytologist* 145: 471–476.

Krouk, G., Lacombe, B., Bielach, A., Perrine-Walker, F., Malinska, K., Mounier, E., Hoyerova, K., Tillard, P., Leon, S., Ljung, K., Zazimalova, E., Benkova, E., Nacry, P., and Gojon, A. (2010) Nitrate-regulated auxin transport by NRT1.1 defines a mechanism for nutrient sensing in plants. *Dev Cell* 18: 927–937.

Léran, S., Garg, B., Boursiac, Y., Corratgé-Faillie, C., Brachet, C., Tillard, P., Gojon, A., and Lacombe, B. (2015) AtNPF5.5, a nitrate transporter affecting nitrogen accumulation in Arabidopsis embryo. *Scientific Reports* 5: 7962.

Lark, R.M., Milne, A.E., Addiscott, T.M., Goulding, K.W.T., Webster, C.P., and O'Flaherty, S. (2004) Scale- and location-dependent correlation of nitrous oxide emissions with soil properties: an analysis using wavelets. *European Journal of Soil Science* 55: 611–627.

Lezhneva, L., Kiba, T., Feria-Bourrellier, A-B., Lafouge, F., Boutet-Mercey, S., Zoufan, P., Sakakibara, H., Daniel-Vedele, F., Krapp, A. (2014) The Arabidopsis nitrate transporter NRT2.5 plays a role in nitrate acquisition and remobilization in nitrogen-starved plants. *The Plant Journal* 80: 230–241.

Li, H., Hu, B., and Chu, C. (2017) Nitrogen use efficiency in crops: lessons from Arabidopsis and rice. *Journal of Experimental Botany* 68: 2477–2488.

Li, J.Y., Fu, Y.L., Pike, S.M., Bao, J., Tian, W., Zhang, Y., Chen, C.Z., Li, H.M., Huang, J., Li, L.G., Schroeder, J.I., Gassmann, W., and Gong, J.M. (2010) The Arabidopsis nitrate transporter NRT1.8 functions in nitrate removal from the xylem sap and mediates cadmium tolerance. *Plant Cell* 22: 1633–1646.

Li, Y., Ouyang, J., Wang, Y.Y., Hu, R., Xia, K., Duan, J., Wang, Y., Tsay, Y.F., and Zhang, M. (2015) Disruption of the rice nitrate transporter OsNPF2.2 hinders root-to-shoot nitrate transport and vascular development. *Sci Rep* 5: 9635.

Lin, S.H., Kuo, H.F., Canivenc, G., Lin, C.S., Lepetit, M., Hsu, P.K., Tillard, P., Lin, H.L., Wang, Y.Y., Tsai, C.B., Gojon, A., and Tsay, YF (2008) Mutation of the Arabidopsis NRT1.5 nitrate transporter causes defective root-to-shoot nitrate transport. *Plant Cell* 20: 2514–2528.

Liu, K.H., Huang, C.Y., and Tsay, Y.F. (1999) CHL1 is a dual-affinity nitrate transporter of Arabidopsis involved in multiple phases of nitrate uptake. *Plant Cell* 11: 865–874.

Liu, K.H., and Tsay, Y.F. (2003) Switching between the two action modes of the dual-affinity nitrate transporter CHL1 by phosphorylation. *EMBO J* 22: 1005–1013.

Luo, B., Chen, J., Zhu, L., Liu, S., Li, B., Lu, H., Ye, G., Xu, G., and Fan, X. (2018) Overexpression of a high-affinity nitrate transporter OsNRT2.1 increases yield and manganese accumulation in rice under alternating wet and dry conditions. 9: 1192. doi: 10.3389/fpls.2018.01192.

Näsholm, T., Kielland, K., and Ganeteg, U. (2009) Uptake of organic nitrogen by plants. *New Phytol* 182: 31–48.

Okamoto, M., Kumar, A., Li, W., Wang, Y., Siddiqi, M.Y., Crawford, N.M., and Glass, A.D. (2006) High-affinity nitrate transport in roots of Arabidopsis depends on expression of the NAR2-like gene AtNRT3.1. *Plant Physiol* 140: 1036–1046.

Orsel, M., Chopin, F., Leleu, O., Smith, S.J., Krapp, A., Daniel-Vedele, F., and Miller, A.J. (2006) Characterization of a two-component high-affinity nitrate uptake system in Arabidopsis. Physiology and protein-protein interaction. *Plant Physiol* 142: 1304–1317.

Smirnoff, N., and Stewart, G.R. (1985) Nitrate assimilation and translocation by higher plants: Comparative physiology and ecological consequences. *Physiologia Plantarum* 64: 133–140.

Tang, Z, Fan, X., Li, Q., Feng, H., Miller, A.J., Shen, Q., and Xu, G. (2012) Knockdown of a rice stelar nitrate transporter alters long-distance translocation but not root influx. *Plant Physiology* 160: 2052.

Taochy, C., Gaillard, I., Ipotesi, E., Oomen, R., Leonhardt, N., Zimmermann, S., Peltier, J.B., Szponarski, W., Simonneau, T., Sentenac, H., Gibrat, R., and Boyer, J.C. (2015) The Arabidopsis root stele transporter NPF2.3 contributes to nitrate translocation to shoots under salt stress. *Plant J* 83: 466–479.

Tsay, Y.F., Schroeder, J.I., Feldmann, K.A., and Crawford, N.M. (1993) The herbicide sensitivity gene CHL1 of Arabidopsis encodes a nitrate-inducible nitrate transporter. Cell 72: 705–713.

Vidal, E.A., Alvarez, J.M., Araus, V., Riveras, E., Brooks, M.D.., Krouk, G, Ruffel, S., Lejay, L., Crawford, N.M., Coruzzi, G.M., and Gutiérrez, R.A. (2020) Nitrate in 2020: Thirty years from transport to signaling networks. *The Plant Cell* 32: 2094.

Wang, W., Hu, B., Yuan, D., Liu, Y., Che, R., Hu, Y., Ou, S., Liu, Y., Zhang, Z., Wang, H., Li, H., Jiang, Z., Zhang, Z., Gao, X., Qiu, Y.R., Meng, X., Liu, Y., Bai, Y., Liang, Y., Wang, Y., Zhang, L., Li, L., Sodmergen, Jing, H., Li, J., and Chu, C. (2018) Expression of the nitrate transporter gene OsNRT1.1A/OsNPF6.3 confers high yield and early maturation in rice. *The Plant Cell* 30: 638–651.

Wang, Y.Y., Cheng, Y.H., Chen, K.E., and Tsay, Y.F. (2018) Nitrate transport, signaling, and use efficiency. *Annu Rev Plant Biol* 69: 85–122.

Wang, Y.Y., Hsu, P.K., and Tsay, Y.F. (2012) Uptake, allocation and signaling of nitrate. *Trends Plant Sci* 17: 458–467.

Wang, Y.Y., and Tsay, Y.F. (2011) Arabidopsis nitrate transporter NRT1.9 is important in phloem nitrate transport. *Plant Cell* 23: 1945–1957.

Watanabe, S., Takahashi, N., Kanno, Y., Suzuki, H., Aoi, Y., Takeda-Kamiya, N., Toyooka, K., Kasahara, H., Hayashi, K-i., Umeda, M., and Seo, M. (2020) The *Arabidopsis* NRT1/PTR family protein NPF7.3/NRT1.5 is an indole-3-butyric acid transporter involved in root gravitropism. *Proceedings of the National Academy of Sciences* 117: 31500.

Xia, X., Fan, X., Wei, J., Feng, H., Qu, H., Xie, D., Miller, A.J., and Xu, G. (2015) Rice nitrate transporter OsNPF2.4 functions in low-affinity acquisition and long-distance transport. *J Exp Bot* 66: 317–331.

Yan, M., Fan, X., Feng, H., Miller, A.J., Shen, Q., and Xu, G. (2011) Rice OsNAR2.1 interacts with OsNRT2.1, OsNRT2.2 and OsNRT2.3a nitrate transporters to provide uptake over high and low concentration ranges. *Plant, Cell & Environment* 34: 1360–1372.

Zhang, G.B., Yi, H.Y., and Gong, J.M. (2014) The Arabidopsis ethylene/jasmonic acid-NRT signaling module coordinates nitrate reallocation and the trade-off between growth and environmental adaptation. *Plant Cell* 26: 3984–3998.

Zhang, J., Liu, Y.-X., Zhang, N., Hu, B., Jin, T., Xu, H., Qin, Y., Yan, P., Zhang, X., Guo, X., Hui, J., Cao, S., Wang, X., Wang, C., Wang, H., Qu, B., Fan, G., Yuan, L., Garrido-Oter, R., Chu, C., and Bai, Y. (2019) NRT1.1B is associated with root microbiota composition and nitrogen use in field-grown rice. *Nature Biotechnology* 37: 676–684.

Zhang, L., Yu, Z., Xu, Y., Yu, M., Ren, Y., Zhang, S., Yang, G., Huang, J., Yan, K., Zheng, C., and Wu, C. (2021) Regulation of the stability and ABA import activity of NRT1.2/NPF4.6 by CEPR2-mediated phosphorylation in Arabidopsis. *Molecular Plant* 14: 633–646]

# 6 Nitrogen Fertilizers and the Environment

*Eric Walling and Celine Vaneeckhaute*

## CONTENTS

6.1 Introduction .................................................................................................. 103
6.2 The Environmental Impact of Nitrogen Fertilizer Production ..................... 105
6.3 The Environmental Impact of Nitrogen Fertilizers Application................... 107
    6.3.1 State of Nitrogen Following Fertilizer Application........................... 107
    6.3.2 Nitrogen Transfer and Impact through Airways................................ 108
        6.3.2.1 Ammonia Emissions ........................................................... 108
        6.3.2.2 Nitrous Oxide Emissions .................................................... 110
        6.3.2.3 Nitric Oxides Emissions ..................................................... 110
        6.3.2.4 Impact on Emissions from Soil Microorganisms................ 111
    6.3.3 Nitrogen Transfer and Impact through Waterways............................ 111
        6.3.3.1 Contamination of Ground and Surface Waters ................... 111
        6.3.3.2 Eutrophication..................................................................... 112
        6.3.3.3 Toxic Algal Blooms ............................................................ 113
        6.3.3.4 Acidification of Waterways................................................. 114
    6.3.4 Nitrogen Transfer and Impact through Soils .................................... 115
        6.3.4.1 Soil Acidification ............................................................... 115
        6.3.4.2 Soil Salinization ................................................................. 117
        6.3.4.3 Soil Erosion......................................................................... 117
        6.3.4.4 Soil Nutrient Leaching and Depletion ............................... 118
        6.3.4.5 Heavy Metal Accumulation, Solubility, and
                 Availability.......................................................................... 118
        6.3.4.6 Loss of Soil Organic Carbon.............................................. 119
    6.3.5 Ecological Impact............................................................................. 120
6.4 Perspective: Toward More Sustainable Agriculture .................................... 122
6.5 Conclusion.................................................................................................... 125

## 6.1 INTRODUCTION

Nitrogen (N) fertilizers are essential to sustaining the human population and have taken on an even more important role since the 1950s, given the boom in population. Indeed, the Haber-Bosch process, through which the ammonia ($NH_3$) used to produce N-fertilizers is synthesized from air and hydrogen, has often been hailed as

DOI: 10.1201/9780429326806-8

**103**

the most important invention of the twentieth century and, as of 2015, is estimated to have enabled the feeding of 3 to 3.5 billion people (Ritchie, 2017). Furthermore, as the global population has continued to grow, so has our reliance on nitrogen fertilizers. Estimates place nearly 50 percent of the world population as being dependent on synthetic nitrogen fertilizers, without considering the contribution from organic sources (Erisman et al., 2008); and this trend does not seem to be waning. Indeed, the International Fertilizer Association (IFA) reported a 46 percent increase in urea, by far the most common N-fertilizer, production between 2003 and 2012 (Heffer and Prud'homme, 2016), while the Food and Agriculture Organization of the United Nations (FAO) anticipated an annual increase in synthetic nitrogen demand of 1.5 percent between 2016 and 2020 (FAOSTAT, 2017). Current N fertilizer needs are only expected to grow as agricultural production is expected to increase by 50 to 100 percent by 2050 (Walling and Vaneeckhaute, 2020b).

However, despite their necessity, our dependency on nitrogen fertilizers has shifted the natural balance, leading to many damaging and long-lasting environmental and ecological consequences. Annually, nearly 120 million tonnes of nitrogen are applied to agricultural fields (FAOSTAT, 2017). This, when compared to natural biological nitrogen fixation of around 190 million tonnes N per year (Brady et al., 2008), demonstrates the breadth of the impact of humanity on the nitrogen cycle. Furthermore, Africa is expected to undergo a major population boom in the coming decades, with its population expected to increase by 1.2 billion people by 2050, representing 63 percent of total population growth between now and then (United Nations, 2017), while it currently only accounts for approximately 3.6 percent of the world N-fertilizer demand (4.3 million tonnes N year$^{-1}$) (FAOSTAT, 2017). Indeed, the average nitrogen application in Africa is around 17 kg N ha$^{-1}$ year$^{-1}$, compared to the world average of 135 kg N ha$^{-1}$ year$^{-1}$ (Harrison, 2020), while generally recommended fertilization rates range between 50 and 300 kg N ha$^{-1}$ year$^{-1}$, depending on the cropping system. Therefore, to sustain this growth, many developing countries, notably in Africa, will have to significantly increase their nitrogen fertilizer consumption (by nearly a thousand percent, assuming Africa reaches expected levels of around 2.4 billion inhabitants by 2050), if they are to reach the standards of the developed world. This is also compounded by the fact that fertilizers, especially N-fertilizers, have been historically over-applied to agricultural lands in the hope of seeing increased crop yield (which do not materialize themselves), only leading to greater nitrogen loss to the environment.

The impact of this can be felt at all levels, affecting the quality of air, waterways, and soil, as well as being a major contributor to global warming, while destabilizing, and potentially destroying entire ecosystems. These effects include the production of fine particulate matter and toxic compounds that are detrimental to human and animal health, emissions of potent greenhouse gases (GHGs), deposition of nitrogen over vast distances, eutrophication of waterways, acidification of both soils and water, soil erosion and nutrient depletion, destruction of habitats, and loss of biodiversity, as well as serious economic and social consequences associated with these phenomena. This chapter will therefore explore in detail the environmental and ecological impact of nitrogen fertilizers, delving into the above-mentioned areas. The chapter is divided as follows: Section 6.2 examines the environmental impact of nitrogen fertilizer

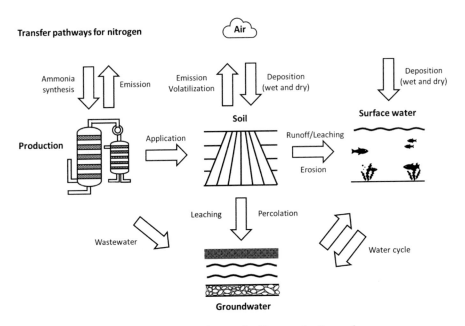

**FIGURE 6.1** Transfer pathways from nitrogen fertilizer production and use.

production, focusing on conventional synthetic N production through the Haber-Bosch process. Section 6.3 investigates the environmental and ecological impact of N-fertilizer application, focusing on the impacts on air (6.3.2), water (6.3.3), soil (6.3.4), and ecology (6.3.5). The final section (section 6.4) offers perspectives on how we can reduce the impact of nitrogen fertilizers, followed by concluding remarks. Figure 6.1 presents a global overview of how nitrogen is transferred to and from the air, soils, and waterways, all of which will be explored in detail throughout sections 6.2 and 6.3, as well as the associated environmental and ecological impact.

## 6.2 THE ENVIRONMENTAL IMPACT OF NITROGEN FERTILIZER PRODUCTION

The primary production pathway for nitrogen fertilizers is through the Haber-Bosch process, which involves reacting nitrogen ($N_2$) from the air with hydrogen ($H_2$), usually from the methane ($CH_4$) in natural gas, at high temperature and pressure, to form ammonia ($NH_3$), with carbon dioxide ($CO_2$) and water ($H_2O$) as the main by-products. The Haber-Bosch process is a behemoth of a procedure, being responsible for over 96 percent of the world's ammonia supply, 88 percent of which is used to produce fertilizers (Smith et al., 2020; USGS, 2020). Consequently, it is responsible for 2 percent of the world's energy use, given the consumption of natural gas as a reagent, and 1-2 percent of global $CO_2$ emissions (IFA, 2009). The ammonia produced through the process can then be directly converted into any variety of N fertilizers, such as urea ($CO(NH_2)_2$), ammonium nitrate ($NH_4NO_3$), ammonium phosphate (($NH_4)_3PO_4$), ammonium bicarbonate ($NH_4HCO_3$), and ammonium sulfate (($NH_4)_2SO_4$), or converted to nitric acid ($HNO_3$) through the Ostwald process, which can then be used to produce nitrates, such as calcium ($Ca(NO_3)_2$) and potassium nitrate ($KNO_3$).

Globally, urea and ammonium nitrate represent approximately 75 percent of straight-N (without P or K) fertilizer consumption, with urea notably being extremely popular in developing countries (85.6 percent of straight N-fertilizer consumption in the developing world), while ammonium phosphate is the most used compound-N fertilizer (NP(K)) (IFA, 2019). Therefore, we will mainly focus on these three when exploring the environmental impact of synthetic N-fertilizer production. Emissions from the production of synthetic N-fertilizers are presented in Table 6.1, adapted from Walling and Vaneeckhaute (2020b). As can be noted by the variability in these emission factors, many components can influence the amount of GHGs emitted by N fertilizer manufacturing processes. These factors notably include the type of fertilizer produced, the maturity of the process in specific countries, and the hydrogen source used for the Haber-Bosch process, energy recovery, and efficiency, among others. Indeed, regarding the type of fertilizer, certain production routes can lead to higher emissions, such as when comparing ammonium nitrate to urea. Ammonium nitrate production requires the production of nitric acid, which can produce significant $N_2O$ emissions, leading to the stark differences observed in Table 6.1, though more and more facilities are implementing catalytic reduction technologies to reduce these emissions, lowering this impact. An even greater impact is the source of energy and hydrogen used for the synthesis of ammonia. For example, in China in 2012, 86 percent of the hydrogen required for ammonia synthesis was provided from coal instead of natural gas, leading to emissions up to 1100 percent higher than a system operating with natural gas (Zhang et al., 2013). Regardless of the specific number, these emissions represent an important contributor to the greenhouse effect and global warming. Production of nitrogen fertilizers is generally considered to be one of the greatest contributors to GHG emissions from crops, alongside in-field emissions following application (Walling and Vaneeckhaute, 2020b); though this is very dependent on the amount of nitrogen emitted from the soil following fertilizer application (Walling and Vaneeckhaute, 2020b). Furthermore, it is important to

**TABLE 6.1**
**Emissions from Synthetic N-Fertilizer Production**

| Fertilizer | Emission factor (kg $CO_2$-eq./kg of N) | Country/Region |
|---|---|---|
| Urea | 1.3–4 | Europe |
|  | 2.7–3.5 | Russia, USA |
|  | 5.5 | China |
| Ammonium nitrate | 3.5–7.2 | Europe |
|  | 8 | Russia, USA |
|  | 10.3 | China |
| Ammonium phosphates | 3.3–4.6 | Europe |
|  | 4.3–16.3 | Russia, USA |
|  | 7.4–22.7 | China |

highlight that focusing purely on emissions does not consider the potential benefits and drawbacks of the different types of N fertilizers during the application, both environmentally and agriculturally.

One important note, however, is that the Haber-Bosch process very often integrates significant carbon-capture measures, recycling the $CO_2$ produced during ammonia synthesis into the production of urea. Furthermore, urea synthesis is an exothermic process, allowing for energy recovery and reintegration throughout the process chain. Though, in the end, this carbon is inevitably returned to the atmosphere in the form of $CO_2$ when urea is applied to the fields. This current method of operation provides a good avenue for $CO_2$ valorization, though this $CO_2$ could easily be acquired from many other industrial sources.

Furthermore, the production of nitrogen fertilizers generates wastewaters rich in nitrogenous compounds, such as ammonia, ammonium, and urea, as well as $CO_2$ and runoffs and residuals from the production process (oils, grease, heavy metals, etc.). Despite generally treating wastewater on location, N-fertilizer plants in developing countries have often been found to greatly exceed environmental norms (Igwe et al., 2016; Laghari et al., 2018), which is only compounded by the considerable amount of water produced by these processes, estimated at around 0.5 tonnes of wastewater per tonne of urea produced (Matijašević et al., 2010).

## 6.3 THE ENVIRONMENTAL IMPACT OF NITROGEN FERTILIZERS APPLICATION

Before delving into the environmental impact of nitrogen fertilizers, it is important to note that, aside from their production, the application of nitrogen fertilizers to soils will mainly have a negative impact when excess fertilizer is applied, relative to the soil's capacity to accept it. Therefore, if all of the nitrogen applied to the soil is taken up by the plants and soil, there would be none left to escape into the environment. However, this is often not the case, with excess nitrogen leaching into the soil and waterways, or volatilizing into the air.

### 6.3.1 State of Nitrogen Following Fertilizer Application

Nitrogen fertilizers provide nitrogen in two main forms: ammonium ($NH_4^+$) and nitrate ($NO_3^-$), given that only these two forms of nitrogen can be used by crops. However, when amended to soils, these chemical species enter the nitrogen cycle and are therefore free to become oxidized/reduced to a variety of different nitrogenous compounds through the nitrification/denitrification pathways. As shown in Figure 6.2, nitrification is the aerobic process by which ammonium is oxidized into nitrite ($NO_2^-$), followed by nitrate ($NO_3^-$), while denitrification is the reduction of nitrate to molecular nitrogen in anoxic conditions, passing through nitrite ($NO_2^-$), nitric oxide (NO), nitrous oxide ($N_2O$), and finally into $N_2$. These processes are biologically driven by soil microorganisms through both nitrifying and denitrifying organisms and can take place simultaneously in close proximity to one another. Furthermore, ammonium can be chemically converted to ammonia ($NH_3$), a highly volatile species, particularly in high pH conditions. As mentioned in the introduction, mankind, notably through modern agriculture driven by fertilizer use, has thrown the natural nitrogen cycle

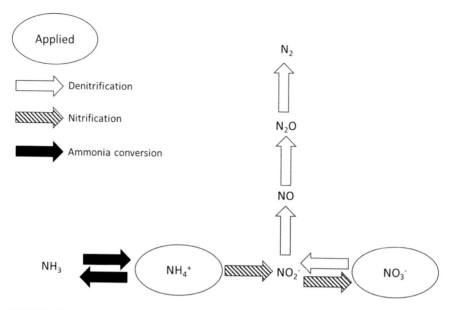

**FIGURE 6.2**  Transformation pathways for applied nitrogen.

out of balance. Fertilizer production and use is not a simple "one-for-one" balance, where nitrogen is removed from air or organic matter and then retransferred to soils to be recycled; it is a transformation of nonreactive nitrogen to reactive forms, such as $NH_3$, $NO_x$, $NO_3^-$, and $N_2O$ (Canfield et al., 2010). Therefore, when applied to soils, especially in cases of over- or improperly timed applications, these various forms of nitrogen can end up in the wider environment, free to react with other compounds or to directly impact the environment, the magnitude and impact of which will be explored in the following sections.

## 6.3.2  Nitrogen Transfer and Impact through Airways

Of the forms of nitrogen mentioned in section 6.3.1, many are released as gaseous compounds. These include ammonia ($NH_3$), nitric oxide (NO), nitrogen dioxide ($NO_2$), and nitrous oxide ($N_2O$) ($N_2$ as well, though it is inert), which can have significant environmental and health consequences, as highlighted in Figure 6.3. Given the widespread use, historic over-application, and the essential nature of nitrogen fertilizers, agriculture and fertilizer use, in particular, have become the largest emitters of some of these gases.

### 6.3.2.1  Ammonia Emissions

Agriculture is responsible for more than 70 percent of anthropogenic $NH_3$ emissions, which originate primarily from the volatilization of ammonium in livestock manure and applied fertilizers (Walling and Vaneeckhaute, 2020b). Ammonia emissions can have significant detrimental impacts, both to human and environmental health.

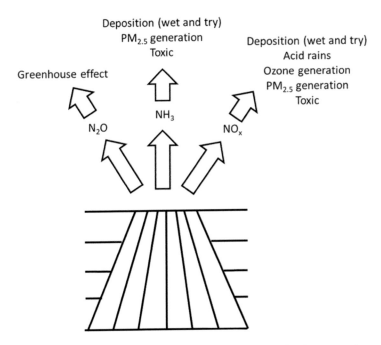

**FIGURE 6.3** Transfer pathways for applied nitrogen through airways and associated environmental impact of these emissions.

Ammonia gas is foul-smelling and corrosive, with exposure to low concentrations leading to skin and eye irritation, while high concentrations can cause respiratory irritation and distress, potentially leading to death, as well as chemical burns and permanent eye damage. Once emitted, ammonia can then return to land through either dry or wet deposition. In the case of dry deposition, atmospheric gases and particles get dumped on surfaces through gravity, wind currents, or given a greater affinity (such as solubility) toward the surface. In the case of wet deposition, aerosol particles are taken up by precipitation and fall down to earth with rain or snow. One of the primary ways that ammonia gets deposited is by binding with other gases, such as sulfur dioxide ($SO_2$), nitric acid, and nitrogen oxides ($NO_x$), the latter of which can also be emitted following application of N-fertilizers through incomplete denitrification, to form fine particulate matter ($PM_{2.5}$) (Hristov, 2011; Wu et al., 2016). These particles ($\geq 2.5$ µm) can remain airborne for a long time and are easily carried through the atmosphere, traveling up to thousands of kilometers away from their emission source (Wang et al., 2017). Beyond the general health impacts of ammonia, $PM_{2.5}$s are particularly detrimental to cardiovascular and respiratory health, leading to increased and worsened cases of heart attacks and strokes, chronic obstructive pulmonary disease (COPD), asthma, and lung cancer. From an environmental perspective, deposition increases nitrogen contents in waterways, soils, and ecosystems, as will be discussed in sections 6.3.3, 6.3.4, and 6.3.5. Although ammonia taken up in the water can decrease the acidity of acid rains, deposition of nitrogen is notably one of the main contributors to the acidification of waters and soils, given its contribution to nitrification, as will be discussed later (sections 6.3.3 and 6.3.4) (Doney et al., 2007; Huang et al., 2014). Deposition, both wet and dry, can also cause the carbon-to-nitrogen (C/N) ratio of soils to decrease and become unbalanced, leading to further nitrogen emissions (Shen et al., 2018).

### 6.3.2.2 Nitrous Oxide Emissions

Ammonia is not the only gas whose primary anthropogenic source is agriculture. Indeed, agriculture is also responsible for over 80 percent of anthropogenic nitrous oxide ($N_2O$) emissions, mainly from fertilizer use (Walling and Vaneeckhaute, 2020b). With a global warming potential of 298 kg $CO_2$-eq./kg $N_2O$ over a 100-year timescale, nitrous oxide is an extremely potent greenhouse gas, while also being the largest ozone-depleting substance emitted by humans (Ravishankara et al., 2009). The contribution of $N_2O$ to global GHG emissions is significant, representing 7 percent of total emissions in the United States (on a $CO_2$-eq. basis), making it the third greatest contributor to global warming (Hockstad and Hanel, 2018). These emissions are produced through incomplete ammonium oxidation and/or (de)nitrification and increase as a function of fertilizer application, though it is unclear whether this relationship is linear or non-linear (Walling and Vaneeckhaute, 2020b). Taking a conservative value of 3 kg $CO_2$-eq./kg of applied N (Walling and Vaneeckhaute, 2020b), global $N_2O$ emissions from nitrogen fertilizer application would amount to approximately 310 million tonnes of $CO_2$-eq. per year, without considering contribution from organic fertilizers, which are expected to emit similar amounts of $N_2O$. When considering emissions from N-fertilizer production, which we will assume at 4 kg $CO_2$-eq./ kg of N as well, based on Table 6.1, a conservative estimate of $N_2O$ emissions from N-fertilizer production and use is approximately 780 million tonnes of $CO_2$-eq. per year, which is equivalent to about 2 percent of global annual $CO_2$ emissions.

### 6.3.2.3 Nitric Oxides Emissions

Nitric oxides ($NO_x$) from fertilizer application can be an important contributor to $NO_x$ budgets, though quantification of its magnitude remains contentious and unclear (Chen et al., 2020). $NO_x$ are detrimental to both health and the environment, given their highly reactive nature, allowing them to react with a variety of compounds and to form undesirable contaminants. As mentioned previously, $NO_x$ can react with ammonia and other molecules to form fine particulate matter ($PM_{2.5}$), which can allow for their transportation over long distances, as well as inherent health impacts of these particles. $NO_x$ in particular can also be harmful on an environmental level, with $NO_x$ and their products being highly phytotoxic (Stevens et al., 2020). Furthermore, in the presence of sunlight, $NO_x$ can react with volatile compounds to form ozone ($O_3$) in the troposphere, another toxic compound for respiratory health. However, this reaction, when occurring in the stratosphere, can have a positive impact on the greenhouse effect by supplementing atmospheric concentrations of ozone, as well as producing hydroxyl radicals as a by-product ($\bullet OH$), which react with methane ($CH_4$) and present the most important sink for this GHG (Zhao et al., 2020). Nevertheless, despite these positive impacts on the greenhouse effect (Fry et al., 2012; Wild et al., 2001),

$NO_x$ are still generally considered as having a negative impact on global warming (Grewe et al., 2019), given their deposition, which leads to further $N_2O$ emissions (Xie et al., 2018), outweighing the benefits of ozone and hydroxyl radical production. Furthermore, a variety of other hazardous compounds can be formed by reactions with $NO_x$, notably mutagenic substances such as nitrosamines and nitroarenes (Fostås et al., 2011). There is also the important contribution of $NO_x$ to acid rains, where they can form nitric acid in the water, increasing the acidity of these precipitations and further contributing to acidification of soil and waterways, as well as leading to significant economic consequences, notably through the accelerated degradation of infrastructure (Zhang et al., 2019b).

### 6.3.2.4  Impact on Emissions from Soil Microorganisms

One area that remains unclear is the impact of nitrogen fertilizers on the emissions from soil microorganisms. As mentioned in section 6.3.1, microorganisms are responsible for the nitrification and denitrification pathways, but they are also responsible for methane emissions from soils, especially in anaerobic conditions. Research has demonstrated that the increased availability of nitrogen can promote the metabolism of certain soil microorganisms and lead to higher methane emissions, though this finding is not always consistent (Banger et al., 2012). This phenomenon can be especially important in semi-aquatic cropping systems, such as rice paddies, which some estimates have placed as contributing up to 17 percent of global anthropogenic methane emissions (Walling and Vaneeckhaute, 2020b). Furthermore, there is an unclear relation between fertilizer application and soil carbon sequestration and respiration (Walling and Vaneeckhaute, 2020b).

### 6.3.3  NITROGEN TRANSFER AND IMPACT THROUGH WATERWAYS

While nitrogen can escape through gaseous form, it can also be lost to the environment through its uptake and transportation by water or aqueous solutions, a process known as leaching. Nitrate, nitrite, ammonia, and ammonium are all water-soluble and can be leached into the environment, notably ending up in waterways and groundwater. This nitrogen leaching is ever more significant when fertilizers are overapplied or when the soils are saturated in nitrogen, given that soils have limited ion exchange capacities, meaning that, when these capacities are exceeded, the ions are not retained in the soil and can be swept out of the system. A general schematic overview of the interactions between the various phenomena described throughout this section is provided in Figure 6.4.

### 6.3.3.1  Contamination of Ground and Surface Waters

Following irrigation of crops and rainfall events, a portion of applied nutrients can make their way into ground and surface waters. In the case of nitrogen fertilizers, this can lead to nitrate contamination, which can pose an important threat to human and animal health, given that these are the primary sources of drinking water across the globe, while ammonium, on the other hand, is not considered as being harmful

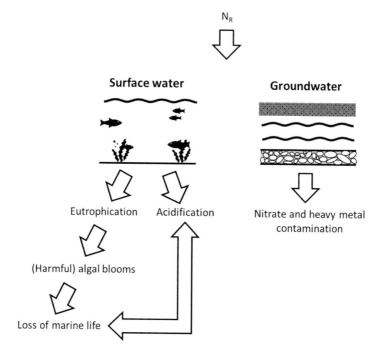

**FIGURE 6.4**  Environmental impacts due to the presence of reactive nitrogen, notably $NH_3/NH_4^+$.

to human health (Schullehner et al., 2017). Nitrate contamination is best known for its impact on infants, leading to methemoglobinemia, though it can also contribute to other health effects and birth defects, such as increasing the risk of developing colorectal, bladder, and breast cancers, thyroid disease, and neural tube defects (Ward et al., 2018). Furthermore, these forms of reactive nitrogen can then make their way into surface waters, through the water cycle, contributing to the following effects that will be explored.

## 6.3.3.2  Eutrophication

One of the primary and best-known environmental impacts of nitrogen pollution of waterways is eutrophication. Eutrophication is a phenomenon through which marine plants and algae grow at an uninhibited rate, in major part due to the abundance of nutrients from fertilizers, such as nitrogen. However, this proliferation of plants and algae comes at a general detriment to the rest of the marine ecosystem. The most well-known of these consequences is the creation of algal blooms, where algae overwhelm an ecosystem through their excessive growth. Indeed, during algal blooms, the high concentration of algae reduces light penetration, consumes dissolved inorganic carbon, and modifies pH, all of which impact the ecosystem in a variety of manners (Amorim and do Nascimento Moura, 2020). Reduced light penetration causes plants in littoral zones (near the shore) to die-off and can reduce the ability of predators to hunt for prey, while depletion of dissolved inorganic carbon, which is a consequence of the elevated rate of photosynthesis from eutrophication, causes an important shift in pH. Indeed, pH ranges of up to 9 have been frequently documented, sometimes increasing up to 11, especially during summer days, due to maximized photosynthetic activity (Raven et al., 2020). The resulting changes in pH thus hamper the survivability of animals that depend on chemosensory abilities (Turner and Chislock, 2010), damages certain species through increased ammonia levels (conversion of ammonium to ammonia), and disrupts reproduction (Ip and Chew, 2010). When the algae die, these dense groups of organisms become an equally dense source of food for bacteria, releasing large quantities of dissolved inorganic carbon into the environment. This has two serious consequences: (1) on a chemical level, the release of this carbon leads to a sharp decrease in pH, causing the acidification of waterways (Cai et al., 2011); while (2), biologically, the decomposition of the remains of the algal bloom requires a significant amount of oxygen, leading to severe depletion of dissolved oxygen in the environment and causing it to become hypoxic (Watson et al., 2016). This, in turn, leads to mass die-offs in these eutrophic waters, often labeled as "dead zones," which are unable to sustain most forms of life. For example, the Gulf of Mexico suffers from seasonal hypoxia, affecting an area up to approximately 23,000 km$^2$ and 140 km$^3$ in volume (Obenour et al., 2013; Rabalais and Turner, 2019). Another example is the Baltic Sea, which contains seven of the world's ten largest dead zones and clearly demonstrates the growing issue of nutrient-driven eutrophication (Carstensen et al., 2014), having grown from an area of 5,000 km$^2$ to more than 60,000 km$^2$ over the past 115 years (Carstensen et al., 2014). As of 2008, 415 dead zones had been identified worldwide, a drastic increase from the 10 cases documented in the 1960s, and a product of the widespread use of N fertilizers (Diaz et al., 2008). By 1993, 54 percent of the Asian Pacific, 53 percent of European, 48 percent of North American, 41 percent of South American, and 28 percent of African lakes were eutrophic (ILEC, 2001). In many of these places, legislation and action plans are being continually deployed to try and reduce eutrophication and aid in the recovery of affected waterways (Bohman, 2018; Preisner et al., 2020); however, this can be a slow process, demonstrating the importance of reducing nutrient pollution at the source (agriculture and wastewater). Despite reductions in the surplus nutrient application and stricter wastewater emission restrictions, eutrophication remains a persistent problem (Van Puijenbroek et al., 2014). For example, recent models have predicted that certain areas in the Baltic Sea will not return to stable (non-eutrophic) levels before 2200 or later, even when the objectives of the Baltic Sea Action Plan are met (Murray et al., 2019).

### 6.3.3.3  Toxic Algal Blooms

Beyond the environmental threats of traditional algal blooms, there are also certain groups known as harmful algal blooms (HABs) that can produce toxic compounds, such as microcystin, saxitoxins, domoic acids, and brevetoxins (Wurtsbaugh et al., 2019). The impact of these harmful blooms is not only limited to what was detailed above in section 6.3.3.2, as they also pose a threat to terrestrial animal life, including humans. Many species can cause HABs, though the best-known are cyanobacteria. Cyanobacterial blooms can cause many unpleasant problems, such as limiting recreational use of waterways, leading to taste and odor problems in drinking water, while also having more serious consequences, such as producing cyanotoxins (Huisman et al., 2018). These toxins, which can affect various organ systems, allowing them to act as hepatotoxins (liver), neurotoxins (nerve tissue), cytotoxins (cells), as well as being skin and gastrointestinal irritants (Codd et al., 2020), are not only a danger if directly consumed (if they are drunk). Indeed, these toxins can reach animals and humans through a variety of pathways (Codd et al., 2020) including (1) through consumption of fish, and particularly shellfish (filter feeders), that have absorbed these contaminants; (2) through inhalation from aerosols, such as in cases where crops are irrigated with contaminated water, from recreational activities (swimming, boating, waterboarding, etc.) (Backer et al., 2010), from dried scum on the shore, and from arid environments following rainfall (Metcalf et al., 2012); (3) through the consumption of crops irrigated with contaminated water, which can absorb the toxins (Lee et al., 2017); (4) through dermal contact; and (5) intravenously (Carmichael et al., 2001). Mild infections can lead to unpleasant symptoms, such as diarrhea, vomiting, fever, and headache, just to name a few, while severe infections can cause blistering of the mouth, pneumonia, confusion, respiratory paralysis, and death (Funari and Testai, 2008). Since their first observation in the 1800s, where the rapid death of stock animals at a freshwater lake in Australia was investigated and attributed to cyanobacteria (Francis, 1878), harmful and toxic blooms have been increasing in number and severity (Gobler, 2020; Lewitus et al., 2012). Though recreational exposure is rarely fatal to humans, it can be particularly toxic to certain animals, such as dogs and birds, which are very susceptible to these toxins (Wood, 2016). However, when consumed orally (drinking, contaminated food), especially through shellfish, which can accumulate high levels of toxins (Negri and Jones, 1995), the toxins released by these harmful blooms can be deadly to humans. Examples of these are plentiful, with the first known case of amnesic shellfish poisoning being recognized in Canada in 1987, where consumption of blue mussels led to the death of three and the acute poisoning of 105 others (Jeffery et al., 2004); though diarrhetic shellfish poisoning had been documented just over a decade earlier (Yasumoto et al., 1978), leading to thousands of yearly infections across the globe (Hallegraeff, 2003). Though the least likely route of exposure, intravenous exposure has also been reported and is extremely dangerous. The first such case was reported in 1996, when patients at a dialysis clinic were treated with contaminated water that was directly responsible for the death of 40 percent (n = 52) of the patients, with 76 percent (n = 100) experiencing subsequent liver-failure (Carmichael et al., 2001).

### 6.3.3.4 Acidification of Waterways

The net impact of the pH changes mentioned in section 6.3.3.2 is the acidification of waterways (lakes, rivers, and coastal oceans) due to the increase in $CO_2$ caused by the decomposition of algal blooms, in addition to acidification from other sources, such as deposition of nitrogen and leaching/runoff from soils. As mentioned in section 6.3.3.2, this can impact the health and lifecycle of many organisms, decreasing survival, growth, development, and abundance of marine organisms (Kroeker et al.,

2013), and can be especially problematic for coral reefs (Silbiger et al., 2018) and for animals that require calcium carbonate ($CaCO_3$) minerals to form their skeletons or shells, given that the carbonate ions will instead form bicarbonate ($HCO_3^-$) in acidic conditions (Fitzer et al., 2016). This impact can also extend to decreasing energy metabolisms for certain organisms, making them weaker, more vulnerable to predation, or less capable to hunt, while also potentially influencing intracellular pH (Lannig et al., 2010; Schalkhausser et al., 2013). This decrease in pH can have disastrous consequences on marine life, favoring the conversion of ammonia to ammonium and leading to central nervous system death by displacing $K^+$ ions and leading to an influx of excessive $Ca^{2+}$ (Ip and Chew, 2010). This added sensitivity of marine life and the loss of coastal protection (degradation of coral reefs) can also have significant social and economic consequences by decreasing yields from fisheries and aquaculture (Hall-Spencer and Harvey, 2019), with losses estimated at US$100 billion by 2100 for loss of mollusc production and US$870 billion from ocean-acidification-induced coral reef loss (Gattuso et al., 2014). Furthermore, this acidification decreases the capacity of oceans, the greatest carbon sinks on the planet, to store $CO_2$ emissions, adding to the contribution to global warming. Potentially, an acidic environment can also significantly decrease ocean nitrification rates, continuing the alteration to the nitrogen cycle and affecting organisms dependent on nitrate for growth.

### 6.3.4 Nitrogen Transfer and Impact through Soils

When not taken up by plants, emitted as gases, or leached out of the soils, applied (reactive) nitrogen will stay in the soil and be free to interact with the environment. This is in addition to the potential deposition of nitrogenous compounds, as discussed in section 6.3.2, which can further contribute to soil nitrogen levels. The impacts of this can be varied and are significant, as will be discussed throughout this section, with a conceptual representation of some of the main contributors and effects provided in Figure 6.5.

#### 6.3.4.1 Soil Acidification

Similar to aquatic settings, the presence of reactive nitrogen from fertilizers can lead to the acidification of the environment. Though fertilizers are themselves rarely acidic, fertilizer nitrogen can interact with the soil in a variety of ways that can lower soil's pH, especially in cases of ammonium-based fertilizers, such as urea, ammonium nitrate, and ammonium sulfate (Han et al., 2015). In soils, nitric acid will directly contribute to acidity, while ammonium, either applied or from deposition, will lead to the release of hydrogen ions, either from biological uptake, nitrification, and/or conversion to ammonia. Indeed, nitrification of excessive N fertilizer (ammonium) is one of the leading causes of soil acidification (Han et al., 2015). Decreases in pH of up to 2 to 3 units over a few decades have been reported following application of nitrogen fertilizers (Guo et al., 2010), such as a long-term trial in Sweden, which found that soil pH decreased from an initial value of 6.5 in 1956 to 4.2 in 2009 when fertilized with ammonium sulfate, while treatments without fertilizer and with calcium nitrate did not significantly impact pH (Kirchmann et al., 2013). The impact of

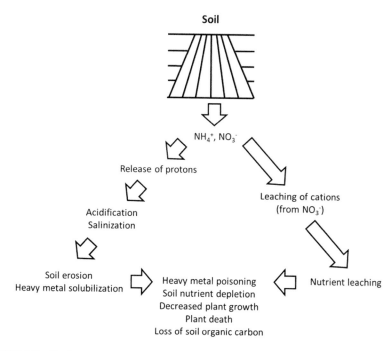

**FIGURE 6.5**   Environmental impacts of excess reactive nitrogen on the soil system.

N fertilizers on soil acidity is widespread, with a meta-analysis undertaken in 2015 finding that, globally, soil pH had decreased by 0.26 on average due to nitrogen addition (Tian and Niu, 2015), with some areas being hit harder, such as China, which observed a decrease of 0.5 between the 1980s and 2000s (Guo et al., 2010).

Soil acidification has multiple environmental consequences: It is harmful to plant and soil organism life, can lead to soil erosion (section 6.3.4.2) and nutrient leaching and depletion (section 6.3.4.3). Regarding the direct impact of soil acidification, the main impact on plant life can be a degradation of root health, leading to smaller and weaker roots, which can significantly inhibit plant growth, while also inhibiting seed germination (Basto et al., 2015; Long et al., 2017). One of the most harmful consequences of soil acidification is, however, the release of toxic elements that become soluble at low pH (<5), such as aluminum. Aluminum is particularly problematic given that it is present in high quantities in soils, is an important component of clay and one of the most abundant elements in the earth's crust (8.1% of its weight) (Bojórquez-Quintal et al., 2017). When released, plants exhibit very high toxicity, leading to impacts on leaves, roots, cell division, respiration, metabolism; released aluminum also interferes with the uptake and use of nutrients such as calcium, magnesium, and phosphorus (Bojórquez-Quintal et al., 2017). Furthermore, acidic water from acid rain – potentially caused by nitrogen emissions, as discussed in section 6.3.2 – can lower the pH of plants, leading them to release important nutrients and minerals into the environment through electrolyte leakage (Long et al., 2017), which can lead to plant death, as well as nutrient leaching and loss. However, plants are not the only ones affected by these low pH conditions. Soil microorganisms are also harmed in acidic conditions. This can lead to a loss of microbial, plant, and animal diversity, the ecological consequences of which are discussed in section 6.3.5.

### 6.3.4.2 Soil Salinization

Similar to acidification, the addition of N fertilizers can contribute to increasing the salinity of soils (Han et al., 2015; Litalien and Zeeb, 2020). The release of protons from N fertilizers, as discussed in the previous section, releases basic cations that accelerate soil salinization (Han et al., 2015). Salinization is very detrimental to plant life, causing severe physiological, morphological, biochemical, and molecular changes that limit their growth, while also changing soil morphology, favoring erosion (section 6.3.4.3) and reducing water infiltration and retention (Litalien and Zeeb, 2020).

### 6.3.4.3 Soil Erosion

One of the primary consequences of soil acidification and salinization is the reduction in root size of plants and trees and loss of soil microorganisms and plant diversity. This leads to another major impact of the application of nitrogen fertilizers: soil erosion. Soil erosion is a form of soil degradation by which topsoil, rich in organic carbon and nutrients, is removed and relocated elsewhere, notably in waterways, leaving behind soil of poor quality and fertility. Although soil erosion is a natural process caused by the erosive nature of water, ice, snow, wind, and animals, the application of fertilizers, notably nitrogen fertilizers, can greatly enhance and quicken this process (Lal, 2007; Zhang et al., 2019a). By acidifying the environment, thus decreasing soil microorganisms and reducing the size and strength of plant roots (section 6.3.4.1), N fertilizers can destroy the structures that are needed to bind the soil together. As a consequence of soil erosion, it is estimated that 23 to 42 billion tonnes of nitrogen is moved per year (Quinton et al., 2010), a significant amount when compared to the 120 million tonnes applied annually through mineral fertilizers. This has significant environmental, economic and social impacts, leading to loss of arable land, desertification, water pollution, vulnerability to flooding, an increased susceptibility to climate change, and threatening the food supply (Pimentel and Burgess, 2013; Sartori et al., 2019). Furthermore, soil erosion can also render the organic carbon and nitrogen in the transported soil easily available for biodegradation (outside of the agricultural system), leading to enhanced GHG emissions in the form of $CO_2$, $CH_4$, and $N_2O$ (Chappell et al., 2016; Lal, 2019).

It is important to note, however, that N-fertilizer application will not necessarily lead to soil erosion. In fact, many studies have demonstrated that nitrogen application can increase soil organic matter content and soil microbial biomass (Geisseler and Scow, 2014; Singh, 2018). However, as mentioned at the beginning of section 6.3, it is the overapplication or unbalanced application of N-fertilizers that can lead to such consequences, which is sadly happening in many places throughout the world (Walling and Vaneeckhaute, 2020b). For example, N application alone, even in N-deficient soils, has been found to have little to no impact on, or even suppresses, microbial diversity and crop yields, especially when pH is below 5, with benefits mainly being observed when applied alongside other nutrients or at higher pH values (Geisseler and Scow, 2014).

### 6.3.4.4   Soil Nutrient Leaching and Depletion

Ironically, the use of nitrogen fertilizers can lead to nutrient depletion in soils. This process, in which nutrients, both macro, and micro, are lost from soils, is mainly caused by soil erosion, leaching, and unbalanced fertilizer application, all of which can be exacerbated by nitrogen fertilizers, as discussed previously.

Direct nutrient depletion can happen through leaching, when nitrate (an anion) is free to move through the soil, where it can tie itself to cations such as calcium and many other nutrients, thus taking these out of the soil and potentially ending up in aquatic systems (Lehmann and Schroth, 2003), further contributing to eutrophication as well (section 6.3.3.2). Furthermore, though not only caused by nitrogen fertilizers, the addition of unbalanced fertilizers, without providing adequate quantity and variety of other macro and micronutrients, can lead to significant soil nutrient depletion. In such situations, the application of nitrogen fertilizers can enhance plant growth, consuming other nutrients as well, such as phosphorus, potassium, calcium, magnesium, nickel, sodium, sulfur, and so forth. However, this increased yield is ephemeral if these nutrients are not returned to the soil in some form (Johnston and Bruulsema, 2014; Mulvaney et al., 2009; Van der Velde et al., 2014). There is also the impact of soil erosion (section 6.3.4.3), which can also lead to significant nutrient loss and depletion, particularly of nitrogen, which is present in the highest concentrations in topsoil (Bashagaluke et al., 2018).

The consequence of soil nutrient depletion is a reduction in soil fertility, as well as a modification of soil microbial communities. In the case of agriculture, this can put the food supply at risk, reducing crop yields and enhancing desertification (Jones et al., 2013). Regarding microbial communities, a lack of nutrients can reduce microbial diversity and activity (Ma et al., 2019; Trivedi et al., 2016), as will be explored in section 6.3.5, although more problematic cases can lead to the emergence of harmful organisms, such as *Burkholderia pseudomallei*, which causes melioidosis and is responsible for nearly 90.000 deaths yearly (Hantrakun et al., 2016).

### 6.3.4.5   Heavy Metal Accumulation, Solubility, and Availability

Nitrogen fertilization can influence the state of heavy metals in soils in a variety of ways, starting with the application of these fertilizers. Heavy metal contamination from mineral fertilizers is most often associated with phosphorus, though arsenic (As) and lead (Pb) have been reported in nitrogen fertilizers (Gambuś and Wieczorek, 2012; Huang and Jin, 2008), as well as a positive correlation between N content and cadmium (Cd), chromium (Cr) and zinc (Zn) in Nitrogen, Phosphorus, and Potassium (NPK) fertilizers (Gambuś and Wieczorek, 2012). However, organic fertilizers, such as manure, composts, sludges, and digestates can have significantly higher heavy metal contents, especially on a nitrogen basis, making them an important source of pollution for soils (Chen et al., 2009; Li and Wu, 2008). If not controlled properly, the application of these fertilizers can lead to a significant accumulation of heavy metals in agricultural soils (Atafar et al., 2010).

Following application and potential accumulation, as detailed in section 6.3.4.1, the change in soil pH can also lead to a transformation in the state and availability of heavy metals, such as Cd, Pb, copper (Cu), Cr, mercury (Hg), As, and Zn (Atafar et al., 2010; Chao et al., 2020). A reduction in pH can increase the solubility of heavy metals, making them available for uptake by plants (Zhou, 2003), thus compounding the effect of accumulation and making food consumption the primary pathway for human exposure to heavy metals (Han et al., 2018; Jaishankar et al., 2014). Furthermore, their increased solubility or a low pH of the leachate can lead to them reaching ground and surface waters, making their way into wider ecosystems and affecting the drinking water supply (Kubier et al., 2020).

Regarding their impact, heavy metals are both environmentally and ecologically damaging. From an environmental perspective, they can be detrimental to plant life and crop growth, causing oxidative stress that leads to cellular damage (Singh et al., 2011). From an ecological perspective, heavy metals are especially dangerous for human and animal health. Lead, which has been most associated with nitrogen fertilizers (Gambuś and Wieczorek, 2012; Huang and Jin, 2008), is detrimental to cardiovascular, renal and reproductive health, being particularly harmful to children and their mental development (Han et al., 2018; Jaishankar et al., 2014). Furthermore, metals such as As and Cd exhibit strong toxicity, even at low concentrations, leading to bladder, lung, and prostate cancers in the case of As (Jaishankar et al., 2014), and cardiovascular, renal, gastrointestinal, neurological, reproductive, and respiratory cancers and diseases for Cd (Jaishankar et al., 2014), from concentrations that can be found in contaminated drinking water or food (Han et al., 2018; Liang et al., 2016; Zhou et al., 2018). For example, a recent and large-scale study of arsenic in agricultural soils across China found that 35.48 percent of soils presented a moderate cancer risk for adults, and 8.06 percent for children, without considering other heavy metals and the potential contribution to groundwater contamination (Zhou et al., 2018), while another study in Taiwan found that 77.7 to 93.3 percent of residents of the Pingtung Plain, an extremely productive agricultural area, were at an increased risk of developing cancer due to As contamination of groundwater (Liang et al., 2016).

### 6.3.4.6 Loss of Soil Organic Carbon

Regardless of whether growth of plant life is hindered or stopped by any of the processes mentioned throughout section 6.3.4, the loss of this life can have significant impacts on soil organic carbon stocks. The impact of soil organic carbon loss goes beyond the domain of agriculture and a loss of fertility; it degrades an important carbon sink, further accelerating global warming and climate change (Chappell et al., 2016; Setia et al., 2013; Wiesmeier et al., 2016). For example, it is estimated that, since becoming saline, saline soils have lost 3.47 tonnes of soil organic carbon per hectare (Setia et al., 2013), while a decrease in agricultural soil organic carbon stocks in Bavaria would increase agricultural $CO_2$ emissions by 4 to 12 percent on a yearly basis between 2000 and 2095 (Wiesmeier et al., 2016).

### 6.3.5 ECOLOGICAL IMPACT

Beyond the direct impacts highlighted in the previous sections, reactive nitrogen can have deep ecological consequences. It is only relatively recently (with the development of nitrogen fertilizers), that reactive nitrogen has become so accessible around the globe. Consequently, most natural and semi-natural ecosystems are not adapted to the increased nitrogen availability, having evolved in low (reactive) nitrogen environments (Sheppard et al., 2011). Therefore, this relatively newfound abundance of nitrogen has caused significant stress to many ecosystems. As discussed in sections 6.3.2 and 6.3.3, nitrogen has the ability to be transported and deposited over great distances, meaning that its impacts are not localized. Consequently, the N cycle in many ecosystems is now driven by exterior, anthropogenic, sources (Erisman et al., 2008; Zhang et al., 2020). In the previous sections, many of the environmental consequences also had associated health consequences for plant and animal life, such as eutrophication, emission of fine particles, acidification of soil and waterways, the impacts of which can cascade throughout an ecosystem, as will be discussed in this section.

The most direct of these consequences is a loss of biodiversity due to N fertilizer application. This loss of biodiversity can impact everything from microorganisms to macrofauna, and also tends to have a more serious impact on rarer and more sensitive species (de Graaff et al., 2019; Ding et al., 2016; Guthrie et al., 2018; Nordin et al., 2005; Wang et al., 2016). This phenomenon happens due to the reactive and easily accessible nitrogen that ends up in ecosystems following nitrogen application, be it from direct application to soils, atmospheric deposition, or transportation through leaching and erosion. Therefore, many species, in both terrestrial and aquatic environments, find themselves unable to compete with more common and homogenous species that are adapted to high nitrogen availability, causing the latter to quickly outgrow and outcompete the prior. This impact can be further compounded by the effects of eutrophication, acidification, and toxicity, as discussed previously. Consequently, certain ecosystems, especially those that are naturally nitrogen-limited, such as alpine and boreal forests, can be particularly sensitive to nitrogen influxes and can witness an extensive loss in biodiversity and even extinction of many species in favor for nitrogen-favoring lifeforms (Bobbink et al., 2010; Dirnböck et al., 2014; Zhang et al., 2018). As such, the nitrogen-tolerant species have expanded rapidly in recent decades, at the expense of the other species (Dirnböck et al., 2014). One does not have to look far or on a global scale to see the impact of nitrogen on biodiversity: This impact can be noted rather starkly when looking at woodlands near major ammonia emitters, such as livestock farms. In such cases, it is not uncommon to find a significantly reduced diversity in close proximity to the emission source, with diversity increasing along with distance (Frati et al., 2007).

Regarding fauna, these impacts can be felt directly at all levels throughout the ecosystem and generally stem from one of six "bottlenecks" that can influence diversity, as identified by Nijssen and colleagues (2017):

(1) chemical stress, (2) [changes in microclimates], (3) decrease in reproductive habitat, (4) changes in food plant quantity, (5) changes in the nutritional quality of

food plants and (6) changes in the availability of prey or host species due to cumulative effects in the food web.

Starting at the microbial level, the addition of nitrogen to ecosystems will most often lead to a decrease in microbial biomass, both bacterial and fungal, microbial respiration, and microbial diversity – effects that increase alongside N application, be it from the direct land application or atmospheric deposition (Zhang et al., 2018). A similar impact is observed for larger soil organisms as well, such as nematodes and earthworms (Blakemore, 2018; Eisenhauer et al., 2012). These negative consequences can be reversed when looking at certain insects, with the application of nitrogen fertilizer leading to increased larval survival rates, longevity, fecundity, and weight of certain species (De Kraker et al., 2000; Throop and Lerdau, 2004). However, this is rarely beneficial, given that it can drastically increase the population of insect pests, affecting agricultural yields, the environment, the economy (notably the agricultural and timber industries), and human health (de Sassi and Tylianakis, 2012). Furthermore, many ground-dwelling animals, especially invertebrates and ectotherms, can find themselves at risk due to an increase in vegetation covers and litter accumulation created by N deposition. These changes in conditions can lead to the creation of microclimates, decreasing sunlight and air circulation near the ground, causing colder and more humid microclimates (Nijssen et al., 2017; Wallisdevries and Van Swaay, 2006). This effect has been noted as being one of the leading causes in loss of certain butterfly species (Öckinger et al., 2006; Wallisdevries and Van Swaay, 2006; Weiss, 1999), though its impact is likely much larger, but understudied. This increase in vegetation and litter can also decrease the reproductive habitat for ground-dwelling species, such as ants, butterflies, grasshoppers, and breeding birds (Nijssen et al., 2017). There is also the impact of acidification which, similar to its impact on waterways (section 6.3.3.4), can weaken and lead to the loss of calcium-rich animals (Pabian et al., 2012), such as snails, isopods, and diplopods, which effects can cascade through the food chain and affect their predators. Known consequences of this have been specially studied for birds, finding a reduction in the strength of bird eggs and bone development (Pabian et al., 2012), though such impacts should be expected in many other predatory species. There is also the notable impact of some of the processes mentioned in the prior sections, such as eutrophication, which can kill off massive quantities of aquatic life; acidification of waterways, which can lead to metabolic issues and potentially the death of marine organisms and insects; and harmful algal blooms, which can be toxic to terrestrial life as well.

Regarding floral ecology, the impacts of nitrogen fertilizers are many: the presence of reactive nitrogen has been noted as enhancing the growth of invasive species, significantly decreasing the biodiversity of lichen communities, altering the forest N and C cycles, especially in sensitive systems such as boreal and alpine forests, promoting forest expansion into grasslands and altering the fire cycle (Fenn et al., 2003). Aside from the impacts mentioned earlier related to soil acidification, salinization, erosion, nutrient depletion, and heavy metal solubilization, plant species can be particularly susceptible to nitrogen deposition. Nitrogen deposition has been found to be toxic to certain plant species by causing physiological changes that either damage the plants directly or cause them to be unable to respond to abiotic and biotic stressors (Bobbink et al., 2010; Sheppard et al., 2011). Nitrogen pollution has also been found to decrease flowering in plants, which impacts both pollination and flower-visiting animals (Hoover et al., 2012; Zhang et al., 2019c). The most sensitive systems tend to be those that evolved in naturally nitrogen-limited conditions, such as boreal forests (Nordin et al., 2005), as well as lichens, which are extremely susceptible to nitrogen (Hauck et al., 2013). However, this does not mean that the impact is non-existent in other ecosystems. For example, a global meta-analysis on the impact of N enrichment on herbaceous communities found decreases in plant species richness of up to 75 percent, with an average reduction of 16 percent (Soons et al., 2017).

On their own, the above-mentioned consequences of nitrogen on ecosystems are problematic enough, but this impact is worsened by how interrelated are actors in ecosystems, allowing for these effects to grow and cascade throughout the ecosystem and on multiple levels. An excellent example of this is provided by Guthrie and colleagues (2018), where the increased growth in vegetation can cause a slower development and longer lifecycles for invertebrates, leading to risks that they will not be able to complete their lifecycle in one season, notably amongst larger insect species (Guthrie et al., 2018). This will therefore favor the development of smaller species that prefer these cold and damp conditions, potentially removing important actors in the food chain. This has been observed and directly attributed to nitrogen pollution in some cases, such as the red-backed shrike, which has seen significant population loss in areas with nitrogen deposition due to a loss of prey but has remained healthy and stable outside of these areas (Kuper et al., 2000). Indeed, this impact of enhanced nitrogen on the food web has been identified as one of the main contributors to the loss of biodiversity in ecosystems (Eisenhauer et al., 2012). A plethora of species has seen their food sources impacted, especially species that are selective feeders, resulting in significant stress. This has included butterflies, pollinators such as bees, grasshoppers, and birds, though the impact of nutrients can be felt throughout the food chain (de Sassi and Tylianakis, 2012).

## 6.4  PERSPECTIVE: TOWARD MORE SUSTAINABLE AGRICULTURE

Despite the environmental consequences of nitrogen fertilizers discussed throughout this chapter, it is important to remember that they are a necessary part of human society, now more than ever. Most of these environmental impacts are due to improper fertilizer management and are not inherent to N-fertilizer production and use. Therefore, implementing and continuously developing better fertilization and mitigation strategies is of utmost importance, especially given how our reliance on nitrogen will only increase with time. Fortunately, work in a variety of fields is aiming at addressing this situation.

Improved and novel methods of ammonia synthesis have seen great interest, aiming at transitioning away from the massive energetic footprint of the Haber-Bosch process. This can include better energy management of the production process through optimized heat recovery and energy efficiency, which could potentially decrease energy requirements by up to 15 percent (Panjeshahi et al., 2008), and transitioning toward greener energy sources and replacing the methane used by the process with renewable sources such as syngas or biogas from biomass gasification or anaerobic digestion, or even hydrogen, further reducing the environmental footprint of the process (Arora, 2017; Tunå et al., 2014). Researchers have also been investigating alternative ammonia synthesis processes, with notable interest surrounding photocatalytic, electrocatalytic, and plasma-assisted processes (Hao et al., 2020; Peng et al., 2018; Soloveichik, 2019).

Though these approaches show great promise, it is unlikely that the Haber-Bosch process will be replaced in the short to medium term. Therefore, one area that can have a positive impact in the short term is the recovery of nitrogen from wastewaters and organic wastes. This can be achieved through nutrient recovery technologies, which allow for the recovery of both organic and mineral fertilizers from these waste streams (Walling et al., 2019). These processes are being implemented more and more throughout the world, and include processes such as ammonia stripping and absorption, ion exchange technologies, and (bio)electrochemical systems, all of which can recover $NH_3$ and be used to produce ammonia fertilizers, alongside a variety of other organic and biobased fertilizers (Walling et al., 2019).

In either case, the transition toward carbon-free ammonia synthesis, which could decrease $CO_2$ emissions from $NH_3$ production by approximately 75 percent (Smith et al., 2020), could also create a massive demand for $CO_2$, assuming that urea demand remains at current levels. This could present a strong driver for further carbon capture and valorization, having a 150 Mt of $CO_2$ gap to fill (Smith et al., 2020).

Regarding fertilizer application, many guidelines exist to limit their environmental impact, such as the 4R Nutrient Stewardship guideline, which promotes the use of the Right fertilizer at the Right time at the Right rate and in the Right place (Johnston and Bruulsema, 2014). Proper fertilizer management is key to ensuring sustainable agriculture, with best management practices regarding nitrogen fertilizers being presented in Table 6.2, based on the information presented throughout this chapter and further inspired by the works of Snyder and colleagues (2009), Johnston and Bruulsema (2014), Snyder (2017), Guthrie and colleagues (2018), and Tei and colleagues (2020). However, social and legislative support and pressure, applied at national and multi-national scales, will be necessary to increase efficient nitrogen use and ensure clean and sustainable fertilization, while also considering the economic, technical, and social realities of agricultural and environmental decision-making (Guthrie et al., 2018; Walling and Vaneeckhaute, 2020a). Though some of these practices can seem as a significant (and thus difficult) change, they can also lead to considerable economic benefits, with a transition toward reduced nitrogen use and machine-assisted deep placement, among other management practices, providing an estimated US$12 billion per year in savings in China alone (Guo et al., 2020). Furthermore, soil systems have shown that they are capable of returning to normal once proper agronomic practices are in place and nitrogen pollution is reversed, though this can process can take time (Stevens, 2016; Zhang et al., 2020) and, as mentioned in section 6.3.3, the situation for waterways is much more complex.

**TABLE 6.2**
**Best Management Practices (BMPs) for Nitrogen Fertilizer Application to Reduce Environmental and Ecological Impact**

| BMP | Specific examples |
|---|---|
| Select the best source of N to meet the needs of the crops | • Consider slow-release fertilizers such as urea-formaldehyde, isobutylenediurea, and struvite.<br>• Consider controlled-release fertilizers which have coatings that limit the rate of solubilization of the fertilizer. |
| Base N application rates on crop and soil requirements | • Establish N fertilizer needs.<br>• Consider soil (accessible) N supply and the N content of all inputs (amendments, fertilizers, irrigation, deposition).<br>• Consider N-fixing from certain plants, notably legumes.<br>• Consider N losses from the field.<br>• Use technologies such as nitrogen sensors to assess soil needs and crop uptake.<br>• Use variable application rates based on crop uptake needs.<br>• Calibrate fertilizer application equipment for optimal delivery (quantity and location). |
| Time N application with crop N uptake demand | • Use technologies such as nitrogen sensors to assess soil needs and crop uptake.<br>• Plan application to coincide with peak uptake and critical growth stages of crops.<br>• Split application of N-fertilizer to favor a more flexible fertilizer program. |
| Ensure appropriate placement of N fertilizers | • Place N sources in or near seed rows during the early season.<br>• Restrict subsurface placement depth based on the N source applied. |
| Ensure adequate soil conditions to receive N | • Avoid application of N to wet or waterlogged soils.<br>• Avoid application of $NO_3$-N sources to highly permeable soils or soils with drainage to avoid contamination of waterways.<br>• Avoid application of $NH_4$-N sources to high pH soils, as well as during warm, dry and windy conditions. |
| Use inhibitors to limit unwanted transformation of reactive nitrogen | • Reduce $NH_3$ emissions through urease inhibitors.<br>• When in an environment with high potential for $NO_3$-N leaching and/or $N_2O$ emissions (humid/wet environments, high rainfall, high $NH_4$-N application rates), use nitrification inhibitors alongside ammoniacal N sources. |
| Provide balanced nutrients | • Focus fertilization strategy on all nutrients, not simply N. N application without appropriate P and K and micronutrients availability will provide little to no benefit. |

## 6.5 CONCLUSION

Moving forward, it is primordial for our continued survival and development that we ensure that soils and crops receive the necessary nutrients to support their health over the long-term. Global warming, soil erosion, acidification of soils and waterways, and destruction of ecosystems, among many others, are some of the current and critical environmental consequences we deal with, in part due to nitrogen fertilizers. However, nitrogen fertilizers on their own are not responsible for most of these issues, and the transition toward sustainable agriculture through responsible fertilization is within reach, as long as we, as a global community, work toward reducing and limiting our impact on the global nitrogen cycle and ensure that the nitrogen we apply remains in soils and crops.

## REFERENCES

Amorim, C.A., do Nascimento Moura, A., 2020. Ecological impacts of freshwater algal blooms on water quality, plankton biodiversity, structure, and ecosystem functioning. *Sci. Total Environ.*, 143605. doi:https://doi.org/10.1016/j.scitotenv.2020.143605.

Arora, P., 2017. Techno-enviro-economic evaluations of biomass gasification for the production of ammonia from synthesis gas utilizing multi-scale modeling. Bombay: The India Institute of Technology-Monash Research Academy.

Atafar, Z., Mesdaghinia, A., Nouri, J., Homaee, M., Yunesian, M., Ahmadimoghaddam, M., Mahvi, A.H., 2010. Effect of fertilizer application on soil heavy metal concentration. *Environ. Monit. Assess.* 160, 83. doi:https://doi.org/10.1007/s10661-008-0659-x.

Backer, L.C., McNeel, S.V., Barber, T., Kirkpatrick, B., Williams, C., Irvin, M., Zhou, Y., Johnson, T.B., Nierenberg, K., Aubel, M., 2010. Recreational exposure to microcystins during algal blooms in two California lakes. *Toxicon* 55, 909–921. doi:https://doi.org/10.1016/j.toxicon.2009.07.006.

Banger, K., Tian, H., Lu, C., 2012. Do nitrogen fertilizers stimulate or inhibit methane emissions from rice fields? *Glob. Change Biol.* 18, 3259–3267. doi:https://doi.org/10.1111/j.1365-2486.2012.02762.x.

Bashagaluke, J.B., Logah, V., Opoku, A., Sarkodie-Addo, J., Quansah, C., 2018. Soil nutrient loss through erosion: Impact of different cropping systems and soil amendments in Ghana. *Plos one* 13, e0208250. doi:https://doi.org/10.1371/journal.pone.0208250.

Basto, S., Thompson, K., Phoenix, G., Sloan, V., Leake, J., Rees, M., 2015. Long-term nitrogen deposition depletes grassland seed banks. *Nat. Commun.* 6, 1–6. doi:https://doi.org/10.1038/ncomms7185.

Blakemore, R.J., 2018. Critical decline of earthworms from organic origins under intensive, humic SOM-depleting agriculture. *Soil Syst.* 2, 33. doi:https://doi.org/10.3390/soilsystems2020033.

Bobbink, R., Hicks, K., Galloway, J., Spranger, T., Alkemade, R., Ashmore, M., Bustamante, M., Cinderby, S., Davidson, E., Dentener, F., 2010. Global assessment of nitrogen deposition effects on terrestrial plant diversity: a synthesis. *Ecol. Appl.* 20, 30–59. doi:https://doi.org/10.1890/08-1140.1.

Bohman, B., 2018. Lessons from the regulatory approaches to combat eutrophication in the Baltic Sea region. *Mar. Pol.* 98, 227–236. doi:https://doi.org/10.1016/j.marpol.2018.09.011.

Bojórquez-Quintal, E., Escalante-Magaña, C., Echevarría-Machado, I., Martínez-Estévez, M., 2017. Aluminum, a friend or foe of higher plants in acid soils. *Front. Plant Sci.* 8, 1767. doi:https://doi.org/10.3389/fpls.2017.01767.

Brady, N.C., Weil, R.R., Weil, R.R., 2008. *The nature and properties of soils*. Upper Saddle River, NJ: Prentice Hall.

Cai, W.-J., Hu, X., Huang, W.-J., Murrell, M.C., Lehrter, J.C., Lohrenz, S.E., Chou, W.-C., Zhai, W., Hollibaugh, J.T., Wang, Y., 2011. Acidification of subsurface coastal waters enhanced by eutrophication. *Nat. Geosci.* 4, 766–770. doi:https://doi.org/10.1038/ngeo1297.

Canfield, D.E., Glazer, A.N., Falkowski, P.G., 2010. The evolution and future of Earth's nitrogen cycle. *Science* 330, 192–196. doi:https://doi.org/10.1126/science.1186120.

Carmichael, W.W., Azevedo, S., An, J.S., Molica, R., Jochimsen, E.M., Lau, S., Rinehart, K.L., Shaw, G.R., Eaglesham, G.K., 2001. Human fatalities from cyanobacteria: chemical and biological evidence for cyanotoxins. *Environ. Health Perspect.* 109, 663–668. doi:https://doi.org/10.1289/ehp.01109663.

Carstensen, J., Andersen, J.H., Gustafsson, B.G., Conley, D.J., 2014. Deoxygenation of the Baltic Sea during the last century. *Proc. Natl. Acad. Sci. U.S.A.* 111, 5628–5633. doi:https://doi.org/10.1073/pnas.1323156111.

Chao, X., Xiang, Q., Qihong, Z., Hanhua, Z., Huang, D., Zhang, Y., 2020. Effect of controlled-release urea on heavy metal mobility in a multimetal-contaminated soil. *Pedosphere* 30, 263–271. doi:https://doi.org/10.1016/S1002-0160(17)60467-3.

Chappell, A., Baldock, J., Sanderman, J., 2016. The global significance of omitting soil erosion from soil organic carbon cycling schemes. *Nat. Clim. Change* 6, 187–191. doi:https://doi.org/10.1038/nclimate2829.

Chen, L., Ni, W., Li, X., Sun, J., 2009. Investigation of heavy metal concentrations in commercial fertilizers commonly used. *J. Zhejiang Sci-Tech Univ.* 26, 223–227.

Chen, Y., Shen, H., Shih, J.S., Russell, A.G., Shao, S., Hu, Y., Odman, M.T., Nenes, A., Pavur, G.K., Zou, Y., 2020. Greater contribution from agricultural sources to future reactive nitrogen deposition in the United States. *Earth's Future*, e2019EF001453. doi:https://doi.org/10.1029/2019EF001453.

Codd, G.A., Testai, E., Funari, E., Svirčev, Z., 2020. Cyanobacteria, cyanotoxins, and human health. doi:https://doi.org/10.1002/9781118928677.ch2.

De Graaff, M.-A., Hornslein, N., Throop, H.L., Kardol, P., van Diepen, L.T., 2019. Effects of agricultural intensification on soil biodiversity and implications for ecosystem functioning: a meta-analysis. In: Sparks, D. (Ed.), *Advances in Agronomy*. Amsterdam and Boston: Academic Press, pp. 1–44. doi:https://doi.org/10.1016/bs.agron.2019.01.001.

De Kraker, J., Rabbinge, R., Van Huis, A., Van Lenteren, J., Heong, K., 2000. Impact of nitrogenous-fertilization on the population dynamics and natural control of rice leaffolders (Lep.: Pyralidae). *Int. J. Pest Manage.* 46, 225–235. doi:https://doi.org/10.1080/096708700415571.

De Sassi, C., Tylianakis, J.M., 2012. Climate change disproportionately increases herbivore over plant or parasitoid biomass. *PLoS One* 7, e40557. doi:https://doi.org/10.1371/journal.pone.0040557.

Diaz, R., Selman, M., Chique-Canache, C., 2008. *Global Eutrophic and Hypoxic Coastal Systems: Eutrophication and Hypoxia–Nutrient Pollution in Coastal Waters*. Washington, DC: World Resource Institute.

Ding, J., Jiang, X., Ma, M., Zhou, B., Guan, D., Zhao, B., Zhou, J., Cao, F., Li, L., Li, J., 2016. Effect of 35 years inorganic fertilizer and manure amendment on structure of bacterial and archaeal communities in black soil of northeast China. *Appl. Soil Ecol.* 105, 187–195. doi:https://doi.org/10.1016/j.apsoil.2016.04.010.

Dirnböck, T., Grandin, U., Bernhardt-Römermann, M., Beudert, B., Canullo, R., Forsius, M., Grabner, M.T., Holmberg, M., Kleemola, S., Lundin, L., 2014. Forest floor vegetation response to nitrogen deposition in Europe. *Global Change Biol.* 20, 429–440. doi:https://doi.org/10.1111/gcb.12440.

Doney, S.C., Mahowald, N., Lima, I., Feely, R.A., Mackenzie, F.T., Lamarque, J.-F., Rasch, P.J., 2007. Impact of anthropogenic atmospheric nitrogen and sulfur deposition on ocean acidification and the inorganic carbon system. *Proc. Natl. Acad. Sci. U.S.A.* 104, 14580–14585. doi:https://doi.org/10.1073/pnas.0702218104.

Eisenhauer, N., Cesarz, S., Koller, R., Worm, K., Reich, P.B., 2012. Global change belowground: impacts of elevated CO2, nitrogen, and summer drought on soil food webs and biodiversity. *Global Change Biol.* 18, 435–447. doi:https://doi.org/10.1111/j.1365-2486.2011.02555.x.

Erisman, J.W., Sutton, M.A., Galloway, J., Klimont, Z., Winiwarter, W., 2008. How a century of ammonia synthesis changed the world. *Nat. Geosci.* 1, 636–639. doi:https://doi.org/10.1038/ngeo325.

FAOSTAT, F.a.A.O.o.t.U.N., 2017. World fertilizer trends and outlook to 2020; available at: http://www.fao.org/3/a-i6895e.pdf.

Fenn, M.E., Baron, J.S., Allen, E.B., Rueth, H.M., Nydick, K.R., Geiser, L., Bowman, W.D., Sickman, J.O., Meixner, T., Johnson, D.W., 2003. Ecological effects of nitrogen deposition in the western United States. *BioSci.* 53, 404–420. doi:https://doi.org/10.1641/0006-3568(2003)053[0404:EEONDI]2.0.CO;2.

Fitzer, S.C., Chung, P., Maccherozzi, F., Dhesi, S.S., Kamenos, N.A., Phoenix, V.R., Cusack, M., 2016. Biomineral shell formation under ocean acidification: A shift from order to chaos. *Sci. Rep.* 6, 21076. doi:https://doi.org/10.1038/srep21076.

Fostås, B., Gangstad, A., Nenseter, B., Pedersen, S., Sjøvoll, M., Sørensen, A.L., 2011. Effects of NOx in the flue gas degradation of MEA. *Energy Procedia* 4, 1566–1573. doi:https://doi.org/10.1016/j.egypro.2011.02.026.

Francis, G., 1878. Poisonous Australian lake. *Nature* 18, 11–12.

Frati, L., Santoni, S., Nicolardi, V., Gaggi, C., Brunialti, G., Guttova, A., Gaudino, S., Pati, A., Pirintsos, S., Loppi, S., 2007. Lichen biomonitoring of ammonia emission and nitrogen deposition around a pig stockfarm. *Environ. Pollut.* 146, 311–316. doi:https://doi.org/10.1016/j.envpol.2006.03.029.

Fry, M.M., Naik, V., West, J.J., Schwarzkopf, M.D., Fiore, A.M., Collins, W.J., Dentener, F.J., Shindell, D.T., Atherton, C., Bergmann, D., 2012. The influence of ozone precursor emissions from four world regions on tropospheric composition and radiative climate forcing. *J. Geophys.l Res. Atmos.* 117. doi:https://doi.org/10.1029/2011JD017134.

Funari, E., Testai, E., 2008. Human health risk assessment related to cyanotoxins exposure. *Crit. Rev. Toxicol.* 38, 97–125. doi:https://doi.org/10.1080/10408440701749454.

Gambuś, F., Wieczorek, J., 2012. Pollution of fertilizers with heavy metals. *Ecol. Chem. Eng. A* 19, 353–360. doi:https://doi.org/10.2428/ecea.2012.19(04)036.

Gattuso, J.P., Brewer, O., Hoegh-Guldberg, O., Kleypas, J.A., Pörtner, H.-O., Schmidt, D.N., 2014. Cross-chapter box on ocean acidification. In: Field, C.B., V.R. Barros, D.J. Dokken, K.J. Mach, M.D. Mastrandrea, T.E. Bilir, M. Chatterjee, K.L. Ebi, Y.O. Estrada, R.C. Genova, B. Girma, E.S. Kissel, A.N. Levy, S. MacCracken, P.R. Mastrandrea, and L.L. White (Ed.), *Climate Change 2014: Impacts, Adaptation, and Vulnerability. Part A: Global and Sectoral Aspects. Contribution of Working Group II to the Fifth Assessment Report of the Intergovernmental Panel on Climate Change.* Cambridge and New York: Cambridge University Press, pp. 129–131.

Geisseler, D., Scow, K.M., 2014. Long-term effects of mineral fertilizers on soil microorganisms–A review. *Soil Biol. Biochem.* 75, 54–63. doi:https://doi.org/10.1016/j.soilbio.2014.03.023.

Gobler, C.J., 2020. Climate change and harmful algal blooms: insights and perspective. *Harmful Algae* 91, 101731. doi:https://doi.org/10.1016/j.hal.2019.101731.

Grewe, V., Matthes, S., Dahlmann, K., 2019. The contribution of aviation NO x emissions to climate change: are we ignoring methodological flaws? *Environ. Res. Lett.* 14. doi:https://doi.org/10.1088/1748-9326/ab5dd7.

Guo, J.H., Liu, X.J., Zhang, Y., Shen, J.L., Han, W.X., Zhang, W.F., Christie, P., Goulding, K., Vitousek, P.M., Zhang, F., 2010. Significant acidification in major Chinese croplands. *Science* 327, 1008–1010. doi:https://doi.org/10.1126/science.1182570.

Guo, Y., Chen, Y., Searchinger, T.D., Zhou, M., Pan, D., Yang, J., Wu, L., Cui, Z., Zhang, W., Zhang, F., 2020. Air quality, nitrogen use efficiency and food security in China are improved by cost-effective agricultural nitrogen management. *Nat. Food* 1, 648–658. doi:https://doi.org/10.1038/s43016-020-00162-z.

Guthrie, S., Giles, S., Dunkerley, F., Tabaqchali, H., Harshfield, A., Ioppolo, B., Manville, C., 2018. *The Impact of Ammonia Emissions from Agriculture on Biodiversity*. Cambridge: RAND Corporation and the Royal Society.

Hall-Spencer, J.M., Harvey, B.P., 2019. Ocean acidification impacts on coastal ecosystem services due to habitat degradation. *Emerging Top. Life Sci.* 3, 197–206. doi:https://doi.org/10.1042/ETLS20180117.

Hallegraeff, G., 2003. Harmful algal blooms: a global overview. In: Hallegraeff, G., Anderson, D., Cembella, A. (Eds), *Manual on Harmful Marine Microalgae*. Paris: UNESCO Publishing, pp. 1–22.

Han, J., Shi, J., Zeng, L., Xu, J., Wu, L., 2015. Effects of nitrogen fertilization on the acidity and salinity of greenhouse soils. *Environ. Sci. Pollut. Res.* 22, 2976–2986. doi:https://doi.org/10.1007/s11356-014-3542-z.

Han, Z., Guo, X., Zhang, B., Liao, J., Nie, L., 2018. Blood lead levels of children in urban and suburban areas in China (1997–2015): Temporal and spatial variations and influencing factors. *Sci. Total Environ.* 625, 1659–1666. doi:https://doi.org/10.1016/j.scitotenv.2017.12.315.

Hantrakun, V., Rongkard, P., Oyuchua, M., Amornchai, P., Lim, C., Wuthiekanun, V., Day, N.P., Peacock, S.J., Limmathurotsakul, D., 2016. Soil nutrient depletion is associated with the presence of Burkholderia pseudomallei. *Appl. Environ. Microbiol.* 82, 7086–7092. doi:https://doi.org/10.1128/AEM.02538-16.

Hao, D., Chen, Z., Figiela, M., Stepniak, I., Wei, W., Ni, B.-J., 2020. Emerging alternative for artificial ammonia synthesis through catalytic nitrate reduction. *J. Mater. Sci. Technol.* doi:https://doi.org/10.1016/j.jmst.2020.10.056.

Harrison, A., 2020. *Africa Fertilizer Map 2020*. UK: Astrategia.

Hauck, M., de Bruyn, U., Leuschner, C., 2013. Dramatic diversity losses in epiphytic lichens in temperate broad-leaved forests during the last 150 years. *Biol. Conserv.* 157, 136–145. doi:https://doi.org/10.1016/j.biocon.2012.06.015.

Heffer, P., Prud'homme, M., 2016. Global nitrogen fertilizer demand and supply: Trend, current level and outlook. International Nitrogen Initiative Conference. Melbourne.

Hockstad, L., Hanel, L., 2018. *Inventory of US Greenhouse Gas Emissions and Sinks*. Environmental System Science Data Infrastructure for a Virtual Ecosystem.

Hoover, S.E., Ladley, J.J., Shchepetkina, A.A., Tisch, M., Gieseg, S.P., Tylianakis, J.M., 2012. Warming, CO2, and nitrogen deposition interactively affect a plant-pollinator mutualism. *Ecol. Lett.* 15, 227–234. doi:https://doi.org/10.1111/j.1461-0248.2011.01729.x.

Hristov, A.N., 2011. Contribution of ammonia emitted from livestock to atmospheric fine particulate matter (PM2. 5) in the United States. *J. Dairy Sci.* 94, 3130–3136. doi:https://doi.org/10.3168/jds.2010-3681.

Huang, J., Mo, J., Zhang, W., Lu, X., 2014. Research on acidification in forest soil driven by atmospheric nitrogen deposition. *Acta Ecol. Sin.* 34, 302–310. doi:https://doi.org/10.1016/j.chnaes.2014.10.002.

Huang, S.-W., Jin, J.-Y., 2008. Status of heavy metals in agricultural soils as affected by different patterns of land use. *Environ. Monit. Assess.* 139, 317. doi:https://doi.org/10.1007/s10661-007-9838-4.

Huisman, J., Codd, G.A., Paerl, H.W., Ibelings, B.W., Verspagen, J.M., Visser, P.M., 2018. Cyanobacterial blooms. *Nat. Rev. Microbiol.* 16, 471–483. doi:https://doi.org/10.1038/s41579-018-0040-1.

IFA (International Fertilizer Association), 2009. *Fertilizers, Climate Change and Enhancing Agricultural Productivity Sustainably.* Paris: IFA.

IFA (International Fertilizer Association), 2019. *Statistics, IFADATA.* Retrieved from: http://ifadata.fertilizer.org/ucSearch.aspx. Consulted on November 30, 2020.

Igwe, K., Uche, K., Obasi, U., 2016. Impact of Discharge Fertilizer Effluents on The Toxicological Profile of Fish Harvested from a Receiving Creek in Okirika, Rivers State. *Waste Technol.* 4, 15–17. doi:https://doi.org/10.14710/4.2.15-17.

ILEC, 2001. *Survey of the State of the World's Lakes, vols. 1-14.* Nairobi: International Lake Environment Committee, Otsu and United Nations Environment Programme.

Ip, A.Y., Chew, S.F., 2010. Ammonia production, excretion, toxicity, and defense in fish: a review. Front. Physiol. 1, 134. doi:https://doi.org/10.3389/fphys.2010.00134.

Jaishankar, M., Tseten, T., Anbalagan, N., Mathew, B.B., Beeregowda, K.N., 2014. Toxicity, mechanism and health effects of some heavy metals. *Interdisc. Toxicol.* 7, 60–72. doi:https://doi.org/10.2478/intox-2014-0009.

Jeffery, B., Barlow, T., Moizer, K., Paul, S., Boyle, C., 2004. Amnesic shellfish poison. *Food Chem. Toxicol.* 42, 545–557. doi:https://doi.org/10.1016/j.fct.2003.11.010.

Johnston, A., Bruulsema, T., 2014. 4R nutrient stewardship for improved nutrient use efficiency. *Procedia Eng.* 83, 365–370. doi:https://doi.org/10.1016/j.proeng.2014.09.029.

Jones, D.L., Cross, P., Withers, P.J., DeLuca, T.H., Robinson, D.A., Quilliam, R.S., Harris, I.M., Chadwick, D.R., Edwards-Jones, G., 2013. Nutrient stripping: the global disparity between food security and soil nutrient stocks. *J. Appl. Ecol.* 50, 851–862. doi:https://doi.org/10.1111/1365-2664.12089.

Kirchmann, H., Schön, M., Börjesson, G., Hamnér, K., Kätterer, T., 2013. Properties of soils in the Swedish long-term fertility experiments: VII. Changes in topsoil and upper subsoil at Örja and Fors after 50 years of nitrogen fertilization and manure application. *Acta Agric. Scand. Sect.* B 63, 25–36. doi:https://doi.org/10.1080/09064710.2012.711352.

Kroeker, K.J., Kordas, R.L., Crim, R., Hendriks, I.E., Ramajo, L., Singh, G.S., Duarte, C.M., Gattuso, J.P., 2013. Impacts of ocean acidification on marine organisms: quantifying sensitivities and interaction with warming. *Global Change Biol.* 19, 1884–1896. doi:https://doi.org/10.1111/gcb.12179.

Kubier, A., Hamer, K., Pichler, T., 2020. Cadmium background levels in groundwater in an area dominated by agriculture. *Integr. Environ. Assess. Manage.* 16, 103–113. doi: https://doi.org/10.1002/ieam.4198.

Kuper, J., Van Duinen, G.-J., Nijssen, M., Geertsma, M., Esselink, H., 2000. *Is the decline of the Red-backed Shrike (Lanius collurio) in the Dutch coastal dune area caused by a decrease in insect diversity?* Ring 22.

Laghari, A.N., Siyal, Z.A., Soomro, M., Bangwar, D.K., Khokhar, A.J., Soni, H.L., 2018. Quality Analysis of Urea Plant Wastewater and its Impact on Surface Water Bodies. *Eng. Technol. Appl. Sci. Res* 8, 2699–2703. doi:https://doi.org/10.48084/etasr.1767.

Lal, R., 2007. Anthropogenic influences on world soils and implications to global food security. *Adv. Agron.* 93, 69–93. doi:https://doi.org/10.1016/S0065-2113(06)93002-8.

Lal, R., 2019. Accelerated soil erosion as a source of atmospheric CO2. *Soil Till. Res.* 188, 35–40. doi:https://doi.org/10.1016/j.still.2018.02.001.

Lannig, G., Eilers, S., Pörtner, H.O., Sokolova, I.M., Bock, C., 2010. Impact of ocean acidification on energy metabolism of oyster, Crassostrea gigas—changes in metabolic pathways and thermal response. *Mar. Drugs* 8, 2318–2339. doi:https://doi.org/10.3390/md8082318.

Lee, S., Jiang, X., Manubolu, M., Riedl, K., Ludsin, S.A., Martin, J.F., Lee, J., 2017. Fresh produce and their soils accumulate cyanotoxins from irrigation water: implications for public health and food security. *Food Res. Int.* 102, 234–245. doi:https://doi.org/10.1016/j.foodres.2017.09.079.

Lehmann, J., Schroth, G., 2003. Nutrient leaching, in: Schroth, G., Sinclair, F. (Eds), *Trees, Crops and Soil Fertility, CABI Publishing, Wallingford.* Wallingford: CAB International, pp. 151–166.

Lewitus, A.J., Horner, R.A., Caron, D.A., Garcia-Mendoza, E., Hickey, B.M., Hunter, M., Huppert, D.D., Kudela, R.M., Langlois, G.W., Largier, J.L., 2012. Harmful algal blooms along the North American west coast region: History, trends, causes, and impacts. *Harmful Algae* 19, 133–159. doi:https://doi.org/10.1016/j.hal.2012.06.009.

Li, D.-P., Wu, Z.-J., 2008. Impact of chemical fertilizers application on soil ecological environment. *J. Appl. Ecol.* 19, 1158–1165.

Liang, C.-P., Wang, S.-W., Kao, Y.-H., Chen, J.-S., 2016. Health risk assessment of groundwater arsenic pollution in southern Taiwan. *Environ. Geochem. Health* 38, 1271–1281. doi:https://doi.org/10.1007/s10653-016-9794-4.

Litalien, A., Zeeb, B., 2020. Curing the earth: A review of anthropogenic soil salinization and plant-based strategies for sustainable mitigation. *Sci. Total Environ.* 698, 134235. doi:https://doi.org/10.1016/j.scitotenv.2019.134235.

Long, A., Zhang, J., Yang, L.-T., Ye, X., Lai, N.-W., Tan, L.-L., Lin, D., Chen, L.-S., 2017. Effects of low pH on photosynthesis, related physiological parameters, and nutrient profiles of citrus. *Front. Plant Sci.* 8, 185. doi:https://doi.org/10.3389/fpls.2017.00185.

Ma, L., Zhao, B., Guo, Z., Wang, D., Li, D., Xu, J., Li, Z., Zhang, J., 2019. Divergent responses of bacterial activity, structure, and co-occurrence patterns to long-term unbalanced fertilization without nitrogen, phosphorus, or potassium in a cultivated vertisol. *Environ. Sci. Pollut. Res.* 26, 12741–12754. doi:https://doi.org/10.1007/s11356-019-04839-2.

Matijašević, L., Dejanović, I., Lisac, H., 2010. Treatment of wastewater generated by urea production. *Resour. Conserv. Recycl.* 54, 149–154. doi:https://doi.org/10.1016/j.resconrec.2009.07.007.

Metcalf, J., Richer, R., Cox, P., Codd, G., 2012. Cyanotoxins in desert environments may present a risk to human health. *Sci. Total Environ.* 421, 118–123. doi:https://doi.org/10.1016/j.scitotenv.2012.01.053.

Mulvaney, R., Khan, S., Ellsworth, T., 2009. Synthetic nitrogen fertilizers deplete soil nitrogen: a global dilemma for sustainable cereal production. *J. Environ. Qual.* 38, 2295–2314. doi:https://doi.org/10.2134/jeq2008.0527.

Murray, C.J., Müller-Karulis, B., Carstensen, J., Conley, D.J., Gustafsson, B.G., Andersen, J.H., 2019. Past, present and future eutrophication status of the Baltic Sea. *Front. Mar. Sci.* 6, 2. doi:https://doi.org/10.3389/fmars.2019.00002.

Negri, A.P., Jones, G.J., 1995. Bioaccumulation of paralytic shellfish poisoning (PSP) toxins from the cyanobacterium Anabaena circinalis by the freshwater mussel Alathyria condola. *Toxicon* 33, 667–678. doi:https://doi.org/10.1016/0041-0101(94)00180-G.

Nijssen, M., WallisDeVries, M., Siepel, H., 2017. Pathways for the effects of increased nitrogen deposition on fauna. *Biol. Conserv.* 212, 423–431. doi:https://doi.org/10.1016/j.biocon.2017.02.022.

Nordin, A., Strengbom, J., Witzell, J., Näsholm, T., Ericson, L., 2005. Nitrogen deposition and the biodiversity of boreal forests: implications for the nitrogen critical load. *Ambio J. Human Environ.* 34, 20–24. doi:https://doi.org/10.1579/0044-7447-34.1.20.

Obenour, D.R., Scavia, D., Rabalais, N.N., Turner, R.E., Michalak, A.M., 2013. Retrospective analysis of midsummer hypoxic area and volume in the northern Gulf of Mexico, 1985–2011. *Environ. Sci. Technol.* 47, 9808–9815. doi:https://doi.org/10.1021/es400983g.

Öckinger, E., Hammarstedt, O., Nilsson, S.G., Smith, H.G., 2006. The relationship between local extinctions of grassland butterflies and increased soil nitrogen levels. *Biol. Conserv.* 128, 564–573. doi:https://doi.org/10.1016/j.biocon.2005.10.024.

Pabian, S.E., Rummel, S.M., Sharpe, W.E., Brittingham, M.C., 2012. Terrestrial liming as a restoration technique for acidified forest ecosystems. Int. J. Forest. Res. 2012. doi:https://doi.org/10.1155/2012/976809.

Panjeshahi, M., Langeroudi, E.G., Tahouni, N., 2008. Retrofit of ammonia plant for improving energy efficiency. *Energy* 33, 46–64. doi:https://doi.org/nergy.2007.08.011.

Peng, P., Chen, P., Schiappacasse, C., Zhou, N., Anderson, E., Chen, D., Liu, J., Cheng, Y., Hatzenbeller, R., Addy, M., 2018. A review on the non-thermal plasma-assisted ammonia synthesis technologies. *J. Clean. Prod.* 177, 597–609. doi:https://doi.org/10.1016/j.jclepro.2017.12.229.

Pimentel, D., Burgess, M., 2013. Soil erosion threatens food production. *Agriculture* 3, 443–463. doi:https://doi.org/10.3390/agriculture3030443.

Preisner, M., Neverova-Dziopak, E., Kowalewski, Z., 2020. An analytical review of different approaches to wastewater discharge standards with particular emphasis on nutrients. *Environ. Manage.* 66, 694–708. doi:https://doi.org/10.1007/s00267-020-01344-y.

Quinton, J.N., Govers, G., Van Oost, K., Bardgett, R.D., 2010. The impact of agricultural soil erosion on biogeochemical cycling. *Nat. Geosci.* 3, 311–314. doi:https://doi.org/10.1038/ngeo838.

Rabalais, N.N., Turner, R.E., 2019. Gulf of Mexico hypoxia: Past, present, and future. *Limnol. Oceanogr. Bull.* 28, 117–124. doi:https://doi.org/10.1002/lob.10351.

Raven, J.A., Gobler, C.J., Hansen, P.J., 2020. Dynamic CO2 and pH levels in coastal, estuarine, and inland waters: Theoretical and observed effects on harmful algal blooms. *Harmful Algae* 91, 101594. doi:https://doi.org/10.1016/j.hal.2019.03.012.

Ravishankara, A., Daniel, J.S., Portmann, R.W., 2009. Nitrous oxide (N2O): the dominant ozone-depleting substance emitted in the 21st century. *Science* 326, 123–125. doi:https://doi.org/10.1126/science.1176985.

Ritchie, H., 2017. How many people does synthetic fertilizer feed?

Sartori, M., Philippidis, G., Ferrari, E., Borrelli, P., Lugato, E., Montanarella, L., Panagos, P., 2019. A linkage between the biophysical and the economic: Assessing the global market impacts of soil erosion. *Land Use Policy* 86, 299–312. doi:https://doi.org/10.1016/j.landusepol.2019.05.014.

Schalkhausser, B., Bock, C., Stemmer, K., Brey, T., Pörtner, H.-O., Lannig, G., 2013. Impact of ocean acidification on escape performance of the king scallop, Pecten maximus, from Norway. *Mar. Biol.* 160, 1995–2006. doi:https://doi.org/10.1007/s00227-012-2057-8.

Schullehner, J., Stayner, L., Hansen, B., 2017. Nitrate, nitrite, and ammonium variability in drinking water distribution systems. Int. *J. Environ. Res. Public Health* 14, 276. doi:https://doi.org/10.3390/ijerph14030276.

Setia, R., Gottschalk, P., Smith, P., Marschner, P., Baldock, J., Setia, D., Smith, J., 2013. Soil salinity decreases global soil organic carbon stocks. *Sci. Total Environ.* 465, 267–272. doi:https://doi.org/10.1016/j.scitotenv.2012.08.028.

Shen, J., Chen, D., Bai, M., Sun, J., Lam, S.K., Mosier, A., Liu, X., Li, Y., 2018. Spatial variations in soil and plant nitrogen levels caused by ammonia deposition near a cattle feedlot. *Atmos. Environ.* 176, 120–127. doi:https://doi.org/10.1016/j.atmosenv.2017.12.022.

Sheppard, L.J., Leith, I.D., Mizunuma, T., Neil Cape, J., Crossley, A., Leeson, S., Sutton, M.A., van Dijk, N., Fowler, D., 2011. Dry deposition of ammonia gas drives species change faster than wet deposition of ammonium ions: evidence from a long-term field manipulation. *Global Change Biol.* 17, 3589–3607. doi:https://doi.org/10.1111/j.1365-2486.2011.02478.x.

Silbiger, N.J., Nelson, C.E., Remple, K., Sevilla, J.K., Quinlan, Z.A., Putnam, H.M., Fox, M.D., Donahue, M.J., 2018. Nutrient pollution disrupts key ecosystem functions on coral reefs. *Proc. R. Soc. B Biol. Sci.* 285, 20172718. doi:https://doi.org/10.1098/rspb.2017.2718.

Singh, B., 2018. Are nitrogen fertilizers deleterious to soil health? *Agronomy* 8, 48. doi:https://doi.org/10.3390/agronomy8040048.

Singh, R., Gautam, N., Mishra, A., Gupta, R., 2011. Heavy metals and living systems: An overview. *Indian J. Pharmacol.* 43, 246. doi:https://doi.org/10.4103/0253-7613.81505.

Smith, C., Hill, A.K., Torrente-Murciano, L., 2020. Current and future role of Haber–Bosch ammonia in a carbon-free energy landscape. *Energy Environ. Sci.* 13, 331–344. doi:https://doi.org/10.1039/C9EE02873K.

Snyder, C.S., 2017. Enhanced nitrogen fertiliser technologies support the '4R' concept to optimise crop production and minimise environmental losses. *Soil Res.* 55, 463–472. doi:https://doi.org/10.1071/SR16335.

Snyder, C.S., Bruulsema, T.W., Jensen, T.L., Fixen, P.E., 2009. Review of greenhouse gas emissions from crop production systems and fertilizer management effects. *Agr. Ecosyst. Environ.* 133, 247–266. doi:https://doi.org/10.1016/j.agee.2009.04.021.

Soloveichik, G., 2019. Electrochemical synthesis of ammonia as a potential alternative to the Haber–Bosch process. *Nat. Catal.* 2, 377–380. doi:https://doi.org/10.1038/s41929-019-0280-0.

Soons, M.B., Hefting, M.M., Dorland, E., Lamers, L.P., Versteeg, C., Bobbink, R., 2017. Nitrogen effects on plant species richness in herbaceous communities are more widespread and stronger than those of phosphorus. *Biol. Conserv.* 212, 390–397. doi:https://doi.org/10.1016/j.biocon.2016.12.006.

Stevens, C., 2016. How long do ecosystems take to recover from atmospheric nitrogen deposition? *Biol. Conserv.* 200, 160–167. doi:https://doi.org/10.1016/j.biocon.2016.06.005.

Stevens, C., Bell, J., Brimblecombe, P., Clark, C., Dise, N., Fowler, D., Lovett, G., Wolseley, P., 2020. The impact of air pollution on terrestrial managed and natural vegetation. *Philos. Trans. R. Soc.* A 378, 20190317. doi:https://doi.org/10.1098/rsta.2019.0317.

Tei, F., De Neve, S., de Haan, J., Kristensen, H.L., 2020. Nitrogen management of vegetable crops. *Agr. Water Manage.* 240, 106316. doi:https://doi.org/10.1016/j.agwat.2020.106316.

Throop, H.L., Lerdau, M.T., 2004. Effects of nitrogen deposition on insect herbivory: implications for community and ecosystem processes. *Ecosystems* 7, 109–133. doi:https://doi.org/10.1007/s10021-003-0225-x.

Tian, D., Niu, S., 2015. A global analysis of soil acidification caused by nitrogen addition. *Environ. Res. Lett.* 10, 024019. doi:https://doi.org/10.1088/1748-9326/10/2/024019.

Trivedi, P., Delgado-Baquerizo, M., Anderson, I.C., Singh, B.K., 2016. Response of soil properties and microbial communities to agriculture: implications for primary productivity and soil health indicators. *Front. Plant Sci.* 7, 990. doi:https://doi.org/10.3389/fpls.2016.00990.

Tunå, P., Hulteberg, C., Ahlgren, S., 2014. Techno-economic assessment of nonfossil ammonia production. *Environ. Prog. Sustain.* 33, 1290–1297. doi:https://doi.org/10.1002/ep.11886.

Turner, A.M., Chislock, M.F., 2010. Blinded by the stink: nutrient enrichment impairs the perception of predation risk by freshwater snails. *Ecol. Appl.* 20, 2089–2095. doi: https://doi.org/10.1890/10-0208.1.

United Nations, D.o.E.a.S.A., Population Division, 2017. *World Population Prospects: The 2017 Revision, Key Findings and Advance Tables.* ESA/P/WP/248.

USGS, 2020. Mineral Commodity Summaries 2020: U.S. Geological Survey. doi:https://doi.org/10.3133/mcs2020.

Van der Velde, M., Folberth, C., Balkovič, J., Ciais, P., Fritz, S., Janssens, I.A., Obersteiner, M., See, L., Skalský, R., Xiong, W., 2014. African crop yield reductions due to increasingly unbalanced Nitrogen and Phosphorus consumption. *Global Change Biol.* 20, 1278–1288. doi:https://doi.org/10.1111/gcb.12481.

Van Puijenbroek, P., Cleij, P., Visser, H., 2014. Aggregated indices for trends in eutrophication of different types of fresh water in the Netherlands. *Ecol. Indic.* 36, 456–462. doi:https://doi.org/10.1016/j.ecolind.2013.08.022.

Walling, E., Babin, A., Vaneeckhaute, C., 2019. Nutrient and carbon recovery from organic wastes. In Schmidt J (Ed.) *Biorefinery – Integrated Sustainable Process for Biomass Conversion to Biomaterials*, Biofuels, and Fertilizers. Springer. doi:https://doi.org/10.1007/978-3-030-10961-5_14.

Walling, E., Vaneeckhaute, C., 2020a. Developing successful environmental decision support systems: Challenges and best practices. *J. Environ. Manage.* 264, 110513–110513. doi:https://doi.org/10.1016/j.jenvman.2020.110513.

Walling, E., Vaneeckhaute, C., 2020b. Greenhouse gas emissions from inorganic and organic fertilizer production and use: A review of emission factors and their variability. *J. Environ. Manage.* 276, 111211. doi:https://doi.org/10.1016/j.jenvman.2020.111211.

Wallisdevries, M.F., Van Swaay, C.A., 2006. Global warming and excess nitrogen may induce butterfly decline by microclimatic cooling. *Global Change Biol.* 12, 1620–1626. doi:https://doi.org/10.1111/j.1365-2486.2006.01202.x.

Wang, J., Zhang, M., Bai, X., Tan, H., Li, S., Liu, J., Zhang, R., Wolters, M.A., Qin, X., Zhang, M., 2017. Large-scale transport of PM 2.5 in the lower troposphere during winter cold surges in China. Sci. Rep. 7, 1–10. doi:https://doi.org/10.1038/s41598-017-13217-2.

Wang, S., Chen, H.Y., Tan, Y., Fan, H., Ruan, H., 2016. Fertilizer regime impacts on abundance and diversity of soil fauna across a poplar plantation chronosequence in coastal Eastern China. *Sci. Rep.* 6, 1–10. doi:https://doi.org/10.1038/srep20816.

Ward, M.H., Jones, R.R., Brender, J.D., De Kok, T.M., Weyer, P.J., Nolan, B.T., Villanueva, C.M., Van Breda, S.G., 2018. Drinking water nitrate and human health: an updated review. *Int. J. Environ. Res. Public Health* 15, 1557. doi:https://doi.org/10.3390/ijerph15071557.

Watson, S.B., Miller, C., Arhonditsis, G., Boyer, G.L., Carmichael, W., Charlton, M.N., Confesor, R., Depew, D.C., Höök, T.O., Ludsin, S.A., 2016. The re-eutrophication of Lake Erie: Harmful algal blooms and hypoxia. *Harmful Algae* 56, 44–66. doi:https://doi.org/10.1016/j.hal.2016.04.010.

Weiss, S.B., 1999. Cars, cows, and checkerspot butterflies: nitrogen deposition and management of nutrient-poor grasslands for a threatened species. *Conserv. Biol.* 13, 1476–1486. doi:https://doi.org/10.1046/j.1523-1739.1999.98468.x.

Wiesmeier, M., Poeplau, C., Sierra, C.A., Maier, H., Frühauf, C., Hübner, R., Kühnel, A., Spörlein, P., Geuß, U., Hangen, E., 2016. Projected loss of soil organic carbon in temperate agricultural soils in the 21 st century: effects of climate change and carbon input trends. *Sci. Rep.* 6, 1–17. doi:https://doi.org/10.1038/srep32525.

Wild, O., Prather, M.J., Akimoto, H., 2001. Indirect long-term global radiative cooling from NOx emissions. Geophys. Res. Lett. 28, 1719–1722. doi:https://doi.org/10.1029/2000GL012573.

Wood, R., 2016. Acute animal and human poisonings from cyanotoxin exposure—A review of the literature. *Environ. Int.* 91, 276–282. doi:https://doi.org/10.1016/j.envint.2016.02.026.

Wu, Y., Gu, B., Erisman, J.W., Reis, S., Fang, Y., Lu, X., Zhang, X., 2016. PM2. 5 pollution is substantially affected by ammonia emissions in China. *Environ. Pollut.* 218, 86–94. doi:https://doi.org/10.1016/j.envpol.2016.08.027.

Wurtsbaugh, W.A., Paerl, H.W., Dodds, W.K., 2019. Nutrients, eutrophication and harmful algal blooms along the freshwater to marine continuum. *Wiley Interdiscip. Rev. Water* 6, e1373. doi:https://doi.org/10.1002/wat2.1373.

Xie, D., Si, G., Zhang, T., Mulder, J., Duan, L., 2018. Nitrogen deposition increases N2O emission from an N-saturated subtropical forest in southwest China. *Environ. Pollut.* 243, 1818–1824. doi:https://doi.org/10.1016/j.envpol.2018.09.113.

Yasumoto, T., Oshima, Y., Sugawara, W., Fukuyo, Y., Oguri, H., Igarashi, T., Fujita, N., 1978. Identification of Dinophysis fortii as the causative organism of diarrhetic shellfish poisoning. *Bull. Jpn. Soc. Sc. Fish.* 46, 1405–1411.

Zhang, S., Yu, J., Wang, S., Singh, R.P., Fu, D., 2019a. Nitrogen fertilization altered arbuscular mycorrhizal fungi abundance and soil erosion of paddy fields in the Taihu Lake region of China. *Environ. Sci. Pollut. Res.* 26, 27987–27998. doi:https://doi.org/10.1007/s11356-019-06005-0.

Zhang, T.a., Chen, H.Y., Ruan, H., 2018. Global negative effects of nitrogen deposition on soil microbes. *ISME J.* 12, 1817–1825. doi:https://doi.org/10.1038/s41396-018-0096-y.

Zhang, W.-f., Dou, Z.-x., He, P., Ju, X.-T., Powlson, D., Chadwick, D., Norse, D., Lu, Y.-L., Zhang, Y., Wu, L., 2013. New technologies reduce greenhouse gas emissions from nitrogenous fertilizer in China. *P. Natl. Acad. Sci. USA* 110, 8375-8380. doi:https://doi.org/10.1073/pnas.1210447110.

Zhang, X., Ward, B.B., Sigman, D.M., 2020. Global Nitrogen Cycle: Critical Enzymes, Organisms, and Processes for Nitrogen Budgets and Dynamics. *Chem. Rev.* 120, 5308-5351. doi:https://doi.org/10.1021/acs.chemrev.9b00613.

Zhang, Y., Gu, L., Li, W., Zhang, Q., 2019b. Effect of acid rain on economic loss of concrete structures in Hangzhou, *China. Int. J. Low-Carbon Technol.* 14, 89–94. doi:https://doi.org/10.1093/ijlct/cty056.

Zhang, Z.-W., Fu, Y.-F., Zhou, Y.-H., Wang, C.-Q., Lan, T., Chen, G.-D., Zeng, J., Chen, Y.-E., Yuan, M., Yuan, S., 2019c. Nitrogen and nitric oxide regulate Arabidopsis flowering differently. *Plant Sci.* 284, 177–184. doi:https://doi.org/10.1016/j.plantsci.2019.04.015.

Zhao, Y., Saunois, M., Bousquet, P., Lin, X., Berchet, A., Hegglin, M.I., Canadell, J.G., Jackson, R.B., Dlugokencky, E.J., Langenfelds, R.L., 2020. Influences of hydroxyl radicals (OH) on top-down estimates of the global and regional methane budgets. *Atmos. Chem. Phys.* 20, 9525–9546. doi:https://doi.org/10.5194/acp-20-9525-2020.

Zhou, Q., 2003. Interaction between heavy metals and nitrogen fertilizers applied to soil-vegetable systems. *Bull. Environ. Contam. Toxicol.* 71, 0338–0344. doi:https://doi.org/10.1007/s00128-003-0169-z.

Zhou, Y., Niu, L., Liu, K., Yin, S., Liu, W., 2018. Arsenic in agricultural soils across China: Distribution pattern, accumulation trend, influencing factors, and risk assessment. *Sci. Total Environ.* 616, 156–163. doi:ws://doi.org/10.1016/j.scitotenv.2017.10.232.

# 7 Nitrate Management by Using Innovative Techniques

*Eleftherios Evangelou and Christos Tsadilas*

## CONTENTS

7.1 Introduction .................................................................................................137
7.2 Soil Properties' Variability in the Field ....................................................138
7.3 Proximal Sensing for Estimation of Soil Properties' Variability..................139
    7.3.1 Measuring Spatial Variability of Electrical Conductivity in the Field.............................................................................................140
7.4 Soil Nitrate Estimation in Real Time .......................................................141
7.5 Managing Soil Properties' Variability in the Field...................................142
7.6 Spatial Variation of Crops' Nitrogen Needs in the Field Level...................143
7.7 Variable Rate Nitrogen Applications........................................................147
7.8 Variable Rate Nitrogen Applications in Real Time. A Case Study from Greece ...............................................................................................149
7.9 Conclusion...................................................................................................154
References.............................................................................................................154

## 7.1 INTRODUCTION

The increasing worldwide demand for food from a growing population, coupled with an increased demand for energy derived from agricultural products, is putting enormous pressure on agricultural productivity (Foley et al., 2011). At the same time, there are growing environmental challenges facing the world and an increasing need for environmental conservation practices. These two rapidly growing needs are extremely difficult to meet simultaneously, and this is multiplying the challenges agriculture is going to face in the current century.

Nitrogen (N) has become one of the major pollutants in agricultural ecosystems (Zhang et al., 2015). On the one hand, the application of nitrogen fertilizer in excessive amounts, often found in high-income areas, leads to environmental problems, such as eutrophication of waters, loss of biodiversity, global warming, and stratospheric ozone depletion, and constitutes the costs of environmental N pollution to exceed the added value from N fertilization (Sutton et al., 2011b). On the other hand, limited access to mineral fertilizers leads to low yields and insufficient food supply,

DOI: 10.1201/9780429326806-9

in many low-income areas (Sutton, et al., 2011a). Worldwide, researchers are trying to improve N management for different types of climates, soils, and crop conditions to increased N use efficiency (NUE) – defined as the grain yield obtained at a certain level of N supplied with fertilizers – and reduce losses to the environment in the short- and long-term. Globally, the application of N fertilizer has increased dramatically in recent decades and is projected to exceed 186 million Mg N yr$^{-1}$ by 2050 (Zhang et al., 2015). Inefficient N management practices have contributed to low NUE and limited N fertilizer recovery (Spiertz, 2010) that could reach 33 percent in cereals (Raun and Johnson, 1999; Cassman et al., 2002). The remaining "unused by cultivating plant N" becomes a potential source of contamination for both atmosphere and water resources, including surface and ground waters (Addiscott, 2005). Nitrate leaching, soil denitrification, and volatilization are the main processes for N-fertilizer excess loss, contributing to environmental pollution (Cameron et al., 2013) – especially nitrate-leaching contaminates in groundwater and other bodies of water, which may contribute to eutrophication. It is estimated that an average 10–30 percent of total N inputs in cropping systems are typically lost due to nitrate leaching (Raun and Schepers, 2008), which has led to environmental contamination and concerns regarding the use of N fertilizers (Li et al., 2015; Rimski-Korsakov et al., 2004).

Many different factors could contribute to low NUE, factors that can ultimately be summarized in three main points, as stated by Shanahan et al. (2008): (a) poor synchrony between soil N supply and crop demand; (b) uniform application rates of N fertilizer to spatially variable landscapes; and (c) failure to account for temporally variable influences on crop N needs.

Traditional farm management uses a whole-field approach, in which each field is treated as a homogeneous area, and the variability in soil, topography, local weather conditions, and land use is not considered (Srinivasan, 2006). Inputs are applied uniformly across the field. This management approach is attractive to growers because it is relatively quick and easy to implement fertilizing, but it implies an inefficient application of inputs. This inefficient application carries hidden environmental costs from nitrogen losses coupled with unwanted, explicit economic costs (Zhang, 2017). Furthermore, the under-application of fertilizers has negative influences on crop growth, and subsequently on yield.

Managing N according to plant needs and to achieve yield potential and limit losses to the environment are challenging due to the temporal and spatial variability in crop N uptake, which affects soil residual – and potentially leachable – N (Delgado, et al., 2005), as soil is an extremely complex and highly variable medium (Phillips, 2001). New innovative technologies can help farmers make better–informed and timely decisions in their nitrogen fertilization programs in order to increase NUE and decrease environmental risks.

## 7.2  SOIL PROPERTIES' VARIABILITY IN THE FIELD

At the field scale, site-specific variation in soil type, texture, soil-structure integrity, soil-moisture content and its availability, and soil nutrient chemistry significantly contribute to the spatial variability in crop yield. Under uniform management in the field, the spatial variability of soil properties greatly affects yield, soil water percolation, and nitrate leaching (Maestrini and Basso, 2018a). Soil variability is the outcome of many processes acting and interacting across a continuum of spatial and temporal scales and is inherently scale-dependent (Parkin, 1993). Generally, the overall variation may increase as the area of study increases. Soil's intrinsic variability is determined by soil-forming factors, including parent material, relief, climate, organisms, and time. On the other hand, management practices – for example, crop rotation, fertilization, and irrigation – can significantly affect dynamic soil properties like nutrient balance, moisture regimes, structural stability, air circulation, and drainage. Similarly, factors like erosion and sedimentation can also influence the field variability to some extent (Ortega et al., 1999).

Variability of soil properties within fields is often described by classical statistical methods, after analyzing in the laboratory soil samplings collected with dense soil-sampling protocols, assuming that variation is randomly distributed within mapping units. Geo-statistics have been described as an effective tool to characterize the spatial variability of soil nutrients (Webster and Oliver, 2007, Liu et al., 2009). Numerous studies have used geostatistics to analyze spatial variability and distribution of soil properties at various scales (Liu et al., 2012a, Elbasiouny et al., 2014, Guan et al., 2017, Usowicz and Lipiec, 2017). Generally, samples collected close to one another are more similar than samples collected at greater distances, although soil properties frequently exhibit spatial dependency. Conventional soil surveys, however, are time-consuming and expensive as a tool to provide efficiently thematic soil data layers to be used in soil assessment for precision fertilization management according to the plant's needs.

Many researchers worldwide have tried to develop sensors and techniques that estimate soil variables "on-the-go" and measure a variety of essential soil properties in real time. In this way, soil data can be provided without the need to collect and analyze samples and can be linked to GPS and computer for on-the-go spatial data collection (Kitchen et al., 2005).

## 7.3 PROXIMAL SENSING FOR ESTIMATION OF SOIL PROPERTIES' VARIABILITY

Proximal soil sensing (PSS) refers to the use of field-based sensors to obtain signals from the soil when the sensor's detector is in contact with, or close to, (within 2 m) the soil (Viscarra-Rossel et al., 2010). The sensors provide soil information because the signals correspond to physical measurements that can be related to soils and their properties. Proximal soil sensors may be described by how they measure (invasively or noninvasively) the source of their energy (active or passive), how they operate (stationary or mobile), and the inference used in the measurement of the target soil property (direct or indirect). During the last several decades, the development of PSS coincided with that of precision agriculture.

Different kinds of soil sensors have been developed for sensing soil attributes such as soil moisture, soil N and apparent electrical conductivity, carbonate rates, and soil porosity. These sensors work with different technologies and involve one of the following measurement methods: (1) electrical and electromagnetic sensors that measure electrical resistivity/conductivity or capacitance affected by the composition of the soil tested; (2) optical and radiometric sensors that use electromagnetic waves to detect the level of energy absorbed/reflected by soil particles; (3) mechanical sensors that measure forces resulting from a tool engaged with the soil; (4) acoustic sensors that quantify the sound produced by a tool interacting with the soil; (5) pneumatic sensors that assess the ability to inject air into the soil; and (6) electrochemical sensors that use ion-selective elements producing a voltage output in response to the activity of selected ions (e.g. hydrogen, potassium, nitrate) like electromagnetic induction, electric conductivity, and ion-selective field-effect transistors (Adamchuk et al., 2004).

### 7.3.1 MEASURING SPATIAL VARIABILITY OF ELECTRICAL CONDUCTIVITY IN THE FIELD

One of the most widely used sensors for the creation of field maps used for site-specific management mapping is the soil's apparent electrical conductivity (ECa) sensor. To distinguish the electrical conductivity (EC) measured by the sensors from the soil-science definition of EC (based upon conductance of a saturated soil paste extract), we call the sensor measured EC as apparent EC (ECa). Measured soil ECa is a function of salinity, clay type and percentage, water content, bulk density, and temperature (Rhoades et al., 1989), and in this way, ECa mapping appears to integrate soil parameters related to productivity to produce a template of potential yield (Jaynes et al., 1993; Sudduth et al., 1995; Kitchen et al., 1999), This is why ECa had been one of the most reliable and frequently used indirect measurements to characterize within-field variability of yield-determining soil factors for implementing a site-specific management plan (Rhoades et al., 1989). Electrical and electromagnetic sensors use electric circuits to measure the capability of soil particles to conduct and/or accumulate electrical charge. When using these sensors, the soil becomes part of an electromagnetic circuit, and the changing local conditions immediately affect the signal recorded by a data logger. Several such sensors are commercially available – for example, sensors produced by Veris Technologies (Salina, KS), Geonics Limited (Mississauga, Ontario), Geocarta (Paris), Geometrics, (San Jose, CA), Dualem (Milton, Ont), and Crop Technology (Bandera, TX). Two types of within-field ECa sensors are commercially available for agricultural purposes: (1) electrode-based sensor requiring soil contact, and (2) non-contact electromagnetic induction sensor. In the electrode-based system, sensors are pulled or rolled across the fields, making direct soil contact, and measurements are recorded simultaneously. A commercial device implementing the electrode-based approach is the Veris 3100, which uses six rolling coulters for electrodes and provides two simultaneous ECa measurements (Lund et al., 1999). A Global Positioning System (GPS) receiver mounted on the Veris unit records the location of each ECa soil measurement point in the field. A field is usually mapped by driving the entire field on parallel paths from 40 to 60 feet apart.

As their name implies, the non-contact sensors do not make direct soil contact for field ECa measurements. The EM-based ECa sensor most often used in agriculture is the EM38 (Geonics, Mississauga, Ont), which was initially developed for root-zone

**FIGURE 7.1** EM 38 system and soil ECa spatial variation map of a field.

salinity assessment (Rhoades and Corwin, 1981). The EM38 induces eddy current loops into the soil with one coil and determines conductivity by measuring the resulting secondary current-induced using another coil (Figure 7.1).

As every soil-sensing technology has strengths and weaknesses, and no single sensor can measure all soil properties, the selection of a complementary set of sensors to measure the required suite of soil properties is important. Integrating multiple proximal soil sensors in a single multisensory platform can provide a number of operational benefits over single-sensor systems. There are a few reports of multisensor systems directed at PSS in the literature. For example, Christy and colleagues (2004) reported the use of a mobile sensor platform that simultaneously measures soil pH and ECa. A near infra-red sensor had also been added to this multisensor platform (Christy, 2008). Taylor and colleagues (2006) reported the development of a multisensor platform consisting of two EMI instruments, ER and pH sensors, a $\gamma$-radiometer, and a differential GPS. Adamchuk and Christenson (2007) described a system that simultaneously measured soil mechanical resistance, optical reflectance, and capacitance. Yurui and colleagues, (2008) reported the development of a multisensor technique for measuring soil physical properties (soil water, mechanical strength, and electrical conductivity).

## 7.4 SOIL NITRATE ESTIMATION IN REAL TIME

Soil nitrate (residual N) concentrations show large variability in both space and time, and its measurement by proximal sensing is not straightforward. Agriculture technologies had been developed for the availability of an economic, automated, on-the-go mapping system that can be used to obtain intensive and accurate "time" data on the levels of nitrate-nitrogen ($NO_3$-N) in the soil. Loreto and Morgan (1996) developed an automated system for on-the-go measurement of soil nitrate using a nitrate ion-selective field-effect transistor as the detector, with good correlations (0.65) with laboratory analysis. Adsett and colleagues (1999) developed a tractor-mounted, automated field-monitoring station for soil nitrate, which produced 10 percent error during a set of simulated field calibration tests. The sensing system consisted of a soil

**FIGURE 7.2**    Soil nitrate mapping system with six sub-units: (1) soil sampler, (2) soil metering and conveying, (3) nitrate extraction and measurement, (4) auto-calibration, (5) control and (6) global positioning system (GPS).

sampler, a nitrate extractor unit, a flow cell, a controller, and a nitrate ion-selective electrode (Figure 7.2).

The field-effect transistor (ISFET) technology sensor chip to measure soil nitrate in a flow injection analysis (FIA) system using low flow rates, short injection times, and rapid rinsing to produce a real-time soil analysis system. The rapid response of the system allowed a sample to be analyzed in 1.25 s, which is satisfactory for real-time soil sensing. Price and colleagues (2003) demonstrated that judicious selection of data analysis techniques could provide nitrate measurements in 2–5 s after injection of the extracting solution into the soil core. Kim and colleagues (2006) further investigated the applicability of multi-ISFET/FIA technology to sense potassium and nitrate content at the same time.

## 7.5   MANAGING SOIL PROPERTIES' VARIABILITY IN THE FIELD

In order to improve farm inputs use efficiency and enhance yield in environmentally friendly conditions, the concept of a management zone (MZ) was introduced. Identification of management zones (MZ) represents a cost-effective method to manage field variability, through field classification into areas of broad similarities (Khosla et al., 2002, Nawar et al., 2017). Therefore, it can be suggested as an alternative method to produce prescription maps for site-specific crop management by identifying areas of similar productivity potential within a field (Hornung et al., 2006). Doerge (1999) defined MZ as sub-regions of a field that express a homogeneous combination of yield-limiting factors. MZs can be considered as homogeneous areas within a field that show similar characteristics in landscape and soil conditions, which should lead to similar yield potential and input use efficiency (Schepers et al., 2004) (Figure 7.3). Early research into the benefits of using MZs to increase yield (Mulla et al., 1992) and/or NUE (Khosla and Alley, 1999) was generally positive and encouraging. Several techniques have been proposed in the literature to delineate MZs, using various soil and crop properties individually or in combination. Topography, bare soil aerial imagery, ECa, farmers' management experiences together with yield maps have been extensively used to define the boundaries of MZ (Khosla et al., 2002, Schepers et al., 2004).

Grain yield data, being a total reflection of all biotic and abiotic factors that can affect crop production, can be combined with other soil variables in order to explain field variability associated with both crop and soil properties (Hornung et al., 2006; Bunselmeyer and Lauer, 2015). Schepers and colleagues (2004) however, observed significant temporal changes in yield spatial patterns, even in irrigated fields. They concluded that use of MZs, for example, in variable-rate applications (VRA) of N, would only have been appropriate in three out of five seasons. In addition, the use of the static soil-based MZ concept that relies on the aggregation of the measured landscape attributes is unlikely to be adequate for VRAs across years exhibiting temporal variability. They suggested that an alternative strategy would be to combine MZs with crop-based, in-season remote sensing systems.

Over years, delineation techniques were developed that used more complex assessments of soil fertility variation and are orientating more and more toward a multivariate approach (e.g., historical yield, soil and crop characteristics, topography, weather, and within-season measures of vigor) (Guastaferro et al., 2010). These developments have improved the performance of MZs, as compared with traditional MZ delineation methods that are based only on soil or crop properties. In addition, the adoption of new sensing technologies for characterizing spatial variability in soil, for example, gamma-ray (i.e., Castrignanò et al., 2012), electromagnetic induction (EMI) (i.e., King et al., 2005), and visible and near-infrared (vis–NIR) spectroscopy (i.e., Mouazen and Kuang, 2016) have gained vast attention among scientists in the last decade. Furthermore, measurement of variability in crop growth proximal to satellite assets to delineate MZs based on spectral indices such as normalized difference vegetation index (NDVI) (i.e., Inman et al., 2008) has meant that a wide range of yield-limiting factors in the soil and crop properties can be measured rapidly at finer resolutions.

## 7.6 SPATIAL VARIATION OF CROPS' NITROGEN NEEDS IN THE FIELD LEVEL

The common way of fertilization by the use of uniform application rates across a field results in side-by-side oversupply and undersupply, as soils are in general heterogeneous in space and time. Variable-rate nitrogen applications is aimed at supplying the correct fertilization rate for satisfying the specific requirements of crops, both in space and time and in this way to improve the NUE (Cao et al., 2018, Rütting et al., 2018).

Soil-based information used alone to manage N fertilization may not always lead to an improvement in NUE. Such an approach fails to account for in-season

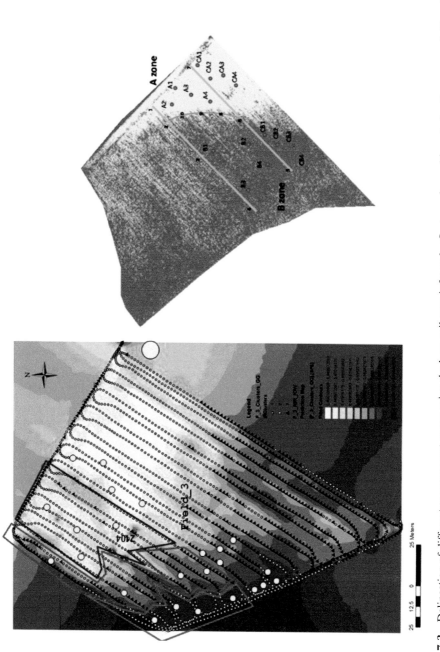

**FIGURE 7.3** Delineation of different management zones using the bare soil near infra red reflectance by multispectral active sensors (left) and by "word-2" satellite data (right), in the HydroSense project (Stamatiadis et al., 2012).

micro-variability (i.e., variability that occurs at a shorter range) associated with crop N status, since the crop response in unstable zones has been demonstrated to be strictly dependent on weather (Maestrini and Basso, 2018b). Consequently, the delineation of MZs alone does not characterize the entire representation for variable N applications (Shanahan et al., 2008). Crop N demand varies spatially and temporally within a field, due to a set of factors that are interrelated as the inherent variations in soil N availability, soil properties, and crop growing conditions across the field (Khosla et al., 2002, Nawar et al., 2017). Crops may not have the physiological ability to uptake and store all of the applied N, with only an estimated 30–50 percent of N fertilizer taken up by cultivated plants (Tilman et al., 2002). Moreover, N supply from the soil varies according to local soil properties and weather conditions, which vary annually and by field site (Rütting et al., 2018).

The last decade's methodologies for the accurate spatial mapping of crops' N needs have been developed in agricultural systems in order to guide variable-rate fertilizer applications for the optimization of crop yield and reduce excessive nitrogen loss.

Crop monitoring, which exploits optical properties of leaf pigments, allows integrating soil, climate, agronomic management, and other environmental factors on crop N status (Shanahan et al., 2008; Muños-Huerta et al., 2013). Leaf N content is often well-correlated with leaf chlorophyll due to the underlying investment of N in chlorophyll molecules (Sage et al., 1987; Evans, 1989; Lamb et al., 2002). A common approach to estimate chlorophyll content, which is easier and more accurate to derive non-destructively, as an indicator of crop N status (Haboudane et al., 2008), is remote sensing through a set of spectral vegetation indices (Gitelson et al., 2005; Wu et al., 2009). Leaf pigments (e.g., carotenoids and anthocyanins) absorb various amounts of light in the visible range of the spectrum. These leaf characteristics influence the reflectance signature of crops. Reflectance measurements made near the crop canopy integrate plant-to-plant variation, while those taken a greater distance from the crop will integrate a larger area. Depending on the field of view of the instrument, measurements taken well above the canopy should make it possible to identify a typical area in a field.

Although following a similar pattern, the leaves of different agricultural crop species exhibit different spectral response patterns (Figure 7.4), as a result of the morphological and physiological characteristics of each species.

Changes in the spectral response patterns can also occur throughout plant growth cycles associated with the impacts of senescence or stress conditions (e.g., due to water or nutrient deficits) on the leaf characteristics (Carter, 1993). Broadly, healthy green vegetation presents high reflectance in the near infra-red (NIR) region, and more NIR energy is reflected as the vegetation canopy increases, while at the same time more red radiant energy is absorbed for photosynthetic purposes. To quantify N-status in plants, different paths have been explored in basic research – for example hyper- and multi-spectral image spectroscopy, which delivers responses as spectral vegetation indices, and radiative transfer models (Féret et al., 2017). It was recently suggested that the relation to proteins is more consistent than the previously assumed relation to chlorophyll (Berger et al., 2020).

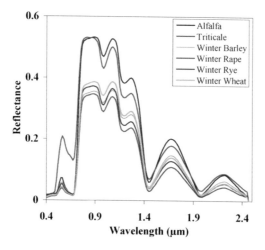

**FIGURE 7.4**  In this figure some typical leaf reflectance curves in the optical region for several agricultural crops are presented based on data from the Leaf Optical Properties Experiment Database (Nidamanuri and Zbell 2011).

Several vegetation indices (Vis) can be determined by combining reflectance data recorded at specific wavelengths (Bajwa et al., 2010). The most widely used vegetation index, the Normalized Difference Vegetation Index (NDVI), calculated as the difference between the NIR and red reflectance divided by the sum of these two values (Shanahan et al., 2008). NDVI values are positively correlated with leaf area index (LAI), green biomass, and leaf N (Shaver et al., 2010). Greater leaf area and green plant biomass translate into higher NDVI values. Since N content of the plant is directly related to leaf area and green plant biomass, higher N content in plants also results in higher NDVI values (Shaver et al., 2011).

Despite its wide use, however, studies have found that NDVI presents some instability under varying soil, sun-view geometry, and atmospheric conditions, and saturates at high biomass conditions (Huete et al., 2002, Huete, 1988). Nevertheless, other VIs, also combining bands in the NIR and red domains, were developed to incorporate soil/atmosphere adjustments and to improve sensitivity to high biomass conditions, for example, the Soil Adjusted Vegetation Index (SAVI) (Huete, 1988), the Enhanced Vegetation Index (EVI) (Huete et al., 1997), and the normalized difference red-edge (NDRE), which are widely used as well (Gitelson et al., 2005).

During the last few decades, the availability from different sources of high spatial, temporal, spectral, and radiometric resolution imagery has been greatly enhanced, widening the agricultural applications further with better precision and accuracy as higher spatial and temporal resolution data facilitates precision management at a smaller spatial scale. Various proximal crop sensors that use canopy reflectance have been developed to assess the total N content of the canopy. Active remote sensors can measure the amount of light reflected from the crop canopy at any time, day or night. This is because these sensors can differentiate between external light sources such as sunlight, and light generated from its own energy, a unique feature of active sensors

(Schepers, 2008). Moreover, they are relatively inexpensive and easy to use. They are also small enough to be handheld or mounted on a tractor (Cabrera-Bosquet et al., 2011, Shaver et al., 2011). Active remote sensing tools such as GreenSeeker (NTech Industries, Ukiah, CA), Crop Circle (Holland Scientific, Lincoln, NE) can measure vegetation indexes such as the NDVI, NDRE, or simple ratio (SR) using light reflectance from crop canopies.

Satellite imagery can cover larger surface areas with up to hundreds of hectares in a single image (Dalla et al., 2015). This capability facilitates rapid data acquisition and, as a result, allows early agronomic management decisions to identify the areas of potential intervention to prevent production losses. Certain satellites have been considered more adapt for regional-scale agricultural studies than others, such as the optical platforms Landsat (Croft et al., 2019, Leslie et al., 2017) and MODIS (Alarcon et al., 2010). New satellites have been suggested to provide more information for the detection of crop parameters, attributable to the support of both the red-edge (RE) region and a higher resolution. One such example is Sentinel-2, launched by the European program Copernicus and the European Space Agency (ESA), showing a high-resolution power and a shorter revisit time (Clevers and Gitelson, 2013). In addition to Sentinel-2, Rapid Eye is an excellent predictor of N content detection (Basso et al., 2016, Magney et al., 2017), possessing a higher spatial resolution (5 meters) and a revisit time of only one day. However, it is common to not acquire a cloud-free image over an area of interest at a critical time in the production cycle of a crop (Tucker and Sellers, 1986). A second challenge for using satellite data is spatial resolutions that are too coarse to resolve important features of interest (Wulder et al., 2004). This is particularly important when the causes of variation in vegetation performance can only be identified using spatial patterns or textures that become visible at relatively fine spatial resolutions. Satellites, therefore, do not provide a complete remote sensing solution to agricultural needs. Other remote sensing platforms, including manned and unmanned aircraft systems, have been proposed to fill the gaps and, in recent years, UAVs (unmanned aerial vehicles commonly referred to as "drones"), have become a practical and widely applied remote sensing tool in agriculture.

Practical applications for drones have progressed significantly in recent years as the technology has improved in tandem with a fall in their cost. Drone-mounted sensors for agricultural applications are typically pointed in a downward direction and detect light that is reflected from surfaces in the direction of the sensor. The low altitudes typically used when deploying drone-based sensors translate into relatively fine spatial resolutions compared to other aerial sensor platforms. This allows information to be related to relatively small objects, in many cases to the level of individual plants or small sections of fields, and it reveals spatial patterns that are often useful in the identification of the causes of anomalies. However, there is an increased risk of data quality degradation due to the use of non-standardized sensors and wide-angle lenses at low altitudes (Rasmussen et al., 2016).

## 7.7  VARIABLE RATE NITROGEN APPLICATIONS

Variable-rate nitrogen application (VRA), a precision management approach to synchronize N input with crop demand, has received increasing attention in recent years. Spatial reallocation of N inputs across various scales has been proposed as an effective way to increase the NUE (Raun et al., 2002, Holland and Schepers, 2010). Unlike the traditional uniform N fertilizer application that neglects within-field spatial variability, VRA applies fertilizers at different rates to meet site-specific demand and therefore can reduce the risks of over- and under-fertilizing plants (Khosla et al., 2002, Holland and Schepers, 2010; Basso et al., 2016). VRA fertilization can be based (1) on static information derived from soil data (e.g., Grisso et al., 2009), remote sensing (Casa et al., 2017) or yield maps from previous years (e.g., Stafford et al., 1999), or (2) on dynamic crop monitoring using real-time, or near-real-time, information (Holland and Schepers, 2010, Nutini et al., 2018).

During the last decades, the development of proximal sensors and variable rate application equipment have made it possible to perform a mid-season assessment of N status and on-the-go application to correct deficiencies. Crop sensing can aid in the identification of areas within fields that require additional N for optimal crop yield and quality, allowing for variable and more precise rate applications (Raun et al., 2002, Tubana et al., 2008), enhancing NUE and reducing spatial variation in end-of-season yield (Stone et al., 1996). Translating vegetation index values into fertilizer N recommendations have followed several paths. Scientists working with wheat at Oklahoma State University developed an algorithm that uses NDVI values and growing-degree days to predict yield potential and then back-calculate crop N requirement (Raun et al., 2005). This approach requires calibration that is based on crop responses and NDVI values from previous years and fields that can be quite different than the situation at hand. Holland and Schepers (2010) developed an algorithm – based on the characteristic quadratic yield response to N fertilizer additions – that does not require external or off-site calibration. This algorithm is sensitive to local conditions because it references an adequately fertilized part of the field at the time of sensing to calculate a sufficiency index, and producers have an opportunity to incorporate their experiences by suggesting an optimum N rate. Most of the algorithms currently in use were developed based on agronomical crop-yield response studies (Lukina et al., 2001, Raun et al., 2002, Raun et al., 2005b, Tubaña et al., 2008, Holland and Schepers, 2010, Bushong et al., 2016) or economic return (Dellinger et al., 2008, Kitchen et al., 2010, Barker and Sawyer, 2010).

VRA has shown to increase NUE by improving grain yields, and net returns while decreasing nutrient overload (Koch et al., 2004, Stamatiadis et al., 2018, Stamatiadis et al., 2019, Evangelou et al., 2020). However, implementing VRA requires high spatial resolution sampling of soil and crops, which can be obtained with both proximal and remote sensing technologies (Mouazen et al., 2020). Analysis and measurement techniques need to be fast, cost-effective, and convenient (Kodaira and Shibusawa, 2013). Shanahan and colleagues (2008) proposed a VRA approach that includes the use of PA tools such as online soil and crop sensors, to enable the development of spatially variable N recommendations based on soil fertility and crop N needs.

A number of commercial systems have been developed in order to support farmers with satellite imagery during the growing season, for example, Farmsat (Geosys, Morges, Switzerland), Satshot (Fargo, ND), Farmstar (Airbus Defence and Space, Toulouse, France), and AgriSat Imberia S.L (Spain). Most of those "fertilizer-oriented" satellite data providers could help farmers creating maps of different fertilization application rates on their farms. Those data could drive suitable fertilizer spreaders for the variable rate applications of nitrogen in the field. In Scandinavia, a free-to-use, non-commercial decision support system based on satellite images (CropSAT), aimed at covering all cropland and assisting farmers in for example, determining optimal N rate for supplementary fertilization of winter wheat, has been in use since 2014 (Söderström et al., 2016). In CropSAT, users can generate a modified version of the soil-adjusted vegetation index maps over selected agricultural fields. It is then possible to manually assign management actions (e.g. how the supplementary N fertilization rate should be varied in relation to the index) and to download prescription files to be used in the fertilizer spreader.

## 7.8  VARIABLE RATE NITROGEN APPLICATIONS IN REAL TIME. A CASE STUDY FROM GREECE

A prototype mechanized VRA fertilizer application system for site-specific in-season N management was developed and tested in field experiments in Greece. The prototype was the result of the work package, "Very-high-resolution variable-rate nitrogen management," in the context of the HORISON2020 project "FATIMA" ("Farming Tools for External Nutrient Inputs and Water Management"). The "Opt-N-Air" integrated active sensor-air delivery system is founded on the ability of ground-based sensors to detect canopy N content, to translate the spatial information into fertilizer N requirement, and to convey a rate signal to a variable-rate spreader for application of granular fertilizer with inter-row precision of placement under real-time conditions.

The VRA prototype shown in Figure 7.5, can be used by any type of tractor and consists of two Crop Circle ACS-430 active crop canopy sensors (Holland Scientific, Lincoln, NE), a GeoScout X data logger (Holland Scientific, Lincoln, NE), a Raven SCS 660 controller (Raven Industries, Inc., Sioux Falls, SD) and a hydraulic motor Gandy Orbit Air 66FSC spreader (Gandy, Owatonna, MN). The data logger processes the geospatial data under real-time conditions to convey an N rate signal to the spreader through the Raven controller. The Raven speed-control compensation system is interfaced with a wheel drive magnetic speed sensor. Nitrogen rate recommendations were made at 1 Hz, which amounts to about every 1.3 m when the tractor with the spreader was traveling at 4–5 km/h. The sensors were mounted in front of the applicator, positioned in a nadir view over 0.6-m from the crop canopy, and measured three optical channels at 670 nm (red), 730 nm (red edge), and 780 nm (NIR).

The sensors monitored the crop at 40,000 Hz and recorded the data at 10 Hz. The data logger was equipped with an internal GPS receiver that collected the geospatial data from the sensors and computed the normalized difference red edge vegetation index (NDRE).

NDRE is similar to the chlorophyll index that accurately estimates chlorophyll contents in contrasting species in terms of LAI, chlorophyll, canopy architecture, and leaf structure for different crops (Gitelson et al., 2005).

**FIGURE 7.5** The "Opt-N-Air" prototype during in season N fertilization in cotton experiments. (1) ACS-430 sensor (2) Gandy Orbit Air 66FSC spreader (3) Speed sensor (4) Air tubes (5) (inside cabin) GeoScout X data logger (left), Raven SCS 660 controller (right).

When set in the VRA mode, the data logger subsequently computed the N application rate based on the algorithm developed by Holland and Schepers (2010):

$$N_{APP} = (N_{OPT}) * \sqrt{\frac{(1-SI)}{\Delta SI}}, \tag{7.1}$$

where Nopt was the optimal in-season N rate to nearly maximize crop yields, SI is the sufficiency index (SI = NDRE sensed/NDRE reference) and DSI is the SI difference parameter (DSI = 0.3). Embedded within Nopt are modifications to the soil N pool from processes that are difficult to quantify like mineralization, immobilization, denitrification, and leaching.

When set in the VRA mode, the prototype system first performed a site-specific auto-calibration procedure using the "virtual strip approach" as described by Holland and Schepers (2013). Based on this approach, the canopy sensors monitored a portion (one strip) of the existing crop that represented the range in crop vigor within the field and then statistically identified plants that are deemed to be non-N limiting by selecting the 95-percentile cumulative value from the histogram of the NDRE values. The data logger then switched to application mode by computing N application rate based on the algorithm of Eq. 1.

The Opt-N-Air was tested on winter wheat, corn, and cotton, three typical arable crops cultivated in the region of central Greece. The experimental design was similar for all experiments, allowing the implementation of VRA N management under full-scale field conditions, by using field strips of specific width to accommodate the operation of each crop harvester, at field-length as experimental units. Under P and K sufficiency and optimal irrigation, real-time VRA N application was compared to uniform farmer N application (single or segmented in-season N application), and an only preplant N control (no in-season N fertilization). Part of each field was divided into four blocks, and treatments were randomly assigned within each block to follow a randomized complete block design.

In 2016, a 2.4 ha winter wheat field with large topographic variability, each treatment was a field strip of 7-m wide running the entire length of the field (200 m). In-season N fertilization was performed at 96 days after planting (DAP) at the advanced tillering stage (Zadoks scale 20-22) on 25 February 2016, just prior to the period of maximum N uptake (Figure 7.6).

In the 2017 3-ha corn field, each replicated treatment was a field strip that averaged 190 m long and 8 rows wide (6.08 m). In-season N fertilization performed 58 DAP in mid-June. In the 2017 5-ha cotton field, each replicated treatment was a field strip of 8 rows wide, with an inter-row spacing of 0.85 m running the block length that averaged 200 m N fertilization performed 82 DAP on July 11 (Figure 7.7).

On average, in the wheat experiment VRA used 72 percent less in-season N or 38 percent less total N than that applied by the farmer without any yield losses. VRA applied more N in areas of lower soil fertility and plant productivity. In areas of low productivity, plants utilized more fertilizer N supplied by VRA, which was the economic optimum nitrogen rate (EONR) for this field. The EONR of VRA resulted in an NUE of 58 percent improvement of 6 percent and 14 percent relative to the preplant and farmer applications, respectively. An economic analysis indicated that VRA provided return over N cost that was €68/ha higher than that of the preplant treatment alone or €118/ha higher than that of the farmer's practice (Stamatiadis et al., 2018) (Table 7.1).

In the 2017 cotton experiment, VRA delivered on average 30 percent less total N than that applied by the farmer without any yield loss. The N response curve was similar to that obtained by Stamatiadis et al. (2018) in winter wheat, where VRA also resulted in an EONR. In comparison to the farmer practice, VRA increased N recovery by an average of 17 percent, reduced energy inputs by 15 percent, and increased energy-use efficiency by 26 percent primarily due to reduced N inputs. VRA provided a return over N cost that was €391/ha higher than that of the preplant treatment alone or €248/ha higher than that of the farmer's practice (Stamatiadis et al., 2019).

In the 2017 corn experiment, the VRA delivered on average 24 percent less total N compared to the farmer's rate, without any yield loss. The reduced N inputs by VRA substantially increased NUE to 56 percent, (21% higher than the farmer's rate) reaching the EONR, but still indicated N losses probably caused by surface run-off. The VRA proximity to EONR translated to net returns over N cost that was €81/ha greater in comparison to the farmer's practice (Evangelou et al., 2020).

The reduced VRA inputs in the wheat experiment resulted in average soil residual nitrate N similar to the pre-plant control and 22.5 percent lower than those in conventional practice. In cotton and corn trials the reduction of residual soil nitrate-nitrogen in comparison to uniform application rates was 15 percent and 12.5 percent respectively (Figure 7.8).

The use of the "Opt–N-Air system" applied N fertilizer based on crop "needs," improved N management by decreasing N inputs, decreased the potential for N leaching, and increased NUE and profitability over fertilizer costs.

Existing studies have also recognized the environmental benefits of more precise N management at the farm level and highlighted the potential of reducing N leaching and

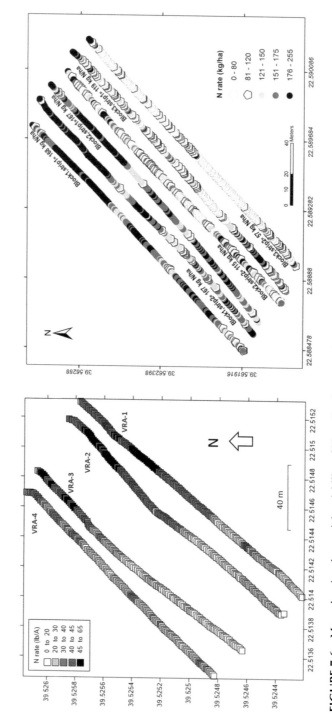

**FIGURE 7.6** Maps showing the spatial variability of N application rates along each VRA strip in wheat (Stamatiadis et al., 2019) (left) and corn (Evangelou et al., 2020) (right) experiments. The data were recorded by the GeoScout X data logger during fertilizer application and processed in five grey-scale classes.

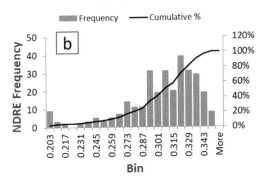

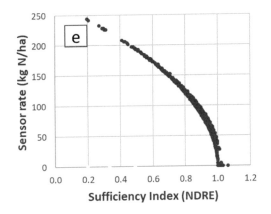

**FIGURE. 7.7** Histogram from cotton trial, showing distribution of canopy NDRE values from which the 95 percentile reference value was extracted (top) and applied N rates in relation to Sufficiency index (Stamatiadis et al., 2019).

**TABLE 7.1**
**Yield, NUE and Economic Performance of the Opt-N-Air System in the Trials**

|  | Treatment | Yield, kg/ha | In season N rate, kg/ha | Total N rate, kg/ha | NUE | Return over N cost |
|---|---|---|---|---|---|---|
| Wheat | Preplant | 5503 | 0 | 100 | 103 | 1197 |
|  | Conventional | 5891 | 112 | 212 | 70 | 1133 |
|  | VRA | 5815 | 31 | 131 | 99 | 1251 |
| Cotton | Preplant | 1501 | 0 | 50 | 20 | 1817 |
|  | Conventional | 1760 | 150 | 200 | 15 | 1960 |
|  | VRA | 1902 | 91 | 141 | 37 | 2208 |
| Corn | Preplant | 9698 | 0 | 80 | 96 | 1553 |
|  | Conventional | 13349 | 190 | 270 | 35 | 1945 |
|  | VRA | 13365 | 125 | 205 | 56 | 2026 |

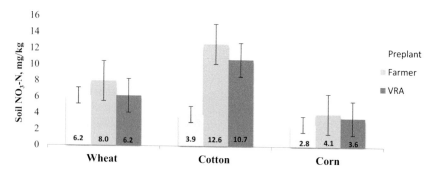

**FIGURE 7.8** Residual NO$_3$-N concentrations after harvest in the 0–40 soil depth of the treatments.

N$_2$O emission (Roberts et al., 2010, Basso et al., 2016, Sela et al., 2019, Gourevitch et al., 2018, Kandulu et al., 2018). A regional assessment of the adoption of the VRA-N could provide useful data for the environmental and economic benefits in larger agricultural areas that could be used by policymakers for the further support of the transition to an N management plan according to the crop needs in time and space.

## 7.9 CONCLUSION

Optimal N management at the field level required to increase NUE, minimize environmental pollution, and decrease cultivation costs. The conventional uniform application of N fertilizers has led to low NUE in arable crops with serious economic and environmental consequences, especially to the nitrate leaching potential. As soil properties at the field level could have a significant spatial variability, intra-field variability of crops N needs has to be taken into account for efficient N fertilization. In the last decades, different technologies and methods have been developed and tested under field conditions in order to adjust N management plans according to spatial and temporal crop N demands. Variable N fertilization rates in different management zones in the field or/and according to a set of canopy indexes have shown to increase the efficiency of N fertilization and decrease the environmental risk by N losses compared with uniform applications. Especially the variable rate applications of N in real time emerge as a technology that addresses in-field variation in soil N availability and crop response, and as such is a tool for more effective site-specific management, with economic and environmental benefits that could be supported by policymakers for its wider adoption by farmers.

## REFERENCES

Adamchuk, V.I., J.W. Hummel, M.T. Morganc, & S.K. Upadhyaya. 2004. On-the-go soil sensors for precision agriculture. *Computers and Electronics in Agriculture* 44: 71–91.

Adamchuk, V.I., & P.T. Christenson. 2007. An instrumented blade system for mapping soil mechanical resistance represented as a second order polynomial. *Soil Tillage and Research* 95(1): 76–83.

Addiscott, T.M. 2005. *Nitrate, Agriculture, and the Environment*. Wallingford, Oxon: CABI International.

Adsett, J.F., J.A. Thotton, & K.J. Sibley. 1999. Development of an automated on the go soil nitrate monitoring system. *Applied Eng. in Agric.* 15(4):351–356.

Alarcon, V., W. McAnally, G. Ervin, & C. Brooks. 2010. Using MODIS land-use/land-cover data and hydrological modeling for estimating nutrient concentrations. In: D. Taniar, O. Gervasi, B. Murgante, E. Pardede, & Apduhan, B.O. (Eds.). Berlin: Springer-Verlag Lecture Notes in Computer Science. *1*(6016), pp. 501–514.

Bajwa, S.G., A.R. Mishra, & R.J. Norman. 2010. Canopy reflectance response to plant nitrogen accumulation in rice. *Precision Agric* 11:488–506.

Barker, D.W. & J.E. Sawyer. 2010. Using active canopy sensors to quantify corn nitrogen stress and nitrogen application rate. *Agronomy Journal*, 102, 964–971.

Basso, B., C. Fiorentino, D. Cammarano, & U. Schulthess. 2016. Variable rate nitrogen fertilizer response in wheat using remote sensing. *Precision Agriculture* 17:168–182.

Berger, K., J. Verrelst, J. Féret, T. Hank, M. Wocher, W.Mauser, & G. Valls. 2020. Retrieval of aboveground crop nitrogen content with a hybrid machine learning method. *International Journal of Applied Earth Observation and Geoinformation* 92:102174.

Bunselmeyer, H.A., & J.G. Lauer. 2015. Using corn and soybean yield history to predict subfield yield response. *Agronomy Journal* 107: 558–562.

Bushong, J., E. Miller, J. Mullock, B. Arnall & W. Raun. 2016. Irrigated and rain-fed maize response to different nitrogen fertilizer application methods. *Journal of Plant Nutrition* 39:1874–1890.

Cabrera-Bosquet, L., G. Molero, M. Stellacci, J. Bort, S. Nogués, & J.L. Araus. 2011. NDVI as a potential tool for predicting biomass, plant nitrogen content and growth in wheat genotypes subjected to different water and nitrogen conditions. *Cereal Res. Commun.* 39:147–159.

Cameron, K.C., H.J. Di, & J.L Moir. 2013. Nitrogen losses from the soil/plant system: a review. *Ann Appl Biol* 162: 145–173.

Cao, P., Lu., C., & Z. Yu. 2018. Historical nitrogen fertilizer use in agricultural ecosystems of the contiguous United States during 1850–2015: Application rate, timing, and fertilizer types. *Earth System Science Data* 10: 969–984.

Carter, G.A. 1993. Responses of leaf spectral reflectance to plant stress. *American Journal of Botany* 80: 239–243.

Casa, R., F. Pelosi, S. Pascucci, F. Fontana, F. Castaldi, S. Pignatti, & M. Pepe. 2017. Early stage variable rate nitrogen fertilization of silage maize driven by multi-temporal clustering of archive satellite data. *Advances in Animal Biosciences*, 8: 288–292.

Cassman, K.G., A.R. Dobermann, & D.T. Walters. 2002. Agroecosystems, nitrogen-use efficiency, and nitrogen management. *Ambio* 31: 132– 140.

Castrignanò, A., M.T.F. Wong, M. Stelluti, D. De Benedetto, & D. Sollitto. 2012. Use of EMI, gamma-ray emission and GPS height as multi-sensor data for soil characterization. *Geoderma* 175–176: 78–89.

Christy, C.D. 2008. Real time measurement of soil attributes using on the go near infrared reflectance spectroscopy. *Comput. Electron.Agric.* 61: 10–19.

Christy, C.D., K. Collings, P. Drummond, & E. Lund. 2004. A mobile sensor platform for measurement of soil pH and buffering: ASAE paper No 041042. *American Society of Agricultural and Biological Engineers*, St Joseph, MI.

Clevers, J.G.P.W., & A.A. Gitelson. 2013. Remote estimation of crop and grass chlorophyll and nitrogen content using red-edge bands on Sentinel-2 and -3. *International Journal of Applied Earth Observation and Geoinformation* 23: 344–351.

Croft, H., J. Arabian, J. Chen, J. Shang, & J. Liu. 2019. Mapping within-field leaf chlorophyll content in agricultural crops for nitrogen management using Landsat-8 imagery. *Precis. Agric.* 21:1–25.

Dalla, M., A. Grifoni, D. Mancini, M. Orlando, F. Guasconi, & F. Orlandini. 2015. Durum wheat in-field monitoring and early-yield prediction: Assessment of potential use of high resolution satellite imagery in a hilly area of Tuscany, Central Italy. *The Journal of Agricultural Science* 153: 68–77.

Delgado, J.A., R. Khosla, W.C. Bausch, D.G. Westfall & D.J. Inman 2005. Nitrogen fertilizer management based on site-specific management zones reduces potential for nitrate leaching. *Journal of Soil and Water Conservation* 60:402–410.

Dellinger, A.E., J.P. Schmidt, & D.B. Beegle. 2008. Developing nitrogen fertilizer recommendations for corn using an active sensor. *Agronomy. Journal.* 10 0:1546–1552.

Doerge, T. 1999. Defining management zones for precision farming. *Crop Insights* 8(21):1–5.

Elbasiouny, H., M. Abowaly, A.A. Alkheir, & A.A. Gad. 2014. Spatial variation of soil carbon and nitrogen pools by using ordinary Kriging method in an area of north Nile Delta, Egypt, *Catena*, 113: 70–78.

Evangelou, E., S. Stamatiadis, J. Schepers, A. Glambedakis, M. Glambedakis, N. Derkas, C. Tsadilas, & T. Nikoli 2020. Evaluation of sensor-based field-scale spatial application of granular N to maize. *Precision Agriculture* 21:1008–1026.

Evans, J.R. 1989. Photosynthesis and nitrogen relationships in leaves of C3 plants. *Oecologia* 78:9–19.

Féret, J.B., A.A. Gitelson, S.D. Noble, & S. Jacquemoud, 2017. Prospect-D: Towards modeling leaf optical properties through a complete lifecycle. *Remote Sensing of Environment* 193:204–215.

Foley, J.A., N. Ramankutty, K.A. Brauman, E.S. Cassidy, J.S. Gerber, M. Johnston, N.D. Mueller, & C. O'Connell, et al. 2011. Solutions for a cultivated planet. *Nature* 478:337–342.

Gitelson, A.A., A. Viña, D.C. Rundquist, V. Ciganda, & T.J. Arkebauer. 2005. Remote estimate of canopy chlorophyll content in crops. *Geophysical Research Letters* 32:1–4.

Gourevitch, J., B. Keeler, & T. Ricketts. 2018. Determining socially optimal rates of nitrogen fertilizer application. *Agriculture, Ecosystems & Environment.* 254. 292–299. 10.1016/j.agee.2017.12.002.

Grisso, R., M. Alley, W.G. Wysor, D. Holshouser, & W. Thomason. 2009. *Precision Farming Tools: Soil Electrical Conductivity.* Virginia Cooperative Extension. Publication.

Guan, F.Y., M.P. Xia, X.L. Tang, & S.H. Fan. 2017. Spatial variability of soil nitrogen, phosphorus and potassium contents in Moso bamboo forests in Yong'an City, China. *Catena* 150:161–172.

Guastaferro, F., A. Castrignano, D. DeBenedetto, D. Sollito, A. Troccoli, & B. Cafarelli. 2010. A comparison of different algorithms for the delineation of management zones. *Precision Agriculture* 11:600–620.

Haboudane, D., N. Tremblay, & J. Miller. 2008. Remote estimation of crop chlorophyll content using spectral indices derived from hyperspectral data. *Transactions on Geoscience and Remote Sensing* 46:423–435.

Holland, K.H. and J.S. Schepers. 2010. Derivation of a variable rate nitrogen application model for in-season fertilization of corn. *Agron. J.* 102:1415–1424.

Holland, K.H. & J.S. Schepers. 2013. Use of a virtual-reference concept to interpret active crop canopy sensor data. *Precis. Agric.* 14:71–85.

Hornung, A., R. Khosla, R.M. Reich, D. Inman, & D.G. Westfall. 2006. Comparison of site-specific management zones: soil-color based and yield-based. *Agronomy Journal* 98:407–415.

Huete, A., K. Didan, T. Miura, E.P. Rodriguez, X. Gao, & L.G. Ferreira. 2002. Overview of the radiometric and biophysical performance of the MODIS vegetation indices. *Remote Sens. Environ* 83:195–213.

Huete, A.R. 1988. A soil-adjusted vegetation index (SAVI). *Remote Sens. Environ.* 25:295–309.

Huete, A.R., H.Q, Liu, K. Batchily, & W. van Leeuwen. 1997. A comparison of vegetation indices over a global set of TM images for EOS-MODIS. *Remote Sensing of Environment* 59:440–451.

Inman, D., R. Khosla, R. Reichand, & D.G. Westfall. 2008. Normalized difference vegetation index and soil color-based management zones in irrigated maize. *Agronomy Journal* 100:60–66.

Jaynes, D.B., T.S. Colvin, & J. Ambuel. 1993. Soil type and crop yield determinations from ground conductivity surveys. Paper 933552. *Am. Soc. Agric. Eng.* St. Joseph, MI.

Kandulu, J., P. Thorburn, J. Biggs, and K. Verburg. 2018. Estimating economic and environmental trade-offs of managing nitrogen in Australian sugarcane systems taking agronomic risk into account. *Journal of Environmental Management* 223, 264–274.

Khosla, R., & M.M. Alley, 1999. Soil-specific nitrogen management on Mid-Atlantic coastal plain soils. *Better Crop.* 83: 6–7.

Khosla R., K. Fleming, J.A. Delgado, T. Shaver, D.G. Westfall. 2002. Use of site-specific management zones to improve nitrogen management for precision agriculture. *Journal of Soil and Water Conservation* 57: 513–518.

Kim, H.J., J.W. Hummel, and K.A. Sudduth. 2006. Sensing nitrate and potassium ions in soil extracts using ion-selective electrodes. *J. Biosystems Eng.* 31:463–473.

King, J.A., P.M.R. Dampney, R.M. Lark, H.C. Wheeler, R.I. Bradley, & T.R. Mayr. 2005. Mapping potential crop management zones within fields: Use of yield-map series and patterns of soil physical properties identified by electromagnetic induction. *Precision Agriculture* 6:167–181.

Kitchen, N.R., K.A. Sudduth, D.B. Myers, S.T. Drummond, & S.Y. Hong. 2005. Delineating productivity zones on claypan soil fields using apparent soil electrical conductivity. *Computer and Electronics in Agriculture* 46:285–308.

Kitchen, N.R., K.A. Sudduth, & S.T. Drummond. 1999. Soil electrical conductivity as a crop productivity measure for claypan soils. *J. Prod. Agric.* 12:607–617.

Kitchen, N.R., K.A. Sudduth, S.T. Drummond, P.C. Scharf, H.L. Palm, D.F. Roberts, & E.D. Vories. 2010. Ground-based canopy reflectance sensing for variable-rate nitrogen corn fertilization. *Agron. J.* 102:71–84.

Koch, B., R. Khosla, W. Frasier, D. Westfall, & D. Inman, (2004). Economic feasibility of variable-rate nitrogen application utilizing site-specific management zones. *Agronomy Journal.* 10.2134/agronj2004.1572.

Kodaira, M., & S. Shibusawa, 2013. Using a mobile real-time soil visible-near infrared sensor for high resolution soil property mapping. Amsterdam: Geoderma, pp. 64–79.

Lamb, D.W., M. Steyn-Ross, P. Schaare, M.M. Hanna, W. Silvester & A. Steyn-Ross. 2002. Estimating leaf nitrogen concentration in ryegrass (Lolium spp.) pasture using the chlorophyll red-edge: Theoretical modelling and experimental observations. *International Journal of Remote Sensing* 23:3619–3648.

Leslie, C.R., L.O. Serbina, & H.M. Miller 2017. Landsat and agriculture – case studies on the uses and benefits of Landsat imagery in agricultural monitoring and production. U.S. *Geol. Surv. Open File Rep* 2017:1034.

Li, H., R. Cong, T. Ren, X. Li, C. Ma & L. Zheng. 2015. Yield response to N fertilizer and optimum N rate of winter oilseed rape under different soil indigenous N supplies. *Field Crops Res* 181:52–59.

Liu, F, A.X, Zhu, B.L. Li, T. Pei, C.Z. Qin, G.H. Liu, Y.J. Wang & C.H. Zhou. 2009. Identification of spatial difference of soil types using land surface feedback dynamic patterns. *Chinese Journal of Soil Science* 40:501–508.

Liu, F., X., Geng, A.X. Zhu & W. Fraser. 2012. Soil texture mapping over low relief areas using land surface feedback dynamic patterns extracted from MODIS. *Geoderma* 171–172: 44–52.

Loreto, A.B., & M.T. Morgan, 1996. *Development of an automated system for field measurement of soil nitrate.* Paper No. 96–1087, ASAE, St. Joseph, MI.

Lukina, E.V., K.W. Freeman, K.J. Wynn, W.E. Thomason, R.W. Mullen, A.R. Klatt, G.V. Johnson, R.L. Elliott, M.L. Stone, J.B. Solie, & W.R. Raun. 2001. Nitrogen fertilization optimization algorithm based on in-season estimates of yield and plant nitrogen uptake. *J. Plant Nutr.* 24:885–898.

Lund, E.D., C.D. Christy, & P.E. Drummond 1999. Practical applications of soil electrical conductivity mapping. pp. 771–779. In J.V. Stafford (Ed.) *Precision agriculture '99. Proc. Eur. Conf. on Precision Agric.,* 2nd, Odense Congress Centre, Odense, Denmark. 11–15 July 1999. SCI, Sheffield, UK.

Maestrini, B. & B. Basso. 2018a. Predicting spatial patterns of within-field crop yield variability. *F. Crop. Res.* 219:106–112.

Maestrini, B., & B. Basso. 2018b. Drivers of within-field spatial and temporal variability of crop yield across the US Midwest. *Sci Rep* 8:14833.

Magney, T.S., C. Frankenberg, J.B. Fisher, Y. Sun, G.B. North, T.S. Davis, A. Kornfeld, & K. Siebke. 2017. Connecting active to passive fluorescence with photosynthesis: a method for evaluating remote sensing measurements of Chlfluorescence. *New Phytologist* 215:1594–1608.

Mouazen, A.M., & B. Kuang. 2016. On-line visible and near infrared spectroscopy for in-field phosphorous management. *Soil Tillage Res* 155:471–477.

Mouazen, A.M., T. Alexandridis, H. Buddenbaum, Y. Cohen, D. Moshou, D. Mulla, S. Nawar, & K.A. Sudduth. 2020. Monitoring. In: A. Castrignanò, G. Buttafuoco, R. Khosla, A.M. Mouazen, D. Moshou, & O. Naud (Eds.). *Agricultural Internet of Things and Decision Support for Precision Smart Farming.* Cambridge, MA: Academic Press, pp. 35–138.

Mulla, D.J., A.U. Bhatti, M.W. Hammond, & J.A. Benson. 1992. A comparison of winter wheat yield and quality under uniform versus spatially variable fertilizer management. *Agric. Ecosy.Environ.* 38:301–311.

Muñoz-Huerta, R.F., R.G. Guevara-Gonzalez, L.M. Contreras-Medina, I. Torres-Pacheco, J. Prado-Olivarez, & R.V. Ocampo-Velazquez. 2013. A review of methods for sensing the nitrogen status in plants: Advantages, disadvantages and recent advances. *Sensors* 13:10823–10843.

Nawar, S., R. Corstanje, G. Halcro, D. Mulla, and A.M. Mouazen. 2017. Delineation of soil management zones for variable-rate fertilization. *Adv. Agron.* 143:175–245.

Nidamanuri, R. R., & B. Zbell. 2011. Normalized spectral similarity score (NS3) as an efficient spectral library searching method for hyperspectral image classification. Selected topics in applied earth observations and remote sensing. *IEEE,* 4(1), pp. 226–240.

Nutini, F., R. Confalonieri, A. Crema, E. Movedi, L. Paleari, D. Stavrakoudis & M. Boschetti. 2018. An operational workflow to assess rice nutritional status based on satellite imagery and smartphone apps. *Computers and Electronics in Agriculture* 154:80–92.

Ortega, R.A., D.G. Westfall, W.J. Gangloff, & G.A. Peterson. 1999. Multivariate approach to N and P recommendations in variable rate fertilizer applications. In: Stafford, J.V. (Ed.), *Proc. 2nd Europ. Conf. Prec. Agric.* Odense, Denmark.

Parkin, T.B. 1993. Spatial variability of microbial processes in soil – a review. *J Environ Quality* 22:409–427.

Phillips, J.D. 2001. The relative importance of intrinsic and extrinsic factors in pedodiversity. *Annals of the Association of American Geographers* 91:609–621.

Price, R.R., J.W. Hummel, S.J. Birrell, & I.S. Ahmad. 2003. Rapid nitrate analysis of soil cores using ISFETs. *Transactions of the ASAE* 46:601–610.

Rasmussen J., G. Ntakos, J. Nielsen, J. Svensgaard, R.N. Poulsen & S. Christensen. 2016. Are vegetation indices derived from consumer-grade cameras mounted on UAVs sufficiently reliable for assessing experimental plots? *Eur. J. Agron.* 74: 75–92.

Raun, W.R., & G.V. Johnson. 1999. Improving nitrogen use efficiency for cereal production. *Agron. J.* 91:357–363.

Raun, W.R., J.B. Solie, G.V. Johnson, M.L. Stone, R.W. Mullen, K.W. Freeman, W.E. Thomason, and E.V. Lukina. 2002. Improving nitrogen use efficiency in cereal grain production with optical sensing and variable rate application. *Agron. J.*, 94:815–820.

Raun, William, J.B. Solie, M. Stone, K.L. Martin, K.W. Freeman, R.W. Mullen, H. Zhang, J. Schepers, & G. Johnson. 2005. Optical sensor-based algorithm for crop nitrogen fertilization. *Communications in Soil Science and Plant Analysis* 36:2759–2781.

Raun, R., & J.S. Schepers. 2008. Nitrogen management for improved use efficiency. In: J.S. Schepers & W.R. Raun (Eds.). *Nitrogen in Agricultural Systems.* Madison, WI: ASA, CSSA, SSSA, pp. 675–694.

Rhoades, J.D., & D.L. Corwin, 1981. Determining soil electrical conductivity-depth relations using an inductive electromagnetic soil conductivity meter. *Soil Sci. Soc. Am. J.* 45:255–260.

Rhoades J.D., N.A. Manteghi, P.J. Shouse, & W.J. Alwes. 1989. Soil electrical conductivity and soil salinity: New formulations and calibrations. *Soil Sci. Soc. Am. J.* 53:433–439.

Rimski Korsakov, H., G. Rubio, & R.S. Lavado. 2004. Potential nitrate losses under different agricultural practices in the Pampas Region, Argentina. *Agricultural Water Management* 65:83–94.

Rütting, T., H. Aronsson, & S. Delin. 2018. Efficient use of nitrogen in agriculture. *Nutr Cycl Agroecosyst* 110:1–5.

Sage R.F., & R.W. Pearcy. 1987. The nitrogen use efficiency of C3 and C4 plants. II Leaf nitrogen effects on the gas exchange characteristics of Chenopodium album L. and Amaranthus retroflexus L. *Plant Physiol* 84:959–963.

Schepers A.R., J.F. Shanahan, M.K. Liebig, J.S. Schepers, S.H. Johnson & A. Luchiari Jr. 2004. Appropriateness of management zones for characterizing spatial variability of soil properties and irrigated corn yields across years. *Agronomy Journal* 96:195–203.

Schepers, J.S. 2008. Active sensor tidbits. Unpublished article. Lincoln, NE.: USDA-ARS.

Sela S., P.B. Woodbury, R. Marjerison and H.M. van Es. 2019. Towards applying N balance as a sustainability indicator for the US corn belt: Realistic achievable targets, spatio-temporal variability and policy implications. *Environmental Research Letters*, Vol. 14, 064015.

Shanahan, J.F., N.R. Kitchen, W.R. Raun, & J.S. Schepers. 2008. Responsive in-season nitrogen management for cereals. *Comput. Electron. Agric.* 61:51–62.

Shaver T.M., R. Khosla & D.G. Westfall. 2011. Evaluation of two crop canopy sensors for nitrogen variability determination in irrigated corn. *Precision Agriculture* 12:892–904.

Shaver, T.M., R. Khosla, & D.G Westfall. 2010. Evaluation of two ground-based active crop canopy sensors in maize: growth stage, row spacing, and sensor movement speed. *Soil Sci. Soc. Am. J.* 74: 2101–2108.

Söderström, M., H. Stadig, J. Martinsson, K. Piikki, & M. Stenberg. 2016. CropSAT – A public satellite-based decision support system for variable-rate nitrogen fertilization in Scandinavia. 13th International Conference on Precision Agriculture (ICPA) St Louis, MI.

Spiertz, J.H.J. 2010. Nitrogen, sustainable agriculture and food security: A review. *Agron. Sustain. Dev.* 30:43–55.

Srinivasan, A. 2006. Precision agriculture: An overview. pp. 3–15. In: A. Srinivasan (Ed.). *Handbook of Precision Agriculture Principles and Applications.* New York: Food Products Press and The Haworth Press.

Stafford, J.V., R.M. Lark, & H.C. Bolam. 1999. Using yield maps to regionalize fields into potential management units. In: P.C. Robert, et al. (Eds.). *Precision Agriculture* (4th Int. Conf.) ASA, CSSA and SSSA. Madison, WI: pp. 225–237.

Stamatiadis, S., J. Schepers, E. Evangelou, C. Tsadilas, A. Glambedakis, M. Glambedakis, N. Derkas, N. Spyropoulos, N. Dalezios, & K. Eskridge. 2018. Variable-rate nitrogen fertilization of winter wheat under high spatial resolution. *Precision Agriculture.* 19: 570–587.

Stamatiadis, S., L. Evangelou, D. Chachalis, & A. Kampas. 2012. *Minimum Data Set for Site-specific Management of Cotton in the HydroSense Project. Final Report.* Deliverable 4.2 for the Action 4 of the Project HydroSense: Innovative Precision Technologies for Optimized Irrigation and Intergraded Crop Management in a Water-limited Agrosystem. LIFE + Environment. pp 50.

Stone, M.L., J.B. Solie, W.R. Raun, R.W. Whitney, S.L. Taylor, & J.D. Ringer. 1996. Use of spectral radiance for correcting in-season fertilizer nitrogen deficiencies in winter wheat. *Trans. ASAE* 39:1623–1631.

Sudduth, K.A., D.F. Hughes, & S.T. Drummond. 1995. Electromagnetic induction sensing as an indicator of productivity on claypan soils. p. 671–681. In P.C. Robert et al. (Ed.) Proc. Int. Conf. on Site-Specific Management for Agricultural Systems, 2nd, Minneapolis, MN. 27–30 March 1994. ASA, CSSA, and SSSA, Madison, WI.

Sutton, M.A., C. Howard, J.W. Erisman, G. Billen, A. Bleeker, P. Grennfelt, H. van Grinsven, & B. Grizzetti. 2011a. Assessing our nitrogen inheritance. In: Sutton, M.A., Howard, C.M., Erisman, J.W., Billen, G., Bleeker, A., Grennfelt, P., van Grinsven, H. & Grizzetti, B. (Eds.). *The European Nitrogen Assessment.* Cambridge, UK: Cambridge University Press, pp 1–6.

Sutton, M.A., G. Billen, A. Bleeker, J.W. Erisman, P. Grennfelt, H. van Grinsven, B. Grizzetti, C.M. Howard, & A. Leip. 2011b. Technical summary. In: Sutton, M.A., Howard, C.M., Erisman, J.W., Billen, G., Bleeker, A., Grennfelt, P., van Grinsven, H., & Grizzetti, B. (Eds.). *The European Nitrogen Assessment.* Cambridge, UK: Cambridge University Press, pp xxxv–li.

Taylor, J.A., A.B. McBratney, R.A. Viscarra-Rossel, B. Minansy, H. Taylor, B. Whelan, & M. Short. 2006. Development of a multi-sensor platform for proximal soil sensing. In: *Proceedings of the 18th World Congress of Soil Science, July 9–15, 2006.* Philadelphia: World Congress of Soil Science.

Tilman, D., K. Cassman, & P. Matson, 2002. Agricultural sustainability and intensive production practices. *Nature* 418:671–677.

Tubana, B., B. Arnall, S. Holtz, J.B. Solie, K. Girma, & W. Raun, 2008. Effect of treating field spatial variability in winter wheat at different resolutions. *Journal of Plant Nutrition* 31:1975–1998.

Tucker, C.J, & P.J. Sellers 1986. Satellite remote sensing of primary production. *International Journal of Remote Sensing* 7:1395–1416.

Usowicz, B., & J. Lipiec. 2017. Spatial variability of soil properties and cereal yield in a cultivated field on sandy soil. *Soil and Tillage Research* 174:241–250.

Viscarra, R., A. Raphael, A.B. McBratney, & B. Minasny. 2010. *Proximal Soil Sensing Series: Progress in Soil Science.* London & New York: Springer.

Webster, R., & M.A. Oliver. 2007. *Geostatistics for Environmental Scientists* (2nd ed.). Chichester: Wiley.

Wu, C., Z. Niu, Q. Tang, W. Huang, B. Rivard, & J. Feng. 2009. Remote estimation of gross primary production in wheat using chlorophyll-related vegetation indices. *Agricultural and Forest Meteorology* 149:1015–1021.

Wulder, M.A., R.J. Hall, N.C. Coops, & S.E. Franklin. 2004. High spatial resolution remotely sensed data for ecosystem characterization. *BioScience* 54:511–521.

Yurui, S., P.S. Lammers, M. Daokun, L. Jianhui, & Z. Qingmeng. 2008. Determining soil physical properties by multi-sensor technique. *Sens. Actuator A* 147:352–357.

Zhang, X. 2017. A plan for efficient use of nitrogen fertilizers. *Nature* 543:322–323.

Zhang, X., E. Davidson, & D. Mauzerall. 2015. Managing nitrogen for sustainable development. *Nature* 528:51–59.

# 8 Estimation of Nitrogen Status in Plants

*Miguel Garcia-Servin, Luis Miguel Contreras-Medina, Irineo Torres-Pacheco, and Ramón Gerardo Guevara-González*

## CONTENTS

8.1 Introduction ..........................................................................................163
8.2 Tissue Analysis......................................................................................164
    8.2.1 The Kjeldahl Digestion ................................................................164
        8.2.1.1 Disadvantages in the Kjeldahl Digestion Method..............165
    8.2.2 Dumas Combustion ......................................................................165
        8.2.2.1 Disadvantages in the Dumas Combustion..........................166
8.3 In Field Systems ....................................................................................166
    8.3.1 Leaf Chlorophyll Meters ..............................................................166
        8.3.1.1 SPAD....................................................................................167
        8.3.1.2 Chlorophyll Fluorescence Meters ......................................168
    8.3.2 Nitrogen Measurement in Plants Using a UAV
        (Unmanned Aerial Vehicle) ..........................................................169
    8.3.3 Biosensors ....................................................................................171
    8.3.4 Nitrate Sap Content and Electrical Variables .............................173
        8.3.4.1 Reflectometers and Nitrate Test Strips...............................173
        8.3.4.2 Ion Selective Electrode .......................................................173
        8.3.4.3 Impedance Measurements....................................................174
8.4 Conclusion.............................................................................................174
References......................................................................................................175

## 8.1 INTRODUCTION

Nitrogen is an essential element in plants due to its key role in the production of primary and secondary metabolites important in plant physiology impacting performance and tolerance to environmental stresses. Moreover, crop yield and biomass are highly affected by nitrogen fertilization. Plants absorb nitrogen as a mineral nutrient mainly from the soil, and it can be taken in the form of ammonium ($NH4^+$) and nitrate ($NO3-$). However, soil nitrogen supply is often limited, thus forcing farmers to increase the amount of nitrogen fertilizers to achieve better crop yield. In agriculture, when nitrogen fertilization is in excess (an issue that commonly occurs), nitrate

DOI: 10.1201/9780429326806-10

leaching, soil denitrification, and volatilization are the main processes for nitrogen fertilizer excess losses, contributing to environmental pollution. To optimize farming practices for sustainable yields and to reduce costs in agricultural and horticultural practices requires applying precision agriculture, which requires the efficient supply of water and nutrients. Precision agriculture techniques require significant knowledge of plant physiology and plant nutrition, as well as electronic instrumentation, in order to design adequate, accurate, and robust devices to measure any plant variable of interest for crop production (in this case, nitrogen status within plants during production of crops).

The present chapter aims to display to readers a current overview of different methodologies to estimate nitrogen in plants, emphasizing the specific features of each method in order to point out strengthens and weaknesses in the various cases.

## 8.2  TISSUE ANALYSIS

A common technique for nitrogen (N) level determination in plants is tissue analysis. This is used as contrast or control in the validation of new methods of N status prediction, as the tissue analysis provides the actual N content in the plant.

### 8.2.1  THE KJELDAHL DIGESTION

The method of N determination in organic compounds that can be considered as the most used is the Kjeldahl digestion proposed by Johan Kjeldahl (Kjeldahl, 1883). This method has been implemented in the N determination in wastewater, beverages, animal and vegetable origin products, and in the agriculture field for soil and plant tissue (Cunniff, 1995; Labconco, 1998; Kalra, 1998; Domini et al., 2009; Sáez-Plaza et al., 2013; Wojciechowski & Barbano, 2015). This procedure consists of three principal stages:

a.  Digestion: the objective of this step is to turn the N of the biological sample into ammonium sulfate, adding an acid, which is usually sulfuric acid, and heating until 373 °C (Persson et al., 2008), so the carbon of the sample is oxidized and separated, obtaining $CO_2$, $H_2O$ and ammonium sulfate. Usually for this stage a Kjeldahl flask is used, which can be described as a spherical glass with a long neck.

b.  Distillation: at this stage, NaOH is added to the solution once its temperature decreases and is subsequently heated again in order to obtain ammonia gas from the ammonium sulfate. Ammonia gas is driven through a condenser and captured in a flask with a solution that can be HCl, $H_3BO_3$, or $H_2SO_4$.

c.  Ammonium quantification: as the amount of N (contained in the biological sample) is proportional to the ammonia concentration obtained in the distillation, the ammonia determination is a crucial step and normally the method of titration is used (Michalowski et al., 2013; Martin et al., 2017).

The Kjeldahl digestion method has many variants and, usually, some other substances are added to improve the process. At the stage of digestion, $K_2SO_4$ is added to the mixture so the boiling point gets higher (Moller, 2010); also, some catalysts as $CuSO_4$ combined with $TiO_2$, Se or HgO and $Na_2SO_4$ with $CuSO_4$ are used to reduce time in the decomposition of organic compounds (American Public Health Association, 1992; Silva & Queiroz, 2002; Silva et al., 2016).

### 8.2.1.1 DISADVANTAGES IN THE KJELDAHL DIGESTION METHOD

With this method it is possible only to measure the N attached to organic molecules and ammonium (Pontes et al., 2009); other forms of N as nitrite or nitrate can be measured only if the sample is treated (so nitrite and nitrate are reduced to ammonium) before the digestion stage. The treatments are based on the addition of substances as salicylic acid ($C_7H_6O_3$), chromium potassium sulfate [$CrK(SO_4)_2$], and phenyl-acetate ($C_8H_8O_2$). It has been reported that the addition of phenyl-acetate with the acid of the usual digestion process to the sample allows for better results in the N measuring for plants' tissue samples (Lee et al., 1996). Some other substance combinations have been reported as salicylic acid diluted in sulfuric acid searching to measure other N forms (Amin & Flowers, 2004) and salicylic acid with sodium thiosulfate ($Na_2S_2O_3$) for the specific treatment of nitrate (Labconco, 1998). Other techniques together with the Kjeldahl digestion for the determination of ammonia have been proposed. Colorimetry used to be the most common method for ammonia determination (Handson & Shelley, 1993); it is used in combination with the indophenol colorimetry method in the Technicon Auto-Analyzer method (Amin & Flowers, 2004). The indophenol colorimetry method was used also complementing the Kjeldahl digestion for ammonia determination in an African grass analysis (Clifton & Clifton, 1991), but as this method has the disadvantage of requiring dangerous substances, other methods, although more expensive, can be used as the ion chromatography (Handson & Shelley, 1993) or the Diffusion conductimetry, which are considered better than the previous ones (Lee et al., 1996; Saha et al., 2012).

Some solutions have been proposed to improve the method itself. In order to reduce time in the digestion stage, microwave and ultrasound energy has been used, achieving a reduction to a quarter of the original time in the classic method (Domini et al., 2009). Ultrasound has been used also in the distillation stage, replacing the classical configuration with a purge and trap one, for a reaction of the alkaline agent with the sample after the Digestion stage (Pontes et al., 2009).

### 8.2.2 DUMAS COMBUSTION

Even the Kjeldahl digestion is a very common method and has gone through several modifications and improvements; it maintains the disadvantage of not being able to measure many N forms and the requirement of toxic reagents (Watson & Galliher, 2001; Lanza et al., 2016). Another method used extensively and that also is a reference for N determination in organic and inorganic samples (Xu et al., 2014; Krotz et al., 2016) is the Dumas combustion, proposed by Jean-Baptiste Dumas (Dumas, 1831). The method can be seen in stages:

a. Heating: after the sample was weighed in a capsule it is heated around 1000 °C in an oxygen atmosphere.
b. Water lost: the heating stage will trigger gas release as water vapor, oxygen, carbon dioxide, and N (as gaseous nitrogen and nitrogen oxides). The water vapor is separated by many methods as using a thermoelectric cooler or a perchlorate trap (Lee et al. 1996; Unkovich et al., 2008)
c. Redox: As some N remains as nitrogen oxides it is made to pass through a reduction heater with pure cooper fillings to obtain gaseous nitrogen.
d. Gas filter: after the previous stages some gases remain as the carbon dioxide and the gaseous nitrogen, but as only the nitrogen is required the carbon dioxide is taken out.
e. N measure: Finally, only gaseous nitrogen is obtained and is measured with a conductivity detector.

### 8.2.2.1 DISADVANTAGES IN THE DUMAS COMBUSTION

Dumas combustion has many advantages over the Kjeldahl digestion, although there are no significant differences between the methods (Lanza et al., 2016) also have negative points. Some N from the sample can be lost if the heating stage is interrupted or not completed; also, the sample weight must be small, around 250 mg (Watson & Galliher, 2001; Unkovich et al., 2008). Another disadvantage that Dumas combustion and Kjeldahl digestion share, is the requirement of invasive sampling. In the case of N measuring in plants, the previous methods require as sample the entire plant or part of the plant (Goffart et al., 2008; Sáez-Plaza et al., 2013), which can represent an affectation in the selection or in the number of samples.

## 8.3 IN FIELD SYSTEMS

The previous methods must be conducted in a laboratory, so considerable time is consumed in addition to the already-mentioned problem with the toxic substances and expensive equipment. This has led to the creation of new techniques for N measuring, looking for noninvasive methods that can be realized directly in the plant without damaging it. Some of the techniques that have been developed for the measurement of N in plants are based on chlorophyll-dependent properties as the transmittance of leaves and leaf chlorophyll fluorescence. Other techniques are based on canopy reflectance measurement (Chen et al., 2012). Most of the developed methods depend on the plant properties that can be measured with optical sensors or analysis because the properties suffer variations according to the state of the plant in relation to pathogen presence, water content, leaf senescence, nutrients in the plant, and N content (Zebarth et al., 2009).

### 8.3.1 LEAF CHLOROPHYLL METERS

Chlorophyll molecules have an important role in photosynthesis, which is the process of capturing and converting solar radiation into chemical energy and protein synthesis, so the content of this molecule in the leaf can be related to the N status (Demotes-Mainard et al., 2008; Taiz & Zeiger, 2010; Ainsworth, 2018). Some devices have been developed for chlorophyll measuring and, due to their portability, fast-result delivery, and low cost (Lin et al., 2010) they have been the best option for N determination in several studies.

### 8.3.1.1 SPAD

Among the most common chlorophyll meters, the SPAD-502 (Spectrum Technologies, Plainfield, IL) is usually used in research and N determination in plants directly in the field (Piekielek et al., 1995; Peng et al., 1996; Bausch & Diker, 2001; Fontes & Araujo, 2006; Perry & Davenport, 2007; Demotes-Mainard et al., 2008; Miao et al., 2009; Lin et al., 2010; Coste et al., 2010; Ling et al., 2011; Xiong et al., 2015; Gromaz et al., 2017). The SPAD meters deliver a reading that is calculated with the value of transmission of red light (650 nm) and infrared light (940 nm), which are the respective values of light absorbance and non-absorbance of chlorophyll (Xiong et al., 2015). A chamber is placed in a section of the leaf that is exposed to the aforementioned lights, and then the amount of light that passes through the leaf is captured by sensors and quantified. The difference in the amount of light is used as the chlorophyll content indicator (Demotes-Mainard et al., 2008). Based on the relation of chlorophyll percentage and N content in the plant, the SPAD has been used in several studies of N content in crops (Yang et al., 2014; Medoza-Tafolla et al., 2019). However, there is not a certain methodology about how to use the equipment for this purpose. Several research works have contributed different points to consider, as the chlorophyll readings may be affected by growth stage, genotypic variations, leaf age, water condition, and others (Xiong et al., 2015). In one-year-old plants of *Lagerstroemia indica, Callicarpa bodinieri*, and *Viburnum tinus* the variation of the SPAD reading due to leaf age was reduced due to the use for the measuring young leaves with the 75 percent of the total size (Demotes-Mainard et al., 2008). In a study with wheat plants for the N determination, it was concluded that a SPAD index had better results than only a SPAD reading, the index was calculated as a ratio of the SPAD reading in an analyzed plant, and the reading of a well-N-supplied plant (Yue et al., 2020). For N measuring in rice plants, better results may be obtained if the readings are taken in the middle of the leaf and in a clear area, about 30 mm far from the midrib. Also, for a standardized procedure on N measuring in rice, four indices were calculated with the average of several readings in fully expanded leaves separated from each other, those indices were related to the N content in the measured leaves, reducing the variation of chlorophyll due to phenological and genotypic variation (Lin et al., 2010). Another study recommends measuring with SPAD in the fourth fully expanded leaf at 2/3 from the base in rice plants due to a chlorophyll concentration in that area (Yuan et al., 2016). The variation in N determination for rice by the geographic area and cultivars was considerably reduced using a leaf-value model that considers the rate of N application, apparent N uptake, the SPAD value, and solid density (Li et al., 2020). In other studies with rice the variation source was irrigation; the N correlation with chlorophyll was not affected when plants were faced with water presence or absence, because the SPAD readings of six fully expanded leaves selected haphazardly were averaged (Cabangon et al., 2011). Also in rice, the SPAD values may be related to grain yield, so could be used as a diagnostic tool, but caution is recommended as N estimation with SPAD vary due to the growing season (Yang et al., 2014); similarly, in wheat plants, the SPAD values were related to the grain yield. nevertheless, there was considerable variation due to cultivars and the growing season (Monostori et al., 2016). Another condition that must be considered at the moment of measuring with SPAD is the exposure of the plant to sunlight (Demotes-Mainard et al., 2008), as it has an effect on the leaf mass per area and it is related to the N content. For the N measuring in leaves of *Prunus persica,* chlorophyll readings were taken with SPAD in 6 leaves from the middle of annual shoots in the tree canopy periphery where sunlight may be moderate (Rubio-Covarrubias et al., 2009). Sunlight was considered also in the N measurement in *Lactuca sativa L. var. longifolia* when the measurements were realized at solar noon (Medoza-Tafolla et al., 2019). In contrast, a study with *Solanum tuberosum* concluded that the chlorophyll readings were barely affected by the hour, between 9:00 and 15:00 (Goffart et al., 2008).

The SPAD equipment has been used together with other techniques for the N quantification and prediction. Using SPAD readings from leaves of *Malus domestica* trees under 4 different N treatments; linear models were created and used to estimate a difference of 25 Kg/ha between the N treatments applied and the estimated readings (Perry & Davenport, 2007). Another study with *Zea mays* related SPAD values for N content with a scale of color that was obtained with image processing, so N content could be estimated with a good degree of accuracy by analyzing the color of leaves (Reyes et al., 2017). Another technique used to improve the chlorophyll meter's performance for the N measuring in several crops is the Nitrogen Sufficiency index (NSI) (Yu et al., 2010; Solari et al., 2010; Tremblay et al., 2011; Yu et al., 2012). The NSI is a value calculated with the proportion between chlorophyll readings in plants that want to be measured for N content, and plants with a good N supply. Several values from 80–100 percent have been chosen to be considered frontier values, although other situations must be considered as soil characteristics, such as the source of the applied N, and the technique for the N application (Tremblay et al., 2011). The value of NSI may vary depending on the plant species, for *Zea mays* a 95 percent value was considered the limit, and under that frontier the plants would be considered to have N deficiency (Yu et al., 2010), although the frontier value for the same plant was set at 97 percent two years later by another study (Yu et al, 2012), in which *Solanum tuberosa* was determined that the frontier value varies across the growing season and, depending on the cultivar, the frontier values should be determined by an equation (Zheng et al., 2015). Something that must be considered while using SPAD for N determination is that the measuring will be affected also by the presence or absence of phosphorus (P) and potassium (K), as seen in plants of *Salvia,* so at the moment of providing N to the plant, P and K must be considered also (Dunn et al., 2018).

### 8.3.1.2   Chlorophyll Fluorescence Meters

The Dualex device is another chlorophyll meter used extensively that measures polyphenolic compounds in leaves. The operation of this device is based on chlorophyll fluorescence: two types of light are applied sequentially to the leaf, UV light (375 nm) and red light (650 nm). The first light is absorbed by polyphenols depending on their concentration; the second light is absorbed by the chlorophyll. The chlorophyll emits fluorescent light (695 nm) after having been exited with the two lights. The Dualex will measure that fluorescent light (Goffart et al., 2008). The SPAD and Dualex devices have been used mainly to achieve better measuring results. When a plant is fertilized with N, the content of photosynthetic proteins that are based on N increases, and the content of polyphenolic components that are based on carbon decreases, the conjunction SPAD/DUALEX present a better response in *Zea mays* crop measuring, with low variation after irrigation (Zhu et al., 2011), and also better results in *Lagerstroemia indica, Callicarpa bodinieri* and *Viburnum tinus,* which are woody plants and for which the SPAD measuring was unreliable (Demotes-Mainard et al., 2008). There are some methods of chlorophyll fluorescence measurement for plant stress and N determination (Thoren & Schmidhalter; 2009; Fernandez-Jaramillo et al., 2012) that could be considered better than the Dualex devices. In *Brassica* plants, N content was determined with a system that measured the plants with a distance from the sensor of 3 or 4 m. The system consisted of a sensor with an optic telescope equipped with filters and beam splitters that detect the chlorophyll´s radiation after a laser has been pulsed into plants (Thoren & Schmidhalter; 2009). The methods based on chlorophyll fluorescence as the Multiplex sensor have advantages over the ones based on chlorophyll reflectance (Yara N-Sensor, GreenSeeker RT 200, Crop Circle ACS-430, CompactSpec dual-channel diode array spectrometer) because it generates 4 wavelengths that excite chlorophyll in 4 different ways, which allows for more exact measurement of N in the absence or presence of sunlight (Tremblay et al., 2011; Friedel et al., 2020).

### 8.3.2 Nitrogen Measurement in Plants Using a UAV (Unmanned Aerial Vehicle)

During the last few years, precision agriculture (PA) has been focused on developing techniques in order to facilitate and make cheaper and faster the measurement of nitrogen (N) in plants for large coverage areas. This kind of monitoring allows, with N management, the estimation of biomass, yield prediction, early detection of disease, and knowing the effect of agricultural management practices. This is being applied to wide varieties of plants. Monitoring nitrogen status of plants in a large coverage area allows for design of models and strategies in order to assure food supply in the region under analysis, and to anticipate possible scarcity of food. However, these systems commonly lack high resolutions, have high costs, and are not easy to operate (Delloye et al., 2018; Sofonia et al., 2019). The conventional way to determine biomass in plants, which is related to nitrogen status, is by using destructive methods, which require harvesting manually and weighing, and in the case of biochemical methods, these increase the drawbacks when the coverage area is large (Han et al., 2019).

Nutritional deficiencies provoked by lack of N show visual symptoms (Cammarano et al., 2014; Osco et al., 2020), and there are several methods of having N monitoring in plants; this is achieved by using, principally, image sensors, which could be CCD (Charge Coupled Device) or CMOS (Complementary Metal Oxide Semiconductor) sensors that cover several regions of the electromagnetic spectrum. These devices give information in several wavelengths that can be employed to obtain vegetation's index (VIs), which are simple and effective algorithms for assessing quantitatively and qualitatively several properties and variables of plants (Xue & Su, 2017).

VIs have the possibility to perceive changes of chemical and physical properties of the plants, which could change according to the state of health or plant nutrition. For example, the Normalized Difference Vegetation Index (NDVI) is the vegetation index most widely used to detect nitrogen deficiencies in plants; however, there is a VIs index used to detect certain phenomena in the plants. In Xue and Su (2017) the authors present more than a hundred VIs and discuss the applicability depending on vegetation to be analyzed, climate, and precision. In Pantazi and colleagues (2016), the authors employ a NDVI vegetation index from images acquired from a UK-DMC-2 satellite. The output images of this satellite are multispectral, including green, red, and near-infrared bands, with 22 m of the resolution, along with this image. This work employs soil information and three machine-learning algorithms to determine yield prediction, which is directly related to the quantity of nitrogen in plants. In Lu and colleagues (2021), the authors used canopy coverage, plant height, and vegetation indexes, this last obtained by using a UAV, a M600, to acquire Red Green Blue (RGB) images to estimate leaf nitrogen in summer maize. One of the main advantages of this project is the low cost; four vegetation indexes were employed: green red ratio vegetation index (GRRI), green red vegetation index (GRVI), atmospherically resistant vegetation index (ARVI) and normalized redness intensity (NRI); and four integrative indexes. Linear models were used to design the relationship between VIs and leaf nitrogen concentration (LNC); the authors prove that, by combining these VIs, canopy height and a canopy covering, the estimation accuracy of the LNC determination might be increased to obtain an $R^2=0.758$ percent and an RMS=0.147 percent. The combination of multispectral and textural information in plants could help to improve the estimation of nitrogen in plants such as in Zheng and colleagues (2018), which employs NDVI, GCI (Green Chlorophyll Index), CIRE (Red Edge Chlorophyll Index), OSAVI (Optimized Soil Adjusted Vegetation Index) and VIopt (Optimal Vegetation Index) to correlate them with nitrogen and chlorophyll concentration. Along with this, gray level co-occurrence matrix, one of the most-used texture algorithms, was employed for texture analysis of the images obtained with UAV.

The aforementioned works show that remote sensing information coming from UAVs is one of the most important areas providing the nitrogen status of plants in large coverage areas, doing so in a fast and non-destructive way. And these methods show that the most common indexes used are red, red edge and near-infrared, which have been shown to be effective estimators of chlorophyll as well as N estimation of plants (Liu et al., 2018). The index to be employed to determine nitrogen in plants vary according to the crops being analyzed. This confirms that there is a need for a lot of work to determine the best index for each species, climatic conditions, and phenological stage, and researchers are working on that continuously.

According to Chlingaryan et al., 2018, several techniques have been used to estimate LNC. These methodologies vary from linear (CR, PLSR) to non-linear (ANN and SVM), and, in concordance with a previous comment, their effectiveness and the selection of vegetative index vary according to the ecological site, growth stage, and variety. The nonlinear algorithms have been gaining popularity in the last years, for example, Han and colleagues (2019), used spectral information from UAV and four machine-learning algorithms (multiple linear regression, SVM, ANN, and Random Forest) to model above-ground biomass (AGB). The results show that the Random Forest algorithm had fewer errors to estimate AGB. AGB could be related to nitrogen quantification in plants. Machine learning has helped in increasing the prediction accuracy of some agronomical variables, and it is easy to conclude that machine learning algorithms could continue helping to develop more cost-effective algorithms for N estimation of plants, situated over large areas, using spectral information from UAV. This information will be used to develop a new index in order to know, in a more quantitative way, the crop condition (Osco et al., 2020).

### 8.3.3 BIOSENSORS

Nowadays, the relevance of biosensors falls into the category of maintaining food safety principally; this industry will pass from USD 17 billion in 2018 to USD 24.6 billion by 2023, according to Giesche and Baeumner (2020). Publications related to biosensors for food safety increased 200 times from 2005 to 2019 and are principally focused on detecting pathogens, pesticides, and genetically modified organisms. Today, biosensors are dealing with in-situ analysis and crop pollutants, and several techniques for biosensing are starting to be used in crops and foods. These techniques include immunoassay, electrochemical impedance spectroscopy (EIS), cyclic voltammetry (CV), spectroscopy, and so forth (Kundo et al., 2019).

Today, the importance of monitoring the stages of food production is vital to ensuring the improvement of the overall yield (Pandey et al., 2018). The principal development of sensors in agriculture is focused on weed control, water content, plant physiology monitoring, and nutrient analysis. For plants, nutrient supply is fundamental to potentiate yield, and of the minerals employed as nutrients by plants, nitrogen is the most important. N is the most expensive compound used in agriculture, and it implies a huge demand for economic and energy resources. The correct and precise monitoring of N in plants will allow producers to save money in fertilizers and reduce the environmental impact of N applications (Barney et al., 2015). Biosensors, due to their selectivity advantage, can help to solve these monitoring requirements.

Classification of agricultural biosensors based on the type of transducers includes electrochemical, optical, piezoelectrical, or magnetic (Kundu et al., 2019) sensors, and among the types of recognition include antibody-antigen, enzyme-coenzyme-substrate, and nucleic acids-complementary sequences; also microorganism such as bacteria or plants and animal cells and tissues can be used. According to Velasco-Garcia and Mottram (2003) and Romero-Galindo and colleagues (2016), biosensors are an alternative to conventional techniques for developing sensors because of their specificity and sensitivity to measuring correctly and with high sensibility several variables in agriculture. Biosensors are also many times cheaper and are used with small devices, compared with the common methodologies.

Today there are several ongoing attempts to improve the production of N and the development of biosensors by using microorganisms; for example, the authors in Barney and colleagues (2015) use bacteria *Azobacter vinelandii* to increase the release of N by disrupting the urease genes that influence the capability of *A. vinelandii* to recycle the area in the cell. In this procedure *lacZ* from *Escherichia coli* MG1655 was added to *A. vinelandii* downstream of the scrX promoter, and the result was a blue phenotype that grows in the existence of X-Gal; and *nifLA* nitrogenase regulatory genes were replaced with tetracycline antibiotic marker. This modification generated a strain that grew well when N was supplied as a component of the medium. In the work presented by Goron and Raizada (2016), the fact of the glutamine increment (Gln) following N application to roots as nitrate or ammonium was used to develop detailed spatial and developmental gradient mapping of maize leaf Gln as consequence of several rates of N rates and uptake durations. Also, the authors tried to find Gln at vein resolution, and both objectives were achieved using GlnLux biosensor technologies. The authors previously had shown in Tessaro and colleagues (2012) that GlnLux luminesces when exogenous free Gln is proportionated, and the authors demonstrated that GlnLux cells are exposed to Gln from maize tissue extracts. Such biosensors can be used to study the dynamics of N assimilation, as can been seen in Goron (2017), where using GlnLux agar, characterized the timing of Gln availability in leaves of maize in the 48 hours following a dose of root N fertilizer. The seedling leaves were monitored at the same time to assess the Gln availability across the organs that could fluctuate in N strength. The authors have shown that the availability time of assimilatory Gln in leaves of maize was 24 h after peak accumulation. The temporal availability was similar in all seedling leaves, but Gln was more dependent on the kind of leaf.

The aforementioned initiatives show several ways of using biosensors to measure N accumulation and N dynamics in plants, and this is based on a microorganism genetically modified. This area seems promising in the development of biosensors to measure N in plants. Nowadays, there are several promoter and reporter genes that can be used to achieve, directly and indirectly, measurements of nutrients in plants, principally N assessments. According to Romero-Galindo and colleagues (2016), there are many promoter genes that can be used to build bacteria for detecting change or compounds in the environment. Those promoters are: *glnA, nirA, GifA, nitA* and *pyeaR*, which can detect several forms of nitrogen, such as ammonium nitrate and nitric oxide. In order to detect and quantify N in plants of in a substance by using genetically modified organisms, in several cases it is necessary to use reporter genes, which commonly employ fluorescence as an output variable. Reporter genes commonly used are *Gfp* (Green fluorescence protein), the operon gene *LacZ, LuxCDABE, dsRed, CobA, Luc, LuxAB, InaZ,* and *GlnLux*.

The disadvantage of biosensors employed in agriculture are the expensive costs, real-time estimation, and *in situ* quantification. Several authors are in concordance with these facts, which need to be solved by developing sensors and biosensors that take into account the real necessities of producers and satisfy the conditions of being easy to use, and need no sample preparation and design for the in-field-handling (Kundu et al., 2019; Griesche & Baeumner, 2020).

### 8.3.4 NITRATE SAP CONTENT AND ELECTRICAL VARIABLES

As noted, the leaves of the plants have optical properties (reflectance, transmittance, and absorptance) that depend heavily on chlorophyll content in tissue (Cabello-Pasini & Macias, 2011). The absence or presence of nutriments and the physiological status of the plant can affect these properties also. As many of the previous methods used for N determination are based on these optical properties, they have the disadvantage of being susceptible to present errors in the measuring due to sunlight variation, the water condition of the soil, and chlorophyll saturation. Nevertheless, the most important disadvantage of the optical methods is the time required between the start of an N deficiency and the reflection in the chlorophyll readings. An N deficiency in *Solanum tuberosum* plants was detected with chlorophyll readings around one month after the problem started, but to the contrary with petiole nitrate sap concentration measurements, the deficiency came out only in two weeks (Wu et al., 2007). Similarly in *Beta vulgaris* and *Brassica oleracea*, chlorophyll readings detected different N treatments after 61 and 108 days of application, while petiole sap nitrate concentration did it in 47 and 87 days, respectively. The measuring of N concentration in the sap of a plant is commonly used to determine the N status in crops (Farneselli et al., 2014; Yosoff et al., 2015; Hu et al., 2020). Instruments normally perform this type of measurement as the nitrate Ion-Selective Electrode (ISE) and a reflectometer is combined with nitrate test strips (Goffart et al., 2008).

### 8.3.4.1 Reflectometers and Nitrate Test Strips

For N measuring, the samples of sap are usually obtained from any petiole of the plant (Goffart et al., 2011), but the process has been reported as being different in several species – from the base of the leaf in *Latuca sativa* (Montemurro, 2010), from the base of the stems in *Zea mays* and *Triticum aestivum* (Lemaire et al., 2008), and from mature but recent leaves in *Cynara scolymus* (Rodrigo et al., 2004). Once the sample is placed, the nitrate test strips will change the color of the two reactive zones into a red-violet combination if nitrate is present in the sample, and the reflectometer will measure that change (Gromaz et al., 2017). Commercial reflectometers that have been reported in the literature are the RQflex (Rodrigo et al., 2008; Parks et al., 2012; Gromaz et al., 2017; San Bautista et al., 2020; Vetrano et al., 2020) and the Nitrachek (Goffart et al., 2008; Montemurro, 2010). The advantage of the use of test strips is that the method does not imply the use of toxic substances; also, the device and operation costs are low (2008). The disadvantage of this method is that the common nitrate concentrations in sap is higher than the maximum concentration that the strips can evaluate, so the sap sample must be diluted before being collocated in the strips, a situation that extebds the required time for the method (Goffart et al., 2008).

### 8.3.4.2 Ion Selective Electrode

The electrical conductivity of the sap can be affected by the content of several ions and forms of N, so the percentage of N in the sap can be measured with an Ion-Selective Electrode (ISE). This device has two electrodes in a container where a sap sample is placed; voltage is generated between electrodes, and the sap sample works as a bridge. The ISE can handle concentrations off 23–2,235 mg·L–1, in contradistinction to the 78 mg·L–1 concentration suggested for the strips (Goffart et al., 2008). An ISE commercial device reported in research is the TwinNO3 – B-341 (Horiba, Kyoto, Japan) (Olson et al., 2012; Tully and Weil, 2014; Carson et al., 2016). The disadvantage of this method is that the measurement by organic compounds and other ions in the sap as nitrite, bicarbonate, and chloride can be affected (Di Gioia et al., 2011).

### 8.3.4.3 Impedance Measurements

Given the ionic content in the sap, some tissues of the plant present electrical properties and conductivity, so the plant has parts that could be seen as electric circuits, and some tissues may act as passive elements with properties as resistance and capacitance (Azzarello et al., 2012). Resistance and capacitance establish the electrical bioimpedance, and it is defined as an opposition of an electrical system to the passage of Alternating Current (AC), which causes variations in amplitude and phase of the signal (Muñoz-Huerta et al., 2013). The electrical impedance in tissues from biological organisms can be affected by conditions of the organism such as water and health condition. In human health evaluations, it is used to estimate substances contained in the body, such as water, fat, and mass (Moon et al., 2008; Meeuwsen et al., 2010). In the case of plant studies, the electrical impedance has been used to determinate crop features as moisture, soluble solids, maturity (Liu, 2006; Guo et al; 2007; Chalermchat et al., 2010; Kuson & Terdwongworakul, 2013; Ando et al., 2014), and as with other physical and chemical properties of the plants the electrical impedance measurements are affected by plant conditions. Plants of *Solanum lycopersicum* presented variations in the electrical impedance measurements due to water stress (He et al., 2011), and in *Solanum melongena* pulp the measurements variated due to the processes of drying and freezing (Wu et al., 2008). Another factor for variations in the electrical impedance measurements is the percentage of water and nutrients content in the plant as potassium, calcium, and phosphorus (Greenham et al., 1982; Wei et al., 1995; Zheng et al., 2015; Meiqing et al., 2016; Jinyang et al., 2016). In plants of *Lactuca sativa L.*, with four N treatments the electrical impedance was measured during specific time periods. The impedance measurement was made with two stainless steel needle electrodes and a LCR HiTESTER 3532-50 (Hioki E. E. Corporation, Nagano, Japan), and the nitrogen content was measured with the Kjeldahl method. A strong correlation was established between the frequency values and the N content (Muñoz-Huerta et al., 2014), with similar results in plants of *Zea mays, Brassica napus, Triticum sp and Glycine max* (Basak et al., 2020), *Solanum lycopersicum* (Li et al., 2017), concluding that the N in a plant can be estimated with the electrical impedance measurement.

## 8.4   CONCLUSION

Taken together with the information aforementioned, the fundamental result is that nitrogen estimation in agricultural and horticultural practices will require more accurate, fast, and low-cost devices based on some or several of the principles mentioned in the strategies included in this chapter. Precision agriculture will require these strategies in order to contribute to a more sustainable vegetable production (Figure 8.1).

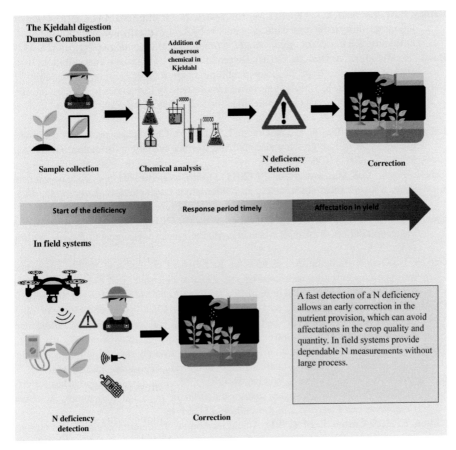

**FIGURE 8.1** The most important difference between the two principal Nitrogen measurement techniques is the time required to identify a N deficiency. Chemical techniques may indicate the lack of N when there is already an affectation on the crop due to the long process required. Otherwise, in field techniques might indicate a problem before the affectation occurs.

## REFERENCES

Ainsworth, E.A. (2018). Photosynthesis. In: Hatfield, J.L., Sivakumar, M.V.K., Prueger, J.H., eds. *Agroclimatology: Linking Agriculture to Climate*. Wiley, 1–23.

American Public Health Association, American Water Works Association, and Water Pollution Control Federation. (1992). *Standard Methods for the Examination of Water and Wastewater. Definition and Procedure for the Determination of Method Detection Limits; 18th Edition*. Washington, DC: American Public Health Association.

Amin, M., & Flowers, T. H. (2004). Evaluation of Kjeldahl digestion method. *Journal of Research in Science Teaching, 15*, 159–179.

Ando, Y., Mizutani, K., Wakatsuki, N. (2014). Electrical impedance analysis of potato tissues during drying. *Journal of Food Engineering*, 121:24–31. https://doi.org/10.1016/j.jfoodeng.2013.08. 008.

Barney, B.M., Eberhart, L.J., Ohlert, J.M., Knutson, C.M., & Plunkett, M.H. (2015). Gene deletions resulting in increased nitrogen release by Azotobacter vinelandii: application of a novel nitrogen biosensor. *Applied and Environmental Microbiology*, *81*(13), 4316–4328.

Basak, R., Wahid, K., & Dinh, A. (2020). Determination of leaf nitrogen concentrations using electrical impedance spectroscopy in multiple crops. *Remote Sensing*, *12*(3), 566.

Bausch, W.C., & Diker, K. (2001). Innovative remote sensing techniques to increase nitrogen use efficiency of corn. *Communications in Soil Science and Plant Analysis*, *32*(7–8), 1371–1390.

Cabangon, R. J., Castillo, E. G., & Tuong, T. P. (2011). Chlorophyll meter-based nitrogen management of rice grown under alternate wetting and drying irrigation. *Field Crops Research*, *121*(1), 136–146.

Cabello-Pasini, A., & Macías-Carranza, V. (2011). Optical properties of grapevine leaves: reflectance, transmittance, absorptance, and chlorophyll concentration. *Agrociencia*, *45*(8), 943–957.

Cammarano, D., Fitzgerald, G. J., Casa, R., & Basso, B. (2014). Assessing the robustness of vegetation indices to estimate wheat N in Mediterranean environments. *Remote Sensing*, *6*(4), 2827–2844.

Carson, L., Ozores-Hampton, M., & Morgan, K. (2016). Correlation of petiole sap nitrate-nitrogen concentration measured by ion selective electrode, leaf tissue nitrogen concentration, and tomato yield in Florida. *Journal of Plant Nutrition*, *39*(12), 1809–1819.

Chalermchat, Y., Malangone, L., & Dejmek, P. (2010). Electropermeabilization of apple tissue: Effect of cell size, cell size distribution, and cell orientation. *Biosystems Engineering*, *105*(3), 357–366.

Chen, H., Wang, P., Li, J., Zhang, J., & Zhong, L. (2012). Canopy spectral reflectance feature and leaf water potential of sugarcane inversion. *Physics Procedia*, *25*, 595–600.

Chlingaryan, A., Sukkarieh, S., & Whelan, B. (2018). Machine learning approaches for crop yield prediction and nitrogen status estimation in precision agriculture: A review. *Computers and Electronics in Agriculture*, *151*, 61–69.

Clifton, K. E., & Clifton, L. M. (1991). A field method for the determination of total nitrogen in plant tissue. *Communications in Soil Science and Plant Analysis*, *22*(9–10), 851–860.

Coste, S., Baraloto, C., Leroy, C., Marcon, É., Renaud, A., Richardson, A.D., … & Hérault, B. (2010). Assessing foliar chlorophyll contents with the SPAD-502 chlorophyll meter: a calibration test with thirteen tree species of tropical rainforest in French Guiana. *Annals of Forest Science*, *67*(6), 607–607.

Cunniff, P., & Association of Official Analytical Chemists. (1995). *Official Methods of Analysis of AOAC International*. Washington, DC: Association of Official Analytical Chemists.

Delloye, C., Weiss, M., & Defourny, P. (2018). Retrieval of the canopy chlorophyll content from Sentinel-2 spectral bands to estimate nitrogen uptake in intensive winter wheat cropping systems. *Remote Sensing of Environment*, *216*, 245–261.

Domini, C., Vidal, L., Cravotto, G., & Canals, A. (2009). A simultaneous, direct microwave/ultrasound-assisted digestion procedure for the determination of total Kjeldahl nitrogen. *Ultrasonics Sonochemistry*, *16*(4), 564–569.

Dumas, J.B.A. (1831). Procedes de l'analyse organique. *Ann. Chim. Phys*, *47*, 198–205.

Dunn, B.L., Singh, H., Payton, M., & Kincheloe, S. (2018). Effects of nitrogen, phosphorus, and potassium on SPAD-502 and leaf sensor readings of Salvia. *Journal of Plant Nutrition*, *41*(13), 1674–1683.

Farneselli, M., Tei, F., & Simonne, E. (2014). Reliability of petiole sap test for N nutritional status assessing in processing tomato. *Journal of Plant Nutrition*, *37*(2), 270–278.

Fontes, P.C.R., & de Araujo, C. (2006). Use of a chlorophyll meter and plant visual aspect for nitrogen management in tomato fertigation. *Journal of Applied Horticulture*, *8*(1), 8–11.

Friedel, M., Hendgen, M., Stoll, M., & Löhnertz, O. (2020). Performance of reflectance indices and of a handheld device for estimating in-field the nitrogen status of grapevine leaves. *Australian Journal of Grape and Wine Research*, *26*(2), 110–120.

Goffart, J.P., Olivier, M., & Frankinet, M. (2011). Crop nitrogen status assessment tools in a decision support system for nitrogen fertilization management of potato crops. *HortTechnology*, *21*(3), 282–286.

Goffart, J. P., Olivier, M., & Frankinet, M. (2008). Potato crop nitrogen status assessment to improve N fertilization management and efficiency: past–present–future. *Potato Research*, *51*(3–4), 355–383.

Goron, T.L., & Raizada, M.N. (2016). Biosensor-based spatial and developmental mapping of maize leaf glutamine at vein-level resolution in response to different nitrogen rates and uptake/assimilation durations. *BMC Plant Biology*, *16*(1), 1–11.

Goron, T.L., & Raizada, M.N. (2020). Biosensor-Mediated in Situ Imaging Defines the Availability Period of Assimilatory Glutamine in Maize Seedling Leaves Following Nitrogen Fertilization. *Nitrogen*, *1*(1), 3–11.

Griesche, C., & Baeumner, A.J. (2020). Biosensors to support sustainable agriculture and food safety. *TrAC Trends in Analytical Chemistry*, 115906.

Gromaz, A., Torres, J.F., San Bautista, A., Pascual, B., López-Galarza, S., & Maroto, J.V. (2017). Effect of different levels of nitrogen in nutrient solution and crop system on nitrate accumulation in endive. *Journal of Plant Nutrition*, *40*(14), 2045–5053.

Guo, W.C., Nelson, S.O., Trabelsi, S., & Stanley, J.K. (2007). Dielectric properties of honeydew melons and correlation with quality. *Journal of Microwave Power and Electromagnetic Energy*, *41*(2), 44–54.

Han, L., Yang, G., Dai, H., Xu, B., Yang, H., Feng, H., … & Yang, X. (2019). Modeling maize above-ground biomass based on machine learning approaches using UAV remote-sensing data. *Plant Methods*, *15*(1), 1–19.

Handson, P. D., & Shelley, B. C. (1993). A review of plant analysis in Australia. *Australian Journal of Experimental Agriculture*, *33*(8), 1029–1038.

He, J.X., Wang, Z.Y., Shi, Y.L., Qin, Y., Zhao, D.J., & Huang, L. (2011). A prototype portable system for bioelectrical impedance spectroscopy. *Sensor Letters*, *9*(3), 1151–1156.

Hu, Y., & Guy, R.D. (2020). Isotopic composition and concentration of total nitrogen and nitrate in xylem sap under near steady-state hydroponics. *Plant, Cell & Environment*, *43*(9), 2112–2123.

Jinyang, L., Meiqing, L., Hanping, M., & Wenjing, Z. (2016). Diagnosis of potassium nutrition level in Solanum lycopersicum based on electrical impedance. *Biosystems Engineering*, *147*, 130–138.

Kalra, Y.P. (1998) Hand Book of Reference Methods for Plant Analysis; CRC Press: Boca Raton, FL, pp. 75–92.

Kjeldahl, J.G.C.T. (1883) Neue methode zur bestimmung des stickstoffs in organischen korpern. *Fresenius' Journal of Analytical Chemistry*, 22, 366–382

Krotz, L., Leone, F., & Giazzi, G. (2016). Nitrogen/protein determination in food and animal feed by combustion method (Dumas) using the Thermo Scientific FlashSmart Elemental Analyzer. *Thermo Fisher Application Note AN42262-EN*, *716*.

Kundu, M., Krishnan, P., Kotnala, R.K., & Sumana, G. (2019). Recent developments in biosensors to combat agricultural challenges and their future prospects. *Trends in Food Science & Technology*, *88*, 157–178.

Kuson, P., & Terdwongworakul, A. (2013). Minimally destructive evaluation of durian maturity based on electrical impedance measurement. *Journal of Food Engineering*, *116*(1), 50–56.

Labconco, C. (1998). A guide to Kjeldahl nitrogen determination methods and apparatus. Houston: Labconco Corporation.

Lanza, J.G., Churión, P.C., & Gómez, N. (2016). Comparación entre el método Kjeldahl tradicional y el método Dumas automatizado (N cube) para la determinación de proteínas en distintas clases de alimentos. *Saber, 28*(2), 245–249.

Lee, D., Nguyen, V., & Littlefield, S. (1996). Comparison of methods for determination of nitrogen levels in soil, plant and body tissues, and water. *Communications in Soil Science and Plant Analysis, 27*(3–4), 783–793.

Lemaire, G., Jeuffroy, M.H., & Gastal, F. (2008). Diagnosis tool for plant and crop N status in vegetative stage: Theory and practices for crop N management. *European Journal of Agronomy, 28*(4), 614–624.

Li, J., Feng, Y., Wang, X., Peng, J., Yang, D., Xu, G., ... & Su, W. (2020). Stability and applicability of the leaf value model for variable nitrogen application based on SPAD value in rice. *PloS One, 15*(6), e0233735.

Lin, M., Jinyang, L., Xinhua, W., & Wenjing, Z. (2017). Early diagnosis and monitoring of nitrogen nutrition stress in tomato leaves using electrical impedance spectroscopy. *International Journal of Agricultural and Biological Engineering, 10*(3), 194–205.

Lin, F.F., Deng, J.S., Shi, Y.Y., Chen, L.S., & Wang, K. (2010). Investigation of SPAD meter-based indices for estimating rice nitrogen status. *Computers and Electronics in Agriculture, 71*, S60–S65.

Ling, Q., Huang, W., & Jarvis, P. (2011). Use of a SPAD-502 meter to measure leaf chlorophyll concentration in Arabidopsis thaliana. *Photosynthesis Research, 107*(2), 209–214.

Liu, S., Li, L., Gao, W., Zhang, Y., Liu, Y., Wang, S., & Lu, J. (2018). Diagnosis of nitrogen status in winter oilseed rape (Brassica napus L.) using in-situ hyperspectral data and unmanned aerial vehicle (UAV) multispectral images. *Computers and Electronics in Agriculture, 151*, 185–195.

Liu, X. (2006). Electrical impedance spectroscopy applied in plant physiology studies. *Master of Engineering*. Royal Melbourne Institute of Technology (RMIT)].

Lu, J., Cheng, D., Geng, C., Zhang, Z., Xiang, Y., & Hu, T. (2021). Combining plant height, canopy coverage, and vegetation index from UAV-based RGB images to estimate leaf nitrogen concentration of summer maize. *Biosystems Engineering, 202*, 42–54.

Martín, J., Fernández Sarria, L., & Asuero, A. G. (2017). The Kjeldahl titrimetric finish: on the ammonia titration trapping in boric acid. In: *Advances in Titration Techniques*. IntechOpen: 23–58.

Meiqing, L., Jinyang, L., Hanping, M., & Yanyou, W. (2016). Diagnosis and detection of phosphorus nutrition level for Solanum lycopersicum based on electrical impedance spectroscopy. *Biosystems Engineering, 143*, 108–118.

Mendoza-Tafolla, R.O., Juarez-Lopez, P., Ontiveros-Capurata, R.E., Sandoval-Villa, M., Iran, A.T., & Alejo-Santiago, G. (2019). Estimating nitrogen and chlorophyll status of romaine lettuce using SPAD and at LEAF readings. *Notulae Botanicae Horti Agrobotanici Cluj-Napoca, 47*(3), 751–756.

Miao, Y., Mulla, D.J., Randall, G.W., Vetsch, J.A., & Vintila, R. (2009). Combining chlorophyll meter readings and high spatial resolution remote sensing images for in-season site-specific nitrogen management of corn. *Precision Agriculture, 10*(1), 45–62.

Michałowski, T., Asuero, A.G., & Wybraniec, S. (2013). The titration in the Kjeldahl method of nitrogen determination: base or acid as titrant? *Journal of Chemical Education, 90*(2), 191–197.

Möller, J. (2010). Protein analysis revisited. *Focus, 34*(2), 22–23.

Monostori, I., Árendás, T., Hoffman, B., Galiba, G., Gierczik, K., Szira, F., & Vágújfalvi, A. (2016). Relationship between SPAD value and grain yield can be affected by cultivar, environment and soil nitrogen content in wheat. *Euphytica, 211*(1), 103–112.

Montemurro, F. (2010). Are organic N fertilizing strategies able to improve lettuce yield, use of nitrogen and N status? *Journal of Plant Nutrition, 33*(13), 1980–1997.

Muñoz-Huerta, R.F., Guevara-Gonzalez, R.G., Contreras-Medina, L.M., Torres-Pacheco, I., Prado-Olivarez, J., & Ocampo-Velazquez, R.V. (2013). A review of methods for sensing the nitrogen status in plants: advantages, disadvantages and recent advances. *Sensors, 13*(8), 10823–10843.

Muñoz-Huerta, R.F., Ortiz-Melendez, A.D.J., Guevara-Gonzalez, R.G., Torres-Pacheco, I., Herrera-Ruiz, G., Contreras-Medina, L.M., … & Ocampo-Velazquez, R.V. (2014). An analysis of electrical impedance measurements applied for plant N status estimation in lettuce (Lactuca sativa). *Sensors, 14*(7), 11492–11503.

Olson, S.M., W.M. Stall, G.E. Vallad, S.E. Webb, S.A. Smith, E.H. Simonne, E.J. McAvoy, and B.M. Santos. (2012). Tomato production in Florida. In: *Vegetable Production Handbook for Florida,* eds. S.M. Olson and B. Santos, pp. 321–344. Vance: Lenexa, KS.

Osco, L.P., Junior, J.M., Ramos, A.P.M., Furuya, D.E.G., Santana, D.C., Teodoro, L.P.R., … & Teodoro, P.E. (2020). Leaf nitrogen concentration and plant height prediction for maize using UAV-based multispectral imagery and machine learning techniques. *Remote Sensing, 12*(19), 3237.

Pandey, R., Teig-Sussholz, O., Schuster, S., Avni, A., & Shacham-Diamand, Y. (2018). Integrated electrochemical Chip-on-Plant functional sensor for monitoring gene expression under stress. *Biosensors and Bioelectronics, 117*, 493–500.

Pantazi, X.E., Moshou, D., Alexandridis, T., Whetton, R.L., & Mouazen, A.M. (2016). Wheat yield prediction using machine learning and advanced sensing techniques. *Computers and Electronics in Agriculture, 121*, 57–65.

Peng, S., Garcia, F.V., Laza, R.C., Sanico, A.L., Visperas, R.M., & Cassman, K.G. (1996). Increased N-use efficiency using a chlorophyll meter on high-yielding irrigated rice. *Field Crops Research, 47*(2–3), 243–252.

Perry, E.M., & Davenport, J.R. (2007). Spectral and spatial differences in response of vegetation indices to nitrogen treatments on apple. *Computers and Electronics in Agriculture, 59*(1–2), 56–65.

Persson, J.Å. (2008). *Handbook for Kjeldahl Digestion: A Recent Review of the Classical Method with Improvements Developed by FOSS.* Hilleroed, Denmark.

Piekielek, W.P., Fox, R.H., Toth, J.D., & Macneal, K.E. (1995). Use of a chlorophyll meter at the early dent stage of corn to evaluate nitrogen sufficiency. *Agronomy Journal, 87*(3), 403–408.

Pontes, F.V., Carneiro, M.C., Vaitsman, D.S., da Rocha, G.P., da Silva, L.I., Neto, A.A., & Monteiro, M.I.C. (2009). A simplified version of the total Kjeldahl nitrogen method using an ammonia extraction ultrasound-assisted purge-and-trap system and ion chromatography for analyses of geological samples. *Analytica Chimica Acta, 632*(2), 284–288.

Reyes, J.F., Correa, C., & Zuniga, J. (2017). Reliability of different color spaces to estimate nitrogen SPAD values in maize. *Computers and Electronics in Agriculture, 143*, 14–22.

Rodrigo, M.C., Ginestar, J., Boix, M., & Ramos, C. (2004, November). Evaluation of rapid methods for nitrate plant sap analysis of globe artichoke grown in sand culture. In *International Symposium on Soilless Culture and Hydroponics, 697* (pp. 393–397).

Romero-Galindo, R., Torres-Pacheco, I., Guevara-Gonzalez, R. G. & Contreras-Medina, L. M. (2016). Biosensors used for quantification of nitrates in plants. *Journal of Sensors,* doi: https://doi.org/10.1155/2016/1630695

Rubio-Covarrubias, O.A., Brown, P.H., Weinbaum, S.A., Johnson, R.S., & Cabrera, R.I. (2009). Evaluating foliar nitrogen compounds as indicators of nitrogen status in Prunus persica trees. *Scientia Horticulturae, 120*(1), 27–33.

Sáez-Plaza, P., Michałowski, T., Navas, M. J., Asuero, A. G., & Wybraniec, S. (2013). An overview of the Kjeldahl method of nitrogen determination. Part I. Early history, chemistry of the procedure, and titrimetric finish. *Critical Reviews in Analytical Chemistry*, *43*(4), 178–223.

Saha, U. K., Sonon, L., & Kissel, D. E. (2012). Comparison of conductimetric and colorimetric methods with distillation–titration method of analyzing ammonium nitrogen in total kjeldahl digests. *Communications in Soil Science and Plant Analysis*, *43*(18), 2323–2341.

San Bautista, A., Gromaz, A., Ferrarezi, R.S., López-Galarza, S., Pascual, B., & Maroto, J.V. (2020). Effect of cropping system and humidity level on nitrate content and tipburn incidence in Endive. *Agronomy*, *10*(5), 749.

Silva, D.D., & Queiroz, A.D. (2002). *Análise de alimentos: métodos químicos e biológicos Viçosa*. MG: Universidade Federal de Vicosa –UFV.

Silva, T.E.D., Detmann, E., Franco, M.D.O., Palma, M.N.N., & Rocha, G.C. (2016). Evaluation of digestion procedures in Kjeldahl method to quantify total nitrogen in analyses applied to animal nutrition. *Acta Scientiarum. Animal Sciences*, *38*(1), 45–51.

Sofonia, J., Shendryk, Y., Phinn, S., Roelfsema, C., Kendoul, F., & Skocaj, D. (2019). Monitoring sugarcane growth response to varying nitrogen application rates: A comparison of UAV SLAM LiDAR and photogrammetry. *International Journal of Applied Earth Observation and Geoinformation*, *82*, 101878.

Solari, F., Shanahan, J.F., Ferguson, R.B., & Adamchuk, V.I. (2010). An active sensor algorithm for corn nitrogen recommendations based on a chlorophyll meter algorithm. *Agronomy Journal*, *102*(4), 1090–1098.

Taiz, L., & Zeiger, E. (2010). *Plant physiology, 5th edn*. Sinauer Associates: Sunderland, MA.

Tessaro, M.J., Soliman, S.S., & Raizada, M.N. (2012). Bacterial whole-cell biosensor for glutamine with applications for quantifying and visualizing glutamine in plants. *Applied and Environmental Microbiology*, *78*(2), 604–606.

Tremblay, N., Fallon, E., & Ziadi, N. (2011). Sensing of crop nitrogen status: Opportunities, tools, limitations, and supporting information requirements. *HortTechnology*, *21*(3), 274–281.

Tully, K. L., & Weil, R. (2014). Ion-selective electrode offers accurate, inexpensive method for analyzing soil solution nitrate in remote regions. *Communications in Soil Science and Plant Analysis*, *45*(14), 1974–1980.

Unkovich, M., Herridge, D., Peoples, M., Cadisch, G., Boddey, B., Giller, K., ... & Chalk, P. (2008). *Measuring Plant-associated Nitrogen Fixation in Agricultural Systems*. Australian Centre for International Agricultural Research (ACIAR).

Velasco-Garcia, M.N., & Mottram, T. (2003). Biosensor technology addressing agricultural problems. *Biosystems Engineering*, *84*(1), 1–12.

Vetrano, F., Miceli, C., Angileri, V., Frangipane, B., Moncada, A., & Miceli, A. (2020). Effect of Bacterial Inoculum and Fertigation Management on Nursery and Field Production of Lettuce Plants. *Agronomy*, *10*(10), 1477.

Watson, M.E., & Galliher, T.L. (2001). Comparison of Dumas and Kjeldahl methods with automatic analyzers on agricultural samples under routine rapid analysis conditions. *Communications in Soil Science and Plant Analysis*, *32*(13–14), 2007–2019.

Wei, Y.Q., Bailey, B.J., & Stenning, B.C. (1995). A wetness sensor for detecting condensation on tomato plants in greenhouses. *Journal of Agricultural Engineering Research*, *61*(3), 197–204.

Wojciechowski, K.L., & Barbano, D.M. (2015). Modification of the Kjeldahl noncasein nitrogen method to include bovine milk concentrates and milks from other species. *Journal of Dairy Science*, *98*(11), 7510–7526.

Wu, J., Wang, D., Rosen, C.J., & Bauer, M.E. (2007). Comparison of petiole nitrate concentrations, SPAD chlorophyll readings, and QuickBird satellite imagery in detecting nitrogen status of potato canopies. *Field Crops Research, 101*(1), 96–103.

Wu, L., Ogawa, Y., & Tagawa, A. (2008). Electrical impedance spectroscopy analysis of eggplant pulp and effects of drying and freezing–thawing treatments on its impedance characteristics. *Journal of Food Engineering, 87*(2), 274–280.

Xiong, D., Chen, J., Yu, T., Gao, W., Ling, X., Li, Y., … & Huang, J. (2015). SPAD-based leaf nitrogen estimation is impacted by environmental factors and crop leaf characteristics. *Scientific Reports, 5*(1), 1–12.

Xu, L., Zhai, L., Liu, Y., & Zhang, L. (2014). Determination of the protein content in milk by Dumas combustion method. *Journal of Food Safety and Quality, 5*(12), 3903–3905.

Xue, J., & Su, B. (2017). Significant remote sensing vegetation indices: A review of developments and applications. *Journal of Sensors, 2017*.

Yang, H., Yang, J., Lv, Y., & He, J. (2014). SPAD values and nitrogen nutrition index for the evaluation of rice nitrogen status. *Plant Production Science, 17*(1), 81–92.

Yosoff, S.F., Mohamed, M.T.M., Parvez, A., Ahmad, S.H., Ghazali, F.M., & Hassan, H. (2015). Production system and harvesting stage influence on nitrate content and quality of butterhead lettuce. *Bragantia, 74*(3), 322–330.

Yu, H., Wu, H-S., & Wang, Z-J. (2010). Evaluation of SPAD and Dualex for in-season corn nitrogen status estimation. *Acta Agronomica Sinica, 36*(5), 840–847.

Yu, W., Miao, Y., Feng, G., Yue, S., & Liu, B. (2012, August). Evaluating different methods of using chlorophyll meter for diagnosing nitrogen status of summer maize. In *2012 First International Conference on Agro-Geoinformatics (Agro-Geoinformatics)* (pp. 1–4). IEEE.

Yuan, Z., Cao, Q., Zhang, K., Ata-Ul-Karim, S.T., Tian, Y., Zhu, Y., … & Liu, X. (2016). Optimal leaf positions for SPAD meter measurement in rice. *Frontiers in Plant Science, 7*, 719.

Yue, X., Hu, Y., Zhang, H., & Schmidhalter, U. (2020). Evaluation of both SPAD reading and SPAD index on estimating the plant nitrogen status of winter wheat. *International Journal of Plant Production, 14*(1), 67–75.

Zebarth, B.J., Drury, C.F., Tremblay, N., & Cambouris, A.N. (2009). Opportunities for improved fertilizer nitrogen management in production of arable crops in eastern Canada: A review. *Canadian Journal of Soil Science, 89*(2), 113–132.

Zheng, H., Cheng, T., Li, D., Yao, X., Tian, Y., Cao, W., & Zhu, Y. (2018). Combining unmanned aerial vehicle (UAV)-based multispectral imagery and ground-based hyperspectral data for plant nitrogen concentration estimation in rice. *Frontiers in Plant Science, 9*, 936.

Zheng, H.-L., Qin, Y.-L., Fan, M.-S., Liu, Y.-C., & Chen, Y. (2015). Establishing dynamic thresholds for potato nitrogen status diagnosis with the SPAD chlorophyll meter. *Journal of Integrative Agriculture, 14*(1), 190–195.

Zheng, L., Wang, Z., Sun, H., Zhang, M., & Li, M. (2015). Real-time evaluation of corn leaf water content based on the electrical property of leaf. *Computers and Electronics in Agriculture, 112*, 102–109.

Zhu, J., Tremblay, N., & Liang, Y. (2011). A corn nitrogen status indicator less affected by soil water content. *Agronomy Journal, 103*(3), 890–898.

# Part 3

## Nitrate in Water

# 9 Nitrate in Fresh Waters

*Daniel F. Gomez Isaza and Essie M. Rodgers*

## CONTENTS

9.1 Introduction .................................................................................................185
9.2 Nitrate in Freshwater and Impacts on Aquatic Life .....................................186
9.3 Nitrate Guidelines.........................................................................................187
9.4 Toxicity of Nitrate to Aquatic Life...............................................................188
9.5 Growth, Development, and Behavior............................................................190
9.6 Histopathological Disruptions to Freshwater Animals..................................192
9.7 Endocrine Disruption and Reproduction......................................................194
9.8 Nitrate and Other Stressors...........................................................................195
    9.8.1 Nitrate and Water Temperature................................................................196
    9.8.2 Nitrate and Low-Water pH .....................................................................196
    9.8.3 Nitrate and Low Oxygen Levels.............................................................197
    9.8.4 Nitrate and Salinity................................................................................197
    9.8.5 Nitrate and Other Pollutants ..................................................................198
9.9 Conclusions and Future Directions ..............................................................199
References.............................................................................................................200

## 9.1 INTRODUCTION

Nitrate is a natural and important component of freshwater ecosystems. Nitrate occurs at "low" background (0–2 mg $NO_3$-N $L^{-1}$) levels that support the productivity of freshwater systems (Galloway et al. 2004). Yet, increased nutrient loadings caused by human activities are a driver of global change and pose significant challenges for managing freshwater ecosystems worldwide (Camargo and Alonso 2006a, Vitousek et al. 1997, Goeller et al. 2019). Many human practices have contributed to the oversaturation of waterbodies with nitrate, including the combustion of fossil fuels, urban wastewater, and aquaculture operations (Camargo and Alonso 2006a, Vitousek et al. 1997). Nitrate concentrations can be of particular concern around agricultural sites, reaching unpresented levels upwards of 50 mg $NO_3$-N $L^{-1}$ in ground and surface water runoff (Camargo and Alonso 2006a, Goeller et al. 2019). These elevated concentrations of nitrate can cause acute toxicity to residing species (Ortiz-Santaliestra et al. 2006, Gomez Isaza, Cramp, and Franklin 2020a). Various governing bodies have imposed the strict management of maximum allowable nitrate concentration that enter freshwaters and, yet, high levels of nitrate remain a concern for some aquatic animals.

DOI: 10.1201/9780429326806-12

When dissolved in freshwater, aquatic animals readily take up nitrate ions via the gill, and this is associated with series of toxic effects. Research on the impact of nitrate on aquatic animals spans several decades and has increased steadily due to the intensification of nitrate in freshwaters (Camargo and Alonso 2006a, Vitousek et al. 1997). High levels of nitrate can be toxic to aquatic animals – that is, causing direct mortality (Camargo and Alonso 2006a, Camargo, Alonso, and Salamanca 2005). But, perhaps more concerningly, high levels of nitrate can lead to more subtle but profound effects, such as developmental delays, behavioural alterations, and histopathological change that together can impact the fitness and survival of aquatic animals (Ortiz-Santaliestra et al. 2006, Gomez Isaza, Cramp, and Franklin 2020a, Hamlin 2006, Chow and Hong 2002). Some species and taxonomic groups are, however, resilient to high levels of nitrate (Yang et al. 2019, McGurk et al. 2006, Baker et al. 2017). Yet, despite this growing interest, the impacts of nitrate on freshwater fauna remain unclear. This is likely because the toxicity of nitrate can vary greatly between freshwater species, and depend on a number of factors, intrinsic (e.g., body size, life stage, sex) and extrinsic (e.g., toxicity modifying factors), and contribute to the current debate surrounding the perceived threat of nitrate on freshwater animals (Ortiz-Santaliestra et al. 2006, Gomez Isaza, Cramp, and Franklin 2020a, Hamlin 2006, Chow and Hong 2002). In the following sections, we aim to provide a summary of the known impacts of nitrate on aquatic animals. We will examine how nitrate interacts with other global threats to affect species' persistence to provide a holistic picture of the threat of elevated nitrate levels on freshwater fauna.

## 9.2   NITRATE IN FRESHWATER AND IMPACTS ON AQUATIC LIFE

In aquatic environments, nitrogen occurs primarily in the form of ammonium ($NH_4^+$), nitrite ($NO_2^-$), and nitrate ($NO_3^-$). These ions are generally present at "low" background levels ($< 2.5$ mg $L^{-1}$) as a result of atmospheric deposition, decomposition of biological matter, and nitrogen fixation by nitrifying prokaryotes (Panno et al. 2006). As part of the nitrogen cycle, ammonium is oxidized to nitrate in a two-step process (ammonium $\rightarrow$ nitrite $\rightarrow$ nitrate) by chemo-autotrophic bacteria (primarily Nitrosomonas and Nitrobacter; Sharma and Ahlert 1977), resulting in nitrate concentrations typically being higher than those of ammonium and nitrite (Glibert 2017). Nitrate (as well as ammonium and nitrite) in the environment can be taken up and assimilated as a nitrogen source by plants, algae, and bacteria (Galloway et al. 2004).

Human activities have significantly altered the global nitrogen cycle by increasing both the availability and mobility of nitrogen across the world (Vitousek et al. 1997). Anthropogenic nitrate pollution has approximately doubled the rate of nitrogen entering aquatic ecosystems. Consequently, in addition to natural sources, nitrate can nowadays enter freshwaters via a combination of inputs, including animal farming, nitrogen-based fertilizers, the combustion of fossil fuels, sewage, mines, and urban run-off (Camargo and Alonso 2006a, Vitousek et al. 1997, Goeller et al. 2019). Nitrate pollution is particularly prominent in areas of high fertilizer use where agricultural land practices (crop and livestock) have increased nitrate loadings in surrounding waters, causing eutrophication, algal blooms, anoxic (oxygen-free) dead zones, altered food webs, and nitrate toxicity to residing species (Camargo, Alonso, and Salamanca 2005, Glibert 2017). For example, after a heavy rainfall, the use of a nitrate-based fertilizer (ammonium nitrate) has caused the concentration of nitrate in surface-water to spike from 2 mg $NO_3 - NL^{-1}$ to 100 mg $NO_3 - NL^{-1}$ in sites adjacent to agriculture plantations (Jaynes et al. 2001). Highly urbanized areas face a similar problem; a positive correlation between nitrate concentration and human population density has been documented for many river systems world-wide (Gu et al. 2013, Mayo et al. 2019, Ouedraogo and Vanclooster 2016), where surface water and groundwater nitrate concentrations have been recorded at persist-ently high levels of ~ 25 mg $NO_3 - NL^{-1}$ and up to $100 \cdot mg \cdot NO_3 - NL^{-1}$, respect-ively (reviewed by Vitousek et al. 1997, Galloway et al. 2004). Such increases in anthropogenic nitrate pollution have shifted levels above historic baselines in fresh-water ecosystems (Panno et al. 2006), and elevated exposure to nitrate may have widespread implications for aquatic life.

## 9.3   NITRATE GUIDELINES

Policy to limit nitrate pollution into freshwater habitats is becoming increasingly recognized as a necessary measure to protect aquatic animals. For example, the European Nitrates Directive (91/676/EEC) was introduced across the European Union in an attempt to limit excess nitrate pollution (e.g., livestock wastewater, industrial wastewater, nitrogen-based fertilizers, and urban runoff) from flowing into freshwaters (European Commission 2018). The directive set "safe" nitrate concentra-tions at $50 \cdot mg \cdot NO_3 - NL^{-1}$. Similarly, the United States Environmental Protection agency set nitrate guidelines at a concentration of $44 \cdot mg \cdot NO_3 - NL^{-1}$ (EPA 2002). The Australian and New Zealand water quality guidelines were introduced to limit water-nitrate concentrations and were derived to offer different levels of ecosystem protection depending on the health and ecological value of the system – for example, high conservation/ecological value systems = 99 percent species protection; slightly to moderately disturbed systems = 95 percent species protection; highly disturbed systems, 80–90 percent species protection (ANZECC 2000). Nitrate guidelines were based on short-term toxicity data (no observed effect concentration; NOEC) of various aquatic animals and were established $72 \cdot mg \cdot NO_3 - NL^{-1}$ for slightly to moderately disturbed environments (offering 95% species protection), $50 \cdot mg \cdot NO_3 - NL^{-1}$ in highly disturbed systems and aquaculture settings (offering 80–90% species protec-tion), and $10 \cdot mg \cdot NO_3 - NL^{-1}$ for recreational purposes (ANZECC 2000). However, ANZECC (2000) guidelines were based on non-native species, before the guidelines were revised in 2013 to include data on native mayfly larvae (*Deleatidium sp.*) and fish (*Galaxias maculatus*) (Hickey 2013). Canada's freshwater nitrate guidelines are set at a maximum of $550 \cdot mg \cdot NO_3 - NL^{-1}$ for short-term exposure, which is based on severe-effects data (such as lethality) and not intended to protect all components of the aquatic ecosystem, while a long-term exposure guideline of 13 mg $NO_3 - NL^{-1}$ has been established based on mostly NOEC data (Canadian Council of Ministers of the Environment 2012). Canadian Water Quality Guidelines also make mention of toxicity modifying factors (see section 7.0 below) yet, the document mainly con-siders water hardness (i.e., the concentration of calcium carbonate, $CaCO_3$) and underestimates other physiochemical parameters (e.g., chloride concentrations, temperature, oxygen levels). Despite these guidelines being established, nitrate concentrations remain elevated in various freshwater systems worldwide (Camargo and Alonso 2006b, Camargo, Alonso, and Salamanca 2005).

## 9.4 TOXICITY OF NITRATE TO AQUATIC LIFE

High levels of nitrate pollution can impair the capacity of aquatic animals (e.g., crustaceans, amphibians, fish, and aquatic insects and molluscs) to survive, grow and reproduce, due to direct toxic effects (Camargo and Alonso 2006a, Camargo, Alonso, and Salamanca 2005). In waters high in nitrate, passive uptake of nitrate ions into the body of organisms occurs across the gill epithelium, where nitrate is then stored in the plasma (Jensen 1996, Stormer, Jensen, and Rankin 1996, Cheng, Tsai, and Chen 2002). Nitrate is also synthesized *in vivo* as part of the nitrate – nitrite – nitric oxide (NO) pathway, but remains at low levels (~ 0.01 – 0.05 μmol/g) in blood and tissue as part of this process (Dolomatov et al. 2011). The main toxic action of nitrate is attributed to the inhibition of oxygen transport capacity within the body of affected organisms (hemoglobin, hemocyanin; Grabda et al. 1974, Monsees et al. 2017, Yang et al. 2019, Cheng and Chen 2002, Gomez Isaza, Cramp, and Franklin 2020b, 2021b). Nitrate penetrates erythrocytes (red blood cells) where it oxidizes the oxygen-transporting molecule, hemoglobin (invertebrates) or hemocynin (invertebrates), converting hemoglobin into methemoglobin, a form that cannot transport oxygen (Jensen 1996, 2003). Under normal conditions, methemoglobin circulates at low levels (<5%) within the body (Williams, Glass, and Heisler 1993). Signs of methemoglobinemia appear as levels exceed 10% methemoglobin or more, although some species are tolerant of elevated methemoglobin concentrations (see Lefevre et al. 2011, van Bussel et al. 2012, Yang et al. 2019). Nitrate also causes damage to peripheral blood and hematopoietic centers (spleen, liver, lymph nodes, and bone marrow; Grabda et al. 1974, Muir, Sutton, and Owens 1991) such that the rate at which red blood cells are replaced is diminished. Methemoglobin formation can be countered by the methemoglobin reductase system, primarily by the activity of NADH-dependent methemoglobin reductase (Jensen 2003, Freeman, Beitinger, and Huey 1983). Excess nitrate is stored in the liver and can be excreted passively through the brachial epithelium or eliminated through urinary losses or bile (Jensen 1996, 2003).

As a biotoxin (i.e., a toxin of biological origin), nitrate has received remarkably little attention. Research to date has primarily consisted of short-term laboratory bioassays to determine the lethal concentrations of nitrate (e.g., lethal concentrations that kill a proportion of the population, 10% = $LC_{10}$; 50% = $LC_{50}$) (e.g., Adelman et al. 2009, Learmonth and Carvalho 2015, Lou et al. 2016, Benítez-Mora et al. 2014, Soucek and Dickinson 2012). These datasets largely conclude that freshwater organisms are, in the short-term, tolerant of elevated nitrate concentrations, with various studies reporting nitrate lethal concentrations upwards of 500 mg $NO_3 - NL^{-1}$ (Figure 9.1A). For example, some fish (e.g., rainbow trout, *Onchorhynchus mykiss*; lake trout, *Salvelinus namaycush*; lake whitefish, *Coregonus clupeaformis*; fat greenling; and *Hexagrammos otakii*) can tolerate nitrate concentrations above 2000 mg $NO_3 - NL^{-1}$ under short-term 96-h lethal toxicity tests (Yang et al. 2019, McGurk

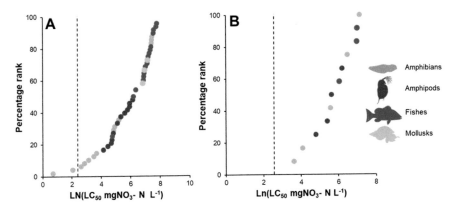

**FIGURE 9.1** Species-sensitivity distributions of acute (A; 96 h) and chronic (B; > 1 week) nitrate toxicity of freshwater animals. Dots are represent median lethal concentration ($LC_{50}$) for four groups of freshwater taxa, including amphibians (green), amphipods (orange), fish (blue) and mollusks (yellow). The broken vertical line is the nitrate guideline (11.3 · mg · $NO_3^- – NL^{-1}$) for the protection of freshwater life.

et al. 2006, Baker et al. 2017). Similarly, the white-clawed crayfish. *Austropotamobius italicus*, can withstand nitrate levels of 2950 mg $NO_3 – NL^{-1}$ over a 96-h lethal toxicity test (Benítez-Mora et al. 2014). Nitrate toxicity, however, increases with longer exposure durations (Figure 9.1B; Camargo, Alonso, and Salamanca 2005, Gomez Isaza, Cramp, and Franklin 2020a). For instance, Adelman et al. (2009) showed that the acute toxicity (96-h $LC_{50}$) of nitrate was 1559 mg $NO_3 – NL^{-1}$ in the freshwater fish, Topeka shiner (*Notropis topeka*), but the chronic toxicity (as determined by the 30-day lowest-observed-effect concentration, LOEC) of nitrate was significantly lower at 268 mg $NO_3 – NL^{-1}$.

Various other factors can influence the toxicity of nitrate, including body size, life stage, diet, and source population (Ortiz-Santaliestra et al. 2006, Gomez Isaza, Cramp, and Franklin 2020a, Hamlin 2006, Chow and Hong 2002). Nitrate toxicity tends to decrease with increases in body size (Camargo and Ward 1992, 1995) due to a higher rate of nitrate accumulation per unit of mass. However, in an interesting study, Hamlin (2006) reported the opposite pattern in Siberian sturgeon (*Acipenser baeri*). Small sturgeon (6.0 g; median $LC_{50}$ = 1028 mg $NO_3 – NL^{-1}$) tolerated much higher nitrate concentrations (about 2.6 times higher) than larger sturgeon (673.8 g; median $LC_{50}$ = 397 mg $NO_3 – NL^{-1}$). Hamlin (2006) suggested that this pattern may be driven by cohort-dependent effects, since $LC_{50}$ values have been shown to be highly variable (Buikema Jr., Niederlehner, and Cairns Jr. 1982). However, the observed effect size between size classes was significant (approximately 2.6-fold greater toxicity in large versus small sturgeon) and may indicate that large Siberian sturgeon are particularly sensitive to nitrate exposure due to some unknown mechanism. Evidence of life-stage specific differences in nitrate toxicity is also available for a variety of aquatic taxa (Camargo and Ward 1992, Ortiz-Santaliestra et al. 2006, Gomez Isaza, Cramp, and Franklin 2020a). Camargo and Ward (1992), for instance, showed that early instar larvae of spinning caddisflies larvae (*Cheumatopsyche pettiti* and *Hydropsyche occidentalis*) were significantly more susceptible to elevated nitrate concentrations (120 h $LC_{0.01}$ values) than last instar larvae. Similarly, early developmental stages (Gosner stages 13–19) of the Iberian painted frog (*Discoglossus galganoi*), the Iberian spadefoot toad (*Pelobates cultripes*), and natterjack toad (*Bufo calamita*) were significantly more susceptible to elevated nitrate concentrations, marked by slower growth rates and higher mortality than later developmental stages (Gosner stages 21–24; Ortiz-Santaliestra et al. 2006). It has been suggested that early life stages are more susceptible to nitrate because they have a higher surface area to volume ratio, meaning that there is a greater rate of absorption of nitrate relative to their body mass (Watts and Jarvis 1997). Early life stages also have a lower metabolic detoxifying ability (Bucciarelli et al. 1999), which may further explain why later life stages are more tolerant of nitrate. Moreover, the early life stages of amphibians have external gills that are in direct contact with the surrounding water before they become internalized during development (Duellman and Trueb 1994). This direct contact between the gills and the surrounding water is predicted to increase the sensitivity of the early life stage to nitrate (Ortiz-Santaliestra et al. 2006). Embryonic life stages, however, appear resilient to nitrate exposure. In a recent meta-analysis, Gomez Isaza and colleagues (2020a) reported that exposure to nitrate does not impact hatching success across amphibians, amphipods, and fish. Furthermore, nitrate exposure during embryonic development has been found to have negligible impacts on hatchling length (Ortiz-Santaliestra, Fernandez, et al. 2011, Lou et al. 2016), days to hatching (Ortiz-Santaliestra, Fernandez, et al. 2011, Ortiz-Santaliestra, Marco, and Lizana 2011), or developmental stage at hatching (Ortiz-Santaliestra, Fernandez, et al. 2011, Ortiz-Santaliestra, Marco, and Lizana 2011), further indicating that embryonic life stages are tolerant of nitrate exposures. Embryonic life stages may be protected from nitrate because the jelly layer of fish and amphibian eggs is impermeable to nitrate. Embryonic life stages may also be resilient to nitrate exposure owing to the incomplete formation of the digestive system, which decreases the formation of methemoglobin in the body (Hecnar 1995, Huey and Beitinger 1982). There is also some evidence of local adaption in response to nitrate pollution (Johansson, Räsänen, and Merilä 2001, Miaud et al. 2011, Hecnar 1995). Johansson, Räsänen, and Merilä (2001) found that common frog (*Rana temporaria*) tadpoles originating from nitrate-polluted sites showed higher growth rates and faster development when reared under elevated nitrate conditions compared to tadpoles originating from unpolluted locations, suggesting that local adaptation may be possible for populations living in chronically nitrate-polluted habitats.

## 9.5   GROWTH, DEVELOPMENT, AND BEHAVIOR

Freshwater animals exposed to elevated levels of waterborne nitrate often face reductions in growth, development, and behavior. Various studies have assessed the impact of nitrate on the growth performance and development of various amphibians (Krishnamurthy and Smith 2010, Wang et al. 2015, Xie et al. 2019), fish (Lou et al. 2016, McGurk et al. 2006, Pereira et al. 2017, van Bussel et al. 2012), and aquatic invertebrates (Baker et al. 2017, Soucek and Dickinson 2012, Douda 2010). Many of these studies report reduced growth and slowed development following periods of nitrate exposure.

Nitrate-exposed animals suffer reduced growth and slowed development due to disruptions of several physiological processes. Most notably, nitrate slows the growth and development of freshwater fauna due to increased concentrations of circulating methemoglobin around the body, causing hypoxemia and a reduced growth capacity (Grabda et al. 1974, Monsees et al. 2017, Gomez Isaza, Cramp, and Franklin 2020b). In addition, nitrate exposure increases basal energy expenditure of aquatic animals (Gomez Isaza, Cramp, and Franklin 2018, 2020c, Meade and Watts 1995) by increasing energetic costs associated with the detoxification (i.e., the conversion of methemoglobin back to hemoglobin by the energy-dependent NADH-methaemoglobin reductase enzyme; Jensen and Nielsen 2018), and removal of nitrate from the body (via urinary losses or bile; Jensen 1996, 2003), and increased energetic costs associated with oxidative damage (Kim et al. 2019) and tissue repair (Sokolova 2013). Growth and development can also be stunted by nitrate exposure owing to alterations in intestinal microbiota diversity and composition (Xie et al. 2020). In their recent study, Xie and colleagues (2020) examined the gut microbiota of *Bufo gargarizans* tadpoles exposed to varying levels of nitrate (0, 20, and 100 mg $NO_3 - NL^{-1}$). Xie and colleagues (2020) found that nitrate-exposed *B. gargarizans* had an altered diversity of gut microbes (facultative anaerobic *Proteobacteria* replaced the obligately anaerobic *Bacteroidetes* and *Fusobacteria*), which lead to the disruption of key metabolic pathways (acid and amino acid metabolic pathways) that affected whole-body fatty acid components of nitrate-exposed tadpoles. These changes in intestinal microbiota and microbe metabolic pathways induced deficient nutrient absorption by the intestine and resulted in malnutrition of *B. gargarizans* tadpoles (Xie et al. 2020). Moreover, exposure to elevated nitrate concentrations can also impact digestion efficiency by increasing the specific dynamic action response (i.e., the energetic costs associated with feeding and digestion), and thereby reduce growth performance (Steinberg et al. 2018). Nitrate has also been shown to lower food intake (Cano-Rocabayera et al. 2019, Krishnamurthy et al. 2006, Schram et al. 2014, van Bussel et al. 2012) and energetic reserves (Cano-Rocabayera et al. 2019, Monsees et al. 2017) of aquatic animals, which further contribute to lowering growth rates. However, some species can withstand exposure to elevated nitrate concentrations, such that growth and development are maintained (Schram et al. 2010, Schram et al. 2014, Cano-Rocabayera et al. 2019, Miaud et al. 2011, Stelzer and Joachim 2010). For instance, Schram and colleagues (2014) reported that pikeperch *Sander lucioperca* can tolerate nitrate concentrations of 1785 mg $NO_3 - NL^{-1}$ without affecting their growth and feeding activity over a 42-day exposure period. Similarly, the tadpoles of the wood frog (*Rana sylvatica*) experienced a period of developmental acceleration upon exposure to nitrate – reaching metamorphosis within a shorter timeframe than their control-unexposed counterparts without affecting growth or survival (Smith et al. 2011). The causes underlying the interspecific variation in growth and development following nitrate exposure are unclear, but nitrate susceptibility may be linked to nitrate uptake rates and gill permeability (Williams et al. 2008), the effectiveness of protective mechanisms – for example, elevated methemoglobin reductase activity (Huey and Beitinger 1982, Jensen and Nielsen 2018) – or nitrate elimination capabilities (Larisch and Goss 2018).

Freshwater fauna face pronounced behavioral alterations following nitrate exposure. Principally, nitrate-exposed animals are characteristically lethargic owing to increased levels of methemoglobin. Burgett and colleagues (2007) noted that *R. sylvatica* tadpoles experienced lethargy when exposed to a nitrate concentration of 78 mg $NO_3 - NL^{-1}$, and showed a 4.3-fold decrease in activity levels compared to unexposed, control tadpoles. Nitrate-exposed animals have also been seen to elicit erratic and disorganized swimming patterns. Rainbow trout (*Oncorhynchus mykiss*) were seen to elicit "side-swimming" behavior (i.e., fish swimming oriented on their side) in response to high nitrate exposure, which was hypothesized to be the result of physiological or morphological disruptions (e.g., imbalance of musculature symmetry, skeletal deformities, or a deviation of swim bladder shape or positioning) caused by nitrate (Davidson et al. 2011, 2014). Nitrate-exposed rainbow trout have also been documented to swim near the surface of the tank and exhibit a "yawning" or "gulping" action, with some even breaking the surface of the water in an attempt to escape (Westin 1974). Similarly, Iberian waterfrog (*Pelophylax perezi*) tadpoles exposed to nitrate (36 and 364 mg $NO_3 - NL^{-1}$) preferred to swim near the water's surface compared to control animals, which swam at the bottom of the test arena (Egea-Serrano, Tejedo, and Torralva 2011). This surface-swimming behavior has been suggested to be a response to nitrate-induced methemoglobinemia, whereby nitrate-exposed animals are forced to swim near the surface of the water in an attempt to increase atmospheric oxygen uptake to meet their oxygen demands (Egea-Serrano, Tejedo, and Torralva 2011). Collectively, these behavioral disruptions in activity and swimming behaviors caused by exposure to nitrate are likely to pose significant fitness constraints on animals living in waters polluted by nitrate, as animals need to move to evade predators, seek out resources and perform repeated locomotor activities (Wolter and Arlinghaus 2003). Nitrate exposure has also been shown to alter other behaviors of aquatic fauna, including foraging-related behaviors (e.g. voracity, latency, foraging efficiency; Cano-Rocabayera et al. 2019, Krishnamurthy et al. 2006, Xu and Oldham 1997, Alonzo and Camargo 2013), mate-guarding behaviors (Pandey, Adams, and Warren 2011), courting behaviors (Secondi et al. 2013), olfactory cues (Secondi et al. 2009), predator recognition (Polo-Cavia, Burraco, and Gomez-Mestre 2016), and predator escape performance (Ortiz-Santaliestra, Fernandes-Beneitez, et al. 2010, García-Muñoz, Guerrero, and Parra 2011), although a broader evaluation of nitrate's impact on various behavioral traits is required to better our understanding of the generality of observed effects (Gomez Isaza, Cramp, and Franklin 2020a).

## 9.6   HISTOPATHOLOGICAL DISRUPTIONS TO FRESHWATER ANIMALS

Nitrate exposure can cause histopathological changes to key organs of aquatic animals. Histopathological changes of the gill epithelium have been the primary focus because the gills are in direct and constant contact with nitrate in polluted

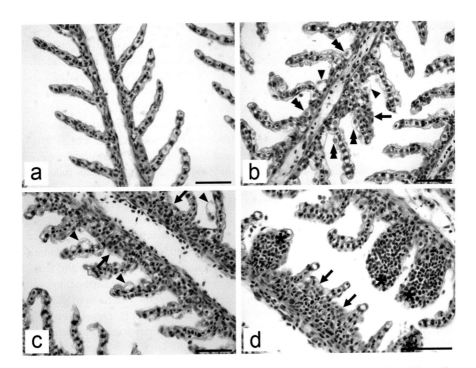

**FIGURE 9.2** Micrographs of gills of control and nitrate-exposed European grayling (*Thymallus thymallus*): A) primary and secondary lamellae without histopathological alterations; B–D) various histopathological lesions responsible for increasing diffusion distance or disturbing circulatory patterns in blood vessels; arrowheads – oedema of secondary epithelium; arrows – hyperplasia of respiratory epithelium; double arrowheads – hypertrophy of epithelial cells; asterisks – presence of aneurisms on secondary lamellae (H&E; ×400; bar = 50 μm).

environments (Pereira et al. 2017). The main pathological changes occurring to the gills following exposure to increased nitrate levels include hyperplasia (increase in cell number) and hypertrophy (increase in cell size) of the epithelial cells, hemorrhaging, oedema of secondary epithelium, aneurisms, and necrosis (Figure 9.2; Davidson et al. 2014, Hrubec, Smith, and Robertson 1996, Pereira et al. 2017, Rodgers et al. 2021, Rodrigues et al. 2011). Research on the impacts of elevated nitrate on the integument (skin) and the digestive tract, which is also in direct contact with nitrate, has been poorly studied, and the available evidence is mixed. For example, Davidson and colleagues (2014) reported no effect of nitrate exposures (up to 100 mg $NO_3 - NL^{-1}$) on the integument (skin) of rainbow trout. Conversely, Pereira and colleagues (2017) noted severe desquamation of the skin on the head and body of zebrafish at nitrate concentrations of 100 mg $NO_3 - NL^{-1}$ and above. Significant histopathological changes have also been documented for the intestine of nitrate-exposed animals. Following exposure to an elevated concentration of nitrate (10 mg $NO_3 - NL^{-1}$), Muir, Sutton, and Owens (1991) reported histological changes in midgut and foregut of the giant tiger shrimp (*Penaeus monodon*). Similarly, Xie and colleagues (2020) noted significant changes to the intestinal histology of *B. gargarizans* tadpoles exposed to elevated nitrate concentrations (20 and 100 mg $NO_3 - NL^{-1}$), including an increase in the cell boundaries of mucosal epithelial cells and an increase in the number of vacuolation of mucosal epithelial cells. No histopathological effect on the intestine was seen at a concentration of 5 mg $NO_3 - NL^{-1}$ in *B. gargarizans* tadpoles, indicating that this is a safe concentration (Xie et al. 2020). Chronic nitrate exposure (100, 200, and 400 mg $NO_3 - NL^{-1}$ for 28 days) was also seen to induce goblet cell hyperplasia, vacuolation, and hypertrophy of the enterocytes (cells of the intestinal lining), as well as villi atrophy in the intestine of zebrafish, *Danio rerio* (Pereira et al. 2017). Histopathological impairment of the liver and kidney has also been documented in response to elevated nitrate exposures. In kidneys, nitrate exposure can induce the mineralization of tubules, increase Bowman's space and tubular lumen, oedemas, glomeruli degeneration, and renal interstitial fibrosis, among other histopathological changes (Davidson et al. 2014, Hrubec, Smith, and Robertson 1996, Pereira et al. 2017, Iqbal, Qureshi, and Ali 2004), whereas hepatocyte vacuolation, foci of necrotic tissue, degraded lipid droplets, and hepatic fibrosis appear to be the most common histopathological changes to the liver (Hrubec, Smith, and Robertson 1996, Pereira et al. 2017, Xie et al. 2019, Iqbal, Qureshi, and Ali 2005). These histopathological changes to key organs are considered as general, non-specific reactions to pollutants aiming to protect animals from toxicity (Rodrigues et al. 2011). Depending on the severity and extent of changes, histopathological damage can compromise the functionality of the affected organs. For example, histopathological changes to the intestine likely cause unwanted effects by decreasing the capacity for nutrient absorption (Xie et al. 2020), whereas histopathological changes to the gills likely impair oxygen uptake (Rodgers et al. 2021).

## 9.7   ENDOCRINE DISRUPTION AND REPRODUCTION

Nitrate has recently garnered attention as an endocrine-disrupting chemical due to its effects on thyroid hormones and reproductive systems (Freitag et al. 2015, Guillette and Edwards 2005, Kellock, Moore, and Bringolf 2018, Edwards and Guillette Jr. 2007). Nitrate is a goitrogen, meaning that it interferes with the production of thyroid hormones by competitively binding to the sodium/iodide symporter on thyroid follicles and decreasing the synthesis of thyroid hormones (De Groef et al. 2006, Guillette and Edwards 2005, Lahti, Harri, and Lindqvist 1985). Periods of prolonged nitrate exposure can also lead to the development of goiter (the enlargement of the thyroid gland). For instance, white-spotted bamboo sharks (*Chiloscyllium plagiosum*) exposed chronically (29-days) to 70 mg $NO_3 - NL^{-1}$ did not show significant reductions in plasma thyroxine levels, but histological analyses revealed moderate hyperplasia and hypertrophy of the thyroid gland (Morris et al. 2011).

Nitrate can exert its influence on amphibian populations by altering the process of metamorphosis (Edwards et al. 2006, Hinther et al. 2012, Wang et al. 2015). Amphibian metamorphosis (e.g., intestine remodeling, hind-limb emergence, and tail resorption) is a process that is completely regulated by thyroid hormones, primarily thyroxine (T4) and 3,5,3′-triiodothyronine (T3) (Buchholz et al. 2007). Because nitrate exposure disrupts thyroid gland function by reducing thyroid hormone levels, competitively inhibiting iodine uptake, and altering thyroid hormone signaling pathways (Edwards et al. 2006, Hinther et al. 2012, Wang et al. 2015), many toad and frog tadpoles exposed to nitrate have shown altered metamorphosis development (Wang et al. 2015, Xu and Oldham 1997, Earl and Whiteman 2009, Ortiz-Santaliestra, Marco, and Lizana 2011). For example, *B. gargarizans* tadpoles that were exposed to nitrate (50 and 100 mg $NO_3 - NL^{-1}$) exhibited a clear reduction in T4 and T3 levels, showed signs of colloid depletion in the thyroid gland follicles, and key enzymes (Dio2 and Dio3 mRNA levels) were up-or down-regulated in different tissues (intestine, limbs, tail), which all reflect disruptions of the thyroid (Wang et al. 2015).

Nitrate can also alter the reproductive physiology of aquatic animals by altering the activity and synthesis of key reproductive hormones. In Siberian sturgeon, exposure to 57 mg $NO_3 - NL^{-1}$ increased plasma testosterone (11-Keto testosterone) and estradiol levels after just 30 days of exposure (Hamlin et al. 2008), whereas Freitag and colleagues (2015) found increased plasma testosterone levels following exposure to concentrations as low as 10.3 mg $NO_3 - NL^{-1}$ in Atlantic salmon (*Salmo salar*). Similarly, nitrate-exposed male fathead minnows (*Pimephales promelas*) had elevated levels of plasma testosterone, and vitellogenin (a yolk nutrient protein stimulated through the activity of the estrogen receptor) levels were elevated in both male and female fathead minnows (Kellock, Moore, and Bringolf 2018). Northern Leopard frogs (*Rana pipiens*) exposed to just 2.26 mg $NO_3 - NL^{-1}$ during early development had testicular oocytes, less-developed testicular tissue, and had female-skewed sex ratios (Orton, Carr, and Handy 2009). Moreover, nitrate exposure has been associated with reduced sperm counts and reduced likelihood of pregnancy in male and female mosquito fish (*Gambusia holbrooki*), respectively (Edwards and Guillette Jr. 2007, Edwards, Miller, and Guillette Jr. 2006), and has been documented to delay reproductive maturity and reproductive output in later life (Alonzo and Camargo 2013, Kolm 2002, Gibbons and McCarthy 1986, Moore and Bringolf 2018). Nitrate has also been shown to exert its influence on secondary sexual traits (e.g., male sexual ornaments, pheromones), male attractiveness, and courting behaviors in palmate newt (*Triturus helveticus*) (Secondi et al. 2009, Secondi et al. 2013). Together, these disruptions to the reproductive physiology may have important implications for the reproductive success and fitness of nitrate-exposed aquatic animals and may scale up to affect whole populations.

## 9.8 NITRATE AND OTHER STRESSORS

Along with increased nitrate concentrations, freshwater fauna must contend with a myriad of co-occurring environmental stressors. The presence of other environmental stressors is likely to mediate the toxicity of nitrate, increasing or decreasing its toxicity and sublethal effects. Co-occurring stressors can include other abiotic factors such as physiochemical parameters of waterbodies (pH, temperature, oxygen levels; Egea-Serrano and Van Buskirk 2016, Ortiz-Santaliestra, Fernandez-Beneitez, et al. 2010, Ortiz-Santaliestra and Marco 2015, Hatch and Blaustein 2000) and other pollutants (Orton, Carr, and Handy 2009, Krishnamurthy and Smith 2011), as well as biotic factors (Ortiz-Santaliestra, Fernandez-Beneitez, et al. 2011, Romansic et al. 2006, Smith et al. 2013).

### 9.8.1 Nitrate and Water Temperature

Water temperature is stipulated to have a key role in governing the toxicity of pollutants, with toxicity tending to increase with temperature increases (Little and Seebacher 2015, Patra et al. 2007, Philippe et al. 2018). Yet, nitrate toxicity has been shown to be largely independent of water temperature (Colt and Tchobanoglous 1976, Egea-Serrano and Van Buskirk 2016). Colt and Tchobanoglous (1976) assessed the short-term toxicity of nitrate to juvenile channel catfish, *Ictalurus punctatus,* at three water temperatures (22, 26, and 30°C), and found no differences in 96 h $LC_{50}$ values (1355, 1423, and 1400 mg $NO_3 - NL^{-1}$, respectively). Similarly, Egea-Serrano and Van Buskirk (2016) assessed the interaction between elevated nitrate (single dose of nitrate 25.7 mg $NO_3 - NL^{-1}$) and water temperature (21 and 24°C) on the growth, development, and behavior of *Rana temporaria* tadpoles and found no differences in any of the metrics measured. Recent research has also shown that long-term exposure to elevated temperatures can offset the negative effects of nitrate (Gomez Isaza, Cramp, and Franklin 2020c, 2021b, Opinion, De Boeck, and Rodgers 2020). Silver perch (*Bidyanus bidyanus*) acclimated to 32°C were able to offset the negative effects of nitrate on the aerobic scope (i.e., maximum–standard metabolic rate), swimming performance, and upper thermal tolerance (i.e., $CT_{MAX}$) when compared to nitrate-exposed fish acclimated to a cooler temperature of 28°C (Gomez Isaza, Cramp, and Franklin 2020c). Similarly, nitrate-exposed European grayling (*Thymallus thymallus*) acclimated to a warmer temperature (22°C) showed an increased aerobic scope compared to fish acclimated to 18°C (Opinion, De Boeck, and Rodgers 2020). The negative effects of nitrate are likely masked at elevated temperatures because warm temperatures induce beneficial cardiorespiratory adjustments (e.g., remodeling of the heart and gill architecture), which are sufficient to overcome nitrate-induced methemoglobinemia (Gomez Isaza, Cramp, and Franklin 2021b). Moreover, erythrocyte methemoglobin reduction is highly thermally sensitive, such that methemoglobin is reduced to hemoglobin more efficiently at warmer temperatures (Jensen and Nielsen 2018, Ha et al. 2019). However, nitrate exposure has been shown to lower the short-term heat tolerance of freshwater fishes (Gomez Isaza, Cramp, and Franklin 2020c, Rodgers et al. 2021), which may make fish more susceptible to heatwaves.

### 9.8.2 Nitrate and Low-Water pH

Environmental pH is one factor that can modify nitrate toxicity in two main ways. First, nitrate uptake is increased in waters of low pH. Gomez Isaza, Cramp, and Franklin (2020b) showed that nitrate uptake was increased in spangled perch (*Leiopotherapon unicolor*) exposed simultaneously to nitrate (50 or 100 mg $NO_3 - NL^{-1}$) and a low pH (4.0) treatment, as evidenced by higher plasma nitrate concentrations compared to fish maintained at circumneutral pH. Similarly, nitrite uptake was increased two-fold following exposure to low pH (pH 5.0) in Coho salmon (*Oncorhynchus kisutch*) (Meade and Perrone 1980). Meade and Perrone (1980) proposed that at low pH, uptake characteristics of ion transporter in the gills are altered, which enhances nitrite toxicity. Second, high levels of plasma nitrate levels promote the formation of methemoglobin. Indeed, spangled perch exposed to elevated nitrate and low pH experienced a two-fold increase in methemoglobin concentrations (Gomez Isaza, Cramp, and Franklin 2020b). Elevated levels of circulating methemoglobin in low pH-exposed fish led to poorer growth and swimming performance, and lowered overall physical condition (Gomez Isaza, Cramp, and Franklin 2020b). In a similar study, Hatch and Blaustein (2000) found a synergistic decrease in survival of tadpoles (*Rana cascadae*) exposed simultaneously to nitrate (5 or 20 mg $NO_3 - NL^{-1}$) and low pH (5.0), while activity levels (proportion of time active) were unaffected. Furthermore, the basal energetic costs were approximately doubled in blue-claw crayfish exposed simultaneously to nitrate (50 or 100 mg $NO_3 - NL^{-1}$) and low pH (5.0) conditions (Gomez Isaza, Cramp, and Franklin 2018), which resulted in poorer growth performance and survival.

### 9.8.3 NITRATE AND LOW OXYGEN LEVELS

The impacts of nitrate on aquatic animals may be compounded under environmental hypoxia (i.e., low oxygen conditions). This is because simultaneous exposure to nitrate and low oxygen conditions results in both environmental hypoxia and internal hypoxemia. Indeed, recent work has shown that nitrate-exposed fish had a lower hypoxia tolerance than un-exposed, control fish (Gomez Isaza, Cramp, and Franklin 2021a, Rodgers et al. 2021). The swimming performance of nitrate-exposed fish was also reduced synergistically when exposed to nitrate and low oxygen conditions (Gomez Isaza, Cramp, and Franklin 2021a). In a similar study, Ortiz-Santaliestra and Marco (2015) exposed natterjack toad tadpoles (*Bufo calamita*) to a combination of elevated nitrate (either 28 or 56 mg $NO_3 - NL^{-1}$) and hypoxia (4.5 mg $L^{-1}$ $O_2$) treatments and found that developmental abnormalities increased synergistically, with abnormalities beginning to appear after only eight-days of exposure. Further, developmental rates were slowed and survival was reduced in tadpoles exposed to a combination of nitrate and hypoxia treatments (Ortiz-Santaliestra and Marco 2015). Such research demonstrates that nitrate exposure increases the risk of species to aquatic hypoxia.

### 9.8.4 NITRATE AND SALINITY

Nitrate toxicity generally decreases with increasing salinity. This is most likely because chloride competitively inhibits nitrate uptake (Stormer, Jensen, and Rankin 1996). Indeed, Tsai and Chen (2002) reported that the 96 h $LC_{50}$ of tiger shrimp (*Penaeus monodon*) was 1449, 1575, and 2316 mg $NO_3 - NL^{-1}$ at salinities of 15, 25, 35 ‰, respectively. Acute nitrate toxicity was also lower at higher salinities (5 versus 10 ‰) in juvenile whiteleg shrimp, *Litopenaeus vannamei* (Alves Neto et al. 2019). Similarly, Soucek and others (2015) found that higher chloride concentrations lowered nitrate toxicity in *Hyalella azteca*, but chloride concentration did not impact the chronic toxicity of nitrate in two freshwater invertebrates (*Ceriodaphnia dubia* and *Hyalella azteca*) (Soucek and Dickinson 2016). Moreover, mortality was increased, rather than decreased, in the Iberian waterfrog (*Pelophylax perezi*) exposed to a combination of nitrate and salinity treatments (Ortiz-Santaliestra, Fernandez-Beneitez, et al. 2010).

## 9.8.5  NITRATE AND OTHER POLLUTANTS

Aquatic animals commonly face elevated levels of nitrate in addition to various herbicides, pesticides, and fertilizers (Krishnamurthy and Smith 2011, Ortiz-Santaliestra, Fernandez, et al. 2011). The combined effects of nitrate and atrazine, a pesticide used to control annual broadleaf and grass weeds (Solomon et al. 1996), have been extensively studied due to their co-occurrence around agricultural sites (Sullivan and Spence 2003, Boone and Bridges-Britton 2006, Allran and Karasov 2000). Atrazine acts by reducing the number of red blood cells (i.e., lower hematocrit), induces developmental deformities, and lowers survival of aquatic taxa (Howe, Gillis, and Mowbray 1998, Solomon et al. 1996, Hussein, El-Nasser, and Ahmed 1996), and is therefore expected to interact with nitrate to reduce fitness. Allran and Karasov (2000) exposed leopard frog (*Rana pipiens*) tadpoles to a combination of nitrate (0, 5, and 30 mg $NO_3 - NL^{-1}$) and atrazine (0, 20, and 200 mg $L^{-1}$) from posthatch through to metamorphosis, and found no significant effect on development rate, time to metamorphosis, survival, mass at metamorphosis, or hematocrit. Boone and Bridges-Britton (2006) later conducted a similar study on the gray treefrog (*Hyla versicolor*), exposing tadpoles to atrazine (0 or 20 µg $L^{-1}$) and nitrate (0 or 10 mg $NO_3 - NL^{-1}$) treatments. They found that tadpoles were not more susceptible to single versus multiple contaminants treatments and hypothesized that combinations of contaminants ameliorate the effects of single contaminants (Boone and Bridges-Britton 2006). However, Sullivan and Spence (2003), which used much higher nitrate (0, 37 and 292 mg $NO_3 - NL^{-1}$) and atrazine (0, 40, and 320 mg $L^{-1}$) reported slowed growth, development, and smaller mass at metamorphosis in African clawed frog (*Xenopus laevis*) exposed to the two contaminants (at either nitrate and atrazine concentrations).

Research on the combined effects of nitrate and malathion (an organophosphate insecticide) has also been conducted. Krishnamurthy and Smith (2010) first examined the interaction between nitrate (0, 2, 4, 8, 16 mg $NO_3 - NL^{-1}$) and malathion (0, 250, 500, and 1000 mg $L^{-1}$) on the growth, development, and survival of American toad tadpoles (*Bufo americanus*). They found that in isolation, malathion increased the frequency of deformities and reduced survival, while nitrate alone reduced the growth and development of tadpoles. However, when combined, nitrate and malathion did not affect the growth, development, or survival of tadpoles (Krishnamurthy and Smith 2010). Similar effects were documented on the wood frog (*Rana sylvatica*) exposed to identical concentrations of nitrate and malathion (Krishnamurthy and Smith 2011). Nitrate and malathion did not jointly influence the survival or metamorph size in *B. americanus* and *R. sylvatica* tadpoles, but nitrate reduced the negative effect of malathion on time to metamorphosis in *R. sylvatica*, such that there was a little delay in metamorphosis compared to control tadpoles, which indicates that the presence of nitrate might ameliorate some of the effects of malathion (Smith et al. 2011). Finally, research on the interactive effects of nitrate and the glyphosate herbicide, Roundup Plus, revealed that combined exposure to these two chemicals during embryonic development increased the size at hatching of the gold-striped salamander (*Chioglossa lusitanica*) without any ill effects (Ortiz-Santaliestra, Fernandez, et al. 2011).

## 9.9 CONCLUSIONS AND FUTURE DIRECTIONS

This chapter provides clear evidence that unnatural increases in nitrate concentrations (mainly due to anthropogenic sources) can pose severe risks to aquatic animals. Nitrate affects aquatic animals primarily due to the formation of methemoglobin in the blood, which reduces the capacity of animals to transport oxygen around the body (Gomez Isaza, Cramp, and Franklin 2021b, Monsees et al. 2017, Yang et al. 2019). Increases in methemoglobin concentrations are associated with several sublethal effects, including reduced growth and development, lethargy, and altered behaviors. High levels of nitrate also induce histopathological changes to key organs (e.g., gills, liver, and intestines) (Pereira et al. 2017, Rodgers et al. 2021, Rodrigues et al. 2011). Moreover, nitrate is being increasingly recognized as an endocrine-disrupting chemical, due to its effects on the thyroid and thyroid hormones (Guillette and Edwards 2005, Kellock, Moore, and Bringolf 2018, Wang et al. 2015). Together, these physiological disruptions render elevated nitrate concentrations a serious risk to aquatic fauna. Current legislative guidelines provide protection to some tolerant freshwater species, though freshwater nitrate concentrations are nowadays regularly exceeding these thresholds. It is therefore recommended that levels be lowered to protect the most sensitive freshwater animals from acute and chronic nitrate pollution.

Additional research on the lethal and sublethal effects of nitrate are required to better understand how species will respond to unnaturally high levels of nitrate being encountered among various freshwater sources. Firstly, data on the responses of native and endangered species to elevated nitrate levels need to be prioritized. To date, data on nitrate toxicity has centered on common (e.g., common arthropods such as daphnia, collembola sp.), or aquaculturally important species (e.g., salmon, trout). It is clear that species responses to elevated nitrate are highly variable (i.e., between life-stages, species, and populations). Therefore, available data may not provide adequate protections to at-risk species that are of conservation concern. Second, there is a paucity of data on the long-term impacts of nitrate on aquatic fauna, despite evidence suggesting that nitrate toxicity increases with longer exposures and that long-term nitrate exposure poses severe fitness consequences to aquatic animals. Hence, long-term data need to be prioritized to determine how freshwater animals will fare in an increasingly nitrate-polluted world. Third, current knowledge surrounding nitrate research is based primarily on laboratory-based experiments where experimental conditions are maintained relatively constant, have low replication, and often use model or domesticated animals. These manipulative laboratory experiments are pivotal in order to gain a cause-and-effect understanding of the impacts on nitrate to aquatic taxa, but it is difficult to predict how such effects might scale up to affect wild populations. To help address these issues, future research should shift focus towards more realistic experiments on wild or semi-wild animals. Numerous examples of natural, field-based experiments exist, including mesocosm studies (Nagrodski et al. 2013, Egea-Serrano and Tejedo 2014) and sampling of natural fish populations along a pollution gradient (McKenzie et al. 2007, Wagenhoff, Townsend, and Matthaei 2012, Adams and Ham 2011). Continuous monitoring of animals living in these natural systems could reveal important responses of animals as an environment becomes more degraded or to assess the effectiveness of rehabilitation and restoration efforts (Adams and Ham 2011, Jeffrey et al. 2015). Fourth, few studies have identified how species respond to the removal (or partial removal) of nitrate. It is known that species have the capacity to recover from periods of stress (Taylor et al. 2000, Hyvärinen, Heinimaa, and Rita 2004), but the timescale and extent of recovery from nitrate exposure remain unclear. For example, it is well documented that organisms can experience compensatory growth (a period of accelerated growth) after the removal of various stressors (e.g. temperature stress, food deprivation, hypoxia; Huang et al. 2008, Tian and Qin 2003, Wei et al. 2008). It has also been established that an individual's history (i.e., previous encounters with stressors) can leave long-lasting consequences on organisms, which can explain their current performance (O'Connor and Cooke 2015). Failure to account for carryover effects and the timescale to recovery might lead to erroneous conclusions about the effectiveness of conservation measures. It is therefore pivotal to understand how the removal (or the partial removal) of nitrate impacts on species' recovery. Lastly, additional studies need to consider how other environmental stressors (e.g., water hardness, suspended sediments, ultraviolet radiation, and other pollutants) moderate the toxicity of nitrate to aquatic animals. Research must also consider how nitrate exposure affects biotic interactions (e.g., competition, predation, and disease) and how these interactions shape ecological and evolutionary trajectories.

## REFERENCES

Adams, S. M., and K. D. Ham. 2011. "Application of biochemical and physiological indicators for assessing recovery of fish populations in a disturbed stream." *Environ. Manag.* 47:1047–1063.

Adelman, I. R., L. I. Kusilek, J. Koehle, and J. Hess. 2009. "Acute and chronic toxicity of ammonia, nitrite and nitrate to the endangered topeka shiner (*Notropis topeka*) and fathead minnows (*Pimephales promelas*)." *Environ Toxicol Chem* 28:2216–2223.

Allran, J. W., and W. H. Karasov. 2000. "Effects of atrazine and nitrate on northern leopard frog (*Rana pipiens*) larvae exposed in the laboratory from posthatch through metamorphosis." *Environ. Toxicol. Chem.* 19:2850–2855.

Alonzo, A., and J. A. Camargo. 2013. "Nitrate causes deleterious effects on the behaviour and reproduction of the aquatic snail *Potamopyrgus antipodarum* (Hydrobiidae, Mollusca)." *Environ. Sci. Pollut. Res.* 20:5388–5396.

Alves Neto, I. E., H. Brandão, P. S. Furtado, and W. Wasielesky Jr. 2019. "Acute toxicity of nitrate in *Litopenaeus vannamei* juveniles at low salinity levels." *Cienc. Rural* 49:e20180439.

ANZECC. 2000. *Australian and New Zealand Guidelines for Fresh and Marine Water Quality - Volume 2: Aquatic Ecosystems — Rationale and Background Information.* Edited by National Water Quality Management Strategy: Australian and New Zealand Environment and Conservation Council.

Baker, J. A., Gilron. G., B. A. Chalmers, and J. R. Elphick. 2017. "Evaluation of the effect of water type on the toxicity of nitrate to aquatic organisms." *Chemosphere* 168: 435–440.

Benítez-Mora, Alfonso, Arantxa Aguirre-Sierra, Alvaro Alonso, and Julio A. Camargo. 2014. "Ecotoxicological assessment of the impact of nitrate (NO3⁻) on the European endangered white-clawed crayfish *Austropotamobius italicus* (Faxon)." *Ecotoxicol. Environ. Saf.* 101:220–225.

Boone, M. D., and C. M. Bridges-Britton. 2006. "Examining multiple sublethal contaminant on the gray treefrog (*Hyla versicolor*): Effects of an insecticide, herbicide and fertiliser." *Environ. Toxicol. Chem.* 25:3261–3265.

Bucciarelli, T., P. Sacchetta, A. Pennelli, L. Cornelio, R. Romagnoli, S. Merino, R. Petruzelli, and C. Di Ilio. 1999. "Characterization of toad glutathione transferase." *Biochim Biophys Acta* 1431:189–198.

Buchholz, D. R., R. A. Heimeier, B. Das, T. Washington, and Y. B. Shi. 2007. "Pairing morphology with gene expression in thyroid hormone-induced intestinal remodeling and identification of a core set of TH-induced genes across tadpole tissues." *Dev. Biol.* 303:576–590.

Buikema Jr., A. L., B. R. Niederlehner, and J. Cairns Jr. 1982. "Biological monitoring part IV—toxicity testing." *Water Res.* 16:239–262.

Burgett, A. A., C. D. Wright, G. R. Smith, D. T. Fortune, and S. L. Johnson. 2007. "Impact of ammonium nitrate on wood frog (*Rana sylvatica*) tadpoles: Effects on survivorship and behavior." *Herpetol. Conser. Biol.* 2:29–34.

Camargo, J. A., and A. Alonso. 2006a. "Ecological and toxicological effects of inorganic nitrogen pollution in aquatic ecosystems: A global assessment." *Environ. Int.* 32:831–849.

Camargo, J. A., and A. Alonso. 2006b. "Ecological and toxicological effects of inorganic nitrogen pollution in aquatic ecosystems: a global assessment." *Environ. Int.* 32:831–849.

Camargo, J. A., A. Alonso, and A. Salamanca. 2005. "Nitrate toxicity to aquatic animals: a review with new data for freshwater invertebrates." *Chemosphere* 58:1255–1267.

Camargo, J. A., and J. V Ward. 1992. "Short-term toxicity of sodium nitrate (NaNO3) to non-target freshwater invertebrates." *Chemosphere* 24:23–28.

Camargo, J. A., and J. V Ward. 1995. "Nitrate (NO3-N) toxicity to aquatic life: A proposal of safe concentrations for two species of nearctic freshwater invertebrates." *Chemosphere* 31:3211–3216.

Canadian Council of Ministers of the Environment. 2012. Canadian water quality guidelines for the protection of aquatic life: Nitrate. In *Canadian Environmental Quality Guidelines*, edited by Canadian Council of Ministers of the Environment. Winnipeg.

Cano-Rocabayera, O., A. de Sostoa, F. Padrós, L. Cárdenas, and A. Maceda-Veiga. 2019. "Ecologically relevant biomarkers reveal that chronic effects of nitrate depend on sex and life stage in the invasive fish *Gambusia holbrooki*." *Plos One* 14:e0211389.

Cheng, S. Y., and J. C. Chen. 2002. "Study on the oxyhemocyanin, deoxyhemocyanin, oxygen affinity and acid-base balance of *Marsupenaeus japonicus* following exposure to combined elevated nitrite and nitrate." *Aquatic Toxicology* 61:181–193.

Cheng, S. Y., S. J. Tsai, and J. C. Chen. 2002. "Accumulation of nitrate in the tissues of *Penaeus monodon* following elevated ambient nitrate exposure after different time periods." *Aquat. Toxicol.* 52:133–146.

Chow, C. K., and C. B. Hong. 2002. "Dietary vitamin E and selenium and toxicity of nitrite and nitrate." *Toxicology* 180:195–207.

Colt, J., and G. Tchobanoglous. 1976. "Evaluation of the short-term toxicity of nitrogenous compounds to channel catfish, *Ictalurus punctatus*." *Aquaculture* 8:209–221.

Davidson, J., C. Good, C. Welsh, and S. T. Summerfelt. 2011. "Abnormal swimming behavior and increased deformities in rainbow trout *Oncorhynchus mykiss* cultured in low exchange water recirculating aquaculture systems." *Aquacult. Eng.* 45:109–117.

Davidson, J., C. Good, C. Welsh, and S. T. Summerfelt. 2014. "Comparing the effects of high vs. low nitrate on the health, performance, and welfare of juvenile rainbow trout *Oncorhynchus mykiss* within water recirculating aquaculture systems." *Aquacult. Eng.* 59:30–40.

De Groef, B., B. R. Decallonne, S. Van der Geyten, V. M. Darras, and R. Bouillon. 2006. "Perchlorate versus other environmental sodium/iodide symporter inhibitors: potential thyroid-related health effects." *Eur. J. Endocrinol.* 155:17–25.

Dolomatov, S. I., P. V. Shekk, W. Zukow, and M. I. Kryukava. 2011. "Features of nitrogen metabolism in fishes." *Rev. Fish. Biol. Fisheries.* 21:733–737.

Douda, K. 2010. "Effects of nitrate nitrogen pollution on Central European unionid bivalves revealed by distributional data and acute toxicity testing." *Aquatic Conserv: Mar. Freshw. Ecosyst.* 20:189–197.

Duellman, W., and L. Trueb. 1994. *Biology of Amphibians.* Baltimore: Johns Hopkins University Press.

Earl, J. E., and H. H. Whiteman. 2009. "Effects of pulsed nitrate exposure on amphibian development." *Environ. Toxicol. Chem.* 28:1331–1337. doi: https://doi.org/10.1897/08-325.1.

Edwards, T. M., and L.J. Guillette Jr. 2007. "Reproductive characteristics of male mosquitofish (*Gambusia holbrooki*) from nitrate-contaminated springs in Florida." *Aquatic Toxicology* 85: 40–47.

Edwards, T.M., H.D. Miller, and L.J. Guillette Jr. 2006. "Water quality influences reproduction in female mosquitofish (*Gambusia holbrooki*) from eight Florida springs." *Environ. Health Perspect.* 114:69–75.

Edwards, T.M., K.A. McCoy, T. Barbeau, M.W. McCoy, J.M. Thro, and L.J. Guillette Jr. 2006. "Environmental context determines nitrate toxicity in southern toad (*Bufo terrestris*) tadpoles." *Aquat. Toxicol.* 78:50–58.

Egea-Serrano, A., and M. Tejedo. 2014. "Contrasting effects of nitrogenous pollution on fitness and swimming performance of Iberian waterfrog, *Pelophylax perezi* (Seoane, 1885), larvae in mesocosms and field enclosures." *Aquat. Toxicol.* 146:144–153.

Egea-Serrano, A., M. Tejedo, and M. Torralva. 2011. "Behavioral responses of the Iberian waterfrog, *Pelophylax perezi* (Seoane, 1885), to three nitrogenous compounds in laboratory conditions." *Ecotoxicology* 20:1246–1257.

Egea-Serrano, A., and J. Van Buskirk. 2016. "Responses to nitrate pollution, warming and density in common frog tadpoles (*Rana temporaria*)." *Amphibia – Reptilia* 37:45–54.

EPA. 2002. *Integrated Risk Information System (IRIS) Database. Nitrate* (CASRN 14797-55-8). Edited by US Environmental Protection Agency. Washington DC.

European Commission. 2018. *On the implementation of Council Directive 91/676/EEC concerning the protection of waters against pollution caused by nitrates from agricultural sources based on Member State reports for the period 2012–2015.* Edited by the Council and the European Parliament. Brussels.

Freeman, L., T.L. Beitinger, and D.W. Huey. 1983. "Methemoglobin reductase activity in phylogenetically diverse piscine species." *Comp. Biochem. Physiol. B* 75:27–30.

Freitag, A. R., L. R. Thayer, C. Leonetti, H. M. Stapleton, and H.J. Hamlin. 2015. "Effects of elevated nitrate on endocrine function in Atlantic salmon, *Salmo salar*." *Aquaculture* 436:8–12.

Galloway, J. N., F. J. Denterner, D. G. Capone, E. W. Boyer, R. W. Haoworth, S. P. Seitzinger, G. P. Asner, C. C. Cleveland, P. A. Green, E. A. Holland, D. M. Karl, A. F. Michaels, J. H. Porter, A. R. Townsend, and C. J. Vörösmarty. 2004. "Nitrogen cycles: past, present, and future." *Biogeochemistry* 70:153–226.

García-Muñoz, E., F. Guerrero, and G. Parra. 2011. "Larval escape behavior in anuran amphibians as a wetland rapid pollution biomarker." *Mar. Freshw. Behav. Physiol.* 44:109–123.

Gibbons, M. M., and T. K. McCarthy. 1986. "The reproductive output of frogs *Rana temporaria* (L.) with particular reference to body size and age." *J. Zool.* 209:579–593.

Glibert, P. M. 2017. "Eutrophication, harmful algae and biodiversity – challenging paradigms in a world of complex nutrient changes." *Mar. Pollut. Bull* 124:591–606.

Goeller, B. C., C. M. Febria, H. J. Warburton, K. L. Hodsden, K. E. Collins, H. S. Devlin, J. S. Harding, and A. R. McIntosh. 2019. "Springs drive downstream nitrate export from artificially-drained agricultural headwater catchments." *Sci. Total Environ.* 671:119–128.

Gomez Isaza, D. F., R. L. Cramp, and C. E. Franklin. 2018. "Negative impacts of elevated nitrate on physiological performance are not exacerbated by low pH." *Aquat. Toxicol.* 200:217–225.

Gomez Isaza, D. F., R. L. Cramp, and C. E. Franklin. 2020a. "Living in polluted waters: A meta-analysis of the effects of nitrate and interactions with other environmental stressors on freshwater taxa." *Environ. Pollut.* 261:114091.

Gomez Isaza, D. F., R. L. Cramp, and C. E. Franklin. 2020b. "Simultaneous exposure to nitrate and low pH reduces the blood oxygen-carrying capacity and functional performance of a freshwater fish." *Conser. Physiol.* 8 (1):coz092. doi: doi:10.1093/conphys/coz092.

Gomez Isaza, D. F., R. L. Cramp, and C. E. Franklin. 2020c. "Thermal acclimation offsets the negative effects of nitrate on aerobic scope and performance." *J. Exp. Biol.* 223:jeb224444. doi: 10.1242/jeb.224444.

Gomez Isaza, D. F., R. L. Cramp, and C. E. Franklin. 2021a. "Exposure to nitrate increases susceptibility to hypoxia in fish." *Physiol. Biochem. Zool.* doi: https://doi.org/10.1086/713252.

Gomez Isaza, D. F., R. L. Cramp, and C. E. Franklin. 2021b. "Thermal plasticity of the cardiorespiratory system provides cross-tolerance protection to fish exposed to elevated nitrate." *Comp. Biochem. Physiol. C.* 240:108920. doi: https://doi.org/10.1016/j.cbpc.2020.108920.

Grabda, E., T. Einszporn-Orecka, C. Felinska, and R. Zbanysek. 1974. "Experimental methemoglobinemia in trout." *Acta Ichthyol. Piscat.* 4: 43–71.

Gu, B., Y. Ge, S. X. Chang, W. Luo, and J. Chang. 2013. "Nitrate in groundwater of China: Sources and driving forces." *Glob. Environ. Chan.* 23:1112–1121.

Guillette, L.J., and T. M. Edwards. 2005. "Is nitrate an ecologically relevant endocrine disruptor in vertebrates?" *Integr. Comp. Biol.* 45:19–27.

Ha, N. T. K., D. T. T. Huong, N. T. Phuong, M. Bayley, and F. B. Jensen. 2019. "Impact and tissue metabolism of nitrite at two acclimation temperatures in striped catfish (*Pangasianodon hypophthalmus*)." *Aquat. Toxicol.* 212:154–161.

Hamlin, H. J. 2006. "Nitrate toxicity in Siberian sturgeon (*Acipenser baeri*)." *Aquaculture* 253:688–693.

Hamlin, H. J., B. C. Moore, T. M. Edwards, I. L. V. Larkin, A. Boggs, W. J. High, K. L. Main, and L. J. Guillette. 2008. "Nitrate-induced elevations in circulating sex steroid concentrations in female Siberian sturgeon (*Acipenser baerii*) in commercial aquaculture." *Aquaculture* 281:118–125.

Hatch, A. C., and A. R. Blaustein. 2000. "Combined effects of UV-B, nitrate, and low pH reduce the survival and activity level of larval Cascades Frogs (*Rana cascadae*)." *Arch. Environ. Contam. Toxicol.* 39:494–499.

Hecnar, S. J. 1995. "Acute and chronic toxicity of ammonium nitrate fertilizer to amphibians from Southern Ontario." *Environ. Toxicol. Chem.* 14:2131–2137.

Hickey, C. W. 2013. *Updating nitrate toxicity effects on freshwater aquatic species.* Hamilton, New Zealand: National Institute of Water & Atmospheric Research.

Hinther, A., T. M. Edwards, L. J. Guillette Jr., and C. C. Helbing. 2012. "Influence of nitrate and nitrite on thyroid hormone responsive and stress-associated gene expression in cultured *Rana catesbeiana* tadpole tail fin tissue." *Front. Genet.* 3:51–58.

Howe, G. E., R. Gillis, and R. C. Mowbray. 1998. "Effect of chemical synergy and larval stage on the toxicity of atrazine and alachlor to amphibian larvae." *Environ. Toxicol. Chem.* 17:519–525.

Hrubec, T. C., S. A. Smith, and J. L. Robertson. 1996. "Nitrate toxicity: A potential problem of recirculating systems." *NRAES* 1:41–48.

Huang, G., L. Wei, X. Zhang, and T. Gao. 2008. "Compensatory growth of juvenile brown flounder Paralichthys olivaceus (*Temminck & Schlegel*) following thermal manipulation." *J. Fish Biol.* 72:2534–2542.

Huey, D. W., and T. L. Beitinger 1982. "A methemoglobin reductase system in channel catfish *Ictalurus punctatus.*" *Can. J. Zool.* 60:483–487.

Hussein, S. Y., M. A. El-Nasser, and S. M. Ahmed. 1996. "Comparative studies on the effects of herbicide atrazine on freshwater fish *Oreochromis niloticus* and *Chrysichthyes auratus* at Assiut, Egypt." *Bull. Environ. Contam. Toxicol.* 57:503–510.

Hyvärinen, P., S. Heinimaa, and H. Rita. 2004. "Effects of abrupt cold shock on stress responses and recovery in brown trout exhausted by swimming." *J. Fish Biol.* 64:1015–1026.

Iqbal, F., I. Z. Qureshi, and M. Ali. 2004. "Histopathological changes in the kidney of common carp, *Cyprinus carpio*, following nitrate exposure." *J. Res. Sci.* 15:411–418.

Iqbal, F., I. Z. Qureshi, and M. Ali. 2005. "Histopathological changes in the liver of a farmed cyprinid fish, *Cyprinus carpio*, following exposure to nitrate." *Pak. J. Zool* 37:297–300.

Jaynes, D. B., T. S. Colvin, D. L. Karlen, C. A. Cambardella, and D. W. Meek. 2001. "Nitrate loss in subsurface drainage as affected by nitrogen fertilizer rate." *J. Environ. Qual.* 30:1305–1314.

Jeffrey, J. D., C. T. Hasler, J. M. Chapman, S. J. Cooke, and C. D. Suski. 2015. "Linking landscape-scale disturbances to stress and condition of fish: Implications for restoration and conservation" *Integr. Comp. Biol.* 55:618–630.

Jensen, F. B. 1996. "Uptake, elimination and effects of nitrite and nitrate in freshwater crayfish (*Astacus astacus*)." *Aquat. Toxicol.* 34:95–104.

Jensen, F. B. 2003. "Nitrite disrupts multiple physiological functions in aquatic animals." *Comp. Biochem. Physiol. A.* 135:9–24.

Jensen, F. B., and K Nielsen. 2018. "Methemoglobin reductase activity in intact fish red blood cells." *Comp. Biochem. Physiol. A* 216:14–19.

Johansson, M., K. Räsänen, and J. Merilä. 2001. "Comparison of nitrate tolerance between different populations of the common frog, *Rana temporaria.*" *Aquat. Toxicol.* 54:1–14.

Kellock, K. A., A. P. Moore, and R. B. Bringolf. 2018. "Chronic nitrate exposure alters reproductive physiology in fathead minnows." *Environ. Pollut.* 232:322–328.

Kim, J. H., Y. J. Kang, K. I. Kim, S. K. Kim, and J. H. Kim. 2019. "Toxic effects of nitrogenous compounds (ammonia, nitrite, and nitrate) on acute toxicity and antioxidant responses of juvenile olive flounder, *Paralichthys olivaceus.*" *Environ. Toxicol. Pharmacol.* 67:73–78.

Kolm, N. 2002. "Male size determines reproductive output in a paternal mouthbrooding fish." *Animal Behav.* 63:727–733.

Krishnamurthy, S. V., D. Meenakumari, H. P. Gurushankara, and R. A. Griffiths. 2006. "Effects of nitrate on feeding and resting of tadpoles of *Nyctibatrachus major* (Anura: Ranidae)." *Aust. J. Ecotoxicol.* 12:123–127.

Krishnamurthy, S. V., and G. R. Smith. 2010. "Growth, abnormalities, and mortality of tadpoles of American toad exposed to combinations of malathion and nitrate." *Environ. Toxicol. Chem.* 29:2777–2782.

Krishnamurthy, S. V., and G. R. Smith. 2011. "Combined effects of malathion and nitrate on early growth, abnormalities, and mortality of wood frog (*Rana sylvatica*) tadpoles." *Ecotoxicology* 20:1361–1367.

Lahti, E., M. Harri, and O. V. Lindqvist. 1985. "Uptake and distribution of radioiodine, and the effect of ambient nitrate, in some fish species." *Comp. Biochem. Physiol. A* 80:337–342.

Larisch, W., and K-U. Goss. 2018. "Uptake, distribution and elimination of chemicals in fish – Which physiological parameters are the most relevant for toxicokinetics?" *Chemosphere* 210:1108–1114.

Learmonth, C., and A. P. Carvalho. 2015. "Acute and chronic toxicity of nitrate to early life stages of zebrafish – setting nitrate safety levels for zebrafish rearing." *Zebrafish* 12:305–311.

Lefevre, S., F. B Jensen, D. T. T. Huong, T. Wang, N. T Phuong, and M. Bayley. 2011. "Effects of nitrite exposure on functional haemoglobin levels, bimodal respiration, and swimming performance in the facultative air-breathing fish *Pangasianodon hypophthalmus*." *Aquat. Toxicol.* 104 (1):86–93.

Little, A. G., and F. Seebacher. 2015. "Temperature determines toxicity: Bisphenol A reduces thermal tolerance in fish." *Environ. Pollut.* 197:84–89.

Lou, S., B. Wu, X. Xiong, and J. Wang. 2016. "Short-term toxicity of ammonia, nitrite, and nitrate to early life stages of the rare minnow (*Gobiocypris rarus*)." *Environ. Toxicol. Chem.* 35:1422–1427.

Mayo, A. L., D. J. Ritter, J. Bruthans, and D. Tingey. 2019. "Contributions of commercial fertilizer, mineralized soil nitrate, and animal and human waste to the nitrate load in the Upper Elbe River Basin, Czech Republic." *Hydro. Res.* 1:25–35.

McGurk, M. D., F. Landry, A. Tang, and C. C. Hanks. 2006. "Acute and chronic toxicity of nitrate to early life stages of lake trout (*Salvelinus namaycush*) and lake whitefish (*Coregonus clupeaformis*)." *Environ. Toxicol. Chem.* 25:2187–2196.

McKenzie, D. J., E. Garofalo, M. J. Winter, S. Ceradini, F. Verweij, N. Day, R. Hayes, R. van der Oost, P. J. Butler, J. K. Chipman, and E. W. Taylor. 2007. "Complex physiological traits as biomarkers of the sub-lethal toxicological effects of pollutant exposure in fishes." *Phil. Trans. R. Soc. B.* 362:2043–2059.

Meade, M. E., and S. A. Watts. 1995. "Toxicity of ammonia, nitrite, and nitrate to juvenile Australian crayfish, *Cherax quadricarinatus*." *J. Shell. Res.* 14:341–346.

Meade, T. L, and S. J. Perrone. 1980. "Effect of chloride ion concentration and pH on the transport of nitrite across the gill epithelia of Coho salmon." *Prog. Fish. Cult.* 42:71–72.

Miaud, C., N. Oromĺ, S. Navarro, and D. Sanuy. 2011. "Intra-specific variation in nitrate tolerance in tadpoles of the Natterjack toad." *Ecotoxicology* 20:1176–1183.

Monsees, H., L. Klatt, W. Kloas, and S. Wuertz. 2017. "Chronic exposure to nitrate significantly reduces growth and affects the health status of juvenile Nile tilapia (*Oreochromis niloticus* L.) in recirculating aquaculture systems." *Aquacult. Res.* 48:3482–3492.

Moore, A. P., and R. B. Bringolf. 2018. "Effects of nitrate on freshwater mussel glochidia attachment and metamorphosis success to the juvenile stage." *Environ. Pollut.* 242:807–813.

Morris, A. L., H. J. Hamlin, R. Francis-Floyd, B. J. Sheppard, and L. J. Guillette. 2011. "Nitrate induced goiter in captive whitespotted bamboo sharks *Chiloscyllium plagiosum*." *J. Aquat. Anim. Health* 23:92–99.

Muir, P. R., D. C. Sutton, and L. Owens. 1991. "Nitrate toxicity to Penaeus monodon protozoea." *Mar. Biol.* 108:67–71.

Nagrodski, A., K. J. Murchie, K. M. Stamplecoskie, C. D. Suski, and S. J. Cooke. 2013. "Effects of an experimental short-term cortisol challenge on the behaviour of wild creek chub *Semotilus atromaculatus* in mesocosm and stream environments." *J. Fish Biol.* 82:1138–1158.

O'Connor, C. M., and S. J. Cooke. 2015. "Ecological carryover effects complicate conservation." *Ambio* 44:582–591.

Opinion, A. G. R., G. De Boeck, and E. M. Rodgers. 2020. "Synergism between elevated temperature and nitrate: Impact on aerobic capacity of European grayling, *Thymallus thymallus* in warm, eutrophic waters." *Aquat. Toxicol.* 226:105563.

Ortiz-Santaliestra, M. E., M. N. Fernandes-Beneitez, A. Marco, and M. Lizana. 2010. "Influence of ammonium nitrate on larval anti-predatory responses of two amphibian species." *Aquat. Toxicol.* 99:198–204.

Ortiz-Santaliestra, M. E., M. J. Fernandez-Beneitez, M. Lizana, and A. Marco. 2010. "Adaptation to osmotic stress provides protection against ammonium nitrate in *Pelophylax perezi* embryos." *Environ. Pollut.* 158:934–940.

Ortiz-Santaliestra, M. E., M. J. Fernandez-Beneitez, M. Lizana, and A. Marco. 2011. "Responses of toad tadpoles to ammonium nitrate fertiliser and predatory stress: differences between popolations on a local scale." *Environ. Toxicol. Chem.* 30:1440–1446.

Ortiz-Santaliestra, M. E., M. J. Fernandez, M. Lizana, and A. Marco. 2011. "Influence of a combination of agricultural chemicals on embryos of the endangered gold-striped salamander (*Chioglossa lusitanica*)." *Arch. Environ. Contam. Toxicol.* 60:672–680.

Ortiz-Santaliestra, M. E., and A. Marco. 2015. "Influence of dissolved oxygen conditions on toxicity of ammonium nitrate to larval natterjack toads." *Arch. Environ. Contam. Toxicol.* 69:95–103. doi: 10.1007/s00244-014-0126-3.

Ortiz-Santaliestra, M. E., A. Marco, M. J. Fernandez, and M. Lizana. 2006. "Influence of developmental stage on sensitivity to ammonium nitrate of aquatic stages of amphibians." *Environ. Toxicol. Chem.* 25:105–111.

Ortiz-Santaliestra, M. E., A. Marco, and M. Lizana. 2011. "Realistic levels of a fertilizer impair Iberian newt embryonic development." *Herpetologica* 67:1–9.

Orton, F., J. A. Carr, and R. D. Handy. 2009. "Effects of nitrate and atrazine on larval development and sexual differentiation in the Northern Leopard Frog *Rana pipiens*." *Environ. Toxicol. Chem.* 25:65–71.

Ouedraogo, I., and M. Vanclooster. 2016. "A meta-analysis of groundwater contamination by nitrates at the African scale." *Hydrol. Earth Syst. Sci. Discuss.* 120:5194.

Pandey, R. B., G. L. Adams, and L. W. Warren. 2011. "Survival and precopulatory guarding behaviour of *Hyalella azteca* (amphipoda) exposed to nitrate in the presence of atrazine." *Environ. Toxicol. Chem.* 30:1170–1177.

Panno, S. V., W. R. Kelly, A. T. Martinsek, and K. C. Hackley. 2006. "Estimating background and threshold nitrate concentrations using probability graphs." *Ground Water.* 44:697–709.

Patra, P. W., J. C. Chapman, R. P. Lim, and P. C. Gehrke. 2007. "The effects of three organic chemicals on the upper thermal tolerances of four freshwater fishes." *Environ. Toxicol. Chem.* 26:1454–1459.

Pereira, A., A. P. Carvalho, C. Cruz, and A. Saraiva. 2017. "Histopathological changes and zootechnical performance in juvenile zebrafish (*Danio rerio*) under chronic exposure to nitrate." *Aquaculture* 473:197–205.

Philippe, C., P. Hautekiet, A. F. Grégoir, E. S. J. Thoré, T. Pinceel, R. Stoks, L. Brendonck, and G. De Boeck. 2018. "Combined effects of cadmium exposure and temperature on the annual killifish (*Nothobranchius furzeri*)." *Environ. Toxicol. Chem.* 37:2361–2371.

Polo-Cavia, N., P. Burraco, and I. Gomez-Mestre. 2016. "Low levels of chemical anthropogenic pollution may threatenamphibians by impairing predator recognition." *Aquat. Toxicol.* 172:30–35.

Rodgers, E.M., A. G. R. Opinion, D. F. Gomez Isaza, B. Rašković, V. Poleksić, and G. De Boeck. 2021. "Double whammy: Nitrate pollution heightens susceptibility to both hypoxia and heat in a freshwater salmonid." *Sci. Total Environ.* 765:142777. doi: https://doi.org/10.1016/j.scitotenv.2020.142777.

Rodrigues, R. V., M. H. Schwarz, B. C. Delbos, E. L. Carvalho, L. A. Romano, and L. A. Sampaio. 2011. "Acute exposure of juvenile cobia *Rachycentron canadum* to nitrate induces gill, esophageal and brain damage." *Aquaculture* 322:223–226.

Romansic, J. M., K. A. Diez, E. M. Higashi, and A. R. Blaustein. 2006. "Effects of nitrate and the pathogenic water mold *Saprolegnia* on survival of amphibian larvae." *Dis. Aquat. Org.* 68:235–243.

Schram, E., J. A. C. Roques, W. Abbink, T. Spanings, P. De Vries, S. Bierman, H. Van de Vis, and G. Flik. 2010. "The impact of elevated water ammonia concentration on physiology, growth and feed intake of African catfish (Clarias gariepinus)." *Aquaculture* 306:108–115.

Schram, E., J. A. C. Roques, W. Abbink, Y. Yokohama, T. Spanings, P. De Vries, S. Bierman, H. Van de Vis, and G. J. Flick. 2014. "The impact of elevated water nitrate concentration on physiology, growth and feed intake of African catfish *clarias gariepinus* (Burchell 1822)." *Aquacult. Res.* 45:1499–1511.

Secondi, J., E. Hinot, S. Djalout, and A. Jadas-Hécart. 2009. "Realistic nitrate concentration alters the expression of sexual traits and olfactory male attractiveness in newts." *Funct. Ecol.* 23:800–808.

Secondi, J., V. Lepetz, G. Cossard, and S. Sourice. 2013. "Nitrate affects courting and breathing but not escape performance in adult newts." *Behav. Ecol. Sociobiol.* 67:1757–1765. doi: DOI 10.1007/s00265-013-1583-9.

Sharma, B., and R. C. Ahlert. 1977. "Nitrification and nitrogen removal." *Water. Res.* 11: 897–925.

Smith, G. R., C. J. Dibble, A. J. Terlecky, C. B. Dayer, A. B. Buner, and M. E. Ogle. 2013. "Effects of invasive Western Mosquitofish and ammonium nitrate on Green Frog tadpoles." *Copeia* 2013:248–253. doi: http://dx.doi.org/10.1643/CE-12-072.

Smith, G. R., S. V. Krishnamurthy, A. C. Burger, and L. B. Mills. 2011. "Differential effects of malathion and nitrate exposure on American toad and wood frog tadpoles." *Arch. Environ. Contam. Toxicol.* 60:327–335.

Sokolova, I. M. 2013. "Energy-limited tolerance to stress as a conceptual framework to integrate the effects of multiple stressors." *Interg. Comp. Biol.* 53:597–608.

Solomon, K. R., D. B. Baker, R. P. Richards, K. R. Dixon, S. J. Klaine, T. W. La Point, R. J. Kendall, C. P. Weisskopf, J. M. Giddings, J. P. Giesy, L. W Hall Jr., and W. M. Williams. 1996. "Ecological risk assessment of atrazine in North American surface waters." *Environ. Toxicol. Chem.* 15:31–76.

Soucek, D. J., and A. Dickinson. 2012. "Acute toxicity of nitrate and nitrite to sensitive freshwater insects, mollusks, and a crustacean." *Arch. Environ. Contam. Toxicol.* 62:233–242.

Soucek, D. J., and A. Dickinson. 2016. "Influence of chloride on the chronic toxicity of sodium nitrate to *Ceriodaphnia dubia* and *Hyalella azteca*." *Ecotoxicology* 25:1406–1416. doi: DOI 10.1007/s10646-016-1691-1.

Soucek, D. J., D. R. Mount, A. Dickinson, J. R. Hockett, and A. R. McEwen. 2015. "Contrasting effects of chloride on growth, reproduction, and toxicant sensitivity in two genetically distinct strains of *Hyalella azteca*." *Environ. Toxicol. Chem.* 34:2354–2362.

Steinberg, K., J. Zimmermann, K. T. Stiller, L. Nwanna, S. Meyer, and C. Schulz. 2018. "Elevated nitrate levels affect the energy metabolism of pikeperch (*Sander lucioperca*) in RAS." *Aquaculture* 497:405–413.

Stelzer, R. S., and B. L. Joachim. 2010. "Effects of elevated nitrate concentration on mortality, growth, and egestion rates of *Gammarus pseudolimnaeus* amphipods." *Arch. Environ. Contam. Toxicol.* 58:694–699.

Stormer, J., F. B. Jensen, and J. C. Rankin. 1996. "Uptake of nitrite, nitrate, and bromide in rainbow trout, *Oncorhynchus mykiss*: Effects on ionic balance." *Can. J. Fish. Aquat. Sci.* 53:1943–1950.

Sullivan, K. B., and K. M. Spence. 2003. "Effects of sublethal concentrations of atrizine and nitrate on metamorphosis of the African clawed frog." *Environ. Toxicol. Chem.* 22: 627–635.

Taylor, L. N., W. J. McFarlane, G. G. Pyle, P. Couture, and D. G. McDonald. 2000. "Use of performance indicators in evaluating chronic metal exposure in wild yellow perch (*Perca flavescens*)." *Aquat. Toxicol.* 67:371–385.

Tian, X., and J. G. Qin. 2003. "A single phase of food deprivation provoked compensatory growth in barramundi Lates calcarifer." *Aquaculture* 224:169–179.

Tsai, S. J., and J. C. Chen. 2002. "Acute toxicity of nitrate on *Penaeus monodon* juveniles at different salinity levels." *Aquaculture* 213:163–170.

Van Bussel, C. G. J., J. P. Schroeder, S. Wuertz, and C. Schulz. 2012. "The chronic effect of nitrate on production performance and health status of juvenile turbot (*Psetta maxima*)." *Aquaculture* 326–329:163–167.

Vitousek, P. M., J. D. Aber, R. W. Howarth, G. E. Linkens, P. A. Matson, D. W. Schnidler, W. H. Schlesinger, and D. G. Tilman. 1997. "Human alteration of the global nitrogen cycle: sources and consequences." *Ecol. Appl.* 7:737–750.

Wagenhoff, A., C. R. Townsend, and C. D. Matthaei. 2012. "Macroinvertebrate responses along broad stressor gradients of deposited fine sediment and dissolved nutrients: a stream mesocosm experiment." *J. Appl. Ecol.* 49:892–902.

Wang, M., L. Chai, H. Zhao, M. Wu, and H. Wang. 2015. "Effects of nitrate on metamorphosis, thyroid and iodothyronine deiodinases expression in *Bufo gargarizans* larvae." *Chemosphere* 139:402–409.

Watts, P. J., and P. Jarvis. 1997. "Survival analysis in palmate newts exposed to ammonium nitrate agricultural fertilize." *Ecotoxicology* 6:355–362.

Wei, L-Z., X-M. Zhang, J. Li, and G-Q. Huang. 2008. "Compensatory growth of Chinese shrimp, *Fenneropenaeus chinensis*, following hypoxic exposure." *Aquacult. Int.* 16:455–470.

Westin, D. T. 1974. "Nitrate and nitrite toxicity to salmonid fishes." *Prog. Fish. Cult.* 36:86–89.

Williams, D. A., M. H. Flood, D. A. Lewis, V. M. Miller, and W. J. Krause. 2008. "Plasma levels of nitrite and nitrate in early and recent classes of fish." *Comp. Med.* 58:431–439.

Williams, E. M., M. L. Glass, and N. Heisler. 1993. "Effects of nitrite-induced methaemoglobinaemia on oxygen affinity of carp blood." *Environ. Biol. Fish.* 37:407–413.

Wolter, C., and R. Arlinghaus. 2003. "Navigation impacts on freshwater fish assemblages: the ecological relevance of swimming performance." *Rev. Fish. Biol. Fish.* 13:63–89.

Xie, L., Y. Zhang, J. Gao, X. Li, and H. Wang. 2020. "Nitrate exposure induces intestinal microbiota dysbiosis and metabolism disorder in *Bufo gargarizans* tadpoles." *Environmental Pollution* 264:114712.

Xie, L., Y. Zhang, X. Li, L. Chai, and H. Wang. 2019. "Exposure to nitrate alters the histopathology and gene expression in the liver of *Bufo gargarizans* tadpoles." *Chemosphere* 217:308–319.

Xu, Q., and R. S. Oldham. 1997. "Lethal and sublethal effects of nitrogen fertilizer ammonium nitrate on common toad (*Bufo bufo*) tadpoles." *Arch. Environ. Contam. Toxicol.* 32:298–303.

Yang, X., L. Peng, F. Hu, W. Guo, E. Hallermen, and Z. Huang. 2019. "Acute and chronic toxicity of nitrate to fat greenling (*Hexagrammos otakii*) juveniles." *J World Aquacult Soc* 50:1016–1025.

# 10 Inorganic Nitrogen Ions of Surface and Ground Waters of Sterea Hellas, Central Greece
## A Case Study

Christos Tsadilas, Miltiadis Tziouvalekas,
Eleftherios Evangelou, Alexandros Tsitouras,
Christos Petsoulas, and Antonios Peppas

## CONTENTS

10.1 Introduction ................................................................................................210
10.2 Materials and Methods ..............................................................................211
    10.2.1 Description of the Area Studied ...................................................211
    10.2.2 Water Sources ..............................................................................211
    10.2.3 Selection of Water-sampling Sites ..............................................214
    10.2.4 Analytical Methods ......................................................................215
10.3 Statistical Analysis ....................................................................................215
10.4 Results and Discussion ..............................................................................215
    10.4.1 River Waters ..................................................................................219
    10.4.2 Lake Waters ..................................................................................224
    10.4.3 Canal Waters ................................................................................225
    10.4.4 Groundwater ................................................................................228
10.5 Concluding Remarks and Proposals ..........................................................231
Acknowledgments ..............................................................................................232
References ...........................................................................................................232

DOI: 10.1201/9780429326806-13

## 10.1 INTRODUCTION

Due to the serious effects of nitrate on the environment and human health, the presence of nitrate in surface and ground waters has received great attention worldwide. Because of its high mobility, nitrate may spread easily to terrestrial and aquatic ecosystems, considerably affecting their quality and health. In aquatic environments, nitrate promotes eutrophication while in terrestrial environments it may leach down, contaminating groundwaters. On the other hand, nitrate is the main form of the nutrient nitrogen which is uptaken through soil by plants, with enormous importance for plant physiology and growth and intimately for food production. Therefore, what is needed, is the adoption of appropriate management of nitrate, so that, on one hand, the significant benefits of their use to be exploited, and on the other hand, the adverse effects on environment and humans to be limited.

The interest in nitrate was noticeably raised when some observations on its possible effect on human health were reported. The first paper published by Comly (1945), in which the author reported the first two cases of cyanosis (methemoglobinemia) in infants who had received well water with high concentrations of nitrate, several other papers were published in American and European journals in which methemoglobinemia was correlated with nitrate content in water. However, at the same time, a debate had started on the correctness of this view. The alternative view was that methemoglobinemia was associated with a high bacterial content of the waters (Parson, 1978), and not immediately with nitrate. This debate still exists, and no definite answer has been given as to whether nitrate is responsible for methemoglobinemia or other diseases, such as stomach cancer. Instead of the view that nitrate causes health problems, there is an opposite opinion stating that nitrate favors human health (L'hirondel and L'hirondel, 2002). From that time onwards, permissible or recommended limits of nitrate concentration in drinking water began to be determined, ranging from country to country and changing over time. At the moment, the upper acceptable limit for drinking water has been set by WHO (2004) and the European Union at 50 mg $NO_3^-$/l.

The water policy of the European Union is addressed in the directive 2000/60/EC, which established a framework for the member states' actions for the protection of the quality of inland surface waters, transitional waters, coastal waters, and groundwaters. Among others, directive 2000/60/EC determines the administrative arrangements to be made in member states for the application of the regulations of this directive. In the formulation processes of strategies and objectives and the management plans of river basins, it is important to take into account the integrated nature of the hydrological and ecological systems. Member states should identify the individual river basins and designate their water districts. In the monitoring of water quality, all water measurements should be scheduled at the river basin scale and not at individual sites' scale.

Agricultural activities, especially intensive practices, play a primary role in water pollution. One of the major problems is the leakage of nutrients into water bodies and seriously affecting water quality. Eutrophication, algae growth, and oxygen reduction in surface water are among the main pollution problems related to nitrate. The Ministry of Rural Development and Foods (MRDF), in accordance with national obligations arising from implementation of Water Directive 2000/60/EK (EC, 2000), and in continuation of similar projects in Northern Greece, assigned a project with similar objectives to the Union of Hellenic Agricultural Organizations, "DEMETER," and ETME Peppas & Associates, aiming for investigation of the quality of freshwater from Sterea Hellas Prefecture, central Greece. More specifically, the purpose of this project was to investigate the quality of surface water and groundwater in an intensively cultivated agricultural area of central Greece, that is, in Sterea Hellas, to estimate the pollution sources and to propose measures of water pollution reduction or remediation, where necessary. In the present chapter, which presents a small part of this project, the main results of a two-year monitoring program of surface and groundwater quality of waters used for irrigation of crops and human consumption are presented, focused mainly on inorganic nitrogen ions, that is, ammonium, nitrite, and nitrate, according to the Water Directive 2000/60 and national regulations.

## 10.2 MATERIALS AND METHODS

### 10.2.1 DESCRIPTION OF THE AREA STUDIED

The study area is located in central Greece and belongs to the Prefecture of Sterea Hellas, which is the second-largest of the country's 13 prefectures, with a total population of about half a million. The prefecture occupies an area of 15,549 $km^2$ corresponding to 11,8 percent of the total country area, with a length 230 km and a width 95 km, and presents diverse geomorphological and socio-economic characteristics. Hydrologically, the area studied consists of two River Water Basins (the eastern EL07 and the western EL04) along with 14 others in the whole country (Figure 10.1).

The eastern part (EL07) is mountainous, including four montane complexes with an altitude up to 2457 m and lowland areas including the values of Sperheios and Kopais. Due to the complex relief of the area, the climate presents a great variety of characteristics. The mean annual rainfall varies from 500 mm in lowlands to 1200 mm in mountains. The average air temperature ranges from 11 °C to 18 °C, depending on the distance from the sea. The water district of the western part (EL04) is mostly mountainous, with very few lowland areas. The altitude of the mountains is 1528 m up to 2461 m. The total precipitation is on average 800–1000 mm with a maximum of 1800 mm, from which, more than 30 percent falls in the higher altitudes. Land use of the eastern part is mainly forests (36%), agriculture (31%), and pastures (27%). In the western part, land use is dominated by forests (about 50% of the area), agriculture (about 10%), and pastures (about 35% of the area). Water use in the eastern part is shared between irrigation of crops (840 $hm^3$, 87%), water supply (76 $hm^{3,}$ 8%), husbandry (10 $hm^3$, 1.1%), and industry (29 $hm^3$, 3.1%). In the western part most of the water is used for irrigation (717 $hm^3$), some of this for human consumption (39 $hm^3$), and the remainder for husbandry (8 $hm^3$) and industry (2 $hm^3$) (NWC, National Waters Committee, 2017a,b).

## 10.2.2 WATER SOURCES

The freshwaters studied in the present work come from rivers, lakes, drainage canals, and aquifers. The study area includes eleven catchments where intensive

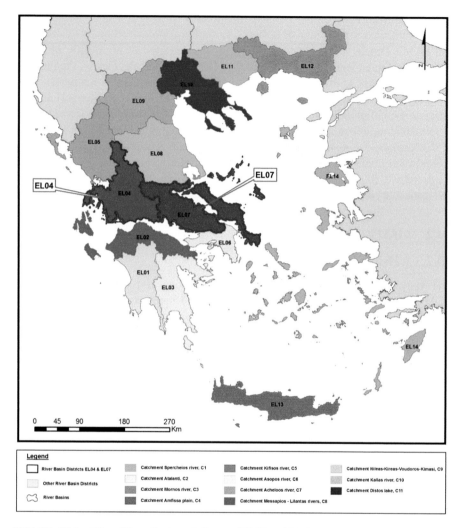

**FIGURE 10.1** Water River Basins of Greece (EL07 and EL04 are the WRBs in which the area studied is located).

agriculture takes place. These catchments are the Spercheios River (EL0718-C1), Atalanti (EL0722-C2), Mornos River (EL0421-C3), Amfissa plain (EL0724-C4), Kifisos River (EL0723-C5), Asopos River (EL0725-C6), Acheloos River (EL0415-C7), Messapios-Lilantas Rivers (EL0719-C8), Nireas-Kireas-Voudoros-Kimasi Rivers (EL0719-C9), Kalas River (EL0719-C10), and Dystos Lake (EL0719-C11) (Figure 10.2).

The rivers that flow through the study area are Spercheios, Mornos, Kifisos, Asopos, Acheloos, Messapios, Lillantas, and a complex of four small rivers Nireas-Kileas-Voudoros-Kimasi, and Kallas.

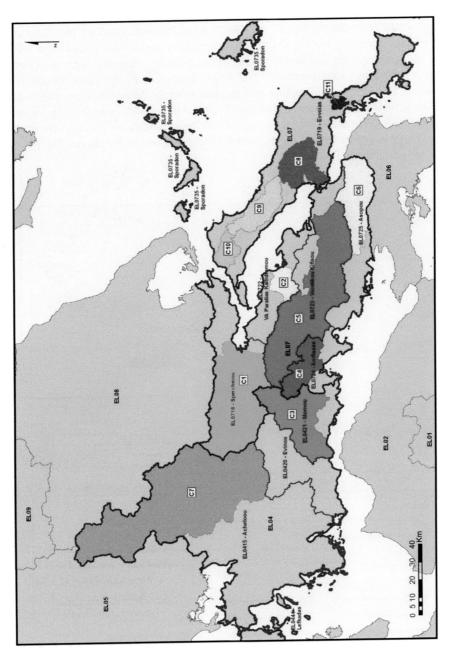

**FIGURE 10.2** The water catchments of the area studied.

## 10.2.3 SELECTION OF WATER-SAMPLING SITES

For two consecutive years (2017, 2018), and a small part of 2019, water samples were selected from the area studied, in which pH, electrical conductivity, and inorganic nitrogen ions (ammonium, nitrate, and nitrite) were determined and are presented in this chapter. Sampling sites in rivers were selected, so that at least one of them was near the water springs and another close to the estuaries. Sampling sites selection criteria in drainage canals included soil relief, kind of crop, and point sources of pollution. In lakes, water sampling was selected from their deepest points and from the lakes' perimeters near the confluence of rivers. Groundwater sampling sites were selected taking into account their geospatial distribution so that the distances between them would be representative for the compilation of isopiezometric curves. Sampling sites are presented in Figure 10.3.

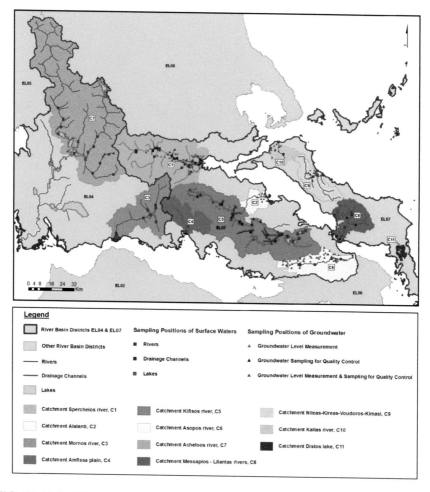

**FIGURE 10.3**   Water sampling sites of the area studied.

#### 10.2.4 ANALYTICAL METHODS

Water samples were analyzed for pH and electrical conductivity using a pH meter and a conductivity meter, respectively. Nitrogen ions, that is, ammonium, nitrite, and nitrate, were measured by using ionic chromatography.

## 10.3 STATISTICAL ANALYSIS

Statistical treatments of the data obtained (average, minimum, maximum, and median values), the standard deviation of the means, and analysis of variance were performed by using the SPSS statistical package. The quality of waters intended for human consumption of pH, electrical conductivity, and nitrogen ions (nitrate, ammonium, and nitrite) was evaluated according to standards set by the European Union and Greek authorities (EC, 2008; EC, 2000; Ministry of Environment, 2013; NWC, 2018). The same waters were evaluated for irrigation use according to the standards proposed by the Food Agricultural Organization – FAO (Ayers and Westcot, 1994).

## 10.4 RESULTS AND DISCUSSION

An overview of the water parameters values from the entire studied area is shown in Table 10.1. The parameter values ranged widely, reflecting the different conditions that prevail in the different water origins, that is, rivers, lakes, groundwater, and drainage canals as well as the special characteristics and conditions that exist in sampling positions. pH ranged between 3.24 and 8.95, electrical conductivity between 171 and 49800 µS/cm, nitrate from traces to 335 mg/l, nitrites from traces up to 1.18, and ammonium from traces up to 116 mg/l. The maximum values are much higher than the permitted values imposed by the respective authorities. However, median values, which describe in a more representative way the real conditions, show that all the properties studied are lying at a normal and acceptable level.

The overall picture of the water properties of the entire study area differs considerably if it is shared among the different water origins, as shown in Table 10.2, from the data of which, it seems that a wide range of values of all the properties studied

**TABLE 10.1**
**Mean, Median, Minimum and Maximum Values of the Water Properties Studied**

|                   | pH   | EC, $\mu$S/cm | $NO_2^-$, mg/l | $NO_3^-$, mg/l | $NH_4^+$, mg/l |
|-------------------|------|---------------|----------------|----------------|----------------|
| Sample numbers, N | 1387 | 1386          | 1380           | 1380           | 1384           |
| Mean              | 7.84 | 995           | 0.05           | 9.54           | 0.33           |
| Median            | 7.85 | 549           | n.d.*          | 2.19           | 0.05           |
| Minimum           | 3.24 | 171           | n.d.           | n.d.           | n.d.           |
| Maximum           | 8.95 | 49800         | 1.18           | 335            | 116            |

* non detectable.

## TABLE 10.2
Mean, Median, Minimum, and Maximum Values of the Properties Studied in
River, Lake, Drainage Canals, and Ground Waters

|  | pH | EC, $\mu$S/cm | NO$_2^-$,mg/l | NO$_3^-$,mg/l | NH$_4^+$,mg/l |
|---|---|---|---|---|---|
| River waters |  |  |  |  |  |
| N* | 415 | 415 | 415 | 415 | 415 |
| Mean | 8.08 | 1191 | 0.05 | 2.42 | 0.14 |
| Median | 8.12 | 429 | n.d. | 1.22 | 0.05 |
| Minimum | 3.24 | 171 | n.d. | n.d. | n.d. |
| Maximum | 8.84 | 49800 | 0.75 | 28 | 4.34 |
| Lake waters |  |  |  |  |  |
| N | 125 | 125 | 124 | 124 | 125 |
| Mean | 8.22 | 395 | n.d. | 0.16 | 0.07 |
| Median | 8.23 | 294 | n.d. | n.d. | n.d. |
| Minimum | 7.10 | 240 | n.d. | n.d. | n.d. |
| Maximum | 8.67 | 2070 | n.d. | 4.00 | 1.00 |
| Drainage canal waters |  |  |  |  |  |
| N | 336 | 335 | 335 | 335 | 335 |
| Mean | 7.88 | 1066 | 0.04 | 3.62 | 0.84 |
| Median | 7.84 | 548 | n.d. | 1.60 | 0.06 |
| Minimum | 6.80 | 301 | n.d. | n.d. | n.d. |
| Maximum | 8.92 | 23600 | 0.66 | 37.00 | 116 |
| Groundwaters |  |  |  |  |  |
| N | 511 | 511 | 506 | 506 | 509 |
| Mean | 7.53 | 936 | 0.07 | 21.61 | 0.23 |
| Median | 7.51 | 708 | n.d. | 8.86 | 0.09 |
| Minimum | 6.66 | 228 | n.d. | 0.01 | n.d. |
| Maximum | 8.95 | 4790 | 1.18 | 335 | 4.71 |

* Samples number

was recorded in the river waters. The pH ranged from 3.24–8.84, electrical conductivity between 171–49800 $\mu$S/cm, nitrites from traces up to 0.75 mgNO$_2$/l, nitrate from traces up to 28 mgNO$_3$/l, and ammonium from traces up to 116 mgNH$_4$/l.

The situation is quite different concerning lake waters, in which the range of the properties values was smaller compared to the river waters. The pH ranged from 7.10–8.67, electrical conductivity from 240–2070 $\mu$S/cm, nitrite concentrations were not detectable, nitrate concentration ranged from traces to 4.00 mgNO$_3$/l, and ammonium from traces up to 1.00 mgNH$_4$/l (Table 10.2).

In the waters of the drainage canals, a wide range was observed in the pH, electrical conductivity and ammonium concentration values (Table 10.2). pH ranged from 6.80 to 8.92, electrical conductivity from 294–23600 $\mu$S/cm, and ammonia from traces to 116 mgNH$_4$/l. Nitrite and nitrate concentrations ranged from traces up to 0.66 mgNO$_2$/l and 116 mgNO$_3$/l, respectively.

In groundwater, extreme values were found mainly in nitrate concentrations and in pH and electrical conductivity (Table 10.2). pH ranged between 6.66 to 8.95,

## TABLE 10.3
## Comparison of the Mean Values of pH, Electrical Conductivity, and Concentrations of Nitrites, Nitrate, and Ammonium of Waters from Different Origins

| Water origin | N* | pH | EC,$\mu$S/cm | $NO_2^-$,mg/l | $NO_3^-$,mg/l | $NH_4^+$,mg/l |
|---|---|---|---|---|---|---|
| Rivers | 415 | 8.08 b | 1191.26 a | 0.046 ab | 2.41 b | 0.13 ab |
| Canals | 336 | 7.88 c | 1066.25 a | 0.039 b | 3.62 b | 0.83 a |
| Groundwaters | 511 | 7.52 d | 936.30 ab | 0.068 a | 21.60 a | 0.22 ab |
| Lakes | 125 | 8.22 a | 395 b | 0.000 c | 0.16 b | 0.07 b |

* observation number, **Different letters in the same column mean statistically significant difference at probability level p<0.05 according to the LSD test.

electrical conductivity from 228 to 4790 $\mu$S/cm, and concentrations of nitrite from traces to 1.18 mgNO$_2$/l, nitrate from 0.01 to 335 mgNO$_3$/l, and ammonium from traces up to 4.71 mgNH$_4$/l.

A comparison between the water from different origins (Table 10.3), showed that the values of the properties studied followed the order: for pH lakes >rivers >canals >ground-waters; for electrical conductivity rivers = canals ≥ ground-waters ≥ lakes; for nitrites ground-waters > canals ≥ rivers > lakes; for nitrate ground-waters > rivers = canals = lakes and for ammonium canals ≥ rivers = ground-waters ≥ lakes. Graphically the comparison between the water properties studied from the various sources is shown in Figure 10.4.

Although water quality assessment has become a critical issue in the last decades, so far a globally accepted water quality index has not been developed – an index that can be applied in all countries and regions. The European Union 200/60 Water Framework Directive (EC, 2000) which established the framework for the protection of the waters in EU countries, proposed the implementation of measures to improve the quality of water in aquatic ecosystems, among which, the first step is to determine ecological and chemical status and include a large number of biological and physiological parameters to be monitored. Then the quality of water is classified in a category such as "high," "good," "moderate," "poor," and "bad" by giving general guidance without providing definite tools on how to do this (Alexakis et al., 2015). Thus, in the present work, the water quality concerning the properties presented in this part of the study is based on the comparison of the properties' values to standards set by the EU and the respective authorities of the country. Given that water use is usually not fully defined, and that it can be used either for irrigation or drinking, the assessment of their quality should take this issue into account. The purpose of this study was to assess the quality of the waters intended both for irrigation and for human consumption and, in this sense, it is discussed here. Concerning the water properties presented in this chapter the standards set by the Council Directive 98/83/EC are used. They are shown in Table 10.4.

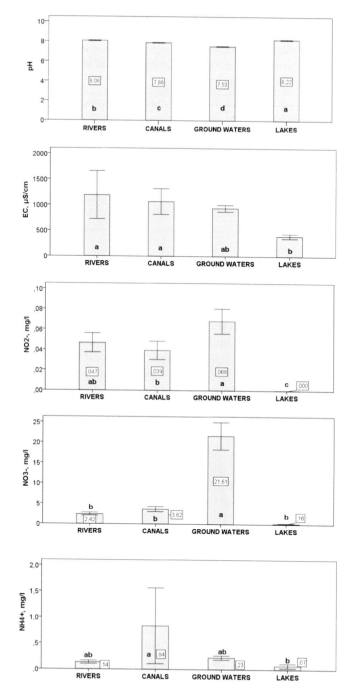

**FIGURE 10.4** Graphical presentation of the comparison between the properties of waters from different origins (different letters mean statistically differences at probability level p<0,05 according to the LSD test. The bars on the column tops indicate the standard error of the means).

**TABLE 10.4**
**Quality Standards Set by the EU for the pH, EC, and Inorganic Nitrogen in Waters Intended for Human Consumption**

| Parameter | Parametric value | Unit |
|---|---|---|
| pH | $\geq 6.5$ and $\leq 9.5$ | - |
| Electrical conductivity, EC | 2500 | $\mu S/cm$ |
| Nitrate, $NO_3^-$ | 50 | $mgNO_3^-/l$ |
| Nitrite, $NO_2^-$ | 0.50 | $mgNO_2^-/l$ |
| Ammonium, $NH_4^+$ | 0.50 | $mgNH_4^+/l$ |

Source: EC (1998).

**TABLE 10.5**
**Guidelines for Interpretation of Selected Water Quality Parameters for Irrigation**

| | | Degree of restriction on use | | |
|---|---|---|---|---|
| Parameter | Unit | None | Slight to moderate | Severe |
| pH | - | Usual range in irrigation water 6.5 – 8.4 | | |
| Salinity | $\mu S/cm$ | <700 | 700-3000 | >3000 |
| Nitrate | $mgNO_3^-/l$ | Usual range in irrigation water 0-50 | | |
| Ammonium | $mgNH_4^+/l$ | Usual range in irrigation water 0-6.5 | | |

Source: Ayers and Westcot (1994).

When the waters are intended for irrigation of crops, the desirable range of the degree of restrictions for the parameters presented in this work, are as in Table 10.5.

## 10.4.1 RIVER WATERS

Table 10.6 presents a comparison between the values of water parameters studied in the rivers from different catchments. pH was recorded very low (3,24) in Asopos River water (catchment C6) only in one sampling site. This is an extreme case caused by anthropogenic action, that is, dumping of industrial wastes. As shown by the color of the water (Photo 10.1), liquid wastes were disposed of into the river, coming from the adjacent industrial activity. This pH is unacceptable for any kind of use. In all the remaining river waters pH values were within the acceptable limits. The lower mean values of pH were recorded in the Asopos River (catchment C6) and the higher ones in the water of the Kallas River (catchment C10).

Electrical conductivity exceeded the permitted values in the catchments of Spercheios (C1), Mornos (C3), and Messapios-Lilantas (C8). The number of samples where the EC exceeded the permissible values for waters intended for human

**TABLE 10.6**
**Minimum and Maximum Values**

| Property/Catchment | Spercheios-C1* N***=93 | Mornos-C3 N=42 | Kifisos-C5 N=45 | Asopos-C6 N=49 | Acheloos-C7 N=96 | Messapios-Lilantas-C8. N=24 | NKBK**-C9 N=48 | Kallas-C10 N=18 |
|---|---|---|---|---|---|---|---|---|
| pH | 7.68-8.84 8.19 ab**** | 7.61-8.49 7.99 bc | 7.24-8.75 8.03 bc | 3.24-8.55 7.79 c | 7.22-88.83 8.15 ab | 7.62-8.78 8.04 ab | 7.55-8.65 8.11 ab | 7.51-8.73 8.23 a |
| Electrical conductivity, $\mu S/cm$ | 174-18630 853 b | 245-10220 916 b | 391-718 500 b | 355-2770 848 b | 171-455 308 b | 367-49800 9317 a | 536-1969 935 b | 619-933 803 b |
| Nitrite ($NO_2^-$), mg/l | n.d.*****-0.26 0.03 bc | n.d.-0.24 0.01 c | n.d.-0.47 0.08 ab | n.d.-0.75 0.09 a | n.d.- 0.01 c | n.d.-0.74 0.08 ab | n.d.-0.30 0.05 abc | n.d.-0.16 0.04 abc |
| Nitrate ($NO_3^-$), mg/l | n.d-9.02 1.57 cd | 0.06-6.87 0.62 d | 0.25-12.22 2.92 c | 0.47-17.29 6.11 b | 0.28-5.68 0.67 d | 2.14-7.76 8.02 a | 0.77-5.31 2.68 c | 0.04-2.31 0.74 d |
| Ammonium ($NH_4^+$), mg/l | n.d-1.05 0.11 ab | 0.02-0.81 0.12 ab | n.d.-0.36 0.05 b | n.d.-4.34 0.30 a | n.d.-0.84 0.08 b | n.d.-2.17 0.32 a | n.d.-1.09 0.12 ab | n.d.-0.52 0.12 ab |

* Catchments, **Nileas, Kireas, Voudoros, Kimasi,, ***Sample numbers, ****Different letters in the same line mean statistically significant difference according to the LSD test at the probability level p<0.05. *****non detectable.

**PHOTO 10.1**    Sampling site of Asopos River (catchment 6) where water pH was 3.24.

consumption, were seven, that is, a value of 8 percent for the Specheios catchment, three (7%) for Mornos catchment (C3) and five (21%) for the catchment of Messapios-Liantas (C8). The higher mean EC value was found in catchment C8 followed by all the others, which had equal mean values. The highest EC values were found in catchment C8 (Messapios-Lilantas, 49800 μS/cm), C1 (Spercheios, 18630 μS/cm), C3 (Mornos, 10220 μS/cm), and C6 (Asopos, 2770 μS/cm). The reasons for these high EC values seem to be mainly the effect of the sea (Figure 10.5). Similar results were reported by others who stated that in areas at the mouth of rivers, where seawater enters, significant amounts of substances in the form of salts are transported, originating mainly from agricultural and secondarily from industrial activity (Piper et al., 1990; Kormas et al., 2003; Vakirtzi et al., 2018).

In the case of C1 (Spercheios) and C8 (Messapios-Lilantas) catchments, the sampling sites were located within the delta of the Spercheios and Messapios Rivers, resulting in seawater intrusion. Besides, it is also noted that, in the case of C1 catchment, at this position two canals receive the effluents from the Wastewater Treatment Plant of Lamia, a city with a population of 76,000, and from the industrial area (Figure 10.5).

According to the first update of River Basin Management Plans of Water River Basins of Eastern Central Greece (EL07) (NWC, 2017a), the point sources of pollution add every year in the catchment of Spercheios (C1) about 90 tons of Biological Oxygen Demand (BOD), coming mostly from sewage treatment plants and secondly from livestock facilities, about 70 tons of N coming mainly from sewage treatments, and in some degree from livestock and industry, and about 50 tons of P coming exclusively from sewage treatment plants. It seems, therefore, those sewage treatments are the main point source of pollution and, in second position, livestock and

**FIGURE 10.5**  Sampling sites in Spercheios (C1), Mornos (C3), and Messapios-Lilants (C8) catchments with maximum EC values of 18630, 10220, and 49800 μS/cm respectively.

industry. The situation is different in Kifisos catchment (C5) concerning the point sources of pollution. In this catchment, most of the BOD and the whole amount of P come from livestock facilities, amounting to up to 300 tons and about 25 tons per year, respectively, while N comes mainly from waste-disposal sites, which amounts up to 280 tons per year. Regarding the non-point pollution sources, BOD comes mostly (82.4%) from animals, N and P come mainly from agriculture (64% and 89%, respectively), and the rest from other non-point pollution sources (diffuse municipal and animals).

Mean values of the nitrogen ions were quite below the acceptable standards (Table 10.6). Slightly higher values than the standard values were recorded for nitrite in the waters of the catchments C6 (Asopos) and C8 (Messapios-Lilantas) in only one sample, but they were quite higher in the case of ammonium. With the exception of the C5 catchment (Kifisos), exceedings from the standard values were recorded in all the other catchments in percentages of 4 percent in the catchment of Spercheios, 2 percent in the catchment Mornos, 12 percent in the catchment Asopos, 1 percent in the catchment Acheloos, 13 percent in the catchment of Messapios-Lilantas, 6 percent in the catchment of the complex Nileas-Kireas-Voudoros-Kimasi (NKVK), and 6 percent in the catchment of Kallas River.

The higher values of ammonium were found in the catchment C6 (Asopos) almost along the entire river length, indicating that the waters of this river receive a heavy load of pollutants from agricultural activities, leading to severe eutrophication. A characteristic case of this type of pollution is presented in Photo 10,2.

Concerning the pollution in the Asopos River, Botsou et al. (2011) reported the Asopos can be considered as a representative case of an intermittent Mediterranean river, in which during the last decades intense industrialization took place with more than three thousand industries, such as metallurgies, and metal furnishing plants, textile and dye production industries and food and chemical industries. Dokou et al. (2016) reported, also, that the Asopos River Basin water bodies face serious water quantity and quality problems. According to the River Basin Management Plan for the District of Eastern Sterea Hellas (EL07), from the total estimated water needs for the Asopos catchment (C6), 82 percent are consumed by the agricultural sector, 11 percent from the industrial sector, and the remaining 7 percent for drinking. The industrial sector hosted in the area of this catchment represents about 20 percent of total national industrial production, due mainly to the proximity to Athens, the capital city. In this catchment effluent and solid wastes from industrial plants have been disposed of legally or illegally for over 45 years. due to a catastrophic adopted

**PHOTO 10.2** A characteristic sampling site in Asopos River with high ammonium concentration and eutrophication.

policy, which provided incentives to industry to relocate their plants from the Athens vicinity to the Asopos catchment area, leading to creation of an unofficial industrial zone (Skouloudis et al., 2017). Furthermore, the Asopos River was designated by the respective authorities as a "receiver of industrial waste," but the discharges of the industrial zone were not regularly monitored, thereby leading to very severe contamination of the water bodies. Since 1969, when an industrial zone was established in the central and eastern parts of this catchment, the river has received an enormous amount of wastewater. Ziogas et al. (2009), reported that around 12,000 m³/day of wastewater, and roughly about 660 kt of solid waste are discharged into the river. This very bad situation was changed after the adoption of the European Water Framework Directive and its transposition into Greek national legislation in 2003. From that point forward, the implementation of environmental law was strict and, from 2004 to 2009, approximately 4.3 million € in fines was imposed on 163 industries operating in the area, due to non-compliance with environmental law (Laoudi et al. 2011). According to the National Water Committee (NWC, 2017a), most of the N load from the point source of pollution (about 250 tons per year) comes from waste disposal sites and BOD (about 125 tons per year) from livestock.

## 10.4.2 LAKE WATERS

In Table 10.7 the minimum, maximum, and a comparison of the mean values of the parameters studied in lake waters, are presented. From the data of this table, it is obvious, that the parameters studied did not exceed the standard values. It means that the water of the four lakes of the area studied is of good quality. This is to be

---

**TABLE 10.7**

**Comparison of the Mean Values of pH, Electrical Conductivity, and Inorganic Nitrogen Forms of Lake Waters (the Numbers above the Mean Values Show the Minimum and Maximum Values)**

| Parameter/Lake | Mornos-C3[*] N[**]=30 | Kifisos-C5 N=24 | Acheloos-C7 N=42 | Lake Dystos-C11 N=5 |
|---|---|---|---|---|
| pH | 7.55-8.30 | 7.58-8.67 | 8.00-8.45 | 7.10-8.44 |
|  | 8.12a[**] | 8.32a | 8.20a | 7.92b |
| Electrical conductivity, $\mu$S/cm | 260-303 | 312-599 | 240-322 | 474-2070 |
|  | 274c[***] | 483b | 277c | 1276a |
| Nitrite ($NO_2^-$), mg/l | - | - | - | n.d.[****] |
|  | 0.42b | 0.49b | 0.90ab | 1.24a |
| Nitrate ($NO_3^-$), mg/l | - | - | 0.05-4.24 | - |
|  | 2.60b | 5.75b | 0.55b | 18a |
| Ammonium ($NH_4^+$), mg/l | - | - | - | - |
|  | 1.35a | 1.33a | 1.56a | 1.49a |

* Catchment, **Sample numbers, *** Different letters in the same line mean statistically significant difference according to the LSD test at the probability level p<0.05. ****non detectable.

expected since all the lakes are in areas (Figure 10.6) where no serious pressures from anthropogenic influences exist.

Mornos is an artificial lake, the ninth-largest in Greece, and constructed during 1972–1979 on the Mornos River's main channel. The dam has a total volume of ~ 17 hm$^3$ and is composed of sand and gravel, with an impermeable clay core. The Mornos Lake watershed has a maximum storage capacity and active water volume of 780 hm$^3$ and 640 hm$^3$ respectively. It receives about 243 hm$^3$ from the Mornos River and its tributaries and another mean annual inflow of 223 hm$^3$ from the artificial lake Evinos through a tunnel 29.4 km long. Mornos Lake is included in the Natura 2000 program (coded GR2450003) (source: EYDAP, Athens Water Supply and Sewerage Company). Mornos Lake provides fresh water to an aqueduct 188 km long (one of the longest in Europe), necessary for about 3.1 million inhabitants of Athens. The water-quality parameters are all within the acceptable limits both for drinking or irrigation, since there are no anthropogenic pressures around its watershed.

Lake Yliki is a key element in the water supply system of Athens. The quantities of water pumped from Yliki, are significant, especially in times of low water supply, as happened in the years 1989–2001, when 81 hm$^3$/year was pumped to avoid the over-exploitation of the Mornos reservoir. Furthermore, the annual water abstraction from lake Yliki, used mainly for irrigation, is about 42 hm$^3$.

Lake Distos (Figure 10.6) is situated in central Evia. Its surface area is 5.07 km$^2$ and its perimeter 11.03 km. There is no significant pressure around this lake, which is covered by reeds and, during the summer months, there is no water. According to NWC (2017a) the annual dilution of the loads of BOD, N and P are 1.73 mgBOD/l, 2.78 mgN/l, and 1.13 mgP/l.

The Acheloos Lake/Reservoir (Kremasta) (Figure 10.7), which became operational in 1965, has a surface area of 80.6 km$^2$, a total storage volume 4,495 hm$^3$ and its watershed covers an area of 3,292 km$^2$. The mean annual precipitation in the watershed is 1433 mm and the mean flow to the reservoir is an estimated 117.1 m$^3$/s, provided mostly from Acheloos River (Nerantzaki et al., 2017). Similar results to our study for pH and the rest properties studied were reported by Economou and colleagues (1991).

## 10.4.3 CANAL WATERS

Mean values of water parameters studied in the canal waters of the entire area were found well below acceptable standards, except for ammonium values in the canals of the C1 (Sperceios) catchment (Table 10.8).

Mean values for pH and EC were greater in the C1 catchment (Spercheios) followed by the C3 (Mornos) and C5 (Kifisos) catchments. High values of EC were recorded in two sampling sites, which were in low areas with poor drainage and which receive soluble salts from the surrounding areas with high inputs, mainly fertilizers (Figure 10.5).

Nitrite concentrations were very low in all the water canals, but nitrate concentration was greater in the catchment C1, also in a low area. Ammonium mean values, however, were greater than the acceptable standards in the canals of catchment C1, followed by catchments C3 and C5, where the ammonium value was low.

**FIGURE 10.6**  Lakes Mornos (up), Yliki and Paralimni (middle), and Distos (down).

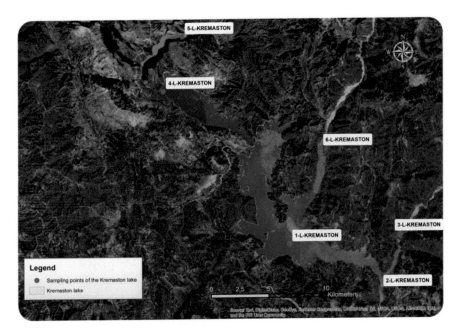

**FIGURE 10.7**    Lake Acheloos (Kremasta).

**TABLE 10.8**
**Comparison of the Mean Values of pH. Electrical Conductivity and Inorganic Nitrogen Forms of Canal Waters (the Numbers above the Mean Values Show the Minimum and Maximum Values)**

| Property/Catchment | Sperchios-C1*<br>N**=130 | Kifisos-C5<br>N=134 | Mornos-C3<br>N=72 |
|---|---|---|---|
| pH | 6.80-8.92<br>7.94a*** | 7.19-8.86<br>7.86b | 7.39-8.48<br>7.81b |
| Electrical conductivity, $\mu$S/cm | 200-**23600**<br>1548a | 245-738<br>604b | 301-12130<br>725b |
| Nitrite ($NO_2^-$), mg/l | n.d****.-0.66<br>0.051a | n.d.-0.48<br>0.043a | n.d.-0.19<br>0.004b |
| Nitrate ($NO_3^-$), mg/l | n.d.-37<br>5.42a | n.d.-23<br>3.35b | n.d.-3.34<br>0.80c |
| Ammonium ($NH_4^+$), mg/l | n.d.-**116**<br>**1.92**a | n.d.-1.39<br>0.17a | n.d.-0.94<br>0.10a |

* Catchments, **Sample numbers, ***Different letters in the same line mean statistically significant difference according to the Least Significant difference at the probability level p<0.05 .****non detectable.

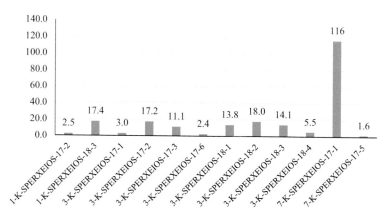

**FIGURE 10.8**  Ammonium concentration in Spercheios canal waters (maximum permissible value 0.5 mg $NH_4^+/l$).

The higher ammonium values were found in sites receiving effluents from the municipal wastewater treatment plant of Lamia city and a cheese factory. Ammonium concentration in these sites are presented in Figure 10.8.

### 10.4.4  GROUNDWATER

In Table 10.9 a comparison between the values of groundwaters parameters between the different catchments is presented.

Mean values of pH were found higher in the catchments C1, C2, C6 followed by the pH values in the catchments C3, C5, C7, C8, and C10 catchments. In all cases, the pH value was within recommended standard values. The same was true concerning the mean values of EC in all catchments except for catchment C4 (Amfissa; Figure 10.9), although there were significant differences between the catchments. In this catchment, groundwater salinity is a major problem seriously affecting crops. In almost all the sampling sites, EC was above the recommended acceptable standards for drinking, as is shown in Figure 10.10. According to the standards for irrigation of crops (Table 10.5), in most cases these waters have a severe degree of restriction. Nitrite, nitrate, and ammonium nitrogen were all well below the upper limits permitted (Table 10.4).

In the plain of the Amfissa catchment, groundwater salinization conditions develop, which seems to extend inside the basin at a great distance from the coastal zone. The results of the present study clearly show salinization as it was highlighted in the provisions of the revised River Basin Management Plans (NWC, 2017a). Salinization is initially attributed, on one hand, to natural conditions (natural salinization), due to the extensive coastal occurrences of carbonate rocks, which surround the plain and, on the other hand, to the over-pumping of groundwater for irrigation of crops. In other words, it seems that the mixing of sea and freshwater, due to the geological structure of surrounding and underlying rocks of the basin, has also affected its formations due to the continuous water pumping. A similar case was reported by Pisimaras et al. (2016) for Sidirochorion area in Northeastern Greece, where EC was found to be very

**TABLE 10.9**
**Comparison of the Mean Values of pH, Electrical Conductivity, and Inorganic Nitrogen Forms of Groundwaters (the Numbers above the Mean Values Show the Minimum and Maximum Values)**

| Property/ Catchment | Sperchios-C1* N=107 | Atalanti-C2 N=23 | Mornos-C3 N=56 | Amfissa-C4 N=29 | Kifisos-C5 N=88 | Asopos-C6 N=78 | Acheloos-C7 N=21 | Messapios-Lilantas-C8 N=15 | NKBK*-C9 N=41 | Kallas-C10 N=26 |
|---|---|---|---|---|---|---|---|---|---|---|
| pH | 6.98-8.40 7.64 a*** | 7.42-8.03 7.66a | 7.04-7.72 7.45b | 6.66-7.64 7.08c | 6.97-8.55 7.48b | 7.07-8.95 7.63a | 7.10-7.70 7.40b | 6.90-7.88 7.49b | 7.22-8.40 7.66a | 7.15-7.67 7.38b |
| Electrical conductivity, $\mu S/cm$ | 367-1976 620 def | 629-1594 928cde | 378-822 533ef | 320-4790 **3158a** | 251-1843 681def | 394-**3400** 1290b | 228-638 464f | 5.23-2200 1013bc | 676-1195 838cde | 610-1117 838cde |
| Nitrite ($NO_2^-$), mg/l | n.d.****-**0.71** 0.06b | n.d.-0.21 0.07 | n.d.-0.24 0.02b | n.d.-0.08 0.06b | n.d.-0.20 0.07b | n.d-**1.10** 0.08ab | n.d.-**1.18** 0.18ab | n.d.-0.74 0.10ab | n.d.-0.24 0.02b | n.d.-0.26 0.05b |
| Nitrate ($NO_3^-$), mg/l | 0.01-94 8.35 b | 6-201 52.39a | 0.06-6.87 1.69b | 0.06-17 7.58b | 0.01-21 16.63b | 1.00-**335** **52.99a** | 0.07-18 5.85b | 5-245 47.01a | 2.71-56 12.81b | 1.89-36 13.78b |
| Ammonium ($NH_4^+$), mg/l | n.d.-4.71 0.32a | n.d.-2.25 0.28a | 0.21-1.74 0.23a | n.d.-0.75 0.18a | n.d.-4.71 0.23a | n.d-2.62 0.31a | n.d.-0.18 0.09a | n.d.-1.45 0.23a | n.d.-1.04 0.19 a | n.d.-064 0.15a |

* Catchments, **Nileas, Kireas, Voudoros, Kallas. ***Different letters in the same line mean statistically significant difference according to the LSD test at the probability level p<0.05. ****non detectable.

**FIGURE 10.9**  Sampling sites in the catchment of Amfissa (C4).

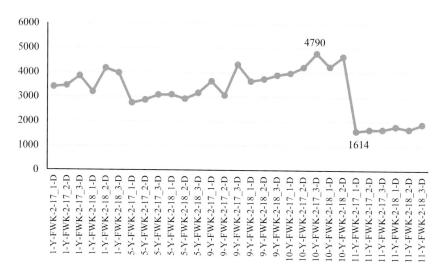

**FIGURE 10.10**  Electrical conductivity ($\mu$S/cm) variation in groundwater of Amfissa catchment.

high. According to NWC (2017a) the total annual loads produced by point pollution sources in the catchment of Amfissa are 48.08 ton BOD, 35.82 ton N, and 6.87 ton P. Most of the pollution comes from sewage treatments plants and, to a lesser degree, from the industry. Diffuse pollution sources coming from urban, agricultural, and livestock farming are quite a bit higher, amounting per year to 142 ton BOD, 102 ton N, and 20 ton P. The source that creates the greatest contamination is livestock farming. The annual supply to the groundwater system of Amfissa catchment is 3.0 $hm^3$, while the annual abstraction amounts to 0.55 $hm^3$, which is used mainly for irrigation. The quantitative situation is characterized as bad.

## 10.5  CONCLUDING REMARKS AND PROPOSALS

The contamination sources on a case-by-case basis appear to be *anthropogenic*: bad agricultural practices and livestock farming, such as unsound inputs management (fertilizers and livestock waste), dumping into drainage canals of untreated municipal or/and industrial (catchments of Spercheios, Kifisos, Asopos, Messapios, and Lilantas). Contamination sources are also natural, such as the seawater intrusion and the chemical composition of the study area's rocks (Amfissa, Asopos, Messapios, Lilantas). Based on the existing soil-water systems (surface and groundwater) of all catchments of the rivers and lakes of Sterea Hellas, nitrate is proposed as an environmental indicator for the pollution of the catchment Spercheios (C1), Amfissa (C4), Asopos (C6), and Messapios-Lilantas (C8).

Suggested measures for facing the contamination problems with salinization and nitrite and nitrate increase in waters are distinguished in the long term and short term. The short-term measures are referred to only in new drillings that are going to be opened, and in which vertical hydraulic protection must be ensured by isolating the surface aquifers (which are the most contaminated with nitrate), to avoid mixing with the deeper aquifers.

Since the increased contents of inorganic nitrogen compounds have been the result of long-term contamination in the last decades, they cannot be eliminated soon or easily. Long-term measures have to be taken which, to perform, will need several years to implement, and a great amount of funds to be spent. The long-term basic measures to be taken are described by the Environmental Goals of article 4 of the Water Framework Directive 2000/60, such as the implementation of the Codes of Good Agricultural Practices, the reconstruction and remediation of wetlands, tackling the problem of water over-pumping, desalinization plants, rehabilitation works of existing infrastructures, especially works on the improvement of infrastructures, of selection, storage, and transport/distribution of water for drinking or irrigation, aiming at the reduction of losses and quality of water, artificial recharge of aquifers, educational works, and, finally, research development, and demonstration works.

All the necessary measures for facing water contamination are clearly described in the recently revised River Basin Management Plans (NWC. 2017 a, b), which briefly refer to

- Tracing of the contamination sources to identify their origin (agricultural, livestock, urban, industrial).

- Implementation of good agricultural practices by applying irrigation methods that reduce nutrient leaching in the aquifer, especially nitrate. Furthermore, innovative techniques, such as precision agriculture practices, should be adopted.
- Continuous monitoring of surface and groundwater through the nitrate indicator.
- Protection measures for water drilling and application of their protection zones (following the River Basin Management Plans proposals).

## ACKNOWLEDGMENTS

The assignment of this project by the Ministry of Rural Development and Food (MRDF) to the Union of Hellenic Agricultural Organization "DEMETER" and ETME Peppas & Associates under the contract no 601/15881/13-2-2017), is greatly appreciated. Special thanks are addressed to the members of the project's monitoring committee of the MRDF for their effective collaboration and help. Many thanks are also addressed to the staff of the ETME Peppas & Associates and to all the technical staff of the Institute of Industrial and Forage Crops of the Hellenic Agricultural Organization DEMETER, especially to Mr. Savas Papadopoulos, Valantis Ntokouzis, Fotis Ntafos, Paraskevi Grypari, and A. Papadoulis for their valuable assistance in sampling and analyzing the water samples.

## REFERENCES

Alexakis, D., V.A. Tsihrintzis, G. Tsakiris, and G.D. Gikas. 2016. Suitability of Water Quality Indices for Application in Lakes in the Mediterranean. *Water Resour. Manage.* 30:1621–1633.

Ayers, R.S. and D.W. Westcot. 1994. Water Quality for Agriculture. FAO Irrigation and Drainage Paper. FAO Rome (www.fao.org/3/T0234E/T0234E00.htm).

Botsou, F., A.P. Karageorgis, E. Dassenakis, and M. Scoullos. 2011. Assessment of heavy metal contamination and mineral magnetic characterization of the Asopos river sediments (Central Greece). *Marine Pollution Bulletin* 62: 547–563.

Comly, H.H. 1945. Cyanosis in infants caused by nitrates in well water. *J. Amer. Med. Assoc.* 129: 112–116.

Dokou, Z., V. Karagiorgi, G.P. Karatzas, N.P. Nikolaidis, and N. Kalogerakis. 2016. Large scale groundwater flow and hexavalent chromium transport modeling under current and future climatic conditions: the case of Asopos River Basin. *Envion. Sci. Pollut. Res.* 23:5307–5321.

Economou, A.N., Ch. Daoulas, and P. Econmou. 1991. Observation on the biology of Leuciscus 'svallize' in the Kremasta reservoir (Greece). *Hydrobiologia* 213: 99–111.

EC. 1998. Council Directive 98/83/EC, 1998 on the quality of water intended for human consumption. L 330/32 5.12.98.

EC. 2000. Directive 2000/60/EC of the European Parliament and of the Council of 23 October 2000 establishing a framework for Community action in the field of water policy. Official Journal of the European Communities L. 327/1.

EC. 2008. Council Directive 98/83/EC of 3 November 1998 on the quality of water intended for human consumption. L 330/32.

Kormas, K.A., A. Nikolaidou, and M. Thessalou-Legaki. 2003. Variability of environmental factors of an eastern Mediterranean Sea river influenced coastal system. *Mediterranean Marine Science* 4(1): 67–77.

Laoudi, A., G. Tentes, and D. Damigos. 2011. "Groundwater damage: A cost-based valuation for Asopos River Basin." In: Proceedings of the 3rd International Conference on Environmental Management, Engineering, Planning and Economics (CEMEPE) and SECOTOX Conference, edited by A. Kungolos, A. Karagiannidis, K. Aravossis, P. Samaras, and K.W. Schramm: 975–982.

L'hirondel, J. and J.-L. L'hirondel. 2002. Nitrate and man. Toxic, harmless or beneficial? Oxford: ABI Publishing, p. 168.

Ministry of Environment. 2013. Common Ministerial Decision No 169280/07-08-2013).

NWC, National Waters Committee. 2017a. Government Gazette, No 4673/29-12-17.

NWC, National Waters Committee. 2017b. Government Gazette, No 4681/29-12-17.

NWC, National Waters Committee. 2018. Government Gazette, No 2686/06-07-18.

Nerantzaki, S.D., G. Giannakis, N. Nikolaidis, I. Zacharias, G. Karatzas, and I.A. Sibetheros. 2016. *Soil Science*, 181(7): 306–314.

Piper, D.J.W., N. Kontopoulos, C. Anabgnostou, G. Chronis, and A.G. Panagos. 1990. *Geo-Marine Letters* 10: 5–12.

Parson, M.L. 1978. Is the nitrate drinking water standard unnecessarily low? Current research indicates that it is. *American Journal of Medical Technology* 44: 952–954.

Pisimaras, V., C. Polychronis, and A. Gemitzi. 2016. Intrinsic groundwater vulnerability determination at the aquifer scale: a methodology coupling travel time estimation and rating methods. *Environ. Earth Sci.* 75: 85. doi:10.100/sl2665-015-4965-7.

Rousakis, G., I.P. Panagiotopoulos, P. Drakopoulou, P. Georgiou, D. Nikolopoulos, N. Mporompokas, V. Kapsimalis, I. Livanos, I.A. Morfis, C. Anagnostou, and M. Koutra. 2018. Sustainability evaluation of Mornos Lake/Reservoir, Greece. Environmental Monitoring Assessment. 190:64 https://doi.org/10.1007/s10661- 017-6431-3.

Skouloudis, A., N. Jones, S. Roumeliotis, D. Isaac, A. Greig, and K. Evangelinos. 2017. Industrial pollution, spatial stigma and economic decline: The case of Asopos river basin through the lens of local small business owners. *Journal of Environmental Planning and Management* 60(9): 1575–1600.

Vakirtzi, I., V. Markogianni, M. Pantazi, K. Pagou, A. Pavlidou, E. Dimitriou. 2018. Effect of river inputs on environmental status and potentially harmful phytoplankton in a coastal area of eastern Mediterranean (Maliakos Gulf, Greece). *Mediterranean Marine Science*. 19(2): 326–343. doi:http://dx.doi.org/10.12681/mms.14591.

WHO, World Health Organization. 2004. *Guidelines for Drinking-water Quality*. 3rd ed. vol. 1. Recommendations, Geneva.

Ziogas, C., C. Theochari, R. Leivadaros, A. Boura, P. Pantelaras, M. Papadopoulou, and A. Stamou. 2009. The Asopos River problem and proposals for its confrontation. TCG, Athens (in Greek).

# 11 Determination of Nitrate in Waters

*Md Eshrat E. Alahi, Fahmida Wazed Tina,*
*and Subhas Mukhopadhyay*

## CONTENTS

11.1 Introduction.................................................................................................235
11.2 Detection Methodologies...........................................................................237
11.3 Electrochemical Detection.........................................................................237
     11.3.1 Potentiometric Detection...............................................................237
     11.3.2 Amperometric Detection................................................................239
     11.3.3 Voltammetric Detection..................................................................242
11.4 Chromatography Detection.........................................................................243
11.5 Biosensors...................................................................................................244
11.6 Flow-Injection Analysis.............................................................................246
11.7 Electromagnetic Sensors............................................................................248
     11.7.1 Planar Electromagnetic Sensor......................................................248
     11.7.2 Planar Interdigital Sensor...............................................................250
11.8 Fiber-Optic Sensor......................................................................................251
11.9 Conclusion..................................................................................................254
References.............................................................................................................255

## 11.1 INTRODUCTION

Nitrogen is considered the most common natural element in the environment, and almost 80 percent of the air we breathe is nitrogen [2]. It can be found in gaseous forms such as Nitrogen ($N_2$), Nitrous oxide ($N_2O$), Nitric oxide (NO), Nitrogen dioxide ($NO_2$), and Ammonia ($NH_3$), and ionic form in water ($NO_3^-$ or $NO_2^-$) [3]. During mixture with rainwater some of these gases produce nitrate, nitrite, and ammonium ions. They can also mix with soil water or groundwater as a solution. The intaking of nitrate ions is necessary for the human body to improve blood flow by reducing the flow. However, excessive intake of nitrate ions through drinking water can be harmful to the human body and create various diseases, such as gastric cancer, and Parkinson's diseases. It is also harmful to infants, as they can suffer the "blue baby syndrome" or methemoglobinemia [4] that reduces the oxygen content of the blood [5, 6]. Six months old infants suffer most from this disease.

DOI: 10.1201/9780429326806-14

In our daily life, nitrate is an essential nutrient, and various nitrogen species are commonly found in the environment, such as in water and soil. In the agricultural field, an excess amount of fertilizers is used for plantation, and mismanagement of natural recourses occurs due to the lack of knowledge. These are the significant reasons to disrupt global and local nitrogen cycles [7]. The excessive and continuous use of nitrate can cause enormous problems and raises numerous concerns [8, 9]. Researchers have identified these problems widely and, as a result, international and government organizations have created frameworks to control the level of contamination. Regulatory bodies have controlled the level of pollution within the environment, industry, and food products [4, 10].

Many sources of nitrate go to ground or surface water. Waste materials, disposal of animal sewage, industrial waste, and excessive use of organic fertilizers in agriculture are the significant sources of nitrate pollution for water [11–13]. In an aquatic medium, nitrate stimulates excessive production of algae and phytoplankton, which leads to eutrophication. In water, they consume more oxygen during the decomposition process, which affects fish and other marine life. Therefore, monitoring the environmental charge of nitrate has gained increasing importance.

The need and desire to monitor nitrate is indisputable, yet tracking the level of nitrate with the presence of other ions can be a substantial challenge to the research community. Similar ions, such as nitrite, ammonium, phosphate, and sulfate are present in natural water. Therefore, improved detection methods are essential to avoid any interference that can be encountered in the environment, industries, food, and industrial activities. A large number of analytical methods and sensing methods have been developed to overcome the peculiarities of the various media. Recently, pollutant removal from water using membrane technology has received significant attention. It has a high adsorption capacity with a low cost and has been integrated into sensor technology when the selectivity towards the targeted ions is enhanced [14]. Appropriate characteristics and selection of membrane material are essential to produce highly selective materials and systems. Many research studies have reported different types of membrane for the detection of nitrate in water. Modified silica polymer, trihexyltetradecyl-phosphonium chloride polymeric membrane [15], and doped polypyrrole, zinc (II) complex polymeric membrane [16], have been used for membrane development. Still, more research is going on to improve the sensitivity of membrane materials and to improve selectivity.

In natural water, there are other environmental impacts such as temperature, pH level, oxygen level, and the presence of other polluting ions, which will affect the precision of the measurements. All these ecological impacts might come naturally or via industrial and experimental waste. Therefore, obtaining reliable data is essential and can be achieved by controlling the measurement conditions. Precision instrumentation and measurement methods are necessary to get highly reliable data. Different methods are available to detect nitrate in water, such as chromatography, flow injection analysis, electrochemical sensors, biosensors, optical fiber sensors, electromagnetic sensors. In recent years, several reviews have been undertaken to investigate the different detection methods [17–22]. The objective of this chapter is to provide a comparative discussion of the reported literature covering all kinds of detection of nitrate, and also to provide a brief assessment of the relative merits of each detection approach.

## 11.2   DETECTION METHODOLOGIES

Usually, two methods are used for nitrate detection: indirect and direct [11]. Indirect methods are complicated and costly as they require chemical reagents, expensive instruments for accurate measurement, and experts to make the measurements. Moreover, indirect methods produce a lot of chemical waste. On the other hand, direct methods are reliable and cost-effective. The disadvantage of using this method is that it provides measurement errors due to interference caused by the presence of other contaminants. Both methods are discussed in subsequent sections.

## 11.3   ELECTROCHEMICAL DETECTION

The electrochemical detection of nitrate has several categories based on their different sensing methods. In an electrochemical cell, nitrate can be detected by using various types of electrodes (i.e., copper, silver, platinum, gold, glassy-carbon electrode (GCE), graphite-epoxy, chitosan/bentonite, graphene, etc.). Sensing methods and the materials of the electrodes determine the sensitivity of the sensor as well as its reusability and LOD. The electrochemical detection method can convert nitrate ions into an impedance, potential difference, or current which can be grouped into impedimetric, potentiometric, and voltammetric, respectively. This method is easy to use, has sufficient sensitivity to nitrate ions in water, has easy miniaturization, and requires low power consumption.

The conventional electrochemical cells that are used for nitrate detection are large enough to be unsuitable as a portable device. Few recent studies [23–25] have reported that nitrate detention was done effectively using small electrochemical cells. Nowadays, developing low-cost portable devices for monitoring the nitrate concentration continuously in a natural aqueous environment is necessary. Moreover, the advanced analytical system should be highly miniaturized and sensitive to nitrate detection over a long period. Therefore, the reusability of any proposed analytical system is essential. In the following subsections, different types of electrochemical detection methods are discussed.

### 11.3.1   POTENTIOMETRIC DETECTION

In the conventional potentiometric order, there are two half-cells, each containing an electrode that is immersed in a solution of ions. The activities of the electrode's potential can be determined in the solution. In the salt bridge of the system, an inert electrolyte – that is, potassium chloride (KCl) – connects the two half-cells. At the end of the salt bridge, porous frits are fixed to allow the electrolyte's ions to move between the salt bridge and half-cells (Figure 11.1).

In late 1976, the potentiometric detection method was first introduced for detecting nitrate ions in water. This method is direct, and it does not need any additional chemical

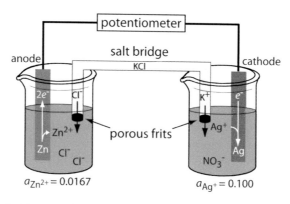

**FIGURE 11.1** Architecture of potentiometry system.

reagent. This method has been improved in the last few years to increase its nitrate selectivity and to reduce the limits of nitrate detection. Several improvements have been made to increase the performance of the method. For example, ion-selective electrodes (ISE) have been integrated with this system to determine the free-ion concentration in water directly. There are many advantages of using ISE: (1) cost-effective; (2) portable device; (3) non-destruction of samples; and (4) pre-treatment of the samples is not required.

In 1976, an organic nitrate and nitramine determination method was reported by Hassan and colleagues [26]. This simple potentiometric method was based on the reaction of a mercury and sulfuric acid mixture. This method was considered highly selective and useful for rapid and accurate detection at the micro and sub-micro levels. The detection limit was 1 to 50 $\mu$M ($\mu$ mol) with a precision of $\pm$ 0.2 percent. A nanobiocomposite was developed as ISE by Mendezo and colleagues [27] for determining nitrate in water, which was based on the intercalation of chitosan in bentonite. The detection limit of this method was 20 mM to 800 mM. A polymeric membrane employing two Zn(II) complexes was developed by Mahajan and colleagues [16]. These two Zn(II) complexes were coordinated by neutral tetradentate ligands, N, N'-ethylene-bis (N-methyl-(S)-alanine methylamide) and N, N'-ethylene-bis (N-methyl-(S)-alanine diethylamide, which was used as anion-selective carriers. The combination of Zn (II) complexes and dioctyl sebacate provided a high sensing selectivity to nitrate ions in water. The detection range was 50 $\mu$ M to 100 mM, which was close to the Nernstian slopes in the wide linear concentration range. Another potentiometric sensor was developed by Li [28] and Nunez [29] who used an Artificial Neural Network (ANN) to determine the nitrate levels in nitrate-contaminated water. According to Bendikov and Harmon [30], doped polypyrrole might be used as a selective membrane in an ISE electrode for the detection of nitrate levels in water. Doped polypyrrole is a highly conductive and stable polymer material that is widely used. Afterward, this doped polypyrrole was used in a potentiometric system for nitrate detection by Zhang and colleagues [31], and it improved the selectivity with a simple recipe procedure. Moreover, using doped

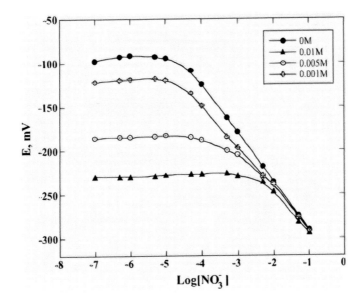

**FIGURE 11.2**  Potentiometric response of an electrode at varying concentrations [1].

polypyrrole is comparatively safe as it has lower toxicity compared to another, conventional membrane – that is, polyvinyl chloride (PVC) [31]. Wardak and colleagues [15] developed an active polymeric membrane using trihexyltetradecylphosphonium chloride (THTDPCl) which enhanced the sensitivity of PVC membrane by reducing electrical resistance.

Since most of the potentiometric sensors are bulky, as they contain an internal reference electrode and reference electrolyte solution with a true-liquid/liquid polymeric membrane, a microfabricated polymeric sensor was developed to solve the problem of bulky potentiometric sensors [1, 32]. This micro-fabrication method has several advantages (1) small-sized, (2) simple design, (3) cost-effective, and (4) possibility of mass production. Until now, different types of materials (e.g., screen-printed thick film, silicon-based transducer, metal-printed flexible film, etc.) have been developed to produce micro-scale potentiometric sensors, and they have demonstrated a good response towards nitrate ions in water. Figure 11.2 shows an example of the reaction from a potentiometric sensor, and Figure 11.3 shows the layout and design of a potentiometric sensor. Table 11.1 summarizes the characteristics of various potentiometric sensors.

## 11.3.2  AMPEROMETRIC DETECTION

Amperometry is known as an electrochemical method in which instrumentation is used to control the potential of the sensing electrode, appearing because of oxidation/reduction. In this method, the current is recorded as an analytical signal, which is measured as a function of time. This method was first introduced in 1976 and was

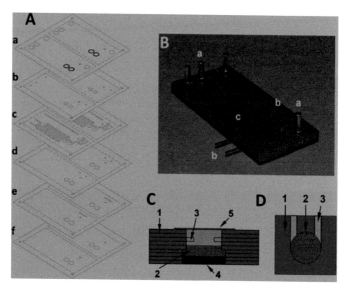

**FIGURE 11.3**  (A) Layout of the biparametric prototype; (B) picture of the final constructed device; (C): front view of the detection-chamber scheme; (D) Top view of the detection-chamber scheme [1].

known as high-performance liquid chromatography (HPLC) [33]. In 1987, a conventional benchtop was introduced [34]. In this method, for any given sample, the magnitude of the generated current is determined by the number of molecules. Calculation of generated current is done using Faraday's law:

$$i_t = \frac{dQ}{dt} = \eta F \frac{dN}{dt} \qquad (1)$$

where, $i_t$ is the current generated at the sensing surface electrode during time t, Q is the charge at the sensing surface, $\eta$ is the number of moles, N is the number of moles of analyte during oxidation/reduction, and F is the Faraday constant (96487 C mol$^{-1}$). The detection limit is attomole or femtomole levels [35, 36]. Therefore, this method (Figure 11.4) is useful for determining nitrates in water.

The amperometric method can be used for nitrate detection in water. N.G. Carpenter and colleagues [38] reported the kind of method where the nitrate concentration range was 0.1–1 mM. This method was able to avoid interference from other similar ions in water. X. Zhang [39] reported how Polypyrrole (PPy) nanowire modified electrodes, based on graphite electrodes, could be used for nitrate detection. In this process, the detection limit was 1.52 μM, and the sensitivity reached 336.28 mA/M cm². It is essential to increase the sensitivity of the detection method by designing the sensing electrodes. Other amperometric detectors [40–49] were reported for nitrate determination in water.

**TABLE 11.1**
**Characteristics of Various Potentiometric Sensors**

| Electrode and membrane | Reference Electrode | Limit of Detection (LOD) | Nernstian Slope (mV/decade) | Response time (s) | Application | Characteristics | Reference |
|---|---|---|---|---|---|---|---|
| Silver bis (bathophenanthroline) nitrate in a plasticized PVC | Ag/AgCl | 0.05 µgm/L | $-55.1 \pm 0.1$ | <15 | Industrial wastewater, Fertilizers and Pharmaceuticals. | Fast response, high sensitivity, long-term stability and excellent selectivity. | [32] |
| Graphite-epoxy and PVC | Ag/AgCl | 4.6 µmol/L | $-68.2$ | Not reported | Photoelectrocatalytic treatment | The simple operation does not require any preparative or pre-treatment stage. | [29] |
| Nitrate polymeric | Ag/AgCl | 9.56 mg/L | $-59.5$ | Not reported | water treatment process plant | Small in size, simultaneous and on-line detection | [1] |
| Glassy carbon electrode and polypyrrole with nitrate doped | Ag/AgCl | 10-4.8 mol/L | $-55.1 \pm 0.1$ | 85 | Monitoring of soil micronutrient | Rapid response, inexpensive cost, simple operation | [31] |
| Chitosan/bentonite and graphite-epoxy | Ag/AgCl | $2 \times 10^{-4}$ M | $-54.6$ | Not reported | Determination of nitrate in water | Low cost, easy to use | [27] |
| Trihexyltetradecylphosphonium chloride, plasticizer and PVC membrane | Ag/AgCl | 2.8 µmol/L | $-60.1$ | 5-10 | Determination of nitrate in water | Simple, fast and cheap | [15] |
| Pencil lead (graphite) and doped polypyrrole | Silver wire | $5 \pm 1 \times 10^{-5}$ M | $-54$ | Not reported | Nitrate determination from tap water, stream or lake water | Simple, easy to prepare and low cost | [30] |
| PVC, N,N'-ethylene-bis (N-methyl-(S)-alanine methylamide) and N,N'-ethylene-bis (N-methyl-(S)-alanine dimethylamide), Zn (II) | Ag/AgCl | $1.0 \times 10^{-3}$ M | $-55.16$ | 25 | Determination of nitrate in water | Fast response, reusability for 3 months. | [16] |

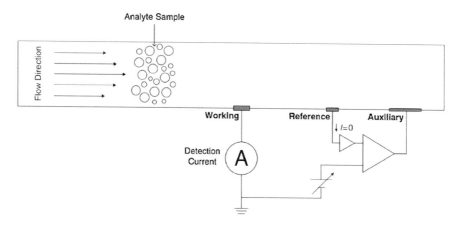

**FIGURE 11.4** Schematic diagram of the amperometric detection cell for lab-on-a-chip application [37].

### 11.3.3 Voltammetric Detection

Cyclic voltammetry (CV) is an effective electrochemical detection method that is used to measure the reduction and oxidation process of any molecular species. Usually, three electrodes (i.e., reference electrode, the working electrode, and counter electrode) are used in this method. The working electrode carries out the electrochemical event of interest, which changes the potential applied to give the desired potential at the reference electrode. During providing the potential to the working electrode, current begins to flow, and the auxiliary electrode completes the electrical circuit to continue the flow of electrons due to the reduction, or oxidation, process.

In 1977, M.E. Bidini and colleagues [50] reported a voltammetric detection method for determining nitrate in irrigation water samples. In this method, the working electrode was designed by depositing copper and cadmium electrochemically on the pyrolytic graphite electrode. The detection range was from 1µM to 1 mM. R.J. Davenport and colleagues [51] reported use of a rotating cadmium disk in the voltammetric method for nitrate and nitrite determination. J. Krista and colleagues [52] reported another determination system using electrodes that were made from a mixture of silver, graphite powder, and methacrylate resin. In this system, the nitrate ion detection range was up to 31 mg/L and LOD was 7 mg/L. Though this system had good reproducibility, a computer-controlled system was required to achieve the results. In situ copper-based electrodes with a low detection limit were reported by S.M. Shahriar and colleagues [53] where simultaneous nitrate determination was done in river water through applying differential pulse voltammetry (DPV). Voltammetric determination of nitrate in drinking water was reported by Neuhold and colleagues and Mareček and colleagues [54, 55], where a carbon-paste electrode was used as a working electrode, and the detection limit was 0.5-60 µg/mL. Application of differential pulse voltammetry (DPV) for nitrate determination in natural water was reported by Solak and colleagues [56]. This method used copper-plated glassy carbon, which

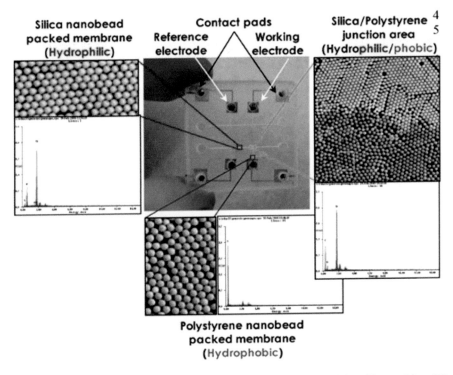

**FIGURE 11.5** Photograph of the dual ion-selective lab chip with self-assembly nBP columns [58].

had a sensitivity of 0.9683 A. L/mol. The nitrate-detection limit was 2.8 µM, and the linear detection range was 2.8 µM–80 µM. Copper-plated glassy carbon electrodes were also used in a square-wave voltammetric method [57] in which a linear range of determination of nitrate was 0.61 µM- 50 µM and LOD was 0.18 µM. A new electrochemical sensing platform with self-assembly nanobeads-packed (nBP) hetero-columns has been developed and reported by Jang and colleagues [58]. Figure 11.5 illustrates the sensing platform reported in that article.

## 11.4 CHROMATOGRAPHY DETECTION

Chromatography detection is used for the separation of different kinds of chemical mixtures. Chromatography is used in the laboratory for various chemical and biological analyses (Figure 11.6). When performing a chemical reaction with an eluent, an anion or cation of the analyte sample is extracted. Several chromatography techniques are used – for example, ion chromatography [59], high-performance liquid chromatography [60], and ultra-performance liquid chromatography [61]. Among these techniques, Ion chromatography is widely used for the detection of nitrate in water [62–68]. Though this technique is popular it is not suitable for in-situ nitrate measurement as it is expensive and bulky. Moreover, it consists of several materials

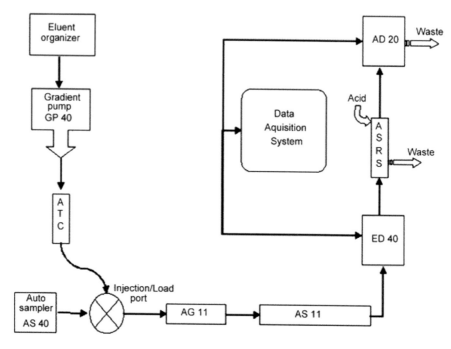

**FIGURE 11.6** Schematic representation of the ion-chromatography setup used for simultaneous determination [63].

such as solvents, solvent degasser, pump, injector, pre-column heat exchanger, guard column, post-column heat exchanger, electrolytic suppressor, and detector. This technique also requires UV spectrometry, fluorimetric, electron capture, and mass spectroscopy [59, 60, 62, 63, 69–71]. The UV detection method is simple and, thus, it is popular. Since the low absorbance of water maintains a low background, it reduces complications during the direct detection of nitrate ions in water.

An HPLC method with improved sensitivity is also available to detect nitrate in water [60]. This method is based on a photochemical reaction and ion-exchange separation. It is easy to handle and has high selectivity towards nitrate ions in water. A simple, fast, and accurate HPLC method was developed by Zuo and colleagues [71] for the detection of nitrate and nitrite ions in lake water, which produced good sensitivity, a high detection range, and a good LOD. Table 11.2 summarizes the type of chromatography methods with their characteristics.

## 11.5  BIOSENSORS

The biosensing method is used as a direct method for nitrate determination in water. In this method, biological materials are employed with a detection system, and a signal-conditioning circuit is used to measure the concentration of the targeted ions in a sampling solution. For measuring the concentration of a sample, the sampling solution is exposed directly to the biosensor and the sensing material. The targeted

**TABLE 11.2**
**The types of chromatography with eluent and LOD**

| Chromatography Type | Detector | Eluent | LOD | Application | References |
|---|---|---|---|---|---|
| Ion Chromatography | Ion exchange and diode-array detection | NaCl | 0.05 mg/L | Detection of nitrate in rain water | [62] |
| Ion Chromatography | low-capacity anion exchange column with amperometric and absorbance. | NaOH | 6 µg/L | Detection of nitrate in water | [63] |
| HPLC | Ion-exchange separation, online photochemical reaction, and luminol chemiluminescence | Borate buffer (pH 10) | $2 \times 10^{-8}$ M | Detection of nitrate in river, rainfall, pond, tap and commercial mineral water | [60] |
| HPLC | UV light absorption | Tetrabutylammonium hydroxide, $Na_2HPO_4$ (pH 3.9), and Acetonitrile | 5 µg/L | Detection of nitrate detection | [71] |
| Liquid Chromatography | Fluorescence detection | Toluene and NaOH | 0.3 pg/L | Biological, food and environmental samples | [69] |

ions interact with the biosensor and the sensing material and provide the required information. This interaction is converted to an electrical signal (i.e., voltage, current, or impedance) that measures the concentration of the sample. This conversion happens based on the biosensing method. Some studies [72–75] reported biosensors and differentiated them based on the nature of the transducer. Various biological components (i.e., DNA, enzymes, immunological systems, receptor proteins, or whole cells) could be used as recognition units. The nature of a transduction component could be acoustic, chemical, electrochemical, microbalance, optical, or piezoelectric, but they should have high sensitivity, specificity, and the ability to work in a wide range of matrices. It would be beneficial if they could be used remotely to make in situ measurements.

Nitrate biosensors have been developed in the last two decades because of the easy availability of appropriate enzymes and their use as a selective material in a substrate. A fluorescence-based fiber-optic biosensor was reported by Zeng and colleagues [76], who used it to trace various contaminants in seawater. In this biosensor, a protein molecule was used as a recognition unit. Another fast, sensitive, and stable conductometric enzyme biosensor has been reported by Xuejiang and colleagues [77], who used it to determine nitrate in water. In the biosensor, a methyl viologen mediator mixed with nitrate reductase (NR) modified the electrodes. This improved biosensor showed a quick response and reached 95 percent of the constant conductance value within 15 s. It had a calibration range from 0.02 to 0.25 mM and a detection limit of 0.005 mM. Cosnier and colleagues [78] entrapped NR in a laponite clay gel, which was cross-linked by glutaraldehyde. This developed biosensor was able to detect nitrate in low concentrations. The use of biosensors could enhance the sensitivity during measuring nitrate in water during in situ remote monitoring. An amperometric nitrate biosensor was reported by Can and colleagues [25], where the film was produced from Polypyrrole (PPy)/ Carbon nanotubes (CNTs). During nitrate measurement, the response of this biosensor was higher compared to the standard methods. The sensitivity of this advanced sensor was 300 nA/mM, and the detection range was 0.44–1.45 mM. Another conductometric biosensor was reported by Zhang and colleagues [79]. Table 11.3 tabulates the type of biological materials, the detection systems, and the limit of detection (LOD).

## 11.6   FLOW-INJECTION ANALYSIS

The flow-based method is a desirable nitrate-detection method, and it can be used as a good alternative to other traditional methods [82–91]. This method is cost-effective and easy to operate. It requires low volumes of reagents and samples, and it shows high throughput analysis. There are four different processes (i.e., distribution, reduction, pre-detection, and detection) in a flow-injection method. In the distribution process, the sample is carried on to the next process with a constant flow rate. The flow-injection method has a pump that allows the supplied sample to proceed endlessly within a specific period. In the reduction process, nitrate ions are converted into active nitrite ions by elevating the nitrate ions inside the sample. In the pre-detection process, the reduction agent reacts with the nitrate ions produced in the earlier process and forms a compound. Afterward, the detection system detects nitrate ions successfully. Usually, Griess-illosvay, acidic hydrogen peroxide, sulfanilamide, and N-(1-naphthyl) ethylenediaminedihydrochloride (NED) are used as detection agents, and zinc column, cadmium column, vanadium (III), titanium (III) chloride, and hydrazine sulfate are used as reduction reagents for nitrate detection in a flow-injection method. It is imperative to choose the correct reduction reagent; otherwise, it takes a long time to complete the injection process. A lengthy process was reported by Hydrazine and colleagues [89] where hazardous reagents were associated as a by-product, which could result in pollution.

Previously, numerous flow-injection methods, along with different types of detectors, have been reported for nitrate detection. Among different types of detection methods, spectrophotometric and chemiluminescence are preferred for nitrate detection. In this detection method, absorbance is the output that is proportional to the sample concentrations. Lambert's law explains this behavior, where absorbance is proportional to the concentration of sample nitrate. For achieving the maximum output of detection, chemical reagent, pH of the solution, flow rate of sample carrier, sample volume, and coil length of the reaction are needed to be related. Moreover,

**TABLE 11.3**
**Types of Biological Materials, Detection Systems, and the Limit of Detection**

| Detection method | Biological material | LOD | Application | Characteristics | Reference |
|---|---|---|---|---|---|
| Conductometric | Methyl viologen mediator with NR | 0.005 mM | Detection of nitrate in wastewater and river water | Response time was 15 s, low detection limit, excellent operational and thermal stability. Life of the sensor not more than two weeks. | [77] |
| Amperometric | CNT/PPy film electrode with NR | 0.17 mM | Measurement of nitrate in water | The developed sensor had better performance than standard analysis, response time was 20 s, and the first 10 days was better. | [25] |
| Cyclic Voltammetry | NaR–SOD1–CNT–PPy–Pt | 200 nM | Nitrate detection in saliva and blood | Extremely sensitive with large sample size required for the measurements. | [80] |
| Amperometric | Escherichia coli with NR | 0.1 mM | Detection of nitrate in drinking water | More extended stability and cheap method to prepare the biosensor. | [81] |
| Cyclic Voltammetry | Laponite clay gel, glutaraldehyde with NR | 7 μM | Determination of nitrate | Low detection limit. | [78] |

these parameters are needed to be controlled to achieve accurate results, sample throughput, and peak shape of the output. Figure 11.7 shows a schematic diagram of the necessary components of a flow-injection analyzer.

Luminol chemiluminescence was reported by Yaqoob and colleagues [84], where nitrate reduction depends on pH. Increasing pH value changes the absorbance level or the detector's response. This process develops the azo dye formation, which increases

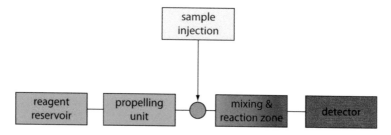

**FIGURE 11.7**   Schematic diagram of necessary components of a simple flow-injection analyzer.

the efficiency of spectrophotometric detection [83]. The flow rate acts as a vital parameter, the significance of which has been reported by Wang and colleagues [85], and the significance of the flow rate has been reported by Wang and colleagues [85]. The flow rate is controlled for the sample, carrier, and reagent solutions to enhance the sensitivity and sample throughput for nitrate ions [88]. Finding the optimum flow rate is essential to stabilize the baseline on the detection area and to improve the output (i.e., intensity or absorbance). Table 11.4 summarizes the various characteristics of flow-injection analysis.

## 11.7   ELECTROMAGNETIC SENSORS

### 11.7.1   PLANAR ELECTROMAGNETIC SENSOR

After conducting a lot of research on impedance-based sensors, it has been proved that they are very good for measuring physical properties and suitable for direct measurement. A Planar Electromagnetic Sensor Array (PESA) is one of those sensors that measures physical properties in terms of impedance. PESA is convenient for in-situ measurement. It is highly reliable, cost-effective, and shows a quick response. During measuring nitrate in water samples, different properties of the sample (i.e., conductivity, dielectric properties, and permeability) can be estimated [14, 99–105]. It is important to enhance the sensor's sensitivity, which can be done through optimizing the sensor design and configuration. A planar electromagnetic sensor used for nitrate detection in an aqueous medium was reported by Yunus and colleagues [99]. This sensor's performance was measured through two different arrangements: series and parallel connections. It was found that series connection showed better performance during nitrate detection in terms of the sensor's sensitivity. Therefore, it is better to use a series connection for electromagnetic sensors. The material of the electrodes is another important thing on which the sensitivity of the sensors depends. In the previously reported work, gold electrodes were used due to the attraction of nitrate ions towards gold. However, for developing a high-sensitivity sensor, it is required to use a high-dielectric substrate to increase the penetration depth of the electric field. Usually, there are three configurations to design the electrodes: parallel, star, and delta. Nor and colleagues [103] reported a sensor array with a thin substrate on a Printed Circuit Board (PCB), which was made by using a conventional PCB

**TABLE 11.4**
**Flow-Injection Methods with Various Characteristics**

| System | Reduction agent and Detection agent | LOD | Detection system | Application | Ref. |
|---|---|---|---|---|---|
| Flow analysis | Zinc granules and N-(1-naphthyl) ethylene diamine (Griess reagent) | 1.3 µg/L | Spectrophotometric | Nitrate measurement in marine and estuarine waters | [92] |
| Flow analysis | Enzymatic assay reagent and Griess-Ilosvay | 0.73 µg/L | Spectrophotometric | Determination of nitrate and nitrite | [93] |
| Automated parametric | Cadmium and Modified Griess-llosvay | 0.0207 mg/L | Spectrophotometric | Determination of nitrate, nitrite, and sulphate in drinking water | [94] |
| Flow injection | Vanadium (III) Chloride and Sulfanilamide and N-(1-naphthyl) ethylenediamine dihydrochloride (NED) | 0.1 µM | Spectrophotometric | Determination of nitrite and nitrate | [95] |
| Flow injection | Cadmium and sulfanilamide and N-(1-naphthyl) ethylenediamine dihydrochloride | 50 µgN/L | Spectrophotometric | Nitrate and nitrite determination in natural waters | [96] |
| Flow injection | Cadmium and sulfanilamide and N-(1-naphthyl) ethylenediamine dihydrochloride | 0.013 µgN/L | Spectrophotometric | Determination of nitrate, nitrite and ammonium in soil | [97] |
| Flow injection | Cadmium and hydrogen peroxide | 0.02 µg N/L | Luminolchemi luminescence | Determination of nitrate and nitrite in freshwater | [98] |

fabrication technique. This designed sensor had several coils or loops of electrodes that were spiral or square. The sensitivity of the sensor also depends on the distance between the coils. The impedance of a sensor can be increased by increasing the distance between the two electrodes and decreasing their area (Wang et al. [106]).

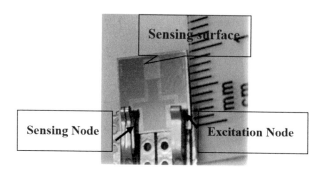

**FIGURE 11.8**   MEMS-based Interdigital sensor [108].

## 11.7.2 Planar Interdigital Sensor

The operating principle planar interdigital sensor follows is the rule of the parallel-plate capacitor, where the electrodes provide a one-sided measurement to the material under test (MUT) [107] (Figure 11.8). From the positive electrode, electric field lines are generated, and they bulge towards the negative electrode through the MUT. The sensor behaves as a capacitor, and the impedance depends on the physical properties of the MUT. Thus, through knowing the impedance behavior of the interdigital sensor, it is possible to know the physical properties of the material.

Some studies [108–114] have been conducted on interdigital sensors to develop a low-cost sensing system for nitrate measurement. In a low-cost sensing system, Electrochemical Impedance Spectroscopy (EIS) is used for measurement in which a small alternating signal is provided to the anode, and an output signal is produced at the cathode. In this method, the impedance measurement is done through an Inductance Capacitance and Resistance (LCR) meter. The sensitivity of the sensor depends on the length, width, and pitch gaps as well as on the electrodes' materials and the sensing area. A smart sensor used for nitrate detection in water was reported by Alahi and colleagues [109, 115]. In this sensor, temperature compensation was added to improve the sensitivity, and the detection limit was 0.01–0.5 mg/L. On the silicon substrate, gold electrodes were used for excitation and sensing. The pitch gap of the electrodes was 25 μm, and the dimension of the sensing surface was 2.5 mm × 2.5 mm. This developed sensor was able to measure nitrate in water and transfer the measured data to the cloud for real-time monitoring. Another study by Alahi and colleagues [110, 116] explained the principle of selective material for nitrate detection. An imprinting polymer technique was used for material development, and the detection range was 1–10 mg/L. This developed material was used as a coating on the interdigital sensor for nitrate detection in lake, stream, river, and canal water. The calibration curve (Figure 11.9) was developed in terms of reactance, and later, it was used to measure unknown nitrate samples.

A graphene-based electrode (Figure 11.10) was used for nitrate detection in water [112, 117] with a detection limit of 1–70 mg/L. The sensitivity of the measurement was improved by printing graphene on Kapton tape. It was also proposed to use temperature compensation for further improvement of the sensitivity of the sensor. This

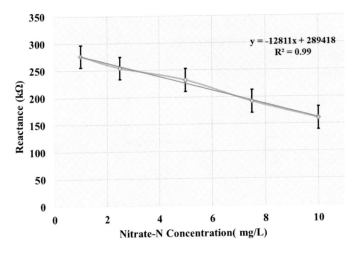

**FIGURE 11.9** Calibration standard for nitrate-N measurement [110].

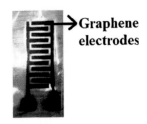

**FIGURE 11.10** Graphene-based interdigital sensor [112].

sensor was robust, cost-effective, easy to operate, and useful for nitrate measurement. Alahi and colleagues [118, 119] developed an interdigital sensor based on FR4 and the Internet of Things for nitrate measurement and found it extremely useful for nitrate detection from creek, stream, lake water in real-time. Table 11.5 summarizes the various characteristics of planar interdigital sensors.

## 11.8 FIBER-OPTIC SENSOR

Several studies have been conducted in the last three decades to develop an optical sensor for determining various parameters in water. This sensor (Figure 11.11) has three essential parts: a source, an optical fiber, and a photodetector for the detection of the optical signal. There are two categories of optical sensors: intrinsic and extrinsic sensors. For the intrinsic sensors, continuous, consistent light sources are required to permit a phase-modulation technique. Only one single-mode fiber is used to design the sensor, and modulation occurs inside the optical fiber. In the case of an external optic sensor, fiber is used as a transmitting channel. This sensor is widely used in remote sensing as well as for the detection of nitrate and nitrite [121] in water as it is small-sized and requires

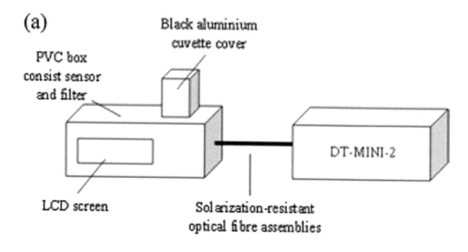

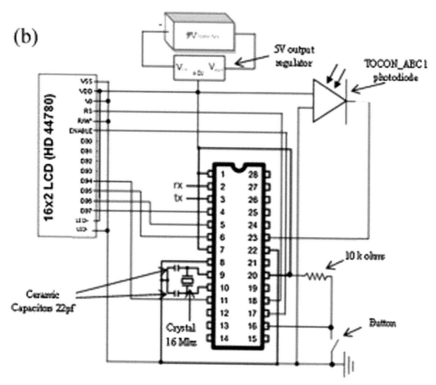

**FIGURE 11.11** (a) Setup of the fiber-optic sensor; and (b) Schematic of the sensor with the microcontroller system [120].

**TABLE 11.5**
**Various Planar Interdigital Sensors with Different Characteristics**

| Type of Sensor | Sensor's response | Limit of Detection (LOD) | Linear range | Application | Ref. |
|---|---|---|---|---|---|
| MEMS sensor | −115543 Ω/ 0.1 ppm | 0.01 mg/L | (0.01–0.5) mg/L | Smart sensing system for agricultural industry | [109] |
| Graphene Sensor | −667.97 Ω/ ppm | 1 mg/L | 1–70 mg/L | IoT-enabled smart sensing system | [112] |
| MEMS-based Coated Sensor | −12811 Ω/ ppm | 1.06 mg/L | 1–10 mg/L | Smart sensing system for water quality measurement | [110] |
| FR4-based sensor | −51.76 Ω/ppm | 1 mg/L | 1–40 mg/L | Water quality monitoring | [118] |
| FR4-based coated sensor | - | - | 5–100 mg/L | Nitrate, phosphate and nickel detection | [21] |

low power. Detection of nitrate concentration was done using Lauth's violet-triacetyl cellulose membrane through absorption spectrophotometry, and the detection range is 10.12 to 1012 ng/mL. B. Mahieuxe [122] modified the fiber-optic sensor using fluorescence emission for nitrate detection in water. In this method, a bathochromic shift and hypochromic effect with different nitrate concentrations were observed.

An external sensor was reported by [123] in which Lophine was used as a sensitive layer on the fiber. As a light source, an ASD FieldSpec 3 Hi-Res Portable Spectroradiometer and a halogen lamp were used. In this method, the detection range was from 1 to 70 mg/L, the detection wavelength was 300–1100 nm, and the response time was 20 ms. Wavelengths from 350 nm to 2500 nm and nitrate-detection range from 0 to 2.50 mg/L were reported by Chong and colleagues [124]. Recently, another new technique for nitrate/nitrite detection has been reported by Moo and colleagues [120]. Different techniques (i.e., transmittance, absorbance, and reflectance) were used for spectroscopic measurement. Channel 0 of the Jaz Spectrometer was used to present all measurements, and Tungsten halogen light was used as a light source. The wavelength was 302 nm to 356 nm with the detection range from 0 mg/L to 50 mg/L. A good linear relationship was observed with a minimum effect of interference from other ions. Johnson and colleagues [125] reported an in-situ ultraviolet spectrophotometry method for long-term nitrate measurements in ocean water. In this method, the measurement of nitrate concentration was done by the absorption of the UV light, which was produced through a deuterium light source. It was guided through fibers

## TABLE 11.6
### Different Optical-Fibre Sensors and Their Characteristics

| Type of Sensor | Analytical Wavelength | Limit of Detection (LOD) | Linear range | Application | Ref. |
|---|---|---|---|---|---|
| Ultraviolet optical fibre sensor | 302 nm | 0.0017 mg/L | 0–50 mg/L | Detection of nitrate and nitrite in water | [120] |
| Ultraviolet Spectrometry | 240 nm | 0.4 µmol/L | 0–45 µmol/L | Ocean-water measurement | [125] |
| Coating-based optical fibre sensor | 300–1100 nm | – | 1–70 mg/L | Nitrate measurement in drinking water | [123] |
| Removal of cladding from optical fiber | 350–2500 nm | – | 0–2.50 mg/L | Nitrate measurement in water | [124] |
| Optical-fiber chemical sensor | 515 nm | – | 0–0.2 µmol/L | Nitrate detection in water | [122] |

in the reflection probe. A microcontroller containing a global positioning system and satellite antenna controlled the in-situ ultraviolet spectrometer. The linear range of nitrate detection was from 0 to10 ppm. This sensor has improved sensitivity, and that is why it is used for various applications [126–131]. Table 11.6 represents the various characteristics of optical-fiber sensors.

## 11.9   CONCLUSION

The chapter has explored the diversified detection methods currently available for the detection of nitrate in water. The merits of these various techniques and their short-comings are discussed. In summary, the electrochemical detection method has good sensitivity, with a simple working principle. Most of the earlier cells are large and are not suitable as portable devices. Chromatography has high sensitivity, but it requires sample preparation, derivatization procedures, and an expert to use the instruments. It also requires specialized equipment, which makes this method expensive compared to other methods. Flow-injection analysis is a popular method in the laboratory and a good alternative to other traditional methods. It has high sensitivity, a low limit of detection, high-throughput analysis, requires low volumes of samples and reagents, is low cost, and easy to operate. Selecting the correct reduction reagent is also an important criterion to complete the flow-injection process in a short time. Biosensors have better sensitivity than electromagnetic sensors. They are also suitable for in-situ measurements for remote water-quality monitoring. Laboratory-based methods are available, which are very sensitive and can be suitable for low-level detection.

However, they produce chemical waste, and training is necessary to run the instruments. They are also expensive. Therefore, biosensors or electromagnetic sensors can be better options to measure nitrate in water, especially for in-situ measurements. Much research is still going on to develop robust sensors that could be utilized continuously. One of the major challenges is to develop sensors that have good repeatability, and more research should be focused on that.

## REFERENCES

[1]  A. Calvo-López, E. Arasa-Puig, M. Puyol, J. M. Casalta, and J. Alonso-Chamarro, "Biparametric potentiometric analytical microsystem for nitrate and potassium monitoring in water recycling processes for manned space missions," *Analytica chimica acta,* vol. 804, pp. 190–196, 2013.

[2]  E. Berner and R. Berner, *The Global Water Cycle.* Englewood Cliffs, NJ: Prentice Hall, 1987.

[3]  J. Gaillard, "February 1995," Lecture on nitrogen cycle.

[4]  W. H. Organization. (2016, 26/04/2018). *Nitrate and nitrite in drinking-water.* Available:    www.who.int/water_sanitation_health/dwq/chemicals/nitrate-nitrite-background-jan17.pdf.

[5]  H. H. Comly, "Cyanosis in infants caused by nitrates in well water," *Jama,* vol. 129, no. 2, pp. 112–116, 1945.

[6]  C. J. Johnson et al., "Fatal outcome of methemoglobinemia in an infant," *Jama,* vol. 257, no. 20, pp. 2796–2797, 1987.

[7]  P. Brimblecombe and D. Stedman, "Historical evidence for a dramatic increase in the nitrate component of acid rain," *Nature,* vol. 298, no. 5873, p. 460, 1982.

[8]  P. F. Swann, "The toxicology of nitrate, nitrite and n-nitroso compounds," *Journal of the Science of Food and Agriculture,* vol. 26, no. 11, pp. 1761–1770, 1975.

[9]  C. S. Bruning-Fann and J. Kaneene, "The effects of nitrate, nitrite and N-nitroso compounds on human health: a review," *Veterinary and human toxicology,* vol. 35, no. 6, pp. 521–538, 1993.

[10]  O. G. Licensee, *The Water Supply (Water Quality) Regulations 2016.* (2016, 2018).

[11]  G. R. Hallberg, "Nitrate in ground water in the United States," in *Developments in Agricultural and Managed Forest Ecology,* vol. 21: Elsevier, 1989, pp. 35–74.

[12]  G. Hallberg and D. Keeney, *Nitrate, Regional groundwater quality.* WJ Alley, Ed," ed: Van Nostrand Reinhold, New York 1993.

[13]  D. Behm, "Ill waters: The fouling of Wisconsin's lakes and streams (Special Report)," *The Milwaukee Journal,* vol. 2, 1989.

[14]  M. A. M. Yunus, S. Ibrahim, W. A. H. Altowayti, G. P. San, and S. C. Mukhopadhyay, "Selective membrane for detecting nitrate based on planar electromagnetic sensors array," in *Control Conference (ASCC), 2015 10th Asian,* 2015, pp. 1–6: IEEE.

[15]  C. Wardak, "Solid contact nitrate ion-selective electrode based on ionic liquid with stable and reproducible potential," *Electroanalysis,* vol. 26, no. 4, pp. 864–872, 2014.

[16]  R. K. Mahajan, R. Kaur, H. Miyake, and H. Tsukube, "Zn (II) complex-based potentiometric sensors for selective determination of nitrate anion," *Analytica chimica acta,* vol. 584, no. 1, pp. 89–94, 2007.

[17]  J. Davis, K. J. McKeegan, M. F. Cardosi, and D. H. Vaughan, "Evaluation of phenolic assays for the detection of nitrite," *Talanta,* vol. 50, no. 1, pp. 103–112, 1999.

[18]  M. Friedberg, M. Hinsdale, and Z. Shihabi, "Analysis of nitrate in biological fluids by capillary electrophoresis," *Journal of Chromatography A,* vol. 781, no. 1–2, pp. 491–496, 1997.

[19]   P. N. Bories, E. Scherman, and L. Dziedzic, "Analysis of nitrite and nitrate in bio-logical fluids by capillary electrophoresis," *Clinical Biochemistry,* vol. 32, no. 1, pp. 9–14, 1999.

[20]   G. Ellis, I. Adatia, M. Yazdanpanah, and S. K. Makela, "Nitrite and nitrate ana-lyses: a clinical biochemistry perspective," *Clinical Biochemistry,* vol. 31, no. 4, pp. 195–220, 1998.

[21]   A. Azmi, A. A. Azman, S. Ibrahim, and M. A. M. Yunus, "Techniques in advancing the capabilities of various nitrate detection methods: A review," *International Journal on Smart Sensing & Intelligent Systems,* vol. 10, no. 2, 2017.

[22]   M. E. E. Alahi and S. C. Mukhopadhyay, "Detection methods of nitrate in water: A review," *Sensors and Actuators A: Physical,* 2018.

[23]   E. Andreoli et al., "Electrochemical conversion of copper-based hierarchical micro/ nanostructures to copper metal nanoparticles and their testing in nitrate sensing," *Electroanalysis,* vol. 23, no. 9, pp. 2164–2173, 2011.

[24]   S. Aravamudhan and S. Bhansali, "Development of micro-fluidic nitrate-selective sensor based on doped-polypyrrole nanowires," *Sensors and Actuators B: Chemical,* vol. 132, no. 2, pp. 623–630, 2008.

[25]   F. Can, S. K. Ozoner, P. Ergenekon, and E. Erhan, "Amperometric nitrate biosensor based on Carbon nanotube/Polypyrrole/Nitrate reductase biofilm electrode," *Materials Science and Engineering: C,* vol. 32, no. 1, pp. 18–23, 2012.

[26]   S. S. Hassan, "Ion-selective electrodes in organic functional group analysis: Microdetermination of nitrates and nitramines with use of the iodide electrode," *Talanta,* vol. 23, no. 10, pp. 738–740, 1976.

[27]   M. O. Mendoza, E. P. Ortega, O. A. de Fuentes, Y. Prokhorov, and J. G. L. Barcenas, "Chitosan/bentonite nanocomposite: preliminary studies of its potentiometric response to nitrate ions in water," in *Sensors (IBERSENSOR), 2014 IEEE 9th Ibero-American Congress on,* 2014, pp. 1–4: IEEE.

[28]   C. Li and L. Li, "Prediction of nitrate and chlorine in soil using ion selective electrode," in *World Automation Congress (WAC),* 2010, pp. 231–234: IEEE.

[29]   L. Nuñez, X. Cetó, M. I. Pividori, M. V. B. Zanoni, and M. Del Valle, "Development and application of an electronic tongue for detection and monitoring of nitrate, nitrite and ammonium levels in waters," *Microchemical Journal,* vol. 110, pp. 273–279, 2013.

[30]   T. A. Bendikov and T. C. Harmon, "A sensitive nitrate ion-selective electrode from a pencil lead. An analytical laboratory experiment," *Journal of Chemical Education,* vol. 82, no. 3, p. 439, 2005.

[31]   L. Zhang, M. Zhang, H. Ren, P. Pu, P. Kong, and H. Zhao, "Comparative investiga-tion on soil nitrate-nitrogen and available potassium measurement capability by using solid-state and PVC ISE," *Computers and Electronics in Agriculture,* vol. 112, pp. 83–91, 2015.

[32]   S. S. Hassan, H. Sayour, and S. S. Al-Mehrezi, "A novel planar miniaturized potentio-metric sensor for flow injection analysis of nitrates in wastewaters, fertilizers and phar-maceuticals," *Analytica Chimica Acta,* vol. 581, no. 1, pp. 13–18, 2007.

[33]   P. T. Kissinger and T. H. Ridgway, "Small-amplitude controlled-potential techniques," *Laboratory Techniques in Electroanalytical Chemistry, Revised and Expanded,* p. 141, 1996.

[34]   R. A. Wallingford and A. G. Ewing, "Capillary zone electrophoresis with electrochem-ical detection," *Analytical Chemistry,* vol. 59, no. 14, pp. 1762–1766, 1987.

[35]   S. Sloss and A. G. Ewing, "Improved method for end-column amperometric detection for capillary electrophoresis," *Analytical Chemistry,* vol. 65, no. 5, pp. 577–581, 1993.

[36]   J. Wang, *Analytical Clectrochemistry.* Wiley, 2006.

[37] T. J. Roussel, D. J. Jackson, R. P. Baldwin, and R. S. Keynton, "Amperometric Techniques," *Encyclopedia of Microfluidics and Nanofluidics*, pp. 1–11, 2013.

[38] N. G. Carpenter and D. Pletcher, "Amperometric method for the determination of nitrate in water," *Analytica Chimica Acta*, vol. 317, no. 1–3, pp. 287–293, 1995.

[39] X.-l. Zhang, J.-x. Wang, Z. Wang, and S.-c. Wang, "Improvement of amperometric sensor used for determination of nitrate with polypyrrole nanowires modified electrode," *Sensors*, vol. 5, no. 12, pp. 580–593, 2005.

[40] J. R. C. da Rocha, L. Angnes, M. Bertotti, K. Araki, and H. E. Toma, "Amperometric detection of nitrite and nitrate at tetraruthenated porphyrin-modified electrodes in a continuous-flow assembly," *Analytica Chimica Acta*, vol. 452, no. 1, pp. 23–28, 2002.

[41] J. E. Newbery and M. P. L. de Haddad, "Amperometric determination of nitrite by oxidation at a glassy carbon electrode," *Analyst*, vol. 110, no. 1, pp. 81–82, 1985.

[42] M. A. Stanley et al., "Comparison of the analytical capabilities of an amperometric and an optical sensor for the determination of nitrate in river and well water," *Analytica Chimica Acta*, vol. 299, no. 1, pp. 81–90, 1994.

[43] J. C. Gamboa, R. C. Pena, T. R. Paixão, and M. Bertotti, "A renewable copper electrode as an amperometric flow detector for nitrate determination in mineral water and soft drink samples," *Talanta*, vol. 80, no. 2, pp. 581–585, 2009.

[44] A. Hulanicki, W. Matuszewski, and M. Trojanowicz, "Flow-injection determination of nitrite and nitrate with biamperometric detection at two platinum wire electrodes," *Analytica Chimica Acta*, vol. 194, pp. 119–127, 1987.

[45] G. A. Sherwood and D. C. Johnson, "A chromatographic determination of nitrate with amperometric detection at a copperized cadmium electrode," *Analytica Chimica Acta*, vol. 129, pp. 101–111, 1981.

[46] S. A. Glazier, E. R. Campbell, and W. H. Campbell, "Construction and characterization of nitrate reductase-based amperometric electrode and nitrate assay of fertilizers and drinking water," *Analytical Chemistry*, vol. 70, no. 8, pp. 1511–1515, 1998.

[47] A. Y. Chamsi and A. G. Fogg, "Oxidative flow injection amperometric determination of nitrite at an electrochemically pre-treated glassy carbon electrode," *Analyst*, vol. 113, no. 11, pp. 1723–1727, 1988.

[48] M. Bertotti and D. Pletcher, "Amperometric determination of nitrite via reaction with iodide using microelectrodes," *Analytica Chimica Acta*, vol. 337, no. 1, pp. 49–55, 1997.

[49] M. A. Alawi, "Determination of nitrate and nitrite in water with HPLC and amperometric detection," *Fresenius' Journal of Analytical Chemistry*, vol. 313, no. 3, pp. 239–240, 1982.

[50] M. E. Bodini and D. T. Sawyer, "Voltammetric determination of nitrate ion at parts-per-billion levels," *Analytical chemistry*, vol. 49, no. 3, pp. 485–489, 1977.

[51] R. J. Davenport and D. C. Johnson, "Voltammetric determination of nitrate and nitrite ions using a rotating cadmium disk electrode," *Analytical Chemistry*, vol. 45, no. 11, pp. 1979–1980, 1973.

[52] J. Krista, M. Kopanica, and L. Novotný, "Voltammetric determination of nitrates using silver electrodes," *Electroanalysis: An International Journal Devoted to Fundamental and Practical Aspects of Electroanalysis*, vol. 12, no. 3, pp. 199–204, 2000.

[53] S. M. Shariar and T. Hinoue, "Simultaneous voltammetric determination of nitrate and nitrite ions using a copper electrode pretreated by dissolution/redeposition," *Analytical Sciences*, vol. 26, no. 11, pp. 1173–1179, 2010.

[54] V. Mareček, H. Jänchenová, Z. Samec, and M. Březina, "Voltammetric determination of nitrate, perchlorate and iodide at a hanging electrolyte drop electrode," *Analytica Chimica Acta*, vol. 185, pp. 359–362, 1986.

[55] C. Neuhold, K. Kalcher, W. Diewald, X. Cai, and G. Raber, "Voltammetric determination of nitrate with a modified carbon paste electrode," *Electroanalysis*, vol. 6, no. 3, pp. 227–236, 1994.

[56] A. O. Solak, P. Gülser, E. Gökm, and F. Gökmeşşe, "A new differential pulse voltammetric method for the determination of nitrate at a copper plated glassy carbon electrode," *Microchimica Acta*, vol. 134, no. 1–2, pp. 77–82, 2000.

[57] A. Osman Solak and P. Çekirdek, "Square wave voltammetric determination of nitrate at a freshly copper plated glassy carbon electrode," *Analytical Letters*, vol. 38, no. 2, pp. 271–280, 2005.

[58] A. Jang, Z. Zou, K. K. Lee, C. H. Ahn, and P. L. Bishop, "Potentiometric and voltammetric polymer lab chip sensors for determination of nitrate, pH and Cd (II) in water," *Talanta*, vol. 83, no. 1, pp. 1–8, 2010.

[59] C. Lopez-Moreno, I. V. Perez, and A. M. Urbano, "Development and validation of an ionic chromatography method for the determination of nitrate, nitrite and chloride in meat," *Food chemistry*, vol. 194, pp. 687–694, 2016.

[60] H. Kodamatani, S. Yamazaki, K. Saito, T. Tomiyasu, and Y. Komatsu, "Selective determination method for measurement of nitrite and nitrate in water samples using high-performance liquid chromatography with post-column photochemical reaction and chemiluminescence detection," *Journal of Chromatography A*, vol. 1216, no. 15, pp. 3163–3167, 2009.

[61] M. R. Siddiqui, S. M. Wabaidur, Z. A. ALOthman, and M. Rafiquee, "Rapid and sensitive method for analysis of nitrate in meat samples using ultra performance liquid chromatography–mass spectrometry," *Spectrochimica Acta Part A: Molecular and Biomolecular Spectroscopy*, vol. 151, pp. 861–866, 2015.

[62] P. Niedzielski, I. Kurzyca, and J. Siepak, "A new tool for inorganic nitrogen speciation study: Simultaneous determination of ammonium ion, nitrite and nitrate by ion chromatography with post-column ammonium derivatization by Nessler reagent and diode-array detection in rain water samples," *Analytica Chimica Acta*, vol. 577, no. 2, pp. 220–224, 2006.

[63] K. Tirumalesh, "Simultaneous determination of bromide and nitrate in contaminated waters by ion chromatography using amperometry and absorbance detectors," *Talanta*, vol. 74, no. 5, pp. 1428–1434, 2008.

[64] M. Tabatabai and W. Dick, "Simultaneous determination of nitrate, chloride, sulfate, and phosphate in natural waters by ion chromatography 1," *Journal of Environmental Quality*, vol. 12, no. 2, pp. 209–213, 1983.

[65] J. A. Morales, L. S. de Graterol, and J. Mesa, "Determination of chloride, sulfate and nitrate in groundwater samples by ion chromatography," *Journal of Chromatography A*, vol. 884, no. 1–2, pp. 185–190, 2000.

[66] I. Dahllöf, O. Svensson, and C. Torstensson, "Optimising the determination of nitrate and phosphate in sea water with ion chromatography using experimental design," *Journal of Chromatography a*, vol. 771, no. 1–2, pp. 163–168, 1997.

[67] E. Kapinus, I. Revelsky, V. Ulogov, and Y. A. Lyalikov, "Simultaneous determination of fluoride, chloride, nitrite, bromide, nitrate, phosphate and sulfate in aqueous solutions at 10– 9 to 10– 8% level by ion chromatography," *Journal of Chromatography B*, vol. 800, no. 1–2, pp. 321–323, 2004.

[68] M. Neal, C. Neal, H. Wickham, and S. Harman, "Determination of bromide, chloride, fluoride, nitrate and sulphate by ion chromatography: comparisons of methodologies for rainfall, cloud water and river waters at the Plynlimon catchments of mid-Wales," *Hydrology and Earth System Sciences*, vol. 11, no. 1, pp. 294–300, 2007.

[69]    M. Akyüz and Ş. Ata, "Determination of low level nitrite and nitrate in biological, food and environmental samples by gas chromatography–mass spectrometry and liquid chromatography with fluorescence detection," *Talanta,* vol. 79, no. 3, pp. 900–904, 2009.

[70]    Y. Li, J. S. Whitaker, and C. L. McCarty, "Reversed-phase liquid chromatography/electrospray ionization/mass spectrometry with isotope dilution for the analysis of nitrate and nitrite in water," *Journal of Chromatography A,* vol. 1218, no. 3, pp. 476–483, 2011.

[71]    Y. Zuo, C. Wang, and T. Van, "Simultaneous determination of nitrite and nitrate in dew, rain, snow and lake water samples by ion-pair high-performance liquid chromatography," *Talanta,* vol. 70, no. 2, pp. 281–285, 2006.

[72]    S. Rodriguez-Mozaz, M. J. L. de Alda, and D. Barceló, "Biosensors as useful tools for environmental analysis and monitoring," *Analytical and Bioanalytical Chemistry,* vol. 386, no. 4, pp. 1025–1041, 2006.

[73]    S. Rodriguez-Mozaz, M. J. L. de Alda, and D. Barceló, "Fast and simultaneous monitoring of organic pollutants in a drinking water treatment plant by a multi-analyte biosensor followed by LC–MS validation," *Talanta,* vol. 69, no. 2, pp. 377–384, 2006.

[74]    B. Roig, I. Bazin, S. Bayle, D. Habauzit, and J. Chopineau, "Biomolecular recognition systems for water monitoring," *Rapid Chemical and Biological Techniques for Water Monitoring,* pp. 175–195, 2009.

[75]    M. Farré, L. Kantiani, S. Pérez, and D. Barceló, "Sensors and biosensors in support of EU Directives," *TrAC Trends in Analytical Chemistry,* vol. 28, no. 2, pp. 170–185, 2009.

[76]    H.-H. Zeng, R. B. Thompson, B. P. Maliwal, G. R. Fones, J. W. Moffett, and C. A. Fierke, "Real-time determination of picomolar free Cu (II) in seawater using a fluorescence-based fiber optic biosensor," *Analytical Chemistry,* vol. 75, no. 24, pp. 6807–6812, 2003.

[77]    W. Xuejiang et al., "Conductometric nitrate biosensor based on methyl viologen/Nafion®/nitrate reductase interdigitated electrodes," *Talanta,* vol. 69, no. 2, pp. 450–455, 2006.

[78]    S. Cosnier, S. Da Silva, D. Shan, and K. Gorgy, "Electrochemical nitrate biosensor based on poly (pyrrole–viologen) film–nitrate reductase–clay composite," *Bioelectrochemistry,* vol. 74, no. 1, pp. 47–51, 2008.

[79]    Z. Zhang et al., "A novel nitrite biosensor based on conductometric electrode modified with cytochrome c nitrite reductase composite membrane," *Biosensors and Bioelectronics,* vol. 24, no. 6, pp. 1574–1579, 2009.

[80]    T. Madasamy, M. Pandiaraj, M. Balamurugan, K. Bhargava, N. K. Sethy, and C. Karunakaran, "Copper, zinc superoxide dismutase and nitrate reductase coimmobilized bienzymatic biosensor for the simultaneous determination of nitrite and nitrate," *Biosensors and Bioelectronics,* vol. 52, pp. 209–215, 2014.

[81]    D. Albanese, M. Di Matteo, and C. Alessio, "Screen printed biosensors for detection of nitrates in drinking water," in *Computer Aided Chemical Engineering,* vol. 28: Elsevier, 2010, pp. 283–288.

[82]    A. Ayala, L. Leal, L. Ferrer, and V. Cerdà, "Multiparametric automated system for sulfate, nitrite and nitrate monitoring in drinking water and wastewater based on sequential injection analysis," *Microchemical Journal,* vol. 100, pp. 55–60, 2012.

[83]    M. Yaqoob, A. Nabi, and P. J. Worsfold, "Determination of nitrite and nitrate in natural waters using flow injection with spectrophotometric detection," *Journal of the Chemical Society of Pakistan,* vol. 34, no. 3, 2013.

[84]    M. Yaqoob, B. Folgado Biot, A. Nabi, and P. J. Worsfold, "Determination of nitrate and nitrite in freshwaters using flow-injection with luminol chemiluminescence detection," *Luminescence,* vol. 27, no. 5, pp. 419–425, 2012.

[85]  S. Wang, K. Lin, N. Chen, D. Yuan, and J. Ma, "Automated determination of nitrate plus nitrite in aqueous samples with flow injection analysis using vanadium (III) chloride as reductant," *Talanta*, vol. 146, pp. 744–748, 2016.

[86]  C. L. Pasquali, A. Gallego-Picó, P. F. Hernando, M. Velasco, and J. D. Alegría, "Two rapid and sensitive automated methods for the determination of nitrite and nitrate in soil samples," *Microchemical Journal*, vol. 94, no. 1, pp. 79–82, 2010.

[87]  C. L. Pasquali, P. F. Hernando, and J. D. Alegria, "Spectrophotometric simultaneous determination of nitrite, nitrate and ammonium in soils by flow injection analysis," *Analytica Chimica Acta*, vol. 600, no. 1–2, pp. 177–182, 2007.

[88]  S. Feng, M. Zhang, Y. Huang, D. Yuan, and Y. Zhu, "Simultaneous determination of nanomolar nitrite and nitrate in seawater using reverse flow injection analysis coupled with a long path length liquid waveguide capillary cell," *Talanta*, vol. 117, pp. 456–462, 2013.

[89]  P. S. Ellis, A. M. H. Shabani, B. S. Gentle, and I. D. McKelvie, "Field measurement of nitrate in marine and estuarine waters with a flow analysis system utilizing on-line zinc reduction," *Talanta*, vol. 84, no. 1, pp. 98–103, 2011.

[90]  A. D. Beaton et al., "Lab-on-chip measurement of nitrate and nitrite for in situ analysis of natural waters," *Environmental Science & Technology*, vol. 46, no. 17, pp. 9548–9556, 2012.

[91]  N. Amini and I. McKelvie, "An enzymatic flow analysis method for the determination of phosphatidylcholine in sediment pore waters and extracts," *Talanta*, vol. 66, no. 2, pp. 445–452, 2005.

[92]  B. Paczosa-Bator, L. Cabaj, M. Raś, B. Baś, and R. Piech, "Potentiometric sensor platform based on a carbon black modified electrodes," *International Journal of Electrochemical Science*, vol. 9, pp. 2816–2823, 2014.

[93]  E. Lindner and B. D. Pendley, "A tutorial on the application of ion-selective electrode potentiometry: an analytical method with unique qualities, unexplored opportunities and potential pitfalls; Tutorial," *Analytica Chimica Acta*, vol. 762, pp. 1–13, 2013.

[94]  A. Stortini, L. Moretto, A. Mardegan, M. Ongaro, and P. Ugo, "Arrays of copper nanowire electrodes: Preparation, characterization and application as nitrate sensor," *Sensors and Actuators B: Chemical*, vol. 207, pp. 186–192, 2015.

[95]  L. T. Duarte, C. Jutten, and S. Moussaoui, "A Bayesian nonlinear source separation method for smart ion-selective electrode arrays," *IEEE Sensors Journal*, vol. 9, no. 12, pp. 1763–1771, 2009.

[96]  P. Ciosek and W. Wróblewski, "Potentiometric electronic tongues for foodstuff and biosample recognition—An overview," *Sensors*, vol. 11, no. 5, pp. 4688–4701, 2011.

[97]  T. Öznülüer, B. Özdurak, and H. Ö. Doğan, "Electrochemical reduction of nitrate on graphene modified copper electrodes in alkaline media," *Journal of Electroanalytical Chemistry*, vol. 699, pp. 1–5, 2013.

[98]  Z. Chang, Y. Zhu, L. Zhang, and S. Du, "Measurement experiment and mathematical model of nitrate ion selective electrode," in *Instrumentation, Measurement, Computer, Communication and Control (IMCCC), 2013 Third International Conference*, 2013, pp. 48–52: IEEE.

[99]  M. A. M. Yunus and S. C. Mukhopadhyay, "Novel planar electromagnetic sensors for detection of nitrates and contamination in natural water sources," *IEEE Sensors Journal*, vol. 11, no. 6, pp. 1440–1447, 2011.

[100]  A. S. M. Nor, M. A. M. Yunus, S. W. Nawawi, and S. Ibrahim, "Low-cost sensor array design optimization based on planar electromagnetic sensor design for detecting nitrate and sulphate," in *Sensing Technology (ICST), 2013 Seventh International Conference*, 2013, pp. 693–698: IEEE.

[101] M. A. M. Yunus, S. Mukhopadhyay, and A. Punchihewa, "Application of independent component analysis for estimating nitrate contamination in natural water sources using planar electromagnetic sensor," in *Sensing Technology (ICST), 2011 Fifth International Conference*, 2011, pp. 538–543: IEEE.

[102] M. A. M. Yunus, S. C. Mukhopadhyay, and S. Ibrahim, "Planar electromagnetic sensor based estimation of nitrate contamination in water sources using independent component analysis," *IEEE Sensors Journal,* vol. 12, no. 6, pp. 2024–2034, 2012.

[103] A. S. M. Nor, M. Faramarzi, M. A. M. Yunus, and S. Ibrahim, "Nitrate and sulfate estimations in water sources using a planar electromagnetic sensor array and artificial neural network method," *IEEE Sensors Journal,* vol. 15, no. 1, pp. 497–504, 2015.

[104] M. M. Yunus, S. Mukhopadhyay, M. Rahman, N. Zahidin, and S. Ibrahim, "The selection of novel planar electromagnetic sensors for the application of nitrate contamination detection," in *Smart Sensors for Real-Time Water Quality Monitoring*: Springer, 2013, pp. 171–195.

[105] M. M. Yunus, S. C. Mukhopadhyay, A. Punchihewa, and S. Ibrahim, "The effect of temperature factor on the detection of nitrate based on planar electromagnetic sensor and independent component analysis," in *Smart Sensing Technology for Agriculture and Environmental Monitoring*: Springer, 2012, pp. 103–118.

[106] X. Wang, Y. Wang, H. Leung, S. C. Mukhopadhyay, M. Tian, and J. Zhou, "Mechanism and experiment of planar electrode sensors in water pollutant measurement," *IEEE Transactions on Instrumentation and Measurement,* vol. 64, no. 2, pp. 516–523, 2015.

[107] A. V. Mamishev, K. Sundara-Rajan, F. Yang, Y. Du, and M. Zahn, "Interdigital sensors and transducers," *Proceedings of the IEEE,* vol. 92, no. 5, pp. 808–845, 2004.

[108] M. E. E. Alahi, L. Xie, S. Mukhopadhyay, and L. Burkitt, "A temperature compensated smart nitrate-sensor for agricultural industry," *IEEE Transactions on Industrial Electronics,* vol. 64, no. 9, pp. 7333–7341, 2017.

[109] M. E. E. Alahi, X. Li, S. Mukhopadhyay, and L. Burkitt, "A temperature compensated Smart nitrate-sensor for agricultural industry," *IEEE Transactions on Industrial Electronics,* 2017.

[110] M. E. E. Alahi, S. C. Mukhopadhyay, and L. Burkitt, "Imprinted polymer coated impedimetric nitrate sensor for real-time water quality monitoring," *Sensors and Actuators B: Chemical,* vol. 259, pp. 753–761, 2018.

[111] M. E. E. Alahi, X. Li, S. Mukhopadhyay, and L. Burkitt, "Application of practical nitrate sensor based on electrochemical impedance spectroscopy," in *Sensors for Everyday Life*: Springer, 2017, pp. 109–136.

[112] M. E. E. Alahi, A. Nag, S. C. Mukhopadhyay, and L. Burkitt, "A temperature-compensated graphene sensor for nitrate monitoring in real-time application," *Sensors and Actuators A: Physical,* vol. 269, pp. 79–90, 2018.

[113] M. E. E. Alahi, L. Xie, A. I. Zia, S. Mukhopadhyay, and L. Burkitt, "Practical nitrate sensor based on electrochemical impedance measurement," in *Instrumentation and Measurement Technology Conference Proceedings (I2MTC), 2016 IEEE International*, 2016, pp. 1–6: IEEE.

[114] M. E. E. Alahi, N. Afsarimanesh, S. C. Mukhopadhyay, and L. Burkitt, "Development of the selectivity of nitrate sensors based on ion imprinted polymerization technique," in *Sensing Technology (ICST), 2017 Eleventh International Conference*, 2017, pp. 1–6: IEEE.

[115] M. E. E. Alahi and S. C. Mukhopadhyay, "Preparation and characterization of the selectivity material of nitrate sensor," in *Smart Nitrate Sensor*: Springer, 2019, pp. 91–113.

[116] M. E. E. Alahi and S. C. Mukhopadhyay, "Temperature compensation for low concentration nitrate measurement," in *Smart Nitrate Sensor*: Springer, 2019, pp. 53–72.

[117] M. E. E. Alahi and S. C. Mukhopadhyay, "Graphite/PDMS capacitive sensor for nitrate measurement," in *Smart Nitrate Sensor*: Springer, 2019, pp. 73–89.

[118] M. E. E. Alahi, N. Pereira-Ishak, S. C. Mukhopadhyay, and L. Burkitt, "An Internet-of-Things enabled Smart Sensing System for nitrate monitoring," *IEEE Internet of Things Journal*, 2018.

[119] M. E. E. Alahi and S. C. Mukhopadhyay, "IoT enabled Smart Sensing System," in *Smart Nitrate Sensor*: Springer, 2019, pp. 115–130.

[120] Y. Moo, M. Matjafri, H. Lim, and C. Tan, "New development of optical fibre sensor for determination of nitrate and nitrite in water," *Optik-International Journal for Light and Electron Optics*, vol. 127, no. 3, pp. 1312–1319, 2016.

[121] N. Amini, M. Shamsipur, M. B. Gholivand, and K. Naderi, "Electrocatalytic and new electrochemical properties of chloropromazine in to silicaNPs/chloropromazine/Nafion nanocomposite: Application to nitrite detection at low potential," *Microchemical Journal*, vol. 131, pp. 43–50, 2017.

[122] B. Mahieuxe, M. Carré, M. Viriot, J. André, and M. Donner, "Fiber-optic fluorescing sensors for nitrate and nitrite detection," *Journal of Fluorescence*, vol. 4, no. 1, pp. 7–10, 1994.

[123] J. Camas-Anzueto, A. Aguilar-Castillejos, J. Castañón-González, M. Lujpan-Hidalgo, H. H. de León, and R. M. Grajales, "Fiber sensor based on Lophine sensitive layer for nitrate detection in drinking water," *Optics and Lasers in Engineering*, vol. 60, pp. 38–43, 2014.

[124] M. Y. Chong, M. Z. M. Jafri, L. H. San, and T. C. Ho, "Detection of nitrate ions in water by optical fiber," in *Computer and Communication Engineering (ICCCE), 2012 International Conference*, 2012, pp. 271–273: IEEE.

[125] K. S. Johnson, L. J. Coletti, H. W. Jannasch, C. M. Sakamoto, D. D. Swift, and S. C. Riser, "Long-term nitrate measurements in the ocean using the In Situ Ultraviolet Spectrophotometer: sensor integration into the Apex profiling float," *Journal of Atmospheric and Oceanic Technology*, vol. 30, no. 8, pp. 1854–1866, 2013.

[126] A. Lalasangi et al., "Fiber Bragg grating sensor for detection of nitrate concentration in water," *Sensors & Transducers*, vol. 125, no. 2, p. 187, 2011.

[127] C. Munkholm, D. R. Walt, and F. P. Milanovich, "A fiber-optic sensor for CO2 measurement," *Talanta*, vol. 35, no. 2, pp. 109–112, 1988.

[128] Y. Zhu and A. Wang, "Miniature fiber-optic pressure sensor," *IEEE Photonics Technology Letters*, vol. 17, no. 2, pp. 447–449, 2005.

[129] S. Zhang, H. Chen, and H. Fu, "Fiber-optic temperature sensor using an optoelectronic oscillator," in *Optical Communications and Networks (ICOCN), 2015 14th International Conference*, 2015, pp. 1–3: IEEE.

[130] F. Delport et al., "Real-time monitoring of DNA hybridization and melting processes using a fiber optic sensor," *Nanotechnology*, vol. 23, no. 6, p. 065503, 2012.

[131] P. Bhatia and B. D. Gupta, "Fabrication and characterization of a surface plasmon resonance based fiber optic urea sensor for biomedical applications," *Sensors and Actuators B: Chemical*, vol. 161, no. 1, pp. 434–438, 2012.

# Part 4

## Nitrate in the Food Chain

# 12 Nitrate in Plant and Animal Foods

*Małgorzata Karwowska*

## CONTENTS

12.1 Introduction.................................................................................................265
12.2 Nitrate in Plant-Based Foods ....................................................................267
    12.2.1 Nitrate in Vegetables.....................................................................267
    12.2.2 Nitrate in Other Plant-Based Foods..............................................270
    12.2.3 Factors Affecting Nitrate/Nitrite Content in Vegetables...............271
12.3 Nitrate in Animal-Based Food ...................................................................273
12.4 Nitrate in Processed Meat..........................................................................274
References.............................................................................................................277

## 12.1 INTRODUCTION

Nitrogen is an essential element for life because it is the main constituent of bio-molecules, including nucleic acids, proteins, chlorophylls, and hormones (Buchanan et al., 2015). In the case of plants, nitrogen is the main limiting nutrient due to the large amount needed to maintain sustained growth, and to its low availability in soil. The main forms of inorganic nitrogen that are available to plants in the soil are ammonium and nitrate (Britto and Kronzucker, 2013). In this context, inorganic nitrate has a strong connection with the food chain and is a naturally occurring compound in foods, especially plant foods and vegetables. In the case of animal-origin food, nitrate/nitrite are also used as additives during processing. According to Reinik et al. (2009) major sources of exogenous nitrate exposure are vegetables and drinking water, therefore more than 80 percent of dietary intake of nitrate is attributed to vegetables. However, the amount of nitrate consumed with vegetables depends on dietary habits and the method of preparing foods. In turn, processed meat and animal origin foods are major nitrite-containing foods. Increased exposure to nitrate/nitrite, due to increased usage of nitrogen fertilizers and additives containing nitrite in processed food products is becoming an important public health issue (Chan, 2011).

Nitrate is relatively non-toxic, but its metabolites – including nitrite, nitric oxide, and N-nitroso compounds – have potentially adverse health implications (Sindelar et al., 2012). Nitrite can increase methemoglobin content by reacting with hemoglobin, transforming ferrous iron to the ferric state, and then arresting or decreasing the oxygenation of blood. High methemoglobin concentrations translate to a reduced

DOI: 10.1201/9780429326806-16

capacity to transport oxygen to tissues, potentially resulting in hypoxia. This is especially dangerous for babies up to ten months of age because during this period the baby's blood contains fetal hemoglobin, which is much easier to oxidize (Mensinga et al., 2003). Nitrite can also generate reactive nitrogen species, including nitric oxide and peroxynitrite anion, in both animals and plants. These reactive nitrogen species can trigger enzyme inactivation, DNA lesions, lipid peroxidation, can damage different organs, and can be considered as one of the most important human dietary carcinogenic factors (Sellimi et al., 2017). In the stomach, nitrite reacts with amines and amides, to form nitrosamines, which are suspected to be carcinogenic (Sindelar et al., 2012).

Many countries have paid attention to the concentrations of nitrate/nitrites in foods, especially in vegetables and baby foods. Hence, many analytical techniques have been used for the determination of nitrate and nitrites in animal and plant food, including spectrophotometry, potentiometry, ion chromatography, polarography, capillary electrophoresis, and high-performance liquid chromatography (Prasad and Chetty, 2008). However, spectroscopic methods are by far the most widely used for nitrate/nitrite determination due to the excellent limits of detection obtained and facile assay-type protocols. As a result, more and more data are available in the literature on the content of nitrate/nitrites in food.

Having more and more knowledge about the content of nitrate/nitrites in food from different counties it is also possible to determine the consumer exposure to these food compounds. Dietary exposure to nitrate and nitrite occurs via three main sources: occurrence in plant-origin foods, food additives in certain processed foods of animal origin, and contaminants in drinking water. Based on data presented by Reinik (2007), the highest exposure to nitrate comes from vegetables and fruits, while the highest dietary exposure to nitrites is in meat products (Figure 12.1). Other dietary

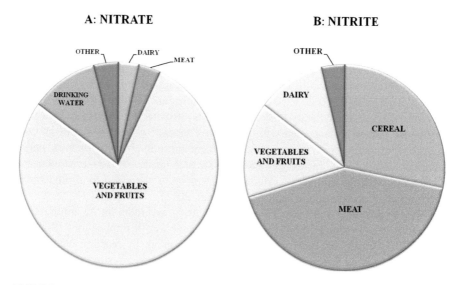

**FIGURE 12.1** General dietary exposure of nitrate and nitrite.

sources of nitrate come from cereals, water, cured meats, and therapeutic treatments for cardiovascular conditions, angina and digital ischemia (Nuῆez de González et al., 2015). According to WHO (2007), humans generally consume between 1.2 and 3.0 mg of nitrite daily. van den Brand and colleagues (2020) have assessed the combined nitrate and nitrite exposure from food and drinking water. This is the first study to perform a combined dietary exposure assessment of nitrate and nitrite from all sources and accounting for the uncertain conversion factor for nitrate to nitrite. A combined exposure was above the acceptable daily intake for nitrite ion (70 µg/kg bw), with the mean exposure varying between 95–114 µg nitrite/kg bw/day. Moreover, the combined exposure was highest (an average of 250 µg nitrite/kg bw/day) in children aged 1 year. The exposure to nitrate and nitrite via contaminants in food was relatively the most contributing source of exposure. Vegetables contributed most to the combined exposure in food (34–41%). Food additive use contributed 8–9 percent to the exposure.

According to the European Commission's former Scientific Committee for Food (SCF) and the Joint FAO/WHO Expert Committee on Food Additives (JECFA), the current acceptable daily intake (ADI) for nitrites is 0.06 and 0.07 milligrams per kilogram of body weight per day, respectively. In the case of nitrate, both organizations establish the ADI at 3.7 mg/kg bw/day.

## 12.2 NITRATE IN PLANT-BASED FOODS

### 12.2.1 NITRATE IN VEGETABLES

Nitrate is a major nitrogen source for plants. It plays an important role in the plant because of its versatile functions in plant nutrition as well as physiological regulations (Wang et al., 2012). As described by Chen et al. (2004) plants supplied with nitrate more than current demand has the ability to accumulate nitrate.

Due to many biological and environmental factors – described in the next part of this chapter – nitrate concentrations of plant food, reported in several studies, varied depending on the country or region. A summary of the nitrate contents of some vegetables and other plant-based food recorded in several studies is depicted in Table 12.1. Studies performed by Raczuk and colleagues (2014) demonstrated that, for nine evaluated vegetables, nitrate concentrations varied between 10 mg kg$^{-1}$ to 4800 mg kg$^{-1}$. The highest nitrate concentrations were found in radishes, butterhead lettuce, beetroots, and iceberg lettuce, values were respectively 2132 mg kg$^{-1}$, 1725 mg kg$^{-1}$, 1306 mg kg$^{-1}$, and 890 mg kg$^{-1}$. The cucumber and tomato we characterized by the lowest concentration of nitrate (32 mg kg$^{-1}$ and 35 mg kg$^{-1}$ respectively). Results presented by Raczuk and colleagues (2014) demonstrated that the consumption of only 100 g of vegetables delivers 40.1 to 96.0 percent of ADI of nitrate to an adult person weighing 60 kg.

In contrast to nitrate, the content of nitrite in vegetables usually falls within the range of 1–2 mg kg$^{-1}$ but in some case, it may exceed 20 mg kg$^{-1}$ and even reach up to 60 mg kg$^{-1}$ in potatoes and 57 mg kg$^{-1}$ for turnip sprouts (Czech and Rusinek, 2005; Santamaria, 2006). In the research by Raczuk and colleagues (2014) the highest concentrations measured were in beetroot (9.2 mg kg$^{-1}$) whilst much smaller amounts

**TABLE 12.1**

**Comparison of Mean Nitrate Contents of Vegetables and Other Plant-Based Foods (mg kg⁻¹ fresh mass) in different countries**

| Vegetable | Country | Nitrate Content | | Analytical Method | References |
|---|---|---|---|---|---|
| | | mean | range | | |
| **Leafy vegetables** | | | | | |
| Lettuce | Poland | 1725 (Butterhead) | 530–4440 | Spectrophotometry | Raczuk et al. (2014) |
| | Poland | 890 (Iceberg) | 125–1580 | Spectrophotometry | Raczuk et al. (2014) |
| | Italy | 1079 | - | Ion chromatography | Roila et al. (2018) |
| Cabbage | Poland | 436 | 116–1260 | Spectrophotometry | Raczuk et al. (2014) |
| | Germany | 3100 | - | Spectrophotometry | Reinik et al. (2009) |
| Spinach | Iran | 1830 | 730.4–2420 | Spectrophotometry | Bahadoran et al. (2016) |
| | Korea | 2123.8 | 17.6–6186.2 | Ion chromatography | Suh et al. (2013) |
| | Italy | 2036 | 96–3559 | Ion chromatography | Roila et al. (2018) |
| **Roots** | | | | | |
| Radishes | Poland | 2132 | 710–4800 | Spectrophotometry | Raczuk et al. (2014) |
| | Iran | 6260 | 3930–8980 | Spectrophotometry | Bahadoran et al. (2016) |
| | Korea | 1494 | ND – 3486.7 | Ion chromatography | Suh et al. (2013) |
| | Italy | 3817 | 3650–3985 | Ion chromatography | Roila et al. (2018) |
| Beetroot | Poland | 1306 | 480–2000 | Spectrophotometry | Raczuk et al. (2014) |
| | Germany | 1630 | - | Spectrophotometry | Reinik et al. (2009) |
| Carrot | Poland | 82.2 | 11–259 | Spectrophotometry | Raczuk et al. (2014) |
| | Iran | 387 | 254–738 | Spectrophotometry | Bahadoran et al. (2016) |
| | Korea | 261.9 | ND – 1005.4 | Ion chromatography | Suh et al. (2013) |

**Other vegetables**

| Food | Country | | | Method | Reference |
|---|---|---|---|---|---|
| Cucumbers | Poland | 8.87 (flesh) | 4.8 – 13.1 (flesh) | Spectrophotometry | Stachniuk et al. (2018) |
| | | 55.7 (peel) | 9.7 – 89.7 (peel) | | |
| | China | 130 | 57–260.0 | Ion chromatography | Chung et al. (2011) |
| | USA | 90 | 17–570 | - | Keeton et. al. (2012) |
| Tomato | Poland | 35 | 10–100 | Spectrophotometry | Raczuk et al. (2014) |
| | Iran | 175 | 84–261 | Spectrophotometry | Bahadoran et al. (2016) |
| Potato | Poland | 54 | 11–90 | Spectrophotometry | Raczuk et al. (2014) |
| | Korea | 206.5 | 26.6–396.1 | Ion chromatography | Suh et al. (2013) |
| | Italy | 96 | 73–223 | Ion chromatography | Roila et al. (2018) |
| **Cereals** | Nigeria | 207.7 (rice) | 120–308 (rice) | Spectrophotometry | Ezeagu (2006) |
| | | 677.5 (maize) | 355–1000 (maize) | | |
| | | 1277.5 (cowpea) | 710 - 1845 (cowpea) | | |
| | Iran | 595.0 (barley) | 320.0–716.0 (barley) | Spectrophotometry | Bahadoran et al. (2016) |
| | | 243.0 (rice) | 144.0–306.0 (rice) | | |
| | | 513.0 (industrial breads) | 391.0–674.0 (industrial breads) | | |
| **Legumes** (soya, cowpea, red bean, chickpea) | Iran | 103.0–322.0 | - | Spectrophotometry | Bahadoran et al. (2016) |
| **Infant food of plant origin** | Italy | 45.5 | 4.82–131.68 | Spectrophotometry | Cortesi et al. (2015) |
| | Portugal | 61.0 | - | Spectrophotometry | Rebelo et al. (2015) |

were present in carrot, cucumbers, iceberg lettuce, white cabbage, tomatoes, and potatoes. According to Menard and colleagues (2008) the maximum content of nitrites in lettuce and spinach in France was 25 mg kg$^{-1}$ and 220 mg kg$^{-1}$ respectively. A survey conducted by Chung and colleagues (2003) to determine the nitrate and nitrite contents of vegetables grown in Korea during different seasons revealed that the average nitrite contents in various vegetables be about 0.6 mg kg$^{-1}$, and the values were not significantly different among most vegetables. Vegetables, which were the material tested in the tests carried out by Bahadoran and colleagues (2016), showed a nitrite content in the range of 2.1–7.4 mg kg$^{-1}$. The study by Correia and colleagues (2010) reports the wider range of nitrite levels of 34 vegetable samples (including different varieties of cabbage, lettuce, spinaches, parsley, and turnips), collected in several locations of an intensive agricultural area in Portugal. Nitrite levels in this study ranged between 1.1 and 57 mg kg$^{-1}$. Similarly, nitrite concentrations of vegetables in a survey in Korea were also reported to a wide range from 1.1 to 138 mg kg$^{-1}$ (Suh et al., 2013). As Ranasinghe and Marapana (2018) describe that in fresh vegetables, kept under proper conditions, nitrite reductase activity rate is in equilibrium with one of the nitrate reductase enzymes, which causes low levels of nitrite. High concentrations of nitrites in vegetables may result from nitrate reduction occurring during improper storage conditions. In particular, the risk of nitrite formation occurs with long storage at temperatures above those recommended or without access to oxygen. Prasad and Chetty (2008) showed that nitrate content decreases during storage under ambient temperature due to its conversion to nitrites, but under refrigerated conditions the nitrite accumulation is limited. Moreover, the nitrite accumulation is inhibited under frozen storage. This explains the different proportions of nitrate and nitrite in commercially available vegetable samples.

## 12.2.2   NITRATE IN OTHER PLANT-BASED FOODS

Nitrate accumulation in fruit mostly depends on the type and dose of fertilizer and herbicide application, physiologically active substances, and the type of soil and environmental conditions, including air humidity (Anjana and Iqbal, 2007; Tahmasi and Ziarati, 2015). Research by Ziarati and colleagues (2017) showed that nitrate content in fresh fruit is significantly lower than in canned products. The authors suppose that the reason for such dependencies is the effect of water for washing fruits during processing. In addition, treating canned apples and pears with CaHPO$_4$ in comparison with untreated canned products revealed a significant reduction in nitrate content. Mohammadi and Ziarati (2016) reported nitrate/nitrite concentration in most familiar fruit juices (mango, pineapple, orange, cherry, grape, tropical) available commercially in the packaged form in the Tehran market. The mean nitrate and nitrite levels in the juices ranged from 14.02 to 43.34 and from 4.23 to 11.82 mg L$^{-1}$, respectively.

Limited data are available regarding nitrate/nitrite content of cereals and cereal products. However, the available studies indicate that they are a poor source of nitrite/nitrate. Studies by Jakszyn and colleagues (2004) reported a relatively low level of nitrate and nitrite in bread (in the range of 1.77–2.50 and 0.13–0.17 mg 100 g$^{-1}$, respectively) as well as wheat flour (0.10–1.98 and 0.10–0.46 mg 100 g$^{-1}$, respectively). Abdulrazak and colleagues (2014) analyzed the level of nitrate/nitrite in selected cereals and found that nitrate ranged from 3.0 to 21.3 mg kg$^{-1}$ and nitrite from 0.03 and 0.15 mg kg$^{-1}$. In a study in Iran (Bahadoran et al., 2016), a relatively high nitrate concentration in bread was observed (50 mg/100 g$^{-1}$). Skendi and colleagues (2020) published the results of a study on nitrate levels in organically grown commercial cereal products (from six different kinds of cereal, including wheat, rye, barley, oat, rice, and maize), available in the Greek market. They indicated that the lowest concentration of nitrate observed in the cereal samples analyzed was in the range of 2.9–102.4 mg kg$^{-1}$, whereas the highest concentration varied from 15.5 to 120.6 mg kg$^{-1}$ with the lowest values observed in rice and the highest in rye samples.

Studies on the content of nitrate also apply to infant and children's food. For infants and children, there is a risk of ingesting a quantity of nitrate greater than ADI because the amount of food to body weight is greater than that consumed by an adult. In the research performed by Cortesi and colleagues (2015), nitrate and nitrite values of infant food ranged between 0.35 and 131.68 mg kg$^{-1}$ and 1.12 and 80.22 mg kg$^{-1}$ respectively. The highest nitrate mean values were found in foods of plant origin (45.5 mg kg$^{-1}$) compared to foods of mixed origin; however, nitrate contents never exceeded the level set by EC Regulation 1881/2006. The maximum level for nitrate in processed cereal-based foods and baby foods for infants and young children according to EC Regulation 1881/2006 is 200 mg kg$^{-1}$. Average and median levels of nitrate in baby foods obtained in research, Rebelo and colleagues (2015), were also lower than the maximum limits established by European Union legislation. Median nitrate values in baby foods were 61, 39, 30, and 15 mg kg$^{-1}$ for vegetable-based baby foods, fish-based baby foods, meat-based baby foods, and fruit-based baby foods, respectively. According to Vasco and Alvito (2011), the nitrate level in baby foods was 108 ± 54 mg kg$^{-1}$ and 15 mg kg$^{-1}$ in vegetable-based foods and fruit-based foods, respectively.

### 12.2.3 FACTORS AFFECTING NITRATE/NITRITE CONTENT IN VEGETABLES

Many papers report that nitrate concentrations in vegetables depend on a number of factors, the most common of which are the following: biological properties of plant culture, light intensity, type of soil, temperature, humidity, density of plants in the field, vegetation period, harvesting time, storage time, and fertilization (Vahed et al. 2015; Costagliola et al. 2014). However, fertilization and light intensity appear to be crucial in determining the amount of nitrate in vegetables. The influence of temperature was indicated by Tamme and colleagues (2010). According to their study vegetables grown in heated greenhouses have higher nitrate content compared to those grown in the open air during the same season, mainly due to lower light intensity and high nitrogen mineralization.

The distribution of nitrate in vegetables is also dependent on the part of the plant studied (Vahed et al. 2015, Chung et al. 2011). Some research showed that higher levels of nitrate tend to be located in leaves, whereas they occur at lower concentrations in seeds and tubers (Ekart et al. 2013; Tamme et al. 2010). Research conducted by Stachniuk and colleagues (2018) showed that the cucumber peel accumulated a nearly threefold higher amount of nitrate than the flesh, regardless of the manner of cultivation (greenhouse or open field). The results obtained by Ebrahimi and colleagues (2020) showed the total amount of nitrate in potato peel was 35 percent higher than in its central parts. Different washing methods reduced the nitrate amount by 7.79 to 14.73 percent. Boiling potatoes also significantly reduced the amount of nitrate, by 59.7 percent.

Heat treatment is another important factor influencing the nitrate content in food. Salehzadeh and colleagues (2020) indicated that the nitrate concentration in cooked vegetables was lower than that in raw vegetables, showing that cooking the vegetables decreased the nitrate levels. Prasad and Chetty (2008) reported that the cooking process reduced the nitrate content of vegetables by 47 to 56 percent. The authors explain that nitrate have a high tendency to dissolve in water. In this context, when vegetables are immersed in water, the nitrate tend to move along the diffusion gradient from inside the vegetable to inside the water in which the vegetable is located. Moreover, increasing temperature and time will favor the diffusion process and the movement of nitrate from the inside of the vegetable into the water. Finally, nitrate levels in the vegetable will reduce.

Processing vegetables by fermentation or acidification may influence nitrate and nitrite concentrations. Research carried out by Ding and colleagues (2018) on the evaluation of nitrate and nitrite contents in pickled fruit and vegetable products indicated that nitrate content in pickled products (beets, cauliflowers, carrots, Brussels sprouts) was generally lower compared to fresh fruits and vegetables. A reduction in nitrate was expected by authors due to the production processes used, including acidification, brining, pasteurization, and shelf-stable. The authors stated that evaluation of nitrite and nitrate concentrations in foods should also include L-ascorbic acid and polyphenolic contents as potentially harmful reactions with nitrite and nitrate may depend on the levels of antioxidants. Their view is supported by the evidence presented by Habermeyer and colleagues (2015). They pointed out that in the conversion of nitrate to nitrite, and subsequent formation of NO, reducing agents such as L-ascorbic acid and polyphenolic compounds facilitate the reduction of nitrite to NO and protect NO from being scavenged once it is produced. Moreover, mentioned compounds have an inhibitory effect on nitrosation reactions. Another study also confirmed that the nitrate content in pickled products was lower than that reported for fresh vegetables (Blekkenhorst et al., 2017). A study by Yang and colleagues (2014) shows that during the fermentation of Chinese cabbage, nitrate decreased while the content of nitrite initially increased and then decreased when the pH was low (<4.5).

In contrast, the study by Salehzadeh and colleagues (2020) demonstrated that the nitrate level in fried vegetables was higher than in raw vegetables. The frying process significantly increased the nitrate level in vegetables over 12 percent, even up to 29 percent. The results obtained by Ebrahimi and colleagues (2020) also confirmed this relationship. Frying potatoes significantly increased the amount of nitrate – by 52 percent.

Research shows that the content of nitrate in vegetables can depend on the sampling season. Pourmoghin and colleagues (2010) showed that the nitrate concentration in tomatoes and potatoes in winter was higher than in the summer. This difference can be attributed to various factors, such as the duration and intensity of light radiation, soil and weather temperature, moisture, and plant age. The authors explain that extreme temperatures reduce the nitrate levels of plants through the assimilation process. Since there are more cloudy days in the fall than in the spring, and the air temperature is lower, this results in lower nitrate assimilation during the fall, so the vegetables have higher nitrate levels.

Another factor that plays an important role in the content of nitrate in vegetables is molybdenum fertilization. Molybdenum is utilized by plants in selected enzymes, including nitrate reductase, xanthine dehydrogenase, aldehyde oxidase, and sulfite oxidase, to carry out redox reactions in particular in processes involving nitrogen metabolism. Its deficiency leads to the accumulation of high concentrations of nitrate in many plants (Moncada et al., 2018). Kaiser and colleagues (2005) indicated that molybdenum raised the nutritional quality of lettuce, escarole, and curly endive by reducing nitrate content down to 1039.2, 1047.3, and 1181.2 mg kg$^{-1}$ fresh weight, respectively.

## 12.3 NITRATE IN ANIMAL-BASED FOOD

Nitrate and nitrites are authorized as food additives in the European Union under Commission Regulation (EU) No 1129/2011. They are used in processed meat and cheese. However, some of these compounds may be present in the raw material mainly as a result of contaminated feed. Zamric (2013) conducted a study of the determination of nitrate and nitrite contents of Syrian white cheese. According to this author, the presence of nitrate/nitrites in milk is caused by contamination of the water with these ions, while the presence of these compounds in cheese results from improper processing during production. Additionally, sodium and potassium nitrate are approved for use in dairy products as antimicrobial agents against *Clostridium tyrobutyricum* and *C. butyricum*. Nitrate in cheese helps to obtain the desired sensory characteristics that may deteriorate as a result of the undesirable microflora. The maximum concentration of nitrite allowed in the regulation for cheese is 150 mg kg$^{-1}$. As reported by Reinik and colleagues (2009), main residues of nitrate and nitrite in dairy products are related to oral administration of potassium nitrate to dairy cows and added water with high nitrate content. Another reason is use of nitrate/nitrite containing food additives. Moreover, nitric oxide production can increase during endotoxin-induced mastitis, which also leads to higher nitrate/nitrite levels in dairy products. According to a study conducted by Bahadoran and colleagues (2016) dairy products were characterized by the mean nitrite and nitrate content in the range of 0.14–0.45 mg 100 g$^{-1}$ and 1.26–5.75 mg 100 g$^{-1}$, respectively. Based on the presented research, it can be concluded that the content of nitrate/nitrites in dairy products is low.

Another group of foods that contain little nitrate is seafood. Chiesa and colleagues (2019) examined seafood, including fish, shrimp, and bivalves from various processing and environmental settings based on ion chromatography with suppressed conductivity. The authors detected nitrite in all samples; however, the presence of nitrate varied from 2.8 (mussels) to 89.3 µg kg$^{-1}$ (clams). The authors suggested that these results were due to the growing system and location (wild and farm). In addition, preservation techniques affect the levels of nitrate in fish. The highest concentration of nitrate was found in smoked salmon samples (median = 60 µg/g), while nitrites were not revealed in any of the samples studied.

Beretta and colleagues (2010) investigated nitrate accumulation in honey of different botanical origins, using a chemiluminescence-based technique (NO-analyzer). All honey (honeydew, chestnut, multiflora, chicory, linden, acacia, eucalyptus, cardoon, lavender, sunflower, orange, licorice, sulla, rhododendron, rosemary, taraxacum, trifolium, pluviar forest, forest, and strawberry) contained appreciable amounts of nitrate ranged from 1.63 to 482.98 mg kg$^{-1}$. The highest level was detected in honeydew honey, 10–40 times than in nectar honey. Low levels of nitrite were found in all samples in the range of 0.01–0.56 mg kg$^{-1}$. Nitrites and nitrate content of animal foods in different countries are presented in Table 12.2.

## 12.4   NITRATE IN PROCESSED MEAT

E249 and E250 are important additives in the meat industry. According to Regulation (EC) No 1333/2008 of the European Parliament and of the Council, the maximum amount of nitrite that can be added to sterilized meat products is 100 mg/kg; for other meat products, the limit is 150 mg/kg. The addition of sodium nitrate is allowed only in uncooked meat, to a maximum amount of 150 mg kg$^{-1}$. Nitrate is added as a precursor to nitrite formed through the action of bacterial nitrate reductase (Bedale et al., 2016). Nitrate and nitrites are added to processed meat due to their multidirectional impact on the quality features of meat products, including inhibition of the growth of bacteria (mainly *Clostridium botulinum*), limiting the process of oxidation, giving the typical reddish color and flavor of the processed meat as well as antioxidative effects. An important aspect of using nitrate compounds in meat processing is the formation of stable red color which is developed in a number of reactions leading to NO-myoglobin formation (Honikel, 2008).

Despite the beneficial effect on the quality characteristics of meat products, nitrate use is associated with negative health effects. In 2018, processed meats, understood as any meat that has been altered through salting, curing, smoking, or other processes, have been classified as a group 1 carcinogen by the International Agency for Research on Cancer (IARC, 2018). IRAC reported that the daily consumption of 50 g of processed meat increases the risk of colorectal cancer by 18 percent. One of the ingredients in processed meat listed as responsible for the cancer-promoting effects is nitrites. Nitrite was classified as a group 2A carcinogen and described as probably carcinogenic (Grosse, 2006). Breda and colleagues (2019) describe the carcinogenic effect of nitrite, pointing to its reactions with N-nitroso compound precursors in the gastrointestinal tract, thereby subsequently forming potentially carcinogenic compounds. Endogenous N-nitroso compound formation is mentioned as one of the main mechanisms underlying the positive relation between colorectal cancer risk and consumption of processed meat. The formation of these endogenous carcinogenic compounds is dependent on some factors, including the presence of nitrosation precursors (mainly amines and amides) and heme iron, which may stimulate their formation. Additionally, it may depend on nitrosation inhibitors present in the diet such, as vitamin C, vitamin E, or polyphenols.

A number of studies report different analytical methods and analysis due to controlling of nitrate/nitrites in meat and meat products. The most commonly used methods in processed meat are chromatographic, capillary electrophoretic, and

**TABLE 12.2**

**Nitrites and Nitrate Content of Animal Foods in Different Countries**

| Samples | Country | Nitrite (mg kg⁻¹) | Nitrate (mg kg⁻¹) | Analytical Method | References |
|---|---|---|---|---|---|
| **Dairy products** | | | | | |
| Milk Iran 1.4–1.5 | | | 12.6–13.2 | Spectrophotometry | Bahadoran et al. (2016) |
| Yogurt Iran 2.4–4.5 | | | 28.6–57.5 | | |
| Cheese Iran 3.4 | | | 29.6 | | |
| Milk | Turkey | 0.12 | 22.43–22.63 | Spectrophotometry | Sungur et al. (2013) |
| Yogurt | Turkey | 0.17–0.33 | 28.70–69.81 | | |
| Cheese | Turkey | 0.12–0.99 | 11.49–323.90 | | |
| **Fish products** | | | | | |
| Canned tuna | Turkey | - | 6.971–22.49 | Spectrophotometry | Kalaycıoğlu and Erim (2016) |
| Canned mackerel | Turkey | - | 17.39–18.19 | | |
| Canned sardine | Turkey | 6.063 | 11.88 | Spectrophotometry | Bahadoran et al. (2016) |
| Fried fish | Iran | 3.39 | 5.56 | | |
| Canned fish | Iran | 2.93 | 6.09 | | |
| **Honey** | | | | | |
| honeydew | Italy | 0.02–0.008 | 162.28–482.98 | Chemiluminescence based technique (NO-analyzer) | Beretta et al. (2010) |
| lavender | Italy | 0.13–0.54 | 4.62–10.78 | | |
| multiflora | Italy | 0.02–0.50 | 4.03–250.22 | | |
| sunflower | Italy | 0.19–0.25 | 2.37–4.41 | | |
| orange | Italy | 0.01–0.31 | 2.43–3.44 | | |
| strawberry | Italy | 0.00–0.14 | 1.63–2.55 | | |
| **Meat and meat products** | | | | | |
| Meat (beef, lamb, ground mixed, chicken) | Iran | 3.72–4.96 | 7.43–13.3 | Spectrophotometry | Bahadoran et al. (2016) |
| Sausages | Iran | 13.9 | 18.8 | | |
| Beef fresh meat | Italy | - | 10.8–15.0 | Chromatographic/IC | Iammarino and Taranto (2012) |
| Pork fresh meat | Italy | - | 10.2–14.8 | | |
| Equine fresh meat | Italy | - | 13.6–36.5 | | |
| Sausage | Belgium | 12.0–34.4 | 23.5–40.4 | Chromatographic/HPLC | Temme et al. (2011) |
| Liver paste | Belgium | 6.6 | 57.3 | | |
| Bacon | Belgium | 4.8 | 85.9 | | |
| Traditional sausages | Polish | 5.79–14.78 | 12.51–34.64 | Spectrophotometry | Halagarda et al. (2018) |
| Conventional sausages | Polish | 9.72–79.21 | 13.79–27.80 | | |

spectrophotometric methods (Table 12.2). Studies available in the literature show that fresh meat has a low nitrate content. Nitrate's presence in fresh meat used in meat processing may have originated from the nitrogen metabolism of animals and feeding. Iammarino and Di Taranto (2012) reported a natural endogenous formation of nitrate with a maximum range of 30–40 mg kg$^{-1}$ in beef and pork and in equine meat, respectively. The differences in the nitrate content were related to feeding regimes. The levels of residual nitrite and nitrate in processed meat products are variable, similar to vegetables, depending on the time and temperature used during processing and storing, the initial addition of nitrite and nitrate, the composition of the meat, pH, addition of antioxidants, and the presence of micro-organisms (Honikel. 2008). According to Sindelar and Milkowski (2012) about 10–20 percent of originally added nitrites, referred to as residual nitrites, are typically present in meat products after production. This amount of residual nitrites additionally declines during the storage period of cured meat products. According to Bahadoran and colleagues (2016), mean nitrate and nitrite concentrations in meats and processed meats were 5.56–19.4 mg 100 g$^{-1}$ and 2.93–13.9 mg 100 g$^{-1}$, respectively. Analysis of commonly consumed processed meat products on the Swedish market (Merino et al., 2016) showed that during the product's early life, there was a decrease in nitrite from the point of sodium nitrite addition. Further reduction of nitrite levels was observed in later stages of product life, however much slower, and at the use-by date, the residual level of nitrite was 5–19 percent of the amount initially added to the products. Interestingly, it has been observed that the depletion of nitrite depended on time and the type of cured meat product. Cured chicken products were characterized by higher residual nitrite levels compared to cured pork/beef products. In this context, the authors are concerned that the Word Health Organization recommendation on limiting consumption of red/processed meat could result in increased nitrite exposure among consumers as a consequence of increased consumption of processed white meat.

Research by Kim and collagues (2017) indicated the effect of using fermented spinach as a natural source of nitrate on the residual nitrite content of cured meat. The residual nitrite content of cured meat with fermented spinach was lower than that of cured meat with sodium nitrite. The authors explain that an acidic environment and abundant antioxidant during the curing process can decrease residual nitrite content because the low pH and antioxidant such as phenols and flavonoids in meat and meat products can easily deplete nitrite to nitric oxide (Kim et al., 2019). Merino and colleagues (2016) examined time-dependent changes in nitrite levels in four Swedish meat products frequently eaten by children (pork/beef sausage, liver paté, and two types of chicken sausage). They also checked how the production process, storage, boiling, and frying affect the initial added nitrite level. The results did not show any effect of boiling on the nitrite content in pork and beef sausage. In contrast, frying reduced nitrite content by about 50 percent. Twenty-four hours after the addition of nitrite, the authors observed a decrease in the amount of nitrite in all meat products tested.

Recently, the European Food Safety Authority has reassessed the safety of nitrite (E 249, E 250) and nitrate (E 251, E 252) as food additives and, based on the evidence gathered, concluded that there is no need to change the previously established dosage limits for these substances in food products. The experts concluded that the exposure to nitrate and nitrites resulting from their use as a food additive does not lead to an excess of ADI for the general population except for a slight excess in children. Considering all sources of dietary nitrite and nitrate exposure, together including food additives, natural presence, and contamination, the ADI would be exceeded in infants, toddlers, and children at the mean level and in people of all ages at the highest exposure (EFSA, 2017).

## REFERENCES

Abdulrazak S., Otie D., & Oniwapele Y.A. (2014). The concentration of nitrate and nitrite in some selected cereals sourced within Kaduna state, Nigeria. *Online Journal of Animal Feed Resources* 4, 37–41.

Anjana S.U., & Iqbal M. (2007). Nitrate accumulation in plants, factors affecting the process, and human health implications. A review. *Agronomy for Sustainable Development* 27, 45–57.

Bahadoran Z., Mirmiran P., Jeddi S., Azizi F., Ghasemi A., & Hadaegh F. (2016). Nitrate and nitrite content of vegetables, fruits, grains, legumes, dairy products, meats, and processed meats. *Journal of Food Composition and Analysis* 2–40.

Bedale W., Sindelar J.J., & Milkowski A.L. (2016). Dietary nitrate and nitrite: Benefits, risks, and evolving perceptions. *Meat Science* 120, 85–92.

Beretta G., Gelmini F., Lodi V., Piazzalunga A., Maffei Facino R. (2010). Profile of nitric oxide (NO) metabolites (nitrate, nitrite and N-nitroso groups) in honeys of different botanical origins: nitrate accumulation as index of origin, quality and of therapeutic opportunities. *Journal of Pharmaceutical and Biomedical Analysis* 2, 53(3), 343–349.

Blekkenhorst L.C., Prince R.L., Ward N.C., Croft K.D., Lewis J.R., Devine A., ... & Bondonno C.P. (2017). Development of a reference database for assessing dietary nitrate in vegetables. *Molecular Nutrition & Food Research* 61(8).

Breda S.G., Mathijs K., Sági-Kiss V., Kuhnle G.G., van der Weer B., Jones R.R., Sinha R., Ward M.H., & de Kok T.M. (2019). Impact of high drinking water nitrate levels on the endogenous formation of apparent N-nitroso compounds in combination with meat intake in healthy volunteers. *Environmental Health* 18, 87, 1–12.

Britto D.T., & Kronzucker H.J. (2013). Ecological significance and complexity of N-source preference in plants. *Annals of Botany* 112, 957–963.

Buchanan B.B., Gruissem W., & Jones R.L. (2015). *Biochemistry and Molecular Biology of Plants*. 2nd ed. John Wiley: Chichester, West Sussex.

Chan T.Y. (2011). Vegetable-borne nitrate and nitrite and the risk of methemoglobinemia. *Toxicology Letters* 200, 107–108.

Chen B.-M., Wang Z.-H., Li S.-X., Wang G.-X., Song H.-X., & Wang X.-N. (2004). Effects of nitrate supply on plant growth, nitrate accumulation, metabolic nitrate concentration, and nitrate reductase activity in three leafy vegetables. *Plant Science* 167, 635–643.

Chiesa L., Arioli F., Pavlovic R., Villa R., & Panseri S. (2019). Detection of nitrate and nitrite in different seafood. *Food Chemistry* 288, 361–367.

Chung S., Kim J., Kim M., Hong M., Lee J., Kim C., & Song I. (2003). Survey of nitrate and nitrite contents of vegetables grown in Korea. *Food Additives and Contaminants* 20, 621–628.

Chung S.W.C., Tran J.C.H., Tong K.S.K., Chen M.Y.Y., Xiao Y., Ho Y.Y., & Chan C.H.Y. (2011). Nitrate and nitrite levels in commonly consumed vegetables in Hong Kong. *Food Additives and Contaminants*, B 4, 34.

Correia M., Barroso M., Barroso F., Soaresa D., Oliveira M.B.P.P., & Delerue-Matosa C. (2010). Contribution of different vegetable types to exogenous nitrate and nitrite exposure. *Food Chemistry* 120, 4, 960–966.

Cortesi M.L., Vollano L., Peruzy M.F., Marrone R., & Mercogliano R. (2015). Determination of nitrate and nitrite levels in infant foods marketed in Southern Italy. *CyTA – Journal of Food* 13, 4, 629–634.

Costagliola A., Roperto F., Benedetto D., Anastasio A., Marrone R., Perillo A., Russo V., Papparella S., & Paciello O. (2014). *Outbreak of fatal nitrate toxicosis associated with consumption of fennels (Foeniculum vulgare) in cattle farmed in Campania region (southern Italy).* Environmental Science and Pollution Research 21(9), 6252–6257.

Czech A., & Rusinek E. (2005). The heavy metals, nitrates and nitrites content in the selected vegetables from Lublin area. *Roczniki Państwowego Zakładu Higieny* 56(3), 229–236 (in Polish).

Ding Z., Johanningsmeier S.D., Price R., Reynolds R., Truong V.-D., Payton S.C., & Breidt F. (2018). Evolution of nitrate and nitrite content in pickled fruit and vegetable products. *Food Contaminants* 90, 304–311.

Ebrahimi R., Ahmadian A., Ferdousi A., Zandi S., Shahmoradi B., Ghanbari R., Mahammadi S., Rezaee R., Safari M., Daraei H., Maleki A., & Yetilmezsoy K. (2020). Effect of washing and cooking on nitrate content of potatoes (cv. Diamant) and implications for mitigating human health risk in Iran. *Potato Research* 63, 449–462.

EFSA (2017). Re-evaluation of potassium nitrite (E 249) and sodium nitrite (E250) as food additives. *EFSA Journal* 15, 4786.

Ekart K., Hmelak Gorenjak A., Madorran E., Lapajne S., & Langerholc T. (2013). Study on the influence of food processing on nitrate levels in vegetables. *EFSA Support Publications*, EN–514.

Ezeagu, I.E. (2006). Contents of nitrate and nitrite in some Nigerian food grains and their potential digestion in the diet – a short report. *Polish Journal of Food and Nutrition Research* 56(3), 283–285.

Grosse Y. (2006). Carcinogenicity of nitrate, nitrite, and cyanobacterial peptide toxins. *Lancet Oncology* 7, 628–629.

Habermeyer M., Roth A., Guth S., Diel P., Engel K. H., Epe B., et al. (2015). Nitrate and nitrite in the diet: How to assess their benefit and risk for human health. *Molecular Nutrition and Food Research* 59(1), 106–128.

Halagarda M., Kędzior W., & Parzyńska E. (2018). Nutritional value and potential chemical food safety hazards of selected Polish sausages as influenced by their traditionality. *Meat Science* 139, 25–34.

Honikel K.O. (2008). The use and control of nitrate and nitrite for the processing of meat products. *Meat Science* 78, 68–76.

Iammarino M., & Di Taranto A. (2012). Nitrite and nitrate in fresh meats: A contribution to the estimation of admissible maximum limits to introduce in directive 95/2/EC. Int. *Journal of Food Science and Technology* 47, 1852–1858.

IARC Working Group (2018). Monographs on the evaluation of carcinogenic risks to humans. *The Lancet Oncology*. Lyon, France.

Jakszyn P., Agudo A., Ibanez R., Garcia-Closas R., Pera G., Amiano P., & Gonzalez C.A. (2004). Development of a food database of nitrosamines, heterocyclic amines: and polycyclic aromatic hydrocarbons. *Journal of Nutrition* 134, 2011–2014.

Kaiser B.N., Gridley K.L., Brady J.N., Phillips T., & Tyerman S.D. (2005). The role of molybdenum in agricultural plant production. *Annals Botany* 96, 745–754.

Kalaycioğlu Z., & Erim F.B. (2016). Simultaneous determination of nitrate and nitrite in fish products with improved sensitivity by sample stacking-capillary electrophoresis. *Food Analytical and Methods* 9, 606–711.

Keeton J.T., Osburn W.N., Hardin M.D., Bryan N.S., & Longnecker T. (2012). A national survey of the nitrite/nitrate concentrations in cured meat products and nonmeat foods available at retail. *Journal of Agriculture and Food Chemistry* 60, 3981–3990.

Kim T.K., Hwang K.E., Song D.H., Ham Y.K., Kim Y.B., Paik H.D., & Choi Y.S. (2019). Effects of natural nitrite source from Swiss chard on quality characteristics of cured pork loin. *Asian-Australasian Journal of Animal Sciences* 32(12), 1933–1941.

Kim T.K., Kim Y.B., Jeon K.H., et al. (2017). Effect of fermented spinach as sources of pre-converted nitrite on color development of cured pork loin. *Korean Journal of Food Science and Animal Resources* 37, 105–113.

Kim T.K., Lee M.A., Sung J.M., Jeon K.H., Kim Y.B., & Choi Y.S. (2019). Combination effects of nitrite from fermented spinach and sodium nitrite on quality characteristics of cured pork loin. *Asian-Australasian Journal of Animal Sciences* 32, 1603–1610.

Menard C., Heraud F., Volatier J.L., & Leblanc J.C. (2008). Assessment of dietary exposure of nitrate and nitrite in France. *Food Additives and Contaminants* 25(8), 971–988.

Mensinga T.T., Speijers G.J., & Meulenbelt J. (2003). Health implications of exposure to environmental nitrogenous compounds. *Toxicology Review* 22(1), 44–51.

Merino L., Darnerud O., Toldrá F., & Ilbäck N.G. (2016). Time-dependent depletion of nitrite in pork/beef and chicken meat products and its effect on nitrite intake estimation. Food Additives & Contaminants Part A, Chemistry, *Analysis, Control, Exposure & Risk Assessment 1*, 33(2), 186–192.

Mohammadi S., & Ziarati, P. (2016). Nitrate and nitrite content in commercially-available fruit juice packaged products. *Journal of Chemical and Pharmaceutical Research* 8(6), 335–341.

Moncada A., Miceli A., Sabatino L., Iapichino G., D'Anna F., & Vetrano F. (2018). Effect of molybdenum rate on yield and quality of lettuce, escarole, and curly endive grown in a floating system. *Agronomy* 8, 171.

Nuñez de González M.T, Osburn W.N., Hardin M.D., Longnecker M., Garg H.K., Bryan N.S., & Keeton J.T. (2015). A survey of nitrate and nitrite concentrations in conventional and organic-labeled raw vegetables at retail. *Journal of Food Science* 80, 5, C942–C949.

Pourmoghim M., Khoshtinat K., Makkei A., Fonod R., Golestan B., & Pirali M. (2010). Determination of nitrate contents of lettuce, tomatoes and potatoes on sale in Tehran central fruit and vegetable market by HPLC. *Iranian Journal of Nutrition Sciences & Food Technology* 5(1), 63–70.

Prasad S., & Chetty A.A. (2008). Nitrate-N determination in leafy vegetables: Study of the effects of cooking and freezing. *Food Chemistry* 106(2), 772–780.

Raczuk J., Wadas W., & Głodaz K. (2014). Nitrates and nitrites in selected vegetables purchased at supermarkets in Siedlce, Poland. *Roczniki Państwowego Zakładu Higieny* 65(1), 15–20.

Ranasinghe R., & Marapana R. (2018). Nitrate and nitrite content of vegetables: A review. *Journal of Pharmacognosy and Phytochemistry* 7(4), 322–328.

Rebelo J.S., Almeida M.D., Vales L., & Almeida C.L.L. (2015). Presence of nitrates in baby foods marketed in Portugal. *Cogent Food & Agriculture* 1, 1.

Reinik, M. (2007). Nitrates, nitrites, N-nitrosamines and polycyclic aromatic hydrocarbons in food: Analytical methods, occurrence and dietary intake. *Dissertation*, University of Tartu.

Reinik M., Tamme M., & Roasto M. (2009). Naturally occurring nitrates and nitrites in food. In: Gilbert G., & Senyuva H.Z. (Eds) *Bioactive compounds in foods*. Blackwell, Oxford, pp. 225–253.

Roila R., Branciari R., Staccini B., Ranucci D., Miraglia D., Altissimo M.S., Mercuri M.L., & Haouet N.M. (2018). Contribution of vegetables and cured meat to dietary nitrate and nitrite intake in Italian population: safe level for cured meat and controversial role of vegetables. *Italian Journal of Food Safety* 7, 7698–7673.

Salehzadeh, H., Maleki, A., Rezaee, R., Shahmoradi, B., & Ponnet, K. (2020). The nitrate content of fresh and cooked vegetables and their health related risks. *PLoS ONE* 15(1), e0227551.

Santamaria P. (2006). Nitrate in vegetables: Toxicity, content, intake and EC regulation (revive). *Journal of the Science of Food and Agriculture* 86, 10–17.

Sellimi S., Ksouda G., Benslima A., Nasri R., Rinaudo M., Nasri M., & Hajji M. (2017). Enhancing colour and oxidative stabilities of reduced-nitrite turkey meat sausages during refrigerated storage using fucoxanthin purified from the Tunisian seaweed Cystoseira Barbata. *Food Chemistry and Toxicology* 107, 620–629.

Sindelar J.J., & Milkowski A.L. (2012). Human safety controversies surrounding nitrate and nitrite in the diet. *Nitric Oxide* 15, 26(4), 259–266.

Sindelar J.J.; & Milkowski A.L. (2012). Human safety controversies surrounding nitrate and nitrite in the diet. *Nitric Oxide* 26, 259–266.

Skendi A., Papageorgiou M., Irakli M., & Katsantonis D. (2020). Presence of mycotoxins, heavy metals and nitrate residues in organic commercial cereal-based foods sold in the Greek market. *Journal of Consumer Protection and Food Safety* 15, 109–119.

Stachniuk A., Szmagara A., & Stefaniak E.A. (2018). Spectrophotometric assessment of the differences between total nitrate/nitrite contents in peel and flesh of cucumbers. *Food Analytical Methods* 11, 2969 – 2977.

Suh J., Paek O., Kang Y.W., Ahn J.E., Jung J.S., An Y.S., Park S.-H., Lee S.-J., & Lee K.-H. (2013). Risk assessment on nitrate and nitrite in vegetables available in the Korean diet. *Journal of Applied Biological Chemistry* 56(4), 205–211.

Sungur S., & Atan M.M. (2013). Determination of nitrate, nitrite and perchlorate anions in meat, milk and their products consumed in Hatay region in Turkey. *Food Additives and Contaminants*, Part B 6, 6–10.

Tahmasi A., & Ziarati P. (2015). Histamnie and chemical composition of canned and frozen green pea (Pisum sativum). *OJC* 9(2), 739–749.

Tamme T., Reinik M., Roasto M., Meremäe K., & Kiis A. (2010). Nitrate in leafy vegetables, culinary herbs, and cucumber grown under cover in Estonia: content and intake. *Food Additives and Contamination* B 3, 108–113.

Temme E.H.M., Vandevijvere S., Vinkx C., Huybrechts I., Goeyens L., & Van Oyen H. (2011). Average daily nitrate and nitrite intake in the Belgian population older than 15 years. *Food Additives and Contaminants*, Part A 28, 1193–1204.

Vahed S., Mosaha L., Mirmohammadi M., & Lakzadeh L. (2015). Effect of some processing methods on nitrate changes in different vegetables. *Food Measures* 9, 241–247.

van den Brand A.D., Beukers M., Niekerk M., van Donkersgoed G., van der Aa M., van de Ven B., Bulder A., van der Voet H., & Sprong, C.R. (2020). Assessment of the combined nitrate and nitrite exposure from food and drinking water: application of uncertainty around the nitrate to nitrite conversion factor. *Food Additives & Contaminants: Part A* 37, 4, 568–582.

Vasco E., & Alvito P. (2011). Occurrence and infant exposure assessment of nitrates in baby foods marketed in the region of Lisbon, *Portugal. Food Additives & Contaminants Part B-Surveillance* 4, 218–225.

Wang Y.Y., Hsu P.K., & Tsay Y.F. (2012). Uptake, allocation and signaling of nitrate. *Trends in Plant Science* 17, 458–467.

World Health Organization (2007). Nitrate and nitrite in drinking water development of WHO guidelines for drinking water quality (pp. 21), Geneva: World Health Organization, 1–21.

Yang Y., Zou H., Qu C., Zhang L., Liu T., Wu H. & Li Y. (2014). Dominant microorganisms during the spontaneous fermentation of Suan Cai, a Chinese fermented vegetable. *Food Science and Technology Research* 20, 5, 915–926.

Zamrik M.A. (2013). Determination of nitrate and nitrite contents of Syrian white cheese. *Pharmacology and Pharmacy* 4, 171–175.

Ziarati P., Sepehr P., Heidari S., & Moslehishad M. (2017). Nitrate reduction in canned apples and pears using calcium hydrogen phosphate (CaHPO4). *Iranian Journal of Toxicology* 11, 5, 53–59.

# 13 Nitrate as Food Additives
## Reactivity, Occurrence, and Regulation

Teresa D'Amore, Aurelia Di Taranto,
Giovanna Berardi, Valeria Vita, and
Marco Iammarino

## CONTENTS

13.1 Introduction.................................................................................281
13.2 Occurrence and Regulation.........................................................283
13.3 Reactivity and Mechanism..........................................................290
13.4 Antimicrobial Function...............................................................292
13.5 Flavoring Function and Antioxidant Properties...........................295
13.6 Colouring Stabilizer and Enhancer.............................................295
Acknowledgments.................................................................................297
References.............................................................................................297

## 13.1 INTRODUCTION

Any substance not normally consumed as a food by itself and not normally used as a typical ingredient of the food, whether or not it has nutritive value, the intentional addition of which to food for a technological (including organoleptic) purpose in the manufacture, processing, preparation, treatment, packing, packaging, transport or holding of such food results, or maybe reasonably expected to result (directly or indirectly), in it or its by-products becoming a component of or otherwise affecting the characteristics of such foods. The term does not include contaminants or substances added to food for maintaining or improving nutritional qualities.

This is the definition of *Food Additive*, reported in the 2019 version of Codex Standard 192-1995 – General Standard for Food Additives (GSFA), refined by Codex Alimentarius Commission (CAC) in collaboration with the Joint Expert Committee on Food Additives (JECFA) of WHO (World Health Organization) and FAO (Food and Agriculture Organization of the United Nations), and generally accepted by the

DOI: 10.1201/9780429326806-17

two major regulators on food additives worldwide, EFSA (European Food Safety Authority) and FDA (Food and Drug Administration of the United States) [1]. From the early twentieth century until today, the usage of chemical additives has increased, to extend shelf-life and maintain or improve the safety, freshness, taste, texture, and appearance of food. Several classifications, based on additives functions, labeling, and numbering, were developed worldwide. However, it is necessary to underline that it is difficult to schedule these molecules in one univocal list, because some agents may have more than one function in foods. The most-used classification is the Codex Alimentarius – GSFA – INS (International Numbering System), which divides all approved additives in categories, assigning a number code, prefixed by "INS". For nitrites and nitrate, it matches with "E-number" assigned by European additives classification [2].

Nitrite ($NO_2^-$) and its congener nitrate ($NO_3^-$), the subjects of this chapter, are numbered as E249, E250, E251, E252, which are referred, in particular, to their potassium and sodium salts. They are the perfect example of additives used for over a century, being also used before their mechanism of action was defined, since they have more than one function in processed food. Scheduled in the list of preservatives, they act as anti-microbial, anti-oxidant, coloring stabilizers, anti-browning, flavoring, and texturing agents. So, they are widely added during the manufacturing of many processed foods, such as in curing processes and to ensure the quality and avoid spoilage of other perishable food (meat and poultry products, fish preparations, and dairy products) [3].

These molecules are both additives and naturally occurring contaminants, since they are an integral part of the Nitrogen cycle. Nitrate is the most stable N fixed oxidized form, so it is widely distributed in the environment, and it may accumulate in several foodstuffs (especially leafy vegetables, such as spinach and lettuce). On the contrary, nitrite is less distributed in nature, since it is quickly oxidized to nitrate in water and soil. Moreover, although in fresh and unprocessed meat and fish products their residual amount is generally low (adding is forbidden), not negligible concentrations of nitrate may be found. A marked influence of this amount rate is due to intensive farming methods, animal feed, and environment pollution [4].

In dietary intake, it has been estimated that 80 percent of nitrate exposure derives from water and vegetable consumption, while meat and other processed food contribute 5–15 percent, depending on age, geographic area, nutritional behavior, and traditions. This percentage is higher for nitrite as an additive, which may contribute to total dietary exposure up to almost 40 percent [5].

The interest in these additives and the monitoring studies grew when it was clarified that nitrate and nitrite may be responsible for some harmful and toxic effects for humans. In particular, uncontrolled ingestion may cause methemoglobinemia, potentially fatal, especially for infants and children, due to the oxidation reaction between a protoporphyrinic group of oxy-hemoglobin and nitrite, producing ferri-hemoglobin. Researchers demonstrated that ingested nitrate/nitrite can react with amino groups of proteins and form nitrosamines (NOCs – N-nitroso compounds), well-known carcinogenic derivates that may increase the risk of gastric, esophageal, and lung cancer. For these toxic effects, red and processed meat products are particularly placed under accusation, because the formation reaction of nitrosamines is favored (cooking methods and food associations may play a notable role) [6]. About this, in 2015 the International Agency for Research on Cancer (IARC) declared "ingested nitrate and nitrite under conditions that result in endogenous nitrosation is probably carcinogenic to humans (Group 2A)" [7, 8].

On the other hand, in the last decade a large amount of data has been brought to the attention of the scientific community to rehabilitate these molecules. The researches correlate the pathway of conversion $NO_3^-/NO_2^-/NO$ with different beneficial effects on the cardio-circulatory system [9]. Furthermore, many studies affirmed that food sources of $NO_2^-/NO_3^-$ play a crucial role. The exposure deriving from processed foods is not similar to that derived from vegetables. These last matrices are rich in beneficial compounds, such as vitamins and polyphenols that inhibit the formation of radicals and NOCs precursors. In this regard, it is reported that there is an increase in production and consumption of meat products "naturally preserved," which are added with various vegetable juices and powders (beetroot powder, celery powder, carrot juice concentrate), naturally rich in nitrate [10, 11].

At any rate, it should be underlined that nitrate is generally considered 10–20 times less toxic than nitrite. A median lethal dose ($LD_{50}$) of 1600–9000 mg of sodium nitrate per kilogram of body weight has been reported in mice, rats, and rabbits. The $LD_{50}$ value for sodium nitrite was 85–220 mg per kilogram of body weight in the same conditions [12].

For all these reasons, the adding of nitrite and nitrate to foodstuffs remains a controversial issue.

A great debate on the overestimation or underestimation of these additives is underway. On the one hand, their replacement in food with natural molecules (essential oils, natural extracts, and powders, etc.) is suggested. They have even been tested as possible supplements for blood pressure lowering, antiplatelet aggregation, and improvement of athletic performances [13].

However, none of the other preservatives have the same manifold functions and a broad spectrum of activity, which may confer unique organoleptic properties to the products (color, texture, and flavor). All these reasons, combined with evidence that nitrite and nitrate are the most active preservatives against the harmful bacterium *Clostridium botulinum,* justify their use in a risk-benefit assessment [14].

## 13.2   OCCURRENCE AND REGULATION

Nowadays, the usage of additives, especially in the food industry, is essential both to prevent contamination and spoilage and to preserve and enhance different sensory properties of foods. Although these compounds have a recognized technological, and safe, role and there is evidence that most foodborne illness derive from microbial contamination events, the regulatory agencies underlined the necessity of regular controls, toxicological and monitoring studies, as well as a periodical re-evaluation of these substances. For example, in Europe, each approved food additive has to demonstrate technological need, but also pass extensive toxicity studies before approval, following the guidelines of Regulation (EC) No 1331/2008, which establishes a uniform authorization procedure for food additives, enzymes, and flavorings [15, 16].

Moreover, the European Food Safety Agency (EFSA), with the support of the European Commission, in 2010 started a program to re-assess the safety of all the existing approved food additives, following Regulation (EU) No 257/2010 [17]. This sequence of re-assessment opinions on all additives authorized before 20 January 2009 was brought to a close in 2020. Nitrate and nitrites salts were included, and two scientific reports were disclosed in 2017. The studies that comprise new toxicological and epidemiological data are useful tools to control and re-confirm the levels of exposure and the acceptable daily intakes (ADIs), which are associated with methods of risk evaluation [18, 19].

In the sodium nitrate and potassium nitrate report, the Panel on Food Additives and Nutrient Sources Added to Food (ANS) underlined that nitrate as food additives contribute less than 5 percent to the overall exposure from all sources in any scenario and for any population group. Also, for nitrites (E 249 and E 250), using the same refined exposure methodology (non-brand-loyal consumer scenario for the general population), the contribution from their use as food additives was established and represented approximately 17 percent (range 1.5–36.0%) of the overall exposure to nitrites. It seems that during recent years the relative contribution of processed food to total dietary intake of nitrite and nitrate has been substantially lowering, in favor of intake from vegetables [11]. Additionally, the necessity of new toxicological studies, especially for thyroids and gastric disease, is explained. In the same manner, the report about the re-evaluation of nitrite disclosed some evidence that links dietary nitrite to gastric and colorectal cancers and stated the urgency of new clarificatory trials.

A long debate was held for the above-mentioned parameter, the ADI, introduced in 1961 by the Joint FAO/WHO Expert Committee on Food Additives (JECFA), which represents the amount of total nitrate that may be ingested daily over a lifetime without an appreciable health risk. For nitrate, the current ADI, which was set in 1995 by the European Commission's former Scientific Committee for Food (SCF) and the JECFA is 3.7 milligrams per kilogram of body weight per day (mg/kg bw/day), equivalent to 222 mg per day for a 60 kg adult. On the contrary, for sodium nitrite and potassium nitrite, two different ADIs were established, equal to 0.07 and 0.06 mg/kg bw/day, respectively. The ANS Panel, although noting that the ADI value may exceed, especially in specific subsets of the population, confirmed SCF and JECFA opinion [20, 21].

The ADI concept is strictly correlated to the NOAEL (No Observed Adverse Effect Level), a parameter used in toxicology, which corresponds to the highest dose or exposure level of a substance or material that produces no noticeable toxic effect. This parameter is derived from long-term animal studies, observations, and experiments. The NOAEL for sodium nitrate, found in the studies for the determination of toxicological parameters, is 500 mg/kg bw/day. It is directly proportional to the ADI and inversely proportional to a safety factor (SF = 100 for $NaNO_3$).

The resulting formula is

$$ADI_{NaNO_3} = {NOAEL}\big/{SF} = 5 mg \, / \, kg \cdot bw \, / \, day$$

Corrected for nitrate molecular weight:

$$ADI_{NO3-} = \frac{ADI_{NaNO3} \times MWN_{NO3-}}{MW_{NaNO3}} = \frac{5\frac{mg}{kg}bw}{84.99\frac{g}{mol}} \times 62\frac{g}{mol} = 3.7 \cdot mg / kg \cdot bw / day$$

At the same manner, for sodium nitrite, the established NOAEL is 10 mg/kg bw/day and the SF is 100. The resulting ADI is 0,1 mg/kg bw/day for $NaNO_2$ which correspond to 0,07 mg/kg bw/day for nitrite ion ($MW_{NaNO2}$= 68.99 g/mol; $MW_{NO2-}$ = 46 g/mol).

In 2002, the FAO/WHO-JECFA Committee confirmed these ADIs and their applicability to all sources of dietary exposure. However, some researchers suggested re-evaluating these values and underlined that some diets, considered healthy, such as DASH (Dietary Approaches to Stop Hypertension), may lead to nitrate intake exceeding the current ADI value by 550 percent; in fact, a single portion of spinach may exceed the ADI for nitrate [22,23].

Other than the ADI and the NOAEL, other important reference parameters for these additives are the maximum permitted levels (MPLs) in different kinds of foodstuffs, that are defined in national and international regulations. Furthermore, official controls have the purpose to verify the compliance of nitrate/nitrite amounts, in authorized foods, with the maximum levels established in the current regulations [24].

The first Regulation in Europe that listed nitrate and nitrite (as potassium and sodium salts) among approved preservatives, and established their MPLs was the No1995/2/EC, amended in 2006 by the Directive No 2006/52/EC. Finally, the Regulation No EC/1333/2008 and its integrations and modifications harmonized the use of food additives in foods in the Community and introduced, through Annex II, the establishment of a Union list of food additives approved for use in foods and their conditions of use. For nitrites and nitrate, MPLs and conditions of use are summed up in Table 13.1 [25–27]. Meat preparations, traditional meat products, and some cheese and fish products are listed with particular specifications and notes. Although the addition of nitrate is permitted only in few dairy products, non-negligible residual amounts may be found in these products due to the presence of additives in animal feed. The same result was detected in fresh meat, in which nitrate is not permitted as a food additive [28].

Certainly, concerning nitrate and nitrites, the most interesting modifications have been made in the Regulations No EC/1129/2011 and EC/601/2014. For example, in the Regulation (EC) No 601/2014 it is specified that member states and/or the meat industry requested and obtained the authorization for novel usages of nitrite as a preservative in certain traditional products at the maximum level of 150 mg/kg, expressed as sodium nitrite (see Table 13.1) [29, 30]. All meat preparations and traditional products that may contain nitrate and nitrites are described in the Regulation No EC/853/2004 [31].

**TABLE 13.1**

Nitrites and Nitrate in Foods as Additives According to the Annex II to Regulation (EC) No 1333/2008-1129/2011–601/2014

| Food category name | Subcategory | Nitrate (E 251-E 252) (MPL mg/kg) | Nitrites (E 249-E 250) (MPLs mg/kg) | Notes |
|---|---|---|---|---|
| Dairy products | Ripened cheese | 150 | - | Only hard, semi-hard and semi-soft cheese, cheese milk. |
| | Whey cheese | | | In the cheese milk or equivalent level if added after removal of whey and addition of water |
| | Dairy analogues, including beverage whiteners | | | |
| Meat products | Non-heat-treated processed meat | 150 | 150 | |
| | Heat-treated processed meat | - | 100 - 150 | Nitrate may be present in some heat-treated meat products resulting from natural conversion of nitrites to nitrate in a low-acid environment |
| | Meat preparations as defined by Regulation (EC) No 853/2004 | - | 150 | Maximum amount is expressed as Sodium nitrite |
| | Traditional immersion cured products (Wiltshire bacon and ham, and similar products) | 250 | 50 -175 | Meat products cured by immersion in a curing solution containing nitrites and/or nitrate, salt and other components |
| | Traditional immersion cured products | 300 | 150 | |
| | Traditional immersion cured products (cured tongue) | 10 | 50 | Immersion cured for at least 4 days and pre-cooked |
| | Traditional dry cured products (bacon, ham, and similar products) | 250 | 50 - 175 | Dry curing process involves dry application of curing mixture containing nitrites and/or nitrate, salt and other components to the surface of the meat followed by a period of stabilization/maturation |
| | Other traditionally cured products (jellied veal and brisket) | 10 | 50 | Immersion and dry cured processes used in combination or where nitrite and/or nitrate is |
| | Other traditionally cured products (salami and kantwurst) | 300 | - | included in a compound product or where the curing solution is injected into the product prior to cooking |
| | Other traditionally cured products | 250 | - | |
| | Other traditionally cured products | - | 180 | |
| Fish | Processed fish and fishery products including molluscs and crustaceans | 500 | - | Only pickled herring and sprat |

Moreover, in these Regulations, two ways of defining the legal limits for the addition of nitrate are outlined: the added amount and the residual amount. During manufacturing, especially in particular matrices such as traditionally cured meat products, the maximum added levels may refer to the maximum amount that may be added during manufacturing. On the contrary, the maximum residual levels refer to the limits at the end of the production process and on the market [32].

This precise distinction is confirmed by many types of research that showed the initially added amount of nitrite/nitrate dramatically depleted from the moment of addition to the point of consumption. In fact, only 10–20 percent of the added nitrite/nitrate remains at the end of the manufacturing process, and the resulting residual levels continue to decline during the storage period, in consequence of a series of enzymatic and non-enzymatic reactions. In most meat products, there is no simple and direct relationship between the initial amount and the residual amount of $NO_3^-/NO_2^-$, which sometimes reaches non-detectable levels [33].

The rate of nitrate/nitrite loss depends on many factors, such as the temperature used during processing and storing, the composition and pH of the product, the addition of ascorbic acid or other reducing agents, and the presence of microorganisms [34]. Moreover, it was suggested that a relevant improvement in manufacturing practice also contributed to these data. For example, in bacon-curing practice, nitrite is used prevalently alone, however, nitrate is often detected. So, note 59 in Regulation (EC) No 1129/2011 specifies "Nitrates may be present in some heat-treated meat products resulting from the natural conversion of nitrites to nitrates in a low-acid environment."

Special attention is also given to the purity criteria of sodium and potassium nitrate/nitrites salts as raw materials. These criteria are defined in the Commission Regulation (EU) No 231/2012 [34].

Often, in order to avoid mis-dosing, tablets, powders, or premixed solutions are recommended: these mixes contain a precise percentage of nitrate and/or nitrite, NaCl, and may contain antioxidants, spices, and other adjuvants.

In the United States the Code of Federal Regulations (CFR) – Title 21, Part 172 – lists all additives permitted for direct addition to food for human consumption, replacing the previous epithet, "Everything Added to Food in the United States" (EAFUS) [35].

For sodium nitrate, a specific paragraph (n. 170) is dedicated, with the following prescriptions:

As a preservative and color fixative, with or without sodium nitrite, in smoked, cured sablefish, smoked, cured salmon, and smoked, cured shad, so that the level of sodium nitrate does not exceed 500 parts per million and the level of sodium nitrite does not exceed 200 parts per million in the finished product.

As a preservative and color fixative, with or without sodium nitrite, in meat-curing preparations for the home curing of meat and meat products (including poultry and wild game), with directions for use which limit the amount of sodium nitrate to not more than 500 parts per million in the finished meat product and the amount of sodium nitrite to not more than 200 parts per million in the finished meat product.

Particular applications for sodium nitrite include color fixation in smoked cured tuna fish products, with an MPLs of 10 mg/kg in the finished product, and inhibition of the

## TABLE 13.2
## Products, Procedures and Concentrations for Nitrite/Nitrate Adding Following USA-CFR-Title 21 and Title 9

### Part A - CFR – Title 21

| Additive | Matrix | Purpose | MPLs (in finished product) mg/kg |
|---|---|---|---|
| Sodium nitrate | meat-curing preparations | preservative and colour fixative | 500 |
| Sodium nitrite | | | 200 |
| Sodium nitrate | smoked, cured sablefish, salmon, and shad | | 500 |
| Sodium nitrite | | | 200 |
| Sodium nitrite | smoked cured tunafish products | color fixative | 10 |
| Sodium nitrite | smoked chub | preservative | 1000 - 200 |
| Potassium nitrate | cod roe | curing agent | 200 |

### Part B - CFR – Title 9 operations for nitrite and nitrate

| Class of substance | Substance | Purpose | Products | Amount |
|---|---|---|---|---|
| Curing Agents | Sodium or potassium nitrate | Source of nitrite | Cured meat products other than bacon. Nitrate may not be used in baby, junior, and toddler foods. Cured, comminuted poultry products | 7 lb to 100-gal pickle; 3 1/2 oz to 100 lb meat or poultry product (dry cure); 2 3/4 oz to 100 lb chopped meat or poultry. |
| | Sodium or potassium nitrite (supplies of sodium nitrite and potassium nitrite and mixtures containing them must be kept | To fix color | Cured meat and poultry products. Nitrites may not be used in baby, junior, or toddler foods | 2 lb to 100-gal pickle at 10 percent pump level; 1 oz to 100 lb meat or poultry product (dry cure); 1/4 oz to 100 lb chopped meat, meat by product or poultry product. The use of nitrites, nitrate or combination shall |

**TABLE 13.2  (Continued)**
**Products, Procedures and Concentrations for Nitrite/Nitrate Adding Following USA-CFR-Title 21 and Title 9**

**Part B - CFR – Title 9 operations for nitrite and nitrate**

| Class of substance | Substance | Purpose | Products | Amount |
|---|---|---|---|---|
| | under the care of a responsible employee of the establishment. | | | not result in more than 200 ppm of nitrite, calculated as sodium nitrite in finished product, except that nitrites may be used in bacon. |
| | Sodium or potassium nitrite | To curing | Bacon | In pumped bacon, immersion cured bacon 120 and 148 ppm for $NaNO_2$ e $KNO_2$ respectively, while in dry cured bacon 200 and 246 ppm for $NaNO_2$ e $KNO_2$ respectively |

outgrowth and toxin formation from *Clostridium botulinum* type E in the commercial processing of smoked chub, with a concentration range of 100–200 mg/kg in the finished loin muscle.

Another use of potassium nitrate may be as a curing agent in the processing of cod roe, with the MPLs of 200 mg/kg.

Moreover, in CFR – Title 9 – Part 424, all operations for food preparation and processing are explained, including specifications of nitrite/nitrate amount in cured meat. In the last part, nitrate are inserted also in "Prohibited uses", in fact, it is specified that "Nitrates shall not be used in curing bacon."

On the contrary, nitrites are permitted in bacon (pumped bacon, immersion cured bacon) following precise recommendations: "sodium nitrite shall be used at 120 parts per million (ppm) ingoing or an equivalent amount of potassium nitrite shall be used (148 ppm ingoing)," while in dry-cured bacon "Sodium nitrite shall not exceed 200 ppm ingoing or an equivalent amount of potassium nitrite (246 ppm ingoing)" [36]. The concomitant use of antioxidants, such as sodium ascorbate or sodium erythorbate is prescribed. Table 13.2 – part A shows up all admitted uses of nitrites and nitrate and the specific products in CFR Title 21, instead of in part B the operations and processing practices (CFR – Title 9) are described.

**TABLE 13.3**
**MPLs for Nitrate and Nitrite in Australia and New Zealand**

FSANZ Code 1.3.1

| Food category name | Nitrates (potassium and sodium salts) (MPL mg/kg) | Nitrites (potassium and sodium salts) (MPLs mg/kg) |
|---|---|---|
| Cheese and cheese products | 50 | - |
| Commercially sterile canned cured meat | - | 50 |
| Dried meat | - | 125 |
| Slow dried cured meat | 500 | 125 |
| Processed comminuted meat, poultry and game products | - | 125 |
| Fermented, uncooked processed comminuted meat products | 500 | - |

In Australia, the local authority, Food Standards Australia New Zealand (FSANZ), Code 1.3.1 lists nitrates and nitrites as permitted additives in cheese and cured meat products. Nitrate is permitted in slow-dried cured meats and fermented uncooked processed comminuted meat products, to a maximum level of 500 mg/kg, and to cheese and cheese products at a maximum level of 50 mg/kg. Table 13.3 lists the products and $NO_2^-/NO_3^-$ limits [37].

In China, in the current national code GB 2760-2014, National Food Safety Standard Food Additive Usage Standard, sodium nitrate and potassium nitrate are used as color fixative and preservative at the maximum dosage of 500 mg/kg. The maximum residual levels shall be calculated as sodium nitrite and must be less than 30 mg/kg. The food list, in which the addition of nitrates is permitted, is shown in Table 13.4 [38].

# 13.3  REACTIVITY AND MECHANISM

$NO_3^-$ is the conjugate base of nitric acid and represents the most oxidized form of nitrogen (o.n.= +5). As a conjugate of a strong mineral acid, it is a very stable anion and slowly reacts with other compounds, depending on the chemical reaction environment. On the contrary, nitrite is partially reduced (o.n. = +3), and, as an intermediate molecule, it is more reactive. As the result, in meat processing – especially in dry sausages and dry-cured hams – although they have the same function, nitrate is often used as a "slow curing agent," instead of nitrite as a "fast curing agent" [39].

For this reason, several researchers affirmed that nitrate is an inactive/inert compound that only serves as a reservoir of nitrite. This statement is only partially true, although nitrate itself has a low and slow activity (bacteriostatic properties), it can be affirmed that nitrate is a sort of "prodrug", that, when intentionally added to foods, is

**TABLE 13.4**
**Nitrate Usage Limits and Products in China**

GB 2760-2014 National Food Safety Standard Food Additive Usage Standard

| Function | Food Category | Maximum dosage/(mg/kg) | Notes |
|---|---|---|---|
| Color fixative Preservative | Cured meat products (e.g. brined meat, preserved pork, preserved duck, Chinese-style ham, Chinese sausage) Thick gravy cooked meat product Smoked, baked or grilled meat Fried meat product Western ham (fire-cure, smudging, stewing ham) Sausages Fermented meat product | 500 | Measured as per sodium nitrite (potassium nitrite), residual amount ≤30mg/kg |

mainly reduced to nitrite by the action of naturally present bacteria or added bacteria with a nitrate reductase activity (bioactivation) [40].

The nitrate-reducing bacteria, so-called "starter cultures" commonly added to meat and dairy products during processing, are *Staphylococci, Micrococci, Streptococci,* and *Lactobacilli.*

Besides, several researchers suggested that nitrate-reduction reaction is not exclusively performed by microorganism enzymes, as a residual nitrate reductase activity in meat was described, as well as the chemical reduction by $Fe_2^+$ and $Cu^+$ [41].

In this context, it should not be forgotten that several pieces of evidence reported that in cheese milk and in fish products, nitrate directly inhibits the growth of some microorganisms, for example, *Clostridium tyrobutyricum* [19].

Although nitrite is considered more active, the experiments aimed at clarifying the precise mechanism of action showed that also $NO_2^-$ needs to be converted in other intermediates: in an acid medium, in its undissociated form (nitrous acid, $HNO_2$, o.n.=+3) and in the reduced form (nitrogen oxide, NO, o.n.=+2). An empirically determined reactivity scale (increasing order) for nitrate-derived-compounds may be:

$$Nitrate\ (NO_3^-) < Nitrite\ (NO_2^-) < Nitrous\ acid\ (HNO_2)$$
$$< nitrogen\ oxide\ (NO) < peroxynitrite\ (ONOO^-)$$

The last radical is the *peroxynitrite*, a structural isomer of nitrate, which is formed from NO and superoxide, $O_2^-$ inside the phagosome ($O_2^{-\cdot} + NO^{\cdot} \rightarrow ONOO^-$).

The inhibitory activity against food pathogens is attributed to these last three molecules since they act as nitrosating agents, strong oxidants, and nucleophiles. They are also called "High reactive nitrogen species" (HRNS) and may be involved in C-, S-, O-, N- nitrosation reactions. [42]. So, they can react with a high range of matrix components, such as lipids, proteins, nucleic acids, vitamins, cofactors (a sensory enhancing mechanism) and, at the same time, interfere with microbial metabolic pathways, binding macromolecules such as key-enzymes, membrane lipids, and nucleic acids, of microorganisms (antimicrobial mechanism) [43]. This affirmation surely makes more comprehensible the above-discussed losses of nitrite and nitrate during manufacturing and storage that may be defined as "apparent losses." It was confirmed by isotope $^{15}N$ labeling studies on cured meat, that most parts of added nitrite and nitrate are not lost during the processing chain, but remain in the product bound to several macromolecules [44]. For example, a study on the fate of $NaNO_2$ in bacon revealed that approximately 10–15 percent of the added nitrite was found bound to the lipids. Proteins, as well, bound a greater part of nitrate/nitrite added (20–30%). Another 1 percent of their added amount was found bound to R-SH compounds. Finally, a small part reacts with myoglobin and hemoglobin forming stable pigment, the nitrosyl myoglobin, and nitrosyl hemoglobin (mechanism of color enhancing) [45].

Figure 13.1 sums up the fate of nitrate when it is added to food and its biochemical targets.

Definitively, it can be said that reducing conditions are fundamental for nitrate/nitrite activity. Moreover, several factors may contribute to increasing or decreasing their reactivity, such as pH, temperature, co-heat treatments, water activity, and salt concentration, oxygen tension, and matrix constituents. For example, it was observed that the bactericidal activity of added nitrite is more effective at low pH values. In a meat medium, the normal range of pH is mild acid (4.8–6.5), so nitrate's and nitrites' conversion in reactive intermediate compounds is relatively fast. In the same manner, the simultaneous addition of acid compounds is therefore regarded as favorable from a food-preservation point of view [46].

Furthermore, an adequate amount of NaCl is important for its drying action, to control osmotic pressure and water activity, and to increase ionic strength. Furthermore, it is reported that Cl$^-$ can react with NO to form nitrosylchloride (NOCl), also a powerful nitrosating agent [47].

As mentioned above, in food chemistry, three main reasons justify the adding of nitrate and its congeners to several processed foodstuffs; the most important is surely the antimicrobial activity, following by color-stabilizers, and flavor enhancer properties.

Although many studies have attempted to explain biochemical and toxicological mechanisms that justify the pleiotropic functions of these additives, many aspects have yet to be clarified.

## 13.4  ANTIMICROBIAL FUNCTION

Preservatives are seen by consumers as unnatural and harmful; however, the technological preservation practices are ancient and often essential. This is the case of $NO_3^-/NO_2^-$, which have been traditionally and secularly used as an ingredient for food curing, though without knowing their mechanism of action. For processed meat, in particular, this practice has a key role in controlling ambient humidity, meat microbial flora, and decomposition. The mixture used for curing mainly contains salts (NaCl),

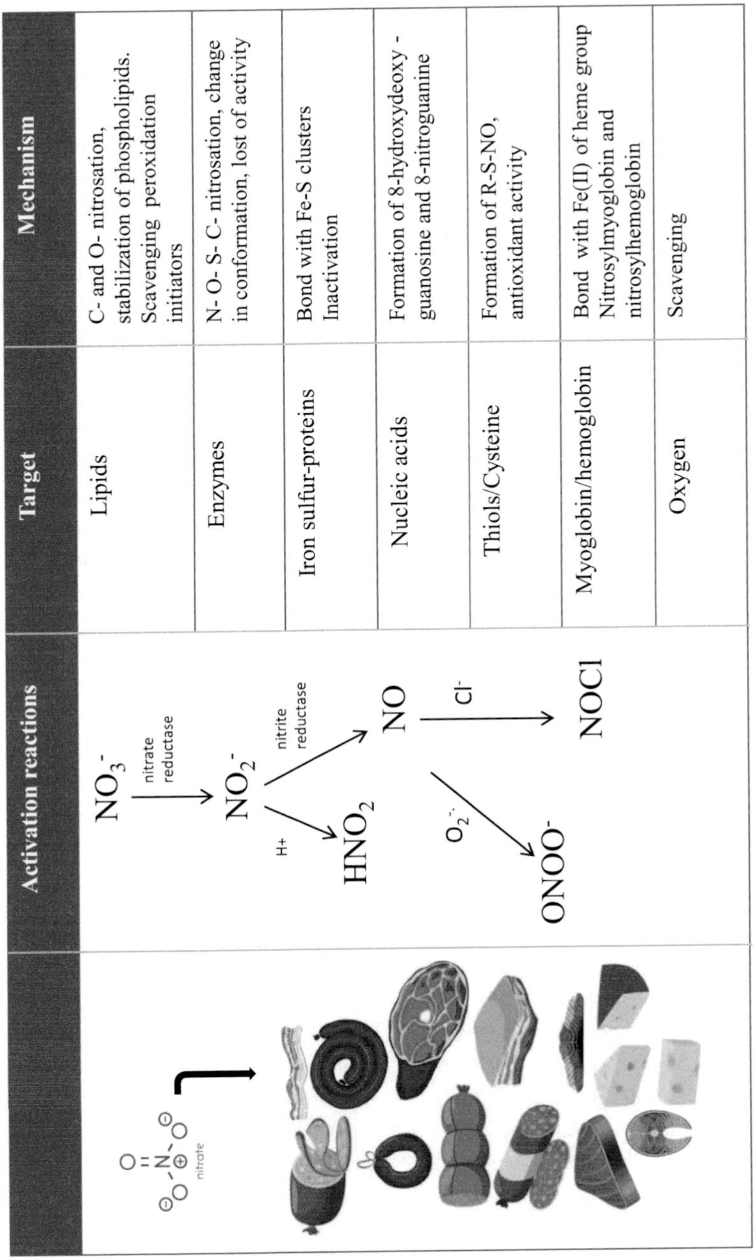

| Activation reactions | Target | Mechanism |
|---|---|---|
| $NO_3^-$ $\xrightarrow{\text{nitrate reductase}}$ $NO_2^-$ | Lipids | C- and O- nitrosation, stabilization of phospholipids. Scavenging peroxidation initiators |
| $NO_2^- \xrightarrow{\text{nitrite reductase}} NO \xrightarrow{Cl^-} NOCl$  $NO_2^- \xrightarrow{H^+} HNO_2$  $NO \xrightarrow{O_2^{\cdot-}} ONOO^-$ | Enzymes | N- O- S- C- nitrosation, change in conformation, lost of activity |
| | Iron sulfur-proteins | Bond with Fe-S clusters Inactivation |
| | Nucleic acids | Formation of 8-hydroxydeoxy - guanosine and 8-nitroguanine |
| | Thiols/Cysteine | Formation of R-S-NO, antioxidant activity |
| | Myoglobin/hemoglobin | Bond with Fe(II) of heme group Nitrosylmyoglobin and nitrosylhemoglobin |
| | Oxygen | Scavenging |

$$O = N \begin{smallmatrix} \ominus \\ O \\ \oplus \\ O^\ominus \end{smallmatrix}$$
nitrate

**FIGURE 13.1** Nitrate bioactivation, HRNS formation and its biochemical targets.

even up to 40 percent, nitrate and/or nitrite, and it may contain sugars and other adjuvants or additives (e.g. antioxidants). The adding of salt is due to the osmotic effect that partially inhibits the microbial growth (indirect inhibition), with a consequent decrease of initially added $NO_3^-/NO_2^-$ amount. Temperature and heat treatments also have a key role [48].

The nitrate and nitrite antimicrobial activity, first described in 1920, is dose-dependant and may be microbicidal, leading to the death of target microorganisms, or microbiostatic, by preventing their growth. Specifically, although most parts of studies regarding the antibacterial function of these preservatives have been focused on the *Clostridium* genus (*Clostridium botulinum* – *Clostridium perfringens* – *Clostridium sporogenes*), nitrates and nitrites show a broad spectrum of activity against both Gram-positive and Gram-negative bacteria [49].In fermented sausages cured with nitrite, complete inhibition of *Escherichia* coli and other *Enterobacteraceae* (*Salmonella*), as well as *Listeriaceae* (*L. monocytogenes*) and *Bacillaceae* (*Bacillus cereus*) was shown [50].

Concerning *Clostridium botulinum*, a marked antibotulinal activity of nitrite against both vegetative cells as well as spores of the pathogen, was shown at a minimal concentration of 50 mg/kg. In dairy products, the inhibition of *Clostridium tyrobutyricum* was investigated, since it may interfere in cheese and milk fermentation by producing small molecules (acetic acid), contributing to excessive cheese blowing; nitrate inhibits this bacterium at a concentration of 150 mg/kg (added amount in cheese milk) [51].

The first complex mechanism of action for nitrite and nitrate in heat-treated meat products was proposed by Perigo in 1960. He observed that the inhibition activity of nitrite against *C. botulinum* was more effective, due to the formation of several compounds called "Perigo-type factor": iron-sulfur-bridged complexes, derived from the reaction of NO with iron (II) salts and nitrosyl/sulfhydryl groups (substances present in a mixture called Roussin black salt) [52].

Unfortunately, it was shown that these factors were only produced during severe heating treatments, so the proposed mechanism did not justify all nitrate and nitrite properties. Nevertheless, this attempt was useful to understand one of the biochemical interference mechanisms of these additives. Macromolecules containing iron and sulfhydryl groups, such as the protein ferredoxin, were put under examination, since they are the most common FeS-cluster proteins in nature and are essential for electron transport and energy production in bacteria, for example during the glycolysis step of conversion pyruvate/acetate, source of ATP.

To support this hypothesis, the accumulation of pyruvate after nitrite treatment and depletion of total ATP was observed in Clostridia bacteria. So, it was suggested that NO and $HNO_2$, bound and inactivated the iron-sulfur proteins (ferredoxin, ferrochelatase, NADH dehydrogenase, hydrogenases, coenzyme Q – cytochrome C reductase, succinate – coenzyme Q reductase, and nitrogenase) [53].

Recently, many studies have demonstrated that the intermediate nitrate and nitrite products, the HRNS, directly inhibit key enzymes involved in the microorganism life cycle, such as microbial dehydrogenase and catalase system, causing oxidative and nitrative stress. Moreover, the inhibition of glutathione peroxidase was observed, which leads to the accumulation of reactive oxygen species, ROS (OH-, $H_2O_2$, OOH-). Also, zinc-finger proteins, protein thiols, membrane, lipids, and iron-sulfur proteins may be oxidized. Moreover, ONOO⁻ may also induce direct damage of DNA, promoting mutations of DNA bases, including the formation of 8-hydroxydeoxyguanosine and 8-nitroguanine [54].

These mechanisms are multiple and complex, and involve several interaction factors, so, they need to be understood more in-depth.

## 13.5  FLAVORING FUNCTION AND ANTIOXIDANT PROPERTIES

The chemical changes responsible for the unique and characteristic flavor of nitrite/nitrate cured meat products are not entirely understood. Surely, flavor-enhancing properties and antioxidant effects are closely related. $NO_2$⁻ e $NO_3$⁻ are involved in the chemistry of lipid oxidation and warmed-over flavor (WOF) in cured meat processes that are also linked, since the oxidative decomposition of lipids results in the development of off-odor and off-flavor compounds (short-chain carbonyl compounds). This is considered to be a major reason for the deterioration of the quality in meat and poultry products.

After being reduced to NO, they act as a suppressor of the rancidity process by stabilizing lipids components of the membranes and inhibiting natural pro-oxidants normally present in meat muscle. Moreover, nitrite scavenges oxygen and binds iron, which are considered the initiators of lipid oxidation.

However, several researchers underlined that these mechanisms do not justify the sensory properties conferred by nitrite/nitrate in cured meat. Other antioxidant additives do not show the same effect. As a result, many sensory studies attempted to identify the compounds involved in this process. For example, during lipid auto-oxidation many volatile compounds, such as aldehydes, alcohols, furans, pyrazines, hydrocarbons are produced, so, experiments of comparison of sensory profiles of cured and non-cured products were carried out. Through Gas Chromatography (GC) profile studies, researchers observed that some volatile compounds, such as the aldehydes n-pentanal, n-hexanal, and malonaldehyde, which are produced during the oxidative rancidity process, are absent in cured meat. In cured ham, 135 volatile compounds were identified, and it was ascertained that a low concentration of $NO_2$⁻/$NO_3$⁻ is sufficient to induce flavor differences [55].

Finally, flavor-producing compounds produced by reactions with NO, such as S-nitrosothiols (R-S-NO) and S-nitrosoaminoacids (S-nitrosocysteine), which exhibit also antioxidant properties, have been identified [56].

## 13.6  COLOURING STABILIZER AND ENHANCER

The appearance of food greatly influences consumers' perceptions and acceptance, since visual characteristics are directly associated with other food-quality attributes. In meat products treated with nitrite and nitrate, the development and fixation of the distinctive red-pinkish color have been demonstrated as the result of chemical interaction between these additives and the protein that gives to meat its color, myoglobin.

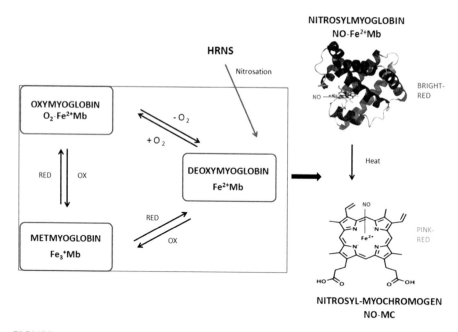

**FIGURE 13.2**   NO, $HNO_2$, $ONOO^-$ react with the reduced form of myoglobin (deoxymyoglobin) forming nitrosylmyoglobin and nitrosylmyochromogen.

Myoglobin is a globular protein, consisting of a single polypeptide chain (153 aminoacids), that bind oxygen, thanks to a prosthetic group, the heme, consisting of a protoporphyrin ring and a central iron (Fe) atom. The Fe ion in oxidation state +2 (ferrous) has the function to coordinate $O_2$, while the $Fe^{3+}$ (ferric) cannot bind oxygen. In the absence of ligands, myoglobin is also called deoxyimoglobin or reduced myoglobin ($Fe^{2+}Mb$), while when myoglobin binds $O_2$ is called oxymyoglobin ($O_2$-$Fe^{2+}Mb$), a bright-red pigment characteristic of fresh meat. Finally, a third form is the oxidized myoglobin or metmyoglobin ($Fe_3^+Mb$), which gives the meat a brown color.

Nitrite and nitrate, after being reduced to NO, are able to bind the ferrous ion of heme group of $Fe^{2+}Mb$, producing a stable red-colored adduct called nitrosylmyoglobin ($NO$-$Fe^{2+}Mb$). In certain severe conditions, such as during heat-curing treatment or at low pH, it can be denatured, forming another adduct, nitroso-myochromogen (NO-MC), a pink pigment typical of heat-treated meat products, which consists in NO-porphyrin ring system alone. In spectral studies of these pigments, it was shown that $NO$-$Fe^{2+}Mb$ has the same $\lambda_{max}$ of $O_2$-$Fe_2^+Mb$ ($\lambda_{max}$= 540nm) [57]. In Figure 13.2 are represented the conversion pathway of myoglobin and the complexes (NO-$Fe^{2+}Mb$; NO-MC) formed after the reaction with HRNS.

The residual hemoglobin remaining in the meat is also converted to NO-haemoglobin (NO-Hb) in the presence of these additives. It is generally accepted that a very little amount of nitrite/nitrate is required to induce a cured color stability in cured meat products (10–15 mg/kg) [58].

## ACKNOWLEDGMENTS

This work was supported by the Italian Ministry of Health, who funded the Project Code GR-2013-02358862.

## REFERENCES

[1] Codex Alimentarius, General Standard for Food Additives (GSFA) – Codex Standard 192-1995, Adopted in 1995, Revision 1997, 1999, 2001, 2003, 2004, 2005, 2006, 2007, 2008, 2009, 2010, 2011, 2012, 2013, 2014, 2015, 2016, 2017, 2018, 2019.

[2] Carocho, M., Barreiro, M.F., Morales, P. and Ferreira, I.C. (2014) Adding molecules to food, pros and cons: A review on synthetic and natural food additives. *Comprehensive Reviews in Food Science and Food Safety*, 13: 377–399.

[3] D'Amore, T., Di Taranto, A., Vita, V., Berardi, G., and Iammarino, M. (2019) Development and validation of an analytical method for nitrite and nitrate determination in meat products by capillary ion chromatography (CIC). *Food Analytical Methods*, 1–10.

[4] Bahadoran, Z. and Mirmiran, P. (2016) Nitrate and nitrite content of vegetables, fruits, grains, legumes, dairy products, meats and processed meats. *Journal of Food Composition and Analysis*, 51, 93–105.

[5] Parvizishad, M., Dalvand A., Mahvi, A.H., and Goodarzi, F. (2017) A review of adverse effects and benefits of nitrate and nitrite in drinking water and food on human health. *Health Scope*. e14164.

[6] Gangolli, S.D., Van den Brandt, P.A., Feron, V.J., Janzowsky, C., Koeman, J.H., Speijers, G.J., Spiegelhalder, B., Walker, R., and Wisnok, J.S. (1994) Nitrate, nitrite and N-nitroso compounds. *Environ. Toxicol. Pharmacol.* 292(1) 1–38.

[7] Iammarino, M., Mangiacotti, M., and Chiaravalle, A.E. (2019) Anion exchange polymeric sorbent coupled to high-performance liquid chromatography with UV diode array detection for the determination of ten N-nitrosamines in meat products: A validated approach. *Int. J. Food Sci. Tech. (in press).* DOI:10.1111/ijfs.14410.

[8] International Agency for Research on Cancer – IARC Working Group (2015) Consumption of red meat and processed meat. *IARC Monogr Eval Carcinog Risks Hum,* 114.

[9] Keeton, J.T. (2011) History of nitrite and nitrate in food. In: *Nitrite and Nitrate in Human Health and Disease*, edited by Nathan Bryan and Joseph Loscalzo, 69–84.

[10] Norman, G., and Conley, M.N. (2017) Regulation of dietary nitrate and nitrite: Balancing essential physiological roles with potential health risks. In: N.S. Bryan and J. Loscalzo (Eds), *Nitrite and Nitrate in Human Health and Disease*, Humana Press, Cham, pp. 153–162.

[11] Hyoung, S. Lee (2018) Exposure estimates of nitrite and nitrate from consumption of cured meat products by the U.S. population, *Food Additives & Contaminants: Part A*, 35:1, 29–39.

[12] World Health Organization (2011) Nitrate and nitrite in drinking water. Background document for development of WHO *Guidelines for Drinking-water Quality*.

[13] Bedale, W., Sindelar, J.J., and Milkowski, A.L. (2016) Dietary nitrate and nitrite: Benefits, risks, and evolving perceptions. *Meat Sci.* 120: 85–92.

[14] Habermeyer, M., Roth, A., Guth, S., Diel, P., Engel, K., Epe, B., Fürst, P., Heinz, V., Humpf, H., Joost, H., Knorr, D., de Kok, T., Kulling, S., Lampen, A., Marko, D., Rechkemmer, G., Rietjens, I., Stadler, R.H., Vieths, S., Vogel, R., Steinberg, P., and Eisenbrand, G. (2015) Nitrate and nitrite in the diet: How to assess their benefit and risk for human health. *Mol. Nutr. Food Res.* 59: 106–128.

[15]  Iammarino M., Di Taranto A., and Cristino, M. (2013) Endogenous levels of nitrites and nitrates in wide consumption foodstuffs: Results of five years of official controls and monitoring. *Food Chem.* 140(4) 763–771.

[16]  European Union (2008) Regulation (EC) No 1331/2008 of the European Parliament and of the Council of 16 December 2008 establishing a common authorisation procedure for food additives, food enzymes and food flavourings.

[17]  European Union (2010) Commission Regulation (EU) No 257/2010 of 25 March 2010 setting up a program for the re-evaluation of approved food additives in accordance with Regulation (EC) No 1333/2008 of the European Parliament and of the Council on food additives.

[18]  EFSA (2017) Panel on food additives and nutrient sources added to food: Re-evaluation of potassium nitrite (E 249) and sodium nitrite (E 250) as food additives. *EFSA Journal,* 15(6) 4786.

[19]  EFSA (2017) Panel on food additives and nutrient sources added to food re-evaluation of sodium nitrate (E 251) and potassium nitrate (E 252) as food additives. *EFSA Journal,* 15(6): 4787.

[20]  Keeton, J., Osburn, W., Hardin, M., Longnecker, M., and Bryan, N. (2009) *A national survey of the nitrite/nitrate concentrations in cured meat products and non-meat foods available at retail.* American Meat Institute Foundations.

[21]  Menard, C., Heraud, F., Volatier, J.L., and Leblanc J.C. (2008) Assessment of dietary exposure of nitrate and nitrite in France. *Food Additives and Contaminants,* 25(8): 971–988.

[22]  Hord, G.N., Tang, Y., and Bryan, N.S. (2009) Food sources of nitrates and nitrites: the physiologic context for potential benefits. *The American Journal of Clinical Nutrition,* 90(1): 1–10.

[23]  Pagliano, E., Meija, J., Campanella, B., Onor, M., Iammarino, M., D'Amore, T., Berardi, G., D'Imperio, M., Parente, A., Mihai, O., and Mester, Z. (2019) Certification of nitrate in spinach powder reference material SPIN-1 by high-precision isotope dilution GC-MS. *Analytical and Bioanalytical Chemistry,* 90(1): 1–11.

[24]  Gorenjak, A.H. (2013) Nitrate in vegetables and their impact on human health. A review. *Acta Alimentaria,* 42 (2): 158–172.

[25]  European Union (1995) European Parliament and Council Directive No 95/2/EC of 20 February 1995 on food additives other than colors and sweeteners.

[26]  European Union (2006) Directive 2006/52/EC of the European Parliament and of the Council of 5 July 2006 amending Directive 95/2/EC on food additives other than colors and sweeteners and Directive 94/35/EC on sweeteners for use in foodstuffs.

[27]  European Union (2008) Regulation (EC) No 1333/2008 of the European Parliament and of the Council of 16 December 2008 on food additives (Text with EEA relevance).

[28]  Iammarino, M. and Di Taranto, A. (2014) Nitrogen fertilizers and nitrite-nitrate accumulation in foodstuffs: A survey. In *Nitrogen Fertilizer: Agricultural Uses, Management Practices and Environmental Effects.* Hauppauge, NY: Nova Science: 1–29.

[29]  European Union (2011) Commission Regulation (EU) No 1129/2011 amending Annex II to Regulation (EC) No 1333/2008 of the European Parliament and of the Council by establishing a Union list of food additives.

[30]  European Union (2014) Commission Regulation (EU) No 601/2014 of 4 June 2014 amending Annex II to Regulation (EC) No 1333/2008 of the European Parliament and of the Council as regards the food categories of meat and the use of certain food additives in meat preparations.

[31]  European Union (2004) Regulation (EC) No 853/2004 of the European Parliament and of the Council of 29 April 2004 laying down specific hygiene rules for food of animal origin.

[32]  Larsson, K., Darnerud, P.O., Ilbäck, N.G., and Merino, L. (2011) Estimated dietary intake of nitrite and nitrate in Swedish children. *Food Additives and Contaminants*, 28: 659–666.

[33]  Temme, E.H., van devijvere, S., et al. (2011) Average daily nitrate and nitrite intake in the Belgian population older than 15 years. *Food Additives and Contaminants Part A*, 28: 1193–204.

[34]  European Union (2012) Commission Regulation (EU) No 231/2012 of 9 March 2012 laying down specifications for food additives listed in Annexes II and III to Regulation (EC) No 1333/2008 of the European Parliament and of the Council.

[35]  FDA Code of Federal Regulation Title 21.

[36]  FDA Code of Federal Regulation Title 9.

[37]  Australia New Zealand Food Standards Code – Standard 1.3.1 – Food Additives.

[38]  China National Regulation (2014) GB 2760-2014, National Food Safety Standard Food Additive Usage Standard.

[39]  Honikel, K.-O. The use and control of nitrate and nitrite for the processing of meat products *Meat Sci.*, 78 (2008) pp. 68–76.

[40]  D'Amore, T., Di Taranto, A., Berardi, G., Vita, V., and Iammarino, M. (2019) Quantification of nitrate levels in spinach and simultaneous determination of nitrite. In: *Agricultural Research Updates*, Vol. 28. Hauppauge, NY: Nova Science: 107–141.

[41]  Merino, L., Örnemark, U., and Toldrá, F. (2017) Analysis of nitrite and nitrate in foods: overview of chemical, regulatory and analytical aspects. *Adv. Food Nutr. Res.*, 81, pp. 65–107.

[42]  Majou, D., Christieans, S., (2018) Mechanisms of the bactericidal effects of nitrate and nitrite in cured meats, *Meat Science*, 145, 273–284.

[43]  Ziarati, P. (2018) Potential health risks and concerns of high levels of nitrite and nitrate in food sources. *SciFed Pharmaceutics Journal*, 1: 3.

[44]  Iammarino, M., and Di Taranto, A. (2012) Nitrite and nitrate in fresh meats: A contribution to the estimation of admissible maximum limits to introduce in directive 95/2/EC. *Int. J. Food Sci. Tech.* 47(9) 1852–1858.

[45]  Woolford, G., and Cassens, R.G. (1977) The fate of sodium nitrite in bacon. *Journal of Food Science.* 42: 586–589.

[46]  Binkerd, E.F., and Kolari, O.E. (1975) The history and use of nitrate and nitrite in the curing of meat, *Food and Cosmetics Toxicology.* 13 (6) 655–661.

[47]  Suresh, G., Xiong, W., Rouissi, T., and Brar, S. (2018) *Nitrates*. Encyclopedia of Food Chemistry, Academic Press, pp. 196–201.

[48]  Govari, M., and Pexara, A. (2015) Nitrates and nitrites in meat products. *Journal of the Hellenic Veterinary Medical Society*, 66, 127–140.

[49]  Roberts, T.A. (1975) The microbiological role of nitrite and nitrate. *J. Sci. Food Agric.*, 26: 1755–1760.

[50]  Scotter, M.J., and Castle, L. (2004) Chemical interactions between additives in foodstuffs: A review. *Food Additives & Contaminants*, 21: 2, 93–124.

[51]  Jung, H., LaDonia, A., Govind, K. and Brou, K. (2017) Curing properties of sodium nitrite in restructured goat meat (chevon) jerky, *International Journal of Food Properties*, 20: 3, 526–537.

[52]  Wedzicha, B.L. (2003) Preservatives – Classifications and Properties. In: *Encyclopedia of Food Sciences and Nutrition* (Second Edition), pp. 4773–4776.

[53]   Kanner, J. (1994) Oxidative processes in meat and meat products: quality implication. *Meat Sci.* 36 673–676.

[54]   Gray, J.I., MacDonald, B., Pearson, A.M., and Morton, I.D. (1981) Role of nitrite in cured meat flavor: *A Review. Journal of Food Protection.* 44. 302–312. 10.4315/ 0362-028X-44.4.302.

[55]   Macdougall, D.B., Mottram, D.S., and Rhodes, D.N. (1975) Contribution of nitrite and nitrate to the colour and flavour of cured meats. *J. Sci. Food Agric.*, 26: 1743–1754.

[56]   Hammes, W.P. (2012) Metabolism of nitrate in fermented meats: The characteristic feature of a specific group of fermented foods, *Food Microbiology*, 29, 2, 151–156.

[57]   Usher, C.D., and Telling, G.M. (1975) Analysis of nitrate and nitrite in foodstuffs: A critical review. *J. Sci. Food Agric.*, 26: 1793–1805.

[58]   Dahle, H.K. (1979) Nitrite as a food additive. *NIPH (National Institute of Public Health) Annals.* 2: 17–24.

# Part 5

---

## Nitrate in the Human Body

# 14 Nitrate and Human Health

## An Overview

*Keith R. Martin and Richard J. Bloomer*

## CONTENTS

14.1 General Overview of Nitrate and Human Health............................................304
    14.1.1 Definition of Nitrate ................................................................304
    14.1.2 Natural Environmental Sources.....................................................304
    14.1.3 Use in Soils..........................................................................306
    14.1.4 Potential Harmful Effects of Nitrate.............................................306
    14.1.5 Potential Health Benefits of Nitrate.............................................309
14.2 Structure and Function of Nitrate and Nitrites.......................................310
14.3 Sources of Nitrate ....................................................................312
    14.3.1 Foods Containing Nitrate...........................................................312
    14.3.2 Nitrate as a Food Additive ........................................................313
    14.3.3 Nitrate as a Dietary Supplement...................................................314
        14.3.3.1 Nitrate and Nitrite Salts .................................................316
        14.3.3.2 Beets and Beetroot Juice.................................................316
        14.3.3.3 Red Spinach ............................................................317
14.4 Potential Harm of Nitrate............................................................318
    14.4.1 Historical View and Rationale .....................................................318
    14.4.2 Current View and Rationale........................................................318
    14.4.3 Toxicity from Ingestion ...........................................................319
14.5 Potential Health Benefits of Nitrate .................................................319
    14.5.1 Increased NO.........................................................................319
    14.5.2 Improved Blood Flow: Response to Hypoxia .........................................320
    14.5.3 Reduction in Blood Pressure .......................................................321
    14.5.4 General Cardioprotection ...........................................................322
    14.5.5 Improved Cognition..................................................................324
    14.5.6 Erectile Dysfunction ...............................................................325
    14.5.7 Improvement in Aerobic Exercise Performance........................................325
    14.5.8 Diabetes, Glycemia, and Insulin Resistance.........................................328
14.6 Directions for Future Research on Nitrate Related to Health.......................328
14.7 Conclusion ..........................................................................329
References................................................................................330

DOI: 10.1201/9780429326806-19

## 14.1 GENERAL OVERVIEW OF NITRATE AND HUMAN HEALTH

The organic nitrite, amyl nitrite ($C_5H_{11}ONO$), was introduced in 1867 as a therapeutic agent in the treatment of angina pectoris, which is characterized by inadequate oxygen supply to cardiac muscle (hypoxia) due to stenosis or blockage of arteries characteristic of coronary artery disease (1). In 1879, glyceryl trinitrate, formerly nitroglycerin, was introduced by William Murrell as an organic treatment for angina pectoris (2). Serendipitous observation had shown that a drop on the tongue could immediately initiate headaches characteristic of marked cranial vasodilation and neural impingement of the trigeminal nerve (2). Subsequently, organic nitrate replaced amyl nitrite because of facilitated delivery and extended duration of action as a vasodilator, although marked side effects resulted, including nitrate tolerance (loss of efficacy) and headache (1). Since these initial discoveries, a considerable amount of information has been gleaned and confirmed regarding the myriad health benefits of nitrate largely on the vasculature, but also in numerous other tissues. Moreover, the bioactive agents responsible and likely mechanism and pathways have been elucidated.

### 14.1.1 DEFINITION OF NITRATE

Nitrate has been used as an anti-ischemic pharmacological agent and exogenous donor of nitric oxide (NO) for more than 130 years without the realization, previously, of the mechanism of action or the specific bioactive entity responsible for its effects. Traditionally, the inorganic anions nitrite ($NO_2^-$) and nitrate ($NO_3^-$) have been considered inert end products of NO metabolism and largely undesired residues in the food chain (3, 4). Recent studies, however, show that naturally occurring nitrate and nitrite are physiologically recycled in blood and tissue to form NO as well as myriad other bioactive nitrogen oxides (5, 6). As a result, nitrate seems to function as storage depots for NO-like bioactivity as an adjunct to the endogenous enzymatic pathways (4). The two major sources of NO are the endogenous, enzymatic NO synthase (NOS) pathway requiring L-arginine and oxygen as substrates and the exogenous consumption of nitrate, primarily in dietary vegetables. For each, the formation of NO requires bioactivation of nitrate via chemical reduction to nitrite, which is largely due to commensal bacteria in the oral cavity (7). Estimates indicate there are >300 species of bacteria with nitrite reductase activity in the oral cavity, which has stimulated considerable research involving the oral microbiome (8, 9). Although these tend to be the primary means of conversion, there are numerous other endogenous molecules (both enzymatic and non-enzymatic) that can produce NO, including hemoglobin, myoglobin, xanthine oxidoreductase, ascorbate, and polyphenols (10).

### 14.1.2 NATURAL ENVIRONMENTAL SOURCES

As previously mentioned, nitrogen represents the largest component of atmospheric air and is the fourth most prevalent element in cellular biomass (11). The cycle between inert atmospheric nitrogen and bioactive nitrogen-encompassing reactions in cellular metabolism and increased biomass is largely controlled by microbial activities and, in fact, is critical for the insertion of nitrogen into genetic material, that is, RNA and DNA, and ultimately protein (11). A specific mandatory process is the ability of bacteria to "fix" or capture $N_2$ converting it to ammonium ($NH_4^+$). Subsequently, this can be further converted via oxidation to form numerous nitrogen oxide entities, including nitrite and nitrate (12). The environmental nitrogen cycle (Figure 14.1) continues with the serial reduction of nitrate to NO, nitrous oxide, and ultimately atmospheric nitrogen.

The environmental nitrogen cycle captures nitrogen and distributes to plants where it may be accumulated and consumed by mammals with ultimate increased human plasma levels rendering nitrate bioavailable. Dietary nitrate intake ultimately depends on the type and amount of vegetables consumed, the concentrations of nitrate in the vegetables (including the nitrate content of fertilizer), and the level of nitrate in the water supply (13). Concentrations can vary considerably and occur in high levels in the petiole (leaf, stem stalk) followed by the leaf, stem, root, influorescence (group or cluster of flowers), tuber, bulb, fruit, and seed (14, 15). For example, the average nitrate content of spinach collected from 3 different markets in Delhi, India, varied from 71 to 429.3 mg/100 g fresh weight (16). The relative accumulation of nitrate also depends on factors such as plant genotype, soil quality, growth environment, and storage and transport conditions (16).

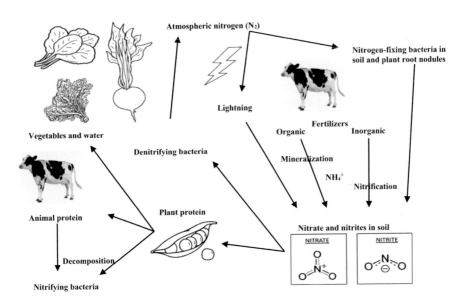

**FIGURE 14.1** Environmental nitrogen cycle. Atmospheric nitrogen is assimilated, or fixed, by symbiotic bacteria associated with plants where it can be transported within the structures of plants, for example, leaves and accumulated. Plants may then be consumed by humans, which increases plasma levels of nitrate. Nitrate can then be reduced to other bioactive nitrogen oxides including the potent vasodilator, nitric oxide.

### 14.1.3  USE IN SOILS

Atmospheric nitrogen, after capture, is essential for successful plant growth, but artificial fertilization (inorganic and organic) as part of agricultural practices can be more efficient for increasing plant biomass and nitrogen composition (17). Both nitrogenous fertilizers can be used in cropping systems. For the former, many agricultural practices include the application of bovine manure (organic matter) containing many nutrients, for example, phosphorous, calcium, and so forth, as well as nitrogenous compounds. However, after the application of organic fertilizers, inorganic nitrogen is released via mineralization and absorbed by plants as inorganic nitrate, the preferential chemical speciation (18). Interestingly, inorganic nitrogen compounds such as nitrate, nitrite, and ammonium represent <5 percent of the total nitrogen in soil but are the prevalent forms used by plants (19). Other factors influence nitrogen accumulation from fertilizers, including agronomic practices, presence of microorganisms, soil properties, ambient temperature, and water content, which may influence translocation to leaves and ultimately accumulation (20). As such, the amounts of nitrate in farmed vegetables, based on the continuing public concern for dietary nitrate and cancer, have justified governmental limitations on allowable nitrate concentrations in farmed vegetables.

### 14.1.4  POTENTIAL HARMFUL EFFECTS OF NITRATE

There are two primary areas of concern regarding toxicities and/or adverse effects from dietary nitrate, including methemoglobinemia and gastric cancer. Methemoglobinemia is a potentially fatal condition in which hemoglobin is oxidized to methemoglobin (>1% of total hemoglobin) with significant reduction, due to oxidation of ferric iron in oxyhemoglobin, in the ability of iron to bind and transport oxygen, leading to hypoxia and cyanosis, or "Blue Baby" syndrome (21–23). Infants <6 months of age may be especially vulnerable when exposed to nitrate from sources such as well water contaminated with bacteria, which reduces and bioactivates nitrate to nitrite (24). As a result, it is vital that potable well water that may be provided to infants directly or indirectly be tested for nitrate/nitrite concentrations (25). Although the bulk of dietary nitrate is derived from vegetables, and infants fed commercially prepared foods with vegetables are not considered to be at risk for excessive nitrate, with the caveat that home-prepared foods from nitrate-rich vegetables such as red spinach, beetroot, squash, and so forth should be avoided until infants are >3 months of age.

Dietary nitrate consumption presents a conundrum during pregnancy. Recommendations to NO-deficient pregnant women for increased consumption of nitrate, for example, beetroot has increasingly been suggested to mitigate hypertension and pre-eclampsia, improve placental blood flow and markedly improve maternal and neonatal health (26). However, caution has also been recommended since potentially lethal outcomes may result in methemoglobinemia, alteration in embryonic cells and malignant transformation, and thyroid disorders. Epidemiologic evidence suggests an association between nitrate-rich water consumption and spontaneous abortions, intrauterine growth restriction, and various birth defects, although the data are limited (27).

$$NO_2^- + H^+ \longrightarrow HNO_2$$

Nitrite     Proton     Nitrous acid

$$HNO_2 + H^+ \longrightarrow H_2NO_2^+$$

Nitrous     Proton     Nitronium
acid

$$H_2NO_2^+ + NO_2^- \longrightarrow N_2O_3 + H_2O$$

Nitronum     Nitrite     Dinitrogen     Water
trioxide

$$N_2O_3 + (R_2)NH \longrightarrow (R)_2N\text{-}N{=}O + NO_2$$

Dinitrogen     Secondary     Nitrosamine     Nitrogen
trioxide     amine                                    dioxide

**FIGURE 14.2** Formation of nitrosoamines from nitrite. Under proton-rich environments with low pH, nitrite can be reduced to nitrous acid and the nitronium ion. Subsequent reaction with nitrite produces dinitrogen trioxide, which can react with secondary amines to form nitrosoamine.

Although a concern, few nitrate and nitrite exposure studies in humans, including either children or adults as subjects, have resulted in the elaboration of methemoglobinemia causing many to proffer alternative explanations for the etiology (21). For example, in one study infants exposed to 175–700 mg nitrate/day did not display methemoglobin concentrations >7.5 percent, which suggests that nitrate alone did not cause methemoglobinemia (28). In a recent study, healthy adults were provided a bolus dose of sodium nitrite (low, 150 to 190 mg or high, 290–380 mg) (29). Methemoglobin concentrations were 12.2 percent and 4.5 percent for the high and low dose, respectively. The data suggest other factors in the etiology of methemoglobinemia such as gastroenteritis or bacteria-induced NO production as an immune response to infection (30, 31).

The second concern with dietary nitrate involves the capacity of nitrate (Figure 14.2) to form carcinogenic nitrosamines ($R_2N\text{-}NO$) at low pH (pH<3) and low $pO_2$ in the gastric lumen (32, 33). Indeed, in a population-based cohort study, there were increased risks for all-cause mortality due to 9 different dietary sources (causes). All were associated with both processed and unprocessed meats via, in part, heme iron, nitrate and nitrites (34).

Although there are ongoing concerns with dietary nitrate-derived cancer, direct evidence from extensive epidemiologic and animal studies has been equivocal and inadequate (31). That is, neither rodent nor human epidemiologic investigations have clearly shown a direct correlation between dietary nitrite exposure and the risk of cancer (35). The basis in large part for the linkage of nitrate/nitrites and gastrointestinal cancer relies on observations that nitrites can empirically and chemically react with secondary amines, for example, proteins, or *N*-alkylamides (bioactive compounds in plants) to generate carcinogenic *N*-nitroso compounds, that is, nitrosoamines as shown in Figure 14.2 (35). Oddly, however, demonstration that nitrosoamines cause cancer in humans has also been inadequate. It is known that enzymatic activation is needed to produce a direct-acting carcinogen and interruption of this process may

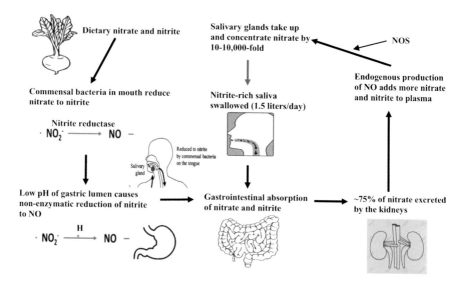

**FIGURE 14.3**    Endogenous salivary cycle illustrating disposition of ingested nitrate. After absorption in the gastrointestinal tract, nitrate enter the bloodstream for systemic circulation. Approximately 75 percent is excreted by the kidneys and 25 percent remains in circulation and enters the enterosalivary cycle where it is reduced to nitrite in the oral cavity by bacteria with nitrite reductase activity. Nitrite-rich saliva is swallowed (~1.5 liters per day) and continues in the cycle.

be a reason for the discrepancies. This has prompted the food industry to include as additives along with nitrate/nitrites, antioxidants that can interrupt this bioactivation process, reducing the potential for nitrosamine formation. Other "green" approaches have been developed substituting other naturally occurring nitrate-rich molecules for the synthetic nitrate.

The occurrence of the enterosalivary cycle known as the nitrate-nitrite-NO pathway, in humans with retention, marked concentration, and conversion of nitrate to nitrite, upon swallowing, can serve as a source for NOS-independent gastric generation of NO and nitrosamines as shown in Figure 14.3 (6, 36, 37).

This occurs after secretion and ionization of hydrochloric acid (HCl), which protonates nitrite in this microenvironment (38, 39). This process can be abolished by proton pump inhibitors such as omeprazole (Prilosec), esomeprasol (Nexium), and so forth, which reduce gastric pH (40). Interestingly, NO levels increase with dietary consumption particularly in the presence of reducing agents such as the antioxidants ascorbic acid and polyphenols, suggesting a detrimental effect (41). However, nitrite in combination with gastric acid is considerably more potent in killing gastric pathogens such as *Helicobacter pylori* than acid alone (42). This was attributed to increased gastric mucosa blood flow and mucus formation protecting the stomach and supports a role for gastric NO as a defense against pathogens (43, 44).

Most salivary nitrite is absorbed systemically, obviating the potential for gastric NO formation. Plasma levels peak around 30 minutes and remain elevated for several

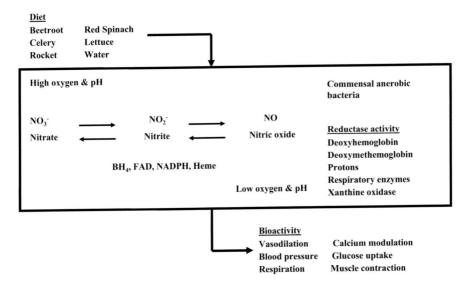

**FIGURE 14.4** Endogenous nitrate-nitite-nitric oxide cycle. Dietary consumption of nitrate-rich vegetables is absorbed and increases plasma nitrate. In the salivary glands, nitrate is reduced to nitrite by commensal bacteria, with further reduction to bioactive nitrogenous compounds, that is, nitric oxide, via other endogenous mechanisms particularly under hypoxic conditions. With oxygen, endogenous enzymatic systems, viz., NOS, nitric oxide may be produced via the oxidation of arginine. (Adapted from reference 139.)

hours, although the half-life is approximately 20–30 minutes, likely due to recirculation of nitrate. In the systemic circulation, myriad proteins, including enzymes, can catalyze one electron reduction of nitrite to NO (Figure 14.4).

Several examples are xanthine oxidoreductase, deoxyhemoglobin, cytochrome P450, mitochondrial proteins, carbonic anhydrase, aldehyde oxidase, and eNOS as well as protons, polyphenols, vitamins E and C (45–47). The mechanism of nitrite reduction ranges from simple protonation to enzymatic activity, and a common observation is enhanced nitrite reduction by the former during hypoxia in the presence of low pH (48). Classical endogenous NOS-dependent NO production requires molecular oxygen and arginine as substrates. Thus, it appears that the nitrate-nitrite-NO pathway functions under homeostasis, whereas nitrite reduction to NO is a backup, or auxiliary, function during periods of ischemia/hypoxia. This redundancy permits NO generation from nitrate as a mechanism of hypoxic vasodilation when oxygen tension falls.

## 14.1.5   POTENTIAL HEALTH BENEFITS OF NITRATE

The concentration of inorganic nitrate in vegetables can serve as a substrate for reduction to nitrite, NO, and other metabolic products ($NO_x$) that produce vasodilation, decrease blood pressure, improve and maintain endothelial function, as well as modulate glucose homeostasis, improve muscle contractility, enhance mitochondrial function, and facilitate respiration. Chronic and acute beetroot juice supplementation has been used as a therapeutic approach for diabetes and insulin hemostasis, renal health, and modulation of the microbiome (49, 50). Thus, collectively these observations have obvious implications for cardiovascular health and exercising muscles, and subsequently athletic performance (51–53). As a result, many experts have suggested increasing vegetable intake, particularly nitrate-rich sources as a heart-healthy strategy and potent ergogenic aid despite the ongoing stigma (54–57).

## 14.2 STRUCTURE AND FUNCTION OF NITRATE AND NITRITES

Nitrate and nitrite are naturally occurring chemical compounds produced from the reaction of diatomic nitrogen and oxygen to form polyatomic anions ($NO_3^-$ and $NO_2^-$, respectively) that can form salts with cations, for example, Na+, K+, to form classes of compounds called nitrates or nitrites, respectively. Both molecules are rich sources of inorganic nitrate (without carbon) in human diets deriving largely from leafy vegetables with differential enrichment depending on the anatomical plant structure. Since almost all nitrates are soluble in water, drinking water is also a significant dietary source. As a result, the WHO has established an upper limit of 10 mg nitrate/L for municipal water supplies (17).

Uptake of nitrate in the salivary glands and subsequent excretion in the saliva is a pivotal, necessary step for conversion of nitrate or nitrite in the oral cavity. Several species of facultative anaerobic bacteria in the crypts of the tongue efficiently convert nitrate to nitrite (58, 59). Due to various intrinsic biological mechanisms, salivary concentrations of both may approach 10–10,000 times that of plasma levels reaching millimolar levels, which may contribute to nitrosamine formation in the low pH environment of the stomach as shown in Figure 14.3 (33, 60–62). As such, there continue to be restrictions on acceptable nitrate levels in farmed vegetables as recommended by the WHO (63).

Several proteins, such as hemoglobin, cytochrome P450 reductase, and cytochrome P450, can catalyze the reduction of nitrite or nitrate to generate NO (10). Interestingly, hemoglobin has enzymatic activity as a nitrite reductase under hypoxic conditions (when hemoglobin is 40–60% saturated) and, moreover, is a sensor and effector of hypoxic vasodilatation (64). Cytochrome P450 reductase also causes NO release by reducing nitrate, which facilitates the production of S-nitrosothiols (65). NO can also be formed via the decomposition of biological molecules to produce reactive products. For example, S-nitrosothiols (R-S-N=O), such as S-nitrosoglutathione and S-nitrosohemoglobin, are also sources of NO when exposed to transition metals, for example, iron, or light (66, 67). S-nitrosothiols can also undergo trans-S-nitrosation with other thiol groups (R-S-H), which can alter cell or protein function of NO via mechanisms analogous to phosphorylation and ubiquitinylation (68, 69).

Endogenous enzymatic production of NO occurs via a family of enzymes called NO synthases (NOS) with endothelial NOS (eNOS) producing ~70 percent in the vascular endothelium (70, 71). This causes vasodilation, blood-pressure regulation, anti-inflammation, and reduced platelet aggregation, which may be pivotal in preventing various types of cardiovascular disease, hypertension, atherosclerosis, and stroke. The two other NO-producing isoforms are neuronal NOS (nNOS) and cytokine-inducible NOS (iNOS). Although eNOS is the major NOS isoform, both nNOS and iNOS exhibit important functional roles in certain tissues and environments (72). For example, vascular injury induces expression of nNOS in smooth muscle cells, and activation of iNOS by proinflammatory cytokines within vascular smooth muscle causes vasodilation in sepsis as part of the innate immune response.

Numerous mechanisms of action have been identified for dietary nitrate. For example, NO induces formation of cyclic guanosine monophosphate (cGMP) by activating soluble guanylyl cyclases in vascular SMC. After this initial step, numerous downstream effectors can be activated such as cGMP-dependent protein kinase (protein kinase G [PKG]), cGMP-gated ion channels, and cGMP sensitive phosphodiesterases (69, 73). PKG induces reuptake of calcium into the sarcoplasmic reticulum (membrane-bound structure in muscle cells that store calcium), extracellular movement of calcium within a cell, and opening of calcium-activated potassium channels. Collectively, these signaling processes cause smooth muscle cell relaxation, vasorelaxation, and improved blood flow.

Although it is becoming clear that NO is critical for homeostasis, another seeming paradox exists. Many argue that free radicals (with an unpaired electron), capable of generating oxidative stress, are causative in the etiology of many, if not most, chronic diseases (74). However, NO is itself a free radical and interacts readily with other free radicals and/or minerals – for example, iron – and can induce nitrosative stress via reactive nitrogen species (75). Moreover, NO can generate noxious, toxic molecules in states of disequilibrium. For example, eNOS may become "uncoupled" producing less NO and more superoxide ($O_2^-$), a reactive oxygen species, which causes endothelial dysfunction found in atherosclerosis, diabetes, hypertension, cigarette smoking, hyperhomocysteinemia, and ischemia/reperfusion injury (65, 76). Major mechanisms of eNOS uncoupling include depletion of its cofactor tetrahydrobiopterin, deficiency of l-arginine, the eNOS substrate, competition with endogenous arginase activity, and/or eNOS S-glutathionylation (77). Uncoupling can lead to the diffusion-controlled reaction of NO with superoxide radical to form highly reactive peroxynitrite (ONOO⁻), which readily kills and damages any molecule within proximity (78). Conversely, cytokine activation of iNOS in the innate immune system significantly increases ONOO- to eliminate infections by bacteria, viruses, or fungi and iNOS-derived NO modulates pathways of glucose and lipid metabolism during inflammation (79).

Nitrite can be bioactivated via numerous non-enzymatic routes (80, 81). NO binds with ferrous iron (a source of electrons) as part of heme – although potentially problematic – and is the basis for many of the signaling activities of NO and a rapid means of modulating NO levels (82). The reduces the half-life of NO to a few seconds with the ultimate formation of nitrate, which can reenter the endogenous nitrogen cycle. NO can also react with Complex I and cytochrome c oxidase (Complex IV), inducing either problematic or protective outcomes, depending on NO, cytochrome, and $O_2$ concentrations (83). NO may also bind to myriad amino acids of diverse proteins to form adducts (two molecules form one) via nitrosation and nitration, which has been shown as a mechanism of altering protein function. Lastly, sulfur-linked nitrosylated proteins can exert NO-like endocrine activity (84, 85). S-nitrosylation involves covalent post-translational modifications of cysteine moieties, which then can propagate NO-triggered signals (86).

## 14.3  SOURCES OF NITRATE

### 14.3.1  FOODS CONTAINING NITRATE

Vegetables accumulate significant amounts of nitrate (~80% of total dietary) from nitrogen-based fertilizers, which are used for rapid and enhanced growth (16). Leafy vegetables, such as lettuce or spinach, and beetroot contain the highest concentrations of nitrate (Table 14.1) (87).

Other examples include radishes, turnips, watercress, bok choy, Chinese cabbage, kohlrabi, chicory leaf, celery, onion, and garlic (7, 88–90). Fruits also contain nitrate,

**TABLE 14.1**
**Nitrate and Nitrite Content of Vegetables**

| Vegetable | Nitrate content (mg/kg) | Nitrite content (mmol/100 g) |
|---|---|---|
| **High** | | |
| Arugula (Rocket) | 2597 | 4.19 |
| Spinach | 2137 | 3.45 |
| Lettuce | 1893 | 3.05 |
| Radish | 1868 | 3.01 |
| Beetroot | 1459 | 2.35 |
| Cabbage | 1388 | 2.24 |
| Average | 1890 | 3.05 |
| **Medium** | | |
| Turnip | 624 | 1.00 |
| Cabbage | 513 | 0.83 |
| Green beans | 496 | 0.80 |
| Leeks | 398 | 0.64 |
| Spring onion | 353 | 0.58 |
| Cucumber | 240 | 0.39 |
| Carrot | 222 | 0.36 |
| Potato | 220 | 0.35 |
| Garlic | 183 | 0.30 |
| Sweet pepper | 117 | 0.19 |
| Green pepper | 111 | 0.18 |
| Average | 316 | 0.51 |
| **Low** | | |
| Onion | 87 | 0.14 |
| Tomato | 69 | 0.11 |
| Average | 78 | 0.13 |
| **Water (250 mL)** | | |
| Tap | 26 | 0.13 |
| Mineral | 3 | 0.01 |

Data were collated from numerous sources. (52, 88, 138).

but at low levels. Examples include watermelon, apples, bananas, grapes, kiwi, pears oranges, and strawberries (91). Interestingly, diets recommended by expert panels that are associated with reduced chronic disease risk tend to be considerably higher in the vegetable intake and include, for example, the Dietary Approaches to Stop Hypertension (DASH) diet and the Mediterranean Diet, with the former estimated to provide >1,200 mg nitrate/d. As a note, current estimates of dietary nitrate consumption in the US ranges from 40–100 mg/d (35).

### 14.3.2 NITRATE AS A FOOD ADDITIVE

Nitrates are routinely added to processed, cured meats as antioxidants, flavor enhancers, color stabilizers (red or pink color in meats) and antimicrobial agents. In fact, they are critical for minimizing or preventing the growth of noxious, disease-causing bacteria such as *Clostridium botulinum*, the causative agent of botulism (92). Examples of processed foods include bacon, bologna, corned beef, hot dogs, ham, luncheon meats, sausages, canned meat, cured meat and hams, all of which are regulated by the FDA and USDA (93). According to the Code of Federal Regulations, the level of sodium nitrite in cured meat products must be $\leq$200 ppm and sodium nitrate must be $\leq$500 ppm. Although generally a public-health concern, food additives are not major contributors to the total estimated intake of nitrate (Table 14.2). Vegetables remain the major contributor, followed by water and other relatively minor sources, including cereals and non-water beverages (94–96).

Nitrate, which can form nitrites, reacts with naturally occurring components of protein, viz., amines, and lead to the formation of nitrosamines, known cancer-causing compounds. Concern over these reactions nearly caused a ban of nitrites and nitrate from food use in the 1970s (97). Paradoxically, vegetables may also naturally contain nitrosamines without artificial addition, although the level is low. An intriguing observation is the perpetuating stigma of food additives (nitrates) as cancer-causing agents, yet individuals frequently consume identical molecules from dietary vegetables. Although cancer risk has been discussed for years, others have recently suggested that nitrate and nitrite be considered essential nutrients because they promote

**TABLE 14.2**
**Nitrate and Nitrite Levels in Processed Foods**

| Product | Nitrate (mg/kg) | Nitrite (mg/kg) |
|---|---|---|
| Sausage (cooked) | 48 | 32 |
| Sausage (smoked) | 56 | 20 |
| Ham | 42 | 24 |
| Salami | 85 | 10 |
| Bacon | 55 | 4 |
| Bacon (nitrite-free) | 30 | 7 |
| Hot dog | 90 | 1 |
| Pork tenderloin | 33 | 0 |
| Cheese | 34 | 10 |

NO production and consequently contribute to cardiovascular health. Nonetheless, this has driven the food industry to seek natural nitrate compounds due to high consumer demand (98). Indeed, alternatives to nitrate and nitrites are the object of many studies, with heightened focus on the addition of vitamins, fruits, chemicals, and natural products containing nitrate (99).

### 14.3.3 NITRATE AS A DIETARY SUPPLEMENT

Inorganic nitrate and nitrite are plant nutrients and legally permitted food additives primarily for processed meats and components of foods. They also may be included in dietary supplements and nutraceuticals since they are associated with blood-pressure–lowering and performance-enhancing effects. It is not surprising that they are immensely popular among consumers with cardiovascular disorders, as well as athletes. Although dietary supplement blends and specific bioactive compounds such as betaine and arginine are generally thought to indirectly produce NO, oddly there is inadequate research to support this assertion (100–103). The most commonly used dietary constituent is nitrate-rich beetroot because of its high nitrate concentration (>250 mg nitrate/100 g [3.5 oz]) coupled with compelling evidence that it does, in fact, increase blood nitrate and/or nitrite concentrations following acute and chronic ingestion as shown in Table 14.3 (104–106).

Others have shown that dietary nitrate supplementation with ~0.1 mmol/kg body mass significantly reduces blood pressure, reduces oxygen cost associated with exercise, improves muscle efficiency, and exerts ergogenic properties (107–110. Valenzuela and colleagues compared dietary supplements, that is, nitrates, and found sufficient evidence for nitrate supporting its acute beneficial effects on muscle strength (111). The mechanism of action is thought to result from enhanced NO bioavailability and subsequent vasodilation leading to improved cardiovascular health and enhanced exercise performance (112).

The results of some clinical trials report that the intake of other nutraceuticals (hawthorn, coenzyme Q10, l-carnitine, d-ribose, carnosine, vitamin D, probiotics, n-3 PUFA, and beet nitrates) might improve self-perceived quality of life and/or functional parameters such as blood flow (left ventricular ejection fraction, stroke volume and cardiac output) in heart failure (HF) patients, with minimal or no side effects (113, 114). This supports the usefulness of supplementation of some nutraceuticals, that is, beet nitrates to improve HF management as an adjunct to evidence-based pharmacological therapy, although corroboration of its efficacy is still needed (115).

There are many other nitrate-rich supplements, either in development or already on the market, that have been shown to be efficacious. For example, the combination of the plant-based ingredients (beetroot, red spinach (*Amaranthus tricolor*), and aronia berry extracts) yields a significant increase in NO metabolites following acute ingestion (100). In other studies based on a meta-analysis, consuming 4–12 mg/kg of nitrate (300–00 mg/d) as a dietary supplement – for example, beetroot juice, beetroot concentrates powders, and/or sodium nitrate – are needed to confer significant cardioprotection (90). Reductions in blood pressure and platelet aggregation have

**TABLE 14.3**
**Nitrate and Nitrite Content of Nitrate-Rich Dietary Supplements**

| Dietary supplement product | Serving size | Nitrate mmol | Nitrite mmol |
|---|---|---|---|
| Redibeets | 100 g | 10.75 | 0.75 |
| Superbeets | 100 g | 20.60 | 2.80 |
| Endurance beets | 100 g | 9.80 | 0.00 |
| BeetBoost | 100 g | 16.20 | 0.00 |
| BeetElite | 100 g | 21.60 | 2.20 |
| PureClean powder | 100 g | 25.60 | 0.00 |
| Beet power | 1 L | 1.02 | 0.00 |
| Beet performer | 1 L | 3.01 | 0.00 |
| Unbeetable Fizz | 1 L | 3.52 | 0.00 |
| Beet performer with passion fruit | 1 L | 3.97 | 0.02 |
| Beet blast | 1 L | 3.96 | 0.06 |
| Ginger beet juice | 1 L | 6.67 | 0.06 |
| Red rush | 1 L | 2.39 | 0.01 |
| Beet performance supplement | 1 L | 2.76 | 0.00 |
| BeetActive | 1 L | 2.76 | 0.00 |
| Beet it organic beetroot shot | 1 L | 5.93 | 0.00 |
| Beet it: Sport elite shot | 1 L | 6.41 | 0.00 |
| Biotta beet juice | 1 L | 4.81 | 0.00 |
| Beet it beet juice | 1 L | 7.55 | 0.00 |
| Beet juice-Knudsen | 1 L | 12.54 | 0.00 |
| Beet juice-Lakewood | 1 L | 18.77 | 0.02 |

Table adapted from (121).

also been observed after supplementation with 4–20 mg/kg (116). There are many other nitrate-rich dietary supplements and nutraceuticals based mostly on beetroot on the market (Table 14.3).

An evolving conundrum regarding nitrate supplementation exists as to whether benefit outweighs detriment with a reasonable intake of dietary nitrate, which may readily exceed recommendations for daily intake. For example, consumption of a single serving of a nitrate-rich food or dietary supplement can exceed the acceptable daily intake (ADI) for nitrate (222 mg/d for a 60-kg adult) per the WHO (35). Moreover, recommendations for dietary supplement intake in exercise are 300–600 mg of nitrate (<10 mg/kg or 0.1 mmol/kg) or 500 mL beetroot juice (3–6 whole beets) within 1.5 h of exercise commencement (117). An additional recommendation is multi-day dosing for around 6 days prior to exercise or an athletic event. Given the potential health benefits and risks for dietary nitrate and nitrite intakes, there is a need for rational dietary guidance regarding nitrate- and nitrite-containing foods and dietary supplements to achieve optimal cardiovascular health and athletic performance, while taking into account the potential negative health risks.

### 14.3.3.1 NITRATE AND NITRITE SALTS

As one might surmise, the notion of supplementing directly with inorganic sodium or potassium nitrate has been considered by many individuals, including endurance athletes. Indeed, both are available either from supermarkets or the internet. However, the indiscriminate use of salts and the assumption of safety have generated considerable concern amongst researchers and experts specifically for nitrite, which has an LD50 (lethal dose) similar to cyanide and can cause death (119). Inorganic nitrate is nontoxic at high doses, but inorganic nitrite can be injurious at much lower doses. The LD50 for nitrite is 100–200 mg/kg body weight, which would be 7–14 g (0.25–0.5 ounces) for a 70 kg (154 pound) human (119). The conversion rate (%) of nitrate to nitrites from dietary sources – that is, vegetables – is relatively low and, thus, even with high dietary nitrate intake, nitrite generation is low, thus safe.

In 2017, the European Food Safety Authority (EFSA) initiated a reevaluation of sodium nitrate (E 251) and potassium nitrate (E 252) as food additives. EFSA calculated nitrate exposure in humans based on salivary secretion rates of nitrate (20–25%), oral conversion rates to nitrite (5–36%), and conversion percentage of 1–9 percent. The ADI range was 1.05–9.4 mg nitrate/kg/d. EFSA compared this to the ADI for nitrite (0.07 mg nitrite/kg body/d) and concluded that intake was within the current, designated ADI of 3.7 mg/nitrate/kg/d (118).

### 14.3.3.2 BEETS AND BEETROOT JUICE

Currently, the most commonly used and most prevalently researched dietary constituent thought to enhance plasma NO levels is beetroot or its juice because of its high nitrate concentration and observations that consumption increases plasma nitrate and/or nitrite subsequent to both acute and chronic ingestion (104, 105, 120). The results have been conflicting, however, and may be due to the wide range of concentrations of nitrate in vegetables and their juices due largely to agronomic practices: for example, nitrate-rich fertilizer use (121, 122). The consensus, however, is that nitrate-rich beetroot juice is efficacious in several aspects of cardiovascular health, as well as physical performance (68). Alternatives, such as the addition of vitamins, fruits, chemicals products, natural products containing nitrite or spices, which have similar properties of nitrites, are in evaluation (99).

Beetroot juice is increasingly used more in athletics due to emerging and cumulative evidence for its role in vasodilation and subsequent reductions in blood pressure and increases in oxygenation of various tissues, particularly exercising muscle that may experience hypoxia. Indeed, others have shown that dietary nitrate can increase muscle efficiency, exercise tolerance, and markedly improve endurance (123). Thus, for athletes the use of dietary nitrate may optimize training leading to significant improvements in athletic performance. Other modulated functions include increased blood flow, improved gas exchange, enhanced mitochondrial biogenesis and efficiency, and strengthening of muscle contraction indicating potential benefit in ergogenic effects and cardiovascular endurance (124). In a meta-analysis of 22 studies, the results showed that beetroot supplementation significantly improved cardiorespiratory endurance, exercise efficiency, athletic performance over a range of distances, time to exhaustion at submaximal intensities, and cardiorespiratory performance at intensities approaching anaerobic threshold and $VO_{2max}$ (124).

Supplementation with dietary nitrate, that is, beetroot juice, is popular amongst not only elite athletes but also recreational exercisers because of its effects on training physiology and exercise performance in both healthy and diseased populations. As a caveat, support for evidence-based sports performance supplements (caffeine, creatine, nitrate/beetroot juice, β-alanine, and bicarbonate) depends on the type of event, the conditions of the event, and the individual responsiveness to the supplement. Indeed, Rothschild and colleagues corroborate these observations, asserting that effects are dependent on dose and duration of the exercise (125). Although conflicting results exist, time to exhaustion seems to increase, and ergogenic benefits may depend on individual aerobic fitness levels. That is, individuals with a lower fitness level may derive greater benefits regarding athletic or exercise performance after nitrate consumption, than well-trained athletes. The purported mechanisms for these effects seem to be improved oxygen cost and consumption during exercise via greater adenosine triphosphate (ATP) production and lesser ATP consumption. In a meta-analysis of beetroot juice studies, nitrate consumption and its relative concentration exerted a substantial effect on athletic or exercise performance when averaged across athletes, non-athletes, and modes of exercise (126). Dietary nitrate supplementation appears to have some effect on training performance in patients with peripheral artery disease, heart failure, and chronic pulmonary obstructive disease. However, larger randomized controlled trials are necessary to determine the overall utility of beetroot as a dietary supplement (127).

### 14.3.3.3 RED SPINACH

Red spinach extract (RSE) is another popular nitrate-rich dietary supplement. In one study, RSE significantly increased plasma nitrate 30 min post-ingestion, with significant acute microvascular (that is, resistance vasculature) reactivity increases in the lower limb of apparently healthy humans (128). Moreover, the acute effect of RSE (1000 mg dose; ~90 mg nitrate) was determined using performance markers during graded exercise testing (GXT). Plasma concentrations of nitrate increased pre and post GXT with RSE, but not with placebo. During GXT, $VO_2$ at the ventilatory threshold was significantly higher with RSE compared to placebo, though time-to-exhaustion and maximal aerobic power (i.e., $VO_2$ peak) were non-significantly lower with RSE. Overall, this suggests that RSE supplementation may exert ergogenic properties by delaying the ventilatory threshold (129). In a study of 17 recreationally active men (n = 9, 22.2 ± 3.8 years) and women (n = 8, 22.8 ± 3.5 years) undergoing 2 randomized testing sessions, subjects supplemented daily with 1 g of RSE or placebo for 7 days prior and 1 hour before completing a 4-km cycling time trial test (130). Compared to placebo, RSE supplementation significantly lowered post-exercise diastolic blood pressure and significantly improved 4-km completion time, average power, relative power, and average speed. Only females displayed significant improvement during RSE trials. Interestingly, it has been proposed that sex differences do, in fact, exist in response to dietary nitrate supplementation, but this proposal lacks firm support because females are underrepresented in research (131). In a different study, RSE supplementation significantly reduced time-to-completion, increased measures of power and speed, and lowered post-exercise diastolic blood pressure during a 4-km cycling time trial without altering subjects' perceived exertion or subjective measures of muscle fatigue.

## 14.4   POTENTIAL HARM OF NITRATE

### 14.4.1   HISTORICAL VIEW AND RATIONALE

Nitrites and nitrates are used as preservatives in cured meats such as bacon, salami, sausages, and hot dogs. The historical basis for concern with dietary nitrate began in the 1950s and 1960s when it was realized that nitrite, which can form from nitrate, react with naturally occurring amines in protein at low pH, such as the gastric lumen, and potentially form carcinogenic nitrosamines. This has prompted many to insist on a ban of synthetic nitrate and nitrites from foods. The potential for methemoglobinemia in infants from nitrate and nitrite exposure from drinking water and some foods has also prompted concern. This stigma surrounding nitrate has continued into recent years, even with the lack of compelling, definitive evidence of harm – coupled with emerging, cumulative evidence of the critical importance of dietary nitrate in health.

### 14.4.2   CURRENT VIEW AND RATIONALE

Emerging evidence regarding the relationship between dietary nitrate/nitrites and health presents a conundrum. Although negatively viewed by many, the high natural nitrate content of beetroot juice has been credited with significantly lowering blood pressure and enhancing exercise performance. Nitrate is also the active ingredient in some medications for angina and have been used for >130 years (amyl nitrite, nitroglycerin). The addition of nitrate and nitrites to processed foods is legal and an effective preservative. The indisputable health benefits of nitrate from dietary vegetable consumption has supported the notion that they are, in fact, dietary nutrients and the benefit surpasses the risk (89, 132).

The rationale for the current view is based on the observation that gastric nitrosation largely occurs via an attack on secondary amines (of proteins), thiols, and phenol groups and the nitric dioxide radical is also formed from nitrite in the stomach. While both NO and nitrite are produced in equimolar amounts, the redox conditions of the microenvironment can shift the reaction in either direction. For example, the presence of reducing agents such as the antioxidants ascorbic acid or polyphenols will synthesize as the principle product, NO. Interestingly, other nitration products can be generated that exert protective effects. For example, nitrate may control the activity of pepsin via nitration or form electrophilic compounds, which react with fatty acids producing anti-inflammatory adducts (133). Moreover, ethanol from alcoholic beverages can be nitrosated by dietary nitrite to form the potent vasodilator ethyl nitrite. Ethyl nitrite, once absorbed, may also release NO at physiological pH (pH=7.4) (134).

In summary, many diverse and potentially reactive nitrogen products are formed in the acidic environment of the stomach. The end products may exert beneficial or detrimental effects depending on the microenvironment and the molecule formed. This argues that the nitrite-producing bacteria of the oral cavity and the acid-producing stomach have pivotal roles in modulating nitrogen oxide signaling throughout the body.

### 14.4.3 TOXICITY FROM INGESTION

Dietary nitrate and nitrites may be associated with cancer risk, but the data are inconsistent. It is, however, estimated that nitrate intake >600 mg/d, nitrite intake >0.2 mg and nitrosamine intake >0.2 µg increase the relative risk. EFSA has recommended an ADI of nitrate for a human adult of 3.7 mg/kg body weight/day, that is for a person with body weight of 60 kg, it is 222.0 mg/day (135). An interesting calculation has been presented by Hord and colleagues, who posit that if nitrite were a carcinogen, the public would be advised to avoid swallowing because saliva contains 50–100 µmol/L nitrite, which can increase to near millimolar levels after a nitrate-rich meal (136). Overall, the data support that normal physiological levels of nitrite and nitrate clearly exceed concentrations considered at-risk. Collectively, this contributes to the conundrum of safety versus toxicity based on regulatory limits. Moreover, in a descriptive cross-sectional study, a total of 90 vegetable samples were collected from nine farms and analyzed for nitrate content. The authors concluded that the amount of nitrate in raw vegetables did not reach the standard limit level for toxicity and, thus, did not cause health problems for consumers (88).

## 14.5 POTENTIAL HEALTH BENEFITS OF NITRATE

Nitrate may exert numerous health benefits. They can produce NO, which readily reduces blood pressure via its vasodilatory effects. Subsequently, the risk for cardiovascular disease, coronary heart disease, myocardial infarction, and stroke are markedly mitigated (137). In a meta-analysis of 34 studies, inorganic nitrate consumption significantly reduced blood pressure, improved endothelial function, reduced arterial stiffness, and reduced platelet aggregation (90, 138). Dietary nitrate consumption also significantly improves and/or reduces the risk for gastric ulcers, renal failure, and metabolic syndrome (135). NO has been implicated in many other physiological functions including neural function and immunity (139, 140).

### 14.5.1 INCREASED NO

Numerous studies have detected plasma nitrate after ingestion. Ingestion of high-dose nitrate, either as synthetic sodium nitrate or natural beetroot juice, in eight young, healthy individuals rapidly increased plasma nitrate concentration up to threefold, a study maintained for two weeks (105). Mayra and colleagues showed in a crossover trial that those consuming high-nitrate leafy salad twice daily for ten days significantly increased fasting plasma nitrate/nitrite concentrations and significantly improved flow-mediated dilation (FMD) by 17 percent (141). Ormesher and colleagues tested in hypertensive pregnant females in a double-blind, placebo-controlled study the effects on blood pressure of daily nitrate consumption (70 mL for 8 days). Dietary nitrate significantly increased plasma and salivary nitrate/nitrite concentrations compared to placebo and was significantly correlated with corresponding reductions of diastolic BP (142). Others have shown, in different experimental models, significant increases in plasma nitrate after consumption of nitrate-rich vegetables and dietary supplements with subsequent beneficial effects. For studies demonstrating protective effects after nitrate consumption, but without measuring plasma nitrate/nitrite levels, there is an assumption that levels did increase to elicit the effect, provided the study was placebo-controlled and well-designed with fairly stringent inclusion/exclusion criteria.

An important source of nitrite is via the enterosalivary NO cycle and the commensal bacteria of the oral cavity. Several studies have shown that elimination of bacteria via the use of antiseptic mouthwash can lead to reduced plasma levels of nitrate and nitrite, increased blood pressure, and lack of gastroprotection as well as contributing to the risk of cardiovascular disease and chronic kidney disease (143–145). It has also been noted that the frequency of tongue cleaning significantly impacts the composition of the human tongue microbiome and enterosalivary circulation of nitrate (146). These observations may be responsible, in part, for reported effects after acute dietary nitrate supplementation, where alterations of the microbiome occur but without subsequent effects on vascular responses (147).

## 14.5.2 IMPROVED BLOOD FLOW: RESPONSE TO HYPOXIA

Due to clear efficacy, nitrate such as nitroglycerin has been valuable in the treatment of cardiovascular disease for decades and continue as an important therapeutic agent in clinical settings. Nitrate causes potent vasodilatation of the capacitance veins and markedly enhance ventricular filling pressure in the heart, and dilate the epicardial (inner layer of the pericardium) coronary arteries, improving coronary blood flow, particularly in ischemic (hypoxic) tissue (148). Bioactivation of organic nitrate generates NO, which induces vasorelaxation via its effect on vascular smooth muscle cells as well as impairs platelet activation and potential aggregation, reducing the risk for subsequent blood clots – for example, thrombi, emboli.

Exercise during hypoxia reduces muscle oxidative function and impairs exercise tolerance (capacity to sustain aerobic exercise). Vanhatalo and colleagues provided beetroot juice (750 ml) to subjects and noted the limit of tolerance was reduced during hypoxia in the placebo group but was restored in those consuming beetroot juice (120). There was also notable attenuation of muscle metabolic perturbation during hypoxic exercise. Moreover, beetroot consumption improved muscle energetics and functional capacity in hypoxic environments. Engan and colleagues provided beetroot juice concentrate to healthy subjects and observed small, yet significant, changes in arterial oxygen saturation between placebo and nitrate after 2 min of static apnea (holding breath while submerged in water) (149). Additionally, maximal apneic duration was prolonged by 11 percent. In contrast, Schiffer and colleagues found no difference after 2 min of apnea, with a lower arterial oxygen saturation after 4 min of static apnea after nitrate intake compared to placebo, but the maximal apneic duration was 5 percent shorter (149).

Kapil and colleagues showed in healthy adults (n=35) that consumption of 250 mL beetroot juice compared to water prevented endothelial dysfunction caused by ischemia-reperfusion (150). This group further showed reduced arterial stiffness after beetroot juice consumption. Asgary and colleagues further showed in hypertensive untreated adults that 500 mL beetroot juice for 15 days significantly reduced endothelial intracellular and vascular cell adhesion molecule expression as well as E-selectin expression, which reduces the risk for atherogenesis (151). Velmurugan and colleagues showed in patients with hypercholesterolemia that consumption of 250 mL beetroot juice per day significantly improved FMD as an indicator of improved vascular response (152). In several studies, acute ingestion of beetroot juice between 2–3 h with doses of 341–1488 mg demonstrated benefits on ischemic reperfusion injury as assessed by FMD as an indicator of endothelial impairment via noxious ischemic reperfusion (150, 153).

### 14.5.3 Reduction in Blood Pressure

Many clinical studies have addressed the capacity for nitrate-rich beetroot juice or some other NO donor to reduce blood pressure (154–157). In 2006 Larsen and colleagues showed that sodium nitrate (8.5 mg/kg/d) consumed for three days increased plasma levels of nitrate and significantly reduced DBP by 4 mmHg in young, healthy volunteers. The dose corresponded to 2–3 beetroots or 200–300 g spinach. A bolus dose of beetroot (500 ml; 1,400 mg) reduced both systolic and diastolic blood pressures by 10 and 8 mmHg, respectively. Subsequent use of a lower dose of beetroot juice (250 ml) caused a 5 mmHg reduction but only in systolic blood pressure. Plasma cGMP was also increased, indicating conversion of soluble guanylyl cyclase involved in NO production and ultimate vasodilation. Several studies have now confirmed the blood-pressure–lowering activity of dietary nitrate mostly in healthy young adults. Similar effects, however, have been shown in older adults, although some report a lack of effect (104, 158). Provision of a single bolus dose of beetroot juice to older adults with peripheral artery disease significantly reduced blood pressure, increased time of claudification pain (leg cramping with exercise caused by arterial obstructions), and prolonged peak walking time. Husmann and colleagues showed that dietary nitrate supplementation mitigated muscle fatigue by alleviating exercise-induced damage to contractile muscle function. Moreover, dietary nitrate reduced the perception of effort and leg muscle pain during exercise (159). Hobbs and colleagues showed that 3 doses of beetroot juice (100, 250, 500 mL) given over 24 h reduced ambulatory blood pressure in healthy adults and was dose-dependent with reduced systolic, but not diastolic blood pressure (160, 161). In the same study, red and white beetroot effects were similar, suggesting that betalains (red and yellow tyrosine-derived pigments) were not responsible for the antihypertensive effect.

Numerous studies have been conducted that demonstrate the hypotensive action of nitrate and nitrate-rich vegetables (141, 162, 163). Keen and colleagues demonstrate in healthy non-smokers in a randomized, double-blind, placebo-controlled study that daily consumption of 70 mL of beetroot juice significantly reduced both mean arterial and diastolic blood pressures (164). Hobbs and colleagues showed in two separate clinical trials, ranging from 100–583 g in healthy adults, significant reductions in systolic and diastolic blood pressure in a dose-dependent fashion as well as subsequent increased endothelium-independent vasodilation (160, 161). In a systematic, meta-analysis, of 22 studies, the data showed that dietary consumption of beetroot juice significantly reduced Systolic Blood Pressure (SBP) and a positive correlation was observed between consumption and resultant mean difference in blood pressure via NO-independent effects (165). In other studies, oral inorganic nitrate and nitrite infusion significantly dilated peripheral arteries leading to increases in forearm blood flow (153, 166). Phase 2 studies in hypertensive patients revealed that dietary inorganic nitrate (beetroot juice) caused sustained blood pressure drops in hypertensive patients (167, 168). The results showed consistent blood pressure drops with the four weeks of the treatment period and therefore no development of tolerance, a frequent side effect of chronic nitrate use. Compelling evidence demonstrates that dietary consumption of synthetic dietary nitrate or nitrate-rich vegetables significantly decreases blood pressure within hours of ingestion and is correlated with dose (165, 169–171). In a study comparing the effects of dietary nitrate of vegetarians and omnivores, the authors noted that vegetarian diets did not alter nitrate or nitrite homeostasis, or the oral microbiome compared to the omnivore diet (170).

From a therapeutic perspective, nitrate-rich beetroot has been routinely used as an adjunct treatment by many for hypertension. In a meta-analysis of 22 studies (2009–2017), the results indicated a significant reduction in both SBP (-3.55 mm Hg) and DBP (-1.32 mm Hg) (165). Moreover, the mean difference of SBP was greater for longer, rather than shorter, time periods (≥14 versus <14 d) and higher, rather than lower (500 versus 70–140 mL/d), doses of beetroot juice (-4.78 compared with -2.37 mm Hg) (165). The beetroot dose was positively correlated with the subsequent blood pressure differences. In a different meta-analysis, studies (2006–2012) with collectively more than 254 subjects (7–30 per study) with durations of 2–15 d were evaluated. The results indicated greater changes in SBP (-4.4 mm Hg] than DBP [-1.1 mm Hg), and there was a clear association between daily nitrate dose and effects on SBP. In a meta-analysis by Bonilla and colleagues, the authors also concluded that beetroot juice supplementation reduced blood pressure in different populations as well as mitigated risk for cardiovascular events and subsequent mortality (169).

### 14.5.4 General Cardioprotection

Cardioprotection includes all mechanisms that collectively contribute to the protection of the heart with mitigation of myocardial injury. Many studies have shown that nitrate consumption reduces infarct size and improves clinical outcomes, but the mechanisms underlying the effect of nitrate against reperfusion injury (RI) are currently unknown. It is, however, well known that nitrate is a NO donor, and NO has a cytoprotective effect via activation of its downstream pathways (172, 173). Since ischemia-reperfusion is characterized as a NO deficit, replacement or restoration of physiological levels of NO should improve vasodilation and mitigate pathologies, although studies are inconsistently demonstrating the need for additional studies (174–176).

Nitrate can favorably influence myocardial infarction, an ischemic, hypoxic event, through several mechanisms (177, 178). Nitrate can reduce infarct size through hemodynamic (blood flow) effects and increased collateral flow of microvessels (arterioles, metartioles, capillaries, venules) typically not open under normal conditions. Moreover, they can accelerate and stabilize reperfusion or prevent adverse remodeling changes, for example, post-injury changes in size, mass, function, in refractory patients. Nitrate also redistributes coronary flow to ischemic regions of the heart that has low oxygen tension. In a study of coronary heart disease (associated with inadequate perfusion of the heart), the highest quintile of nitrate consumption compared to the lowest, there was a protective association for CHD (relative risk 0.77) (179). However, this effect was reduced after adjusting for largely modifiable lifestyle factors, that is, smoking, physical activity, and so forth.

NO is essential also in maintaining the function of the vascular endothelium largely via its vasodilatory, antiaggregatory, and antiadhesive effects. Divakaran and colleagues showed that FMD in healthy individuals was acutely improved after oral sodium nitrate, but the response to intravenous nitroglycerin did not alter the vascular smooth muscle response (1). It did improve endothelial function. Webb and colleagues demonstrated that a 20-minute ischemic insult to the forearm followed by a 20-minute perfusion reduced the FMD response by 60 percent in control subjects. Consumption of beetroot juice (500 ml) completely abrogated the effect indicating preservation of endothelial integrity and function. Moreover, the result aligns with the observation that, in humans, plasma levels of inorganic nitrate closely correlate with normal endothelial function and, as a result, vascular homeostasis (160, 180). This was further corroborated by the observations that daily dietary intake of inorganic nitrate improved endothelial dysfunction (155). In contrast, other studies have shown equivocal results with no apparent improvement to endothelial function, although plasma levels of nitrate significantly increased. The disparities in research results require further research to fully understand the reasons.

In patients with congestive heart failure (CHF), peripheral abnormalities are apparent, including a high degree of vasoconstriction relative to a maximally dilated state and ultrastructural changes to the cellular architecture of skeletal muscle. Recent studies in patients with HF with preserved ejection fraction (% blood ejected divided by maximally filled left ventricle) show that beetroot juice and potassium nitrate improved exercise capacity, presumably via reduced vascular resistance, increased muscle power, and markedly improved vascular compliance (the relationship between blood volume of vessel and BP that is generated). Others have shown that synthetic sodium nitrite provided either by inhalation or intravenous infusion exerts significant effects on biventricular central filling pressures, cardiac output, and improved exercise capacity. NO-generating nitroglycerin was the first and most frequently used organic nitrate for clinical treatment of angina pectoris because it causes vasodilatation of the capacitance veins and improves ventricular filling pressure as well as dilating the epicardial coronary arteries, improving coronary blood flow, particularly in ischemic zones, reducing infarct size and improving clinical outcomes.

Nitrate via synthetic or dietary means – that is, green leafy vegetables – is pivotal in cardioprotection. Cardiac anomalies leading to hypoxia disrupt energetics and mitochondrial function of the heart (181, 182). Thus, the provision of dietary nitrate may reverse hypoxia-induced effects on respiration, mitochondrial complex I levels and activity, and oxidative stress that occurs concomitant with hypoxia (183). Since hypoxia and arginase deficiency (leading to reduced NO) are key features of heart failure, dietary nitrate may confer protection, particularly since the non-enzymatic pathway of NO production is preferred at hypoxic sites (184).

### 14.5.5 IMPROVED COGNITION

Cerebral blood flow regulation exerts an important role in cognitive function and ischemia and/or energy depletion (185). NO is critical in regulating cerebral blood flow and the coupling of neural activity to perfusion in the brain (186). In a study by Presley and colleagues, older adults were given high-dose and low-dose nitrate. The former significantly increased regional cerebral perfusion in the frontal lobe of the brain in regions involved in executive functioning (working memory, flexible thinking, self-control) (187). Oral nitrate supplementation differentially altered cerebral arterial blood velocity and subsequent prefrontal oxygenation under normoxic conditions but not hypoxic conditions (188, 189). In a study by Kelly and colleagues, older subjects were given beetroot juice for 2.5 days. However, no significant effects were noted using a panel of cognitive tests or in concentrations of brain metabolites (190).

The increased blood flow and regional perfusion in the brain due to beetroot consumption suggests a means of improving mental function and reducing the progression of age-related cognitive decline as well as dementia. In a study by Wightman and colleagues, healthy adults were recruited to assess the effect of dietary nitrate consumption on cognitive performance and cerebral blood-flow to the prefrontal cortex (191). Forty healthy adults were randomized to groups for either placebo or 450 ml beetroot juice (~5.5 mmol nitrate). After the 90-minute consumption period, participants completed an array of cognitive tasks known to activate the frontal cortex. The bioconversion of nitrate to nitrite was confirmed in plasma. Moreover, dietary nitrate modulated the hemodynamic response associated with task performance, with an initial increase in cerebral blood flow tapering off for the last of the three tests. Cognitive performance was also improved. Collectively, the results demonstrate that a bolus dose of nitrate can modulate cerebral blood flow during task performance and potentially improve cognition. Other studies have shown that dietary nitrate improves oxygenation and cerebral flow during hypoxia (186, 192).

Previous studies have shown that both acute and prolonged supplementation of dietary nitrate in older adults can significantly improve oxygen uptake and agility in exercise, and increase time to fatigue (delayed tiredness), thus promoting improved exercise performance (171). Furthermore, nitrate supplementation can significantly improve cognitive performance, as shown by enhanced reaction times in older adults, although the potential benefits of dietary nitrate supplementation are limited. Studies have also shown that, in older adults, consuming nitrate reduced blood pressure and improved blood flow to the brain and muscle, suggesting cardiovascular and cerebrovascular benefits. Older adults frequently display reduced vasodilation (reduced blood flow), cardiovascular function, cognitive function, and mood. As a result, dietary nitrate supplementation or increased nitrate-rich vegetable intake may be particularly efficacious for older adults.

### 14.5.6 Erectile Dysfunction

Erectile dysfunction is a common, multifactorial disorder associated with aging and a range of organic, hormonal, and psychogenic conditions and is considered a marker for cardiovascular disease (193). Given the vascular involvement in ED, NO deficiency is involved in the etiology, since it is a key vasoactive neurotransmitter of penile tissue (194). NO is secreted by neural and endothelial cells of the corpora cavernosa, where it activates, as described earlier, soluble guanylyl cyclase, which increases cGMP levels releasing calcium from intracellular stores in smooth muscle cells (57). This can also interact with vasorelaxation-inducing contractile proteins. As one might surmise, the absence or impairment of NO bioactivity and its vasorelaxing properties are major contributors to erectile dysfunction. The efficacy of such drugs as sildenafil, viz., Viagra, illustrates the importance of the NO-cGMP pathway. Sildenafil is a PDE-5 inhibitor that prevents the degradation of NO-generated cGMP (195). With this knowledge, other drugs are being investigated that function as activators of guanylyl cyclase, donors of NO, and so forth. Recent evidence suggests that neuronal and endothelial NOS (nNOS and eNOS, respectively) play major roles also in causing NO bioactivity necessary for erectile function. Moreover, S-nitrosylation/denitrosylation has been shown to regulate eNOS activity via S-nitrosoglutathione reductase, contributing directly and indirectly to erectile function/dysfunction (196).

Demonstration that blood-flow-dependent generation of NO involves phosphorylation of penile eNOS questions the current paradigm of NO-dependent erectile mechanisms. Regulation of erectile function may not be mediated exclusively by neurally derived NO, but by fluid shear stress in the penile vasculature. This stress stimulates phosphatidyl-inositol 3-kinase to phosphorylate PKB which, in turn, phosphorylates eNOS to generate NO as discussed earlier (197). Thus, working in tandem, nNOS may initiate cavernosal tissue relaxation, while eNOS initiates and sustains full erection.

### 14.5.7 Improvement in Aerobic Exercise Performance

Increasingly, there has been considerable interest in the beneficial effects of dietary nitrate supplementation on athletic performance and exercise in general. In the last 5 years, >180 publications have been published on dietary nitrate and exercise, more than double that published in the 10 years before that (117). Indeed, dietary nitrate has been touted as an ergogenic aid and potential exercise therapeutic (139). Others have reviewed supplements and their use and the myriad beneficial effects to endurance athletes, physical performance, and exercise performance (108–110).

Dietary nitrate supplementation may exert enhanced effects on aerobic exercise performance and improved exercise tolerance, as reported in numerous studies (198–202). Many studies have also shown that dietary nitrate exerts ergogenic effects associated with lower oxygen cost during submaximal exercise (203–205). Mechanisms have been posited as greater production of mitochondrial ATP and thriftiness in ATP use during the work of skeletal muscles. Many effects are also based on the capacity to significantly enhance vascular function, but also modulation of both metabolism and muscle function (206). Acute dietary nitrate supplementation (five days) has also been shown to reduce muscle fatigue primarily caused by lower exercise-induced dysfunction in contractile capacity (207).

Dietary nitrate can improve performance during high-intensity exercise. Dominguez reports in a meta-analysis of nine studies on the effects of beetroot juice supplementation and high-intensity exercise. Beetroot juice given as a single dose or over a few days improved muscle fatigue and exercise performance at intermittent, high-intensity efforts with short rest periods in between (124). A potential mechanism proffered is enhanced phosphocreatine (an energy store in muscle) resynthesis mitigating the rate of depletion. Wylie and colleagues have shown that nitrate concentration in skeletal muscle is indeed considerably higher than blood concentrations, and is further increased by dietary nitrate ingestion (208). Further, the authors show that high-intensity exercise reduces the skeletal nitrate store following dietary supplementation, presumably via conversion of nitrite to NO. Nitrate-rich beetroot juice also improved muscle power output, presumably due to more rapid muscle shortening velocity (124). Kelly and colleagues report the effects of dietary nitrate in nine recreationally active male subjects in a double-blind, randomized, crossover study by measuring and calculating critical power and the curvature constant of power duration, which describe the relationship of tolerance to severe-intensity exercise. Dietary nitrate significantly enhanced endurance in recreationally active subjects (209). The data, however, are inconsistent, and some studies do not report beneficial effects. For example, Bescos and colleagues report that sodium nitrate supplementation did not enhance performance in endurance athletes (188).

In runners, Thompson and colleagues have shown that dietary nitrate improves sprint and high-intensity running performance, as well as cognitive function (210, 211). Margaret and colleagues also showed that nitrate-rich whole beetroot consumption acutely improved running performance (212). As a result, many nitrate-containing supplements and regimens have been proposed for optimal function and performance enhancement of track and field athletes (127).

In rowers, high-dose (8.4 mmol) dietary nitrate provided 2 hours prior to rowing has been shown to improve performance in well-trained rowers in simulated 2,000-meter rowing ergometer tests (108). Bond and colleagues also showed that nitrate supplementation, as beetroot juice, improved maximal rowing-ergometer repetitions (213). Muggeridge and colleagues showed that a single dose of beetroot concentrate can improve performance in trained flatwater kayakers (214).

In a different study, 6 d of nitrate supplementation (beetroot concentrate, 140 ml/ d; 8 mmol) reduced $VO_2$ during submaximal exercise and improved time-trial performance in trained cyclists (215). Regarding endurance, dietary sodium nitrate (10 mg/kg) supplementation to 11 cyclists reduced the $VO_2$ peak without compromising maximal exercise performance (216). RSE supplementation (1 g/d; 7 d and 1 hour prior) was provided to 17 males and females prior to a 4-km cycling time-trial. RSE significantly reduced time-to-completion, increased power and speed, and lowered diastolic BP (130). Conversely, in an acute study, a single dose of beetroot juice (500 ml beetroot juice) was provided prior to a 50-mile time-trial performance in 8 well-trained male cyclists. The data indicated no significant improvement in performance (217). Garnacho-Castano and colleagues gave 70 mL beetroot juice (6.5 mmol) to 12 well-trained male triathletes and conducted endurance tests on cycle ergometers and noted no differences between groups on cardiovascular efficiency/economy, $VO_2$ time-trial, energy expenditure, carbohydrate oxidation, or fat oxidation (218). Both authors suggest that the training status of subjects may have blunted the physiological and performance responses.

Wylie reports that dietary nitrate supplementation improves team sport-specific intense intermittent exercise performance (219). Vanhatale and colleagues further elaborates on both the acute and chronic effects in 8 healthy subjects of dietary nitrate (beetroot juice, 500 mL/d 4–6 days). The results indicated that dietary nitrate acutely reduced BP and the oxygen cost of submaximal exercise, which was maintained for 15 days with continuous consumption (106). Kramer and colleagues supplemented athletes with potassium nitrate (8 mmol/d for 6 d), followed by subject participation in a heavily power-dependent CrossFit sport regimen. The authors concluded that peak power improved during one test, viz., Wingate, but not the others testing strength or endurance (220). The results showed that plasma nitrate and the $O_2$ cost of moderate-intensity exercise were altered dose-dependently with beetroot juice, but there was no additional improvement in exercise tolerance after doubling the dose (16.8 mmol) (221). Moreover, positive effects occur even with chronic disease, such as peripheral artery disease, and treated, but uncontrolled, hypertension, and thus is important in maintaining cardiovascular health (222–224).

Regarding tolerance, beetroot juice amplifies oxygen uptake kinetics and significantly improves exercise tolerance during severe-intensity exercise with elevated metabolic rates (225). Nitrate also reduces skeletal muscle metabolism perturbation under conditions of hypoxia (199). In a meta-analysis by Van de Walle and colleagues the effect of nitrate on exercise tolerance and performance was compared. A total of 29 studies using the time to exhaustion (TTE) as the outcome variable revealed a significant effect of nitrate supplementation on exercise compared with placebo, although it revealed no significant effect of nitrate supplementation on exercise performance compared with placebo. Thus, it appears that nitrate supplementation improves exercise tolerance and capacity, which may subsequently improve exercise performance (226).

In a meta-analysis by McMahon and colleagues, data from 76 trials were reviewed and analyzed for time-trial performance, TTE, and graded-exercise test (GXT) protocols. The results indicated small but non-significant effects for the time and GXT trials and TTE data displayed small to moderate statistically significant effects. The data suggest nitrate supplementation elicits a positive outcome when testing endurance exercise capacity, but is less effective for time-trial performance (227). In a meta-analysis by Campos and colleagues, 54 studies were analyzed for the effects of dietary nitrate supplementation on human performance. Nitrate supplementation was ergogenic in non-athletes using long-duration, open-ended tests, but, overall, did not enhance the performance of athletes. Thus, the present study suggests that dietary nitrate supplementation improves physical performance in non-athletes but others were non-refractory (228).

Regarding oxygen use, dietary nitrate supplementation has been shown to significantly reduce the oxygen cost of exercise of walking and running (200, 229). It also reduced maximal oxygen consumption while maintaining work performance in maximal exercise (204). Bailey and colleagues have shown that dietary nitrate supplementation reduces oxygen cost of low-intensity exercise and enhances tolerance to high-intensity exercise (203). Nitrate from red spinach consumption has increased the ventilatory threshold during graded exercises (129). Masschelein and colleagues show as well that dietary nitrate improves skeletal muscle oxygenation during exercise in hypoxia (192). Cermak and colleagues, however, report that nitrate-rich beetroot did not improve endurance performance after a single dose (215). Bescos and colleagues have also reported that sodium nitrate supplementation does not enhance the performance of endurance athletes (188).

### 14.5.8 Diabetes, Glycemia, and Insulin Resistance

There is compelling evidence that NO modulates carbohydrate metabolism, and lack of NO contributes to the development of type 2 diabetes (230). A review of five human studies was analyzed and reported as significantly reducing blood glucose levels and beneficially affecting both glycemic and insulin responses (231). In 16 healthy adults, participants were given 225 mL beetroot juice. Postprandial insulin response was reduced in the early phase (0–60 min) and the glucose response was reduced in the 0–30 min phase (232). Subjects were provided three beverages, including (1) beetroot with lemon, (2) beetroot with glucose, fructose, and sucrose and (3) beetroot juice with added glucose. A positive correlation was observed with beetroot juice plus lemon, but not the two other beverages, and glycemic response was lower with both the first two glycemic responses significantly lower than the third. In a second study, there was a trend for 35 percent reduction in plasma glucose in 30 subjects consuming the beverage over longer periods and 10 percent for 4 weeks, suggesting that chronic consumption must be maintained (233, 234). In a study of 57 individuals, co-ingestion of beetroot juice and glucose caused greater elevation of glucose in obese versus non-obese up to 90 min. The authors concluded that obesity with an intrinsic higher risk for developing insulin resistance may receive greater benefit than non-obese (235) in a meta-analysis of 173 (11) studies (2004–2019). Interestingly, preliminary experimental findings strongly support the hypothesis that $NO_3$ can be considered as a natural anti-obesity agent (231).

## 14.6 DIRECTIONS FOR FUTURE RESEARCH ON NITRATE RELATED TO HEALTH

Given the relatively recent appreciation of the capacity of dietary nitrate to generate NO, and the clear beneficial effects of NO, there are many areas of research that require attention. First, elucidation of the lowest effective, efficacious dose and the duration needed to significantly lower blood pressure and improve cardioprotection are needed from dietary intervention studies. Moreover, the demonstration of the potential for these effects in target "at-risk" populations or those with overt chronic disease (hypertensive) is needed. More specific biomarkers with greater sensitivity and validity are needed, since plasma NOx are currently measured colorimetrically as nitrite and nitrate, which can be metabolized and/or interconverted. The clear capacity for other dietary components to interact both positively and negatively requires further investigation to elicit dichotomous effects and the microenvironments that may contribute to their action. The ongoing conundrum of whether dietary nitrate cause gastric cancer requires resolution, thus a reliable risk-benefit analysis would be helpful. Future cancer research should focus on at-risk populations, such as tobacco smokers, supplement users (particularly sodium and potassium nitrate and nitrite), and specific dietary components that might yield pro-carcinogenic and/or mutagenic molecules. By extension, a comparison of health or disease status to overall nitrate-rich vegetable intake (amounts and types) and water would be helpful in establishing risk versus benefit. Other recommendations have been made, including the development and/or expansion of dietary databases derived from epidemiological studies to assist with more accurate estimates of dietary consumption patterns and amounts. As such, the cardiovascular benefits may be correlated to intakes. Furthermore, studies of total intake coupled with excretion or at least urinary and plasma concentrations should be incorporated into studies to clarify the disposition of nitrate.

Given the effects of nitrate based largely on the assumption that NO is specifically produced from nitrate and nitrites, and that the vasculature is a specific target, then most pathologies and/or conditions that are based in part on vasoconstriction and/or altered hemodynamics should merit additional research. Regarding exercise and sports performance, there is a need for elaborating the impact of dietary nitrate on anaerobic exercise, identifying interactions with other dietary components, determining efficacious, non-toxic doses, and optimal time of supplementation. Collectively, this may help unravel the many noted discrepancies in research results.

## 14.7  CONCLUSION

Dietary vegetables represent a nitrate-rich source for consumption by humans. Both dietary nitrate and nitrites can be bioactivated endogenously to form NO, as well as other nitrated bioactive molecules, with subsequent reabsorption of 25 percent, which enters the nitrate-nitrite-NO pathway or enterosalivary cycle and is concentrated 10–10,000-fold. The relative critical importance of NO and its vasodilatory function, among many others, has garnered considerable research interest despite the ongoing stigma that nitrate can form carcinogenic nitrosamines at low pH. Indeed, nitrate can be problematic regarding methemoglobinemia, and perhaps some cancers, but data are inconsistent. Nonetheless, there is considerable interest in exogenous dietary nitrate for endogenous protective functions such as reducing the risk for cerebrovascular incident, myocardial infarction, cardioprotection, angina pectoris, hypertension, erectile dysfunction, athletic performance, gastric ulcers, and so forth. Dietary consumption of nitrate and nitrite, although low for the latter, do not cause harm in humans. Dietary supplementation with purified salts, in particular, and certain nutraceutical formulations may be problematic regarding safety. Given the considerable myriad protective effects, dietary nitrate and nitrite are pivotal in physiological health and homeostasis largely via the formation of NO either enzymatically via NOS or non-enzymatically. With the clear observations of benefit – tempered, however, with cautious concerns for safety – further investigation is needed regarding biological functions, efficacy, dosing, and duration, as well as therapeutic applications of dietary nitrate.

## REFERENCES

1. Divakaran S, Loscalzo J. 2017. The role of nitroglycerin and other nitrogen oxides in cardiovascular therapeutics. J Am Coll Cardiol. 70(19):2393–410. doi: 10.1016/j.jacc.2017.09.1064.
2. Steinhorn BS, Loscalzo J, Michel T. Nitroglycerin and nitric oxide – A rondo of themes in cardiovascular therapeutics 2015. 373:277–80.
3. Gladwin MT, Schechter AN, Kim-Shapiro DB, Patel RP, Hogg N, Shiva S, Cannon RO, Kelm M, Wink DA, Espey MG, Oldfield EH, Pluta RM, Freeman BA, Lancaster J R Jr, Feelisch M, Lundberg JO. The emerging biology of the nitrite anion. Nature Chemical Biology 1:308–14.
4. Lundberg JO, Carlström M, Weitzberg E. Metabolic effects of dietary nitrate in health and disease. Cell Metabolism 2018. 28:9–22.
5. Lundberg JO, van Faassen, EEH Gladwin MT, Ahluwalia A, Benjamin N. 2009. Nitrate and nitrite in biology, nutrition and therapeutics. Nature Chemical Biology 5(12):865–9.
6. Lundberg JO, Gladwin MT, Weitzberg E. 2008. The nitrate-nitrite-nitric oxide pathway in physiology and therapeutics. Nature Reviews Discovery 7(2):156–67.
7. Blekkenhorst LC, Prince RL, Ward NC, Croft KD, Lewis JR, Devine A, Shinde S, Woodman RJ, Hodgson JM, Bondonno CP. 2017. Development of a reference database for assessing dietary nitrate in vegetables. Molecular Nutrition and Food Research 61(8):1600982.
8. Hyde ER, Andrade F, Vaksman Z, Parthasarathy K, Jiang H, Parthasarathy DK, Torregrossa AC, Tribble G, Kaplan HB, Petrosino JF, and Bryan N. 2014. Metagenomic analysis of nitrate-reducing bacteria in the oral cavity: Implications for nitric oxide homeostasis. PloS ONE 9(3):e88645.
9. Koch CD, Gladwin MT, Freeman BA, Lundberg JO, Weitzberg E, Morris A. 2017. Enterosalivary nitrate metabolism and the microbiome: Intersection of microbial metabolism, nitric oxide and diet in cardiac and pulmonary vascular health. Free Radical Biology and Medicine 105:48–67.
10. Oliveira-Paula GH, Pinheiro LC, Tanus-Santos JE. 2019. Mechanisms impairing blood pressure responses to nitrite and nitrate. Nitric Oxide – Biology and Chemistry 85:35–43.
11. Stein LY, Klotz MG. 2016. The nitrogen cycle. Current Biology 26(3):R94–8.
12. Kuypers MMM, Marchant HK, Kartal B. 2018. The microbial nitrogen-cycling network. Nature Reviews Microbiology 16(5):263–76.
13. Pennington JAT. 1998. Dietary exposure models for nitrates and nitrites. Food Control 9(6):385–95.
14. Santamaria P. 2006. Nitrate in vegetables: toxicity, content, intake and EC regulation. Journal of the Science of Food and Agriculture 86(1):10–17.
15. Guadagnin SG, Rath S, Reyes FGR. 2005. Evaluation of the nitrate content in leaf vegetables produced through different agricultural systems. Food Additives and Contaminants 22(12):1203–8.
16. Anjana SU, Iqbal M. 2007. Nitrate accumulation in plants, factors affecting the process, and human health implications. A review. *Agronomy for Sustainable Development* 27(1):45–57.

17. Hakeem KR, Sabir M, Ozturk M, Akhtar MS, Ibrahim FH. 2017. Erratum to: Nitrate and nitrogen oxides: Sources, health effects and their remediation. In: de Voogt P. (eds) Reviews of Environmental Contamination and Toxicity 242 Reviews of Environmental Contamination and Toxicology (Continuation of Residue Reviews), vol 242. Springer, Cham. https://doi.org/10.1007/978-3-319-51243-3_15.

18. Ginting D, Kessavalou A, Eghball B, Doran JW. 2003. Greenhouse gas emissions and soil indicators four years after manure and compost applications. Journal of Environmental Quality 32(1):23–32.

19. Renseigné, Umar S, Iqbal M. 2007. Nitrate accumulation in plants, factors affecting the process, and human health implications. A review Agronomy for Sustainable Development 27(1):45–57.

20. Dessureault-Rompré J, Zebarth BJ, Burton DL, Sharifi M, Cooper J, Grant CA, Drury CF. 2010. Relationships among mineralizable soil nitrogen, soil properties, and climatic indices. Soil Science Society of America 74(4):1218–27.

21. Bryan NS, van Grinsven H. 2013. The role of nitrate in human health. Advances in Agronomy119:153.

22. Fan AM, Steinberg VE. 1996. Health implications of nitrate and nitrite in drinking water: An update on methemoglobinemia occurrence and reproductive and developmental toxicity. Regulatory Toxicology and Pharmacology 23(1):35–43.

23. Cosby, K., Partovi KS, Crawford J.H., Parte RP, Reiter CD, Martyt S, Yang BK, Waclawiw MA, Zalos G, Xu X, Huang KT, Shields H, Kim-Shapipo DB, Schechter AN, Canno RO, Gladwin MT. 2003. Nitrite reduction to nitric oxide by deoxyhemoglobin vasodilates the human circulation. Nature Medicine 9, 1498–1505 (2003). https://doi.org/10.1038/nm954 9:1498–505.

24. Ward M, Jones R, Brender J, de Kok T, Weyer P, Nolan B, Villanueva C, van Breda S. 2018. Drinking Water Nitrate and Human Health: An Updated Review. International Journal of Environmental Research 15(7):1557.

25. Greer FR, Shannon M, Committee on Nutrition, Committee on Environmental Health. Infant Methemoglobinemia: The role of dietary nitrate in Food and Water. Pediatrics 116(3):784–6. doi: 10.1542/peds.2005-1497.

26. Bahadoran Z, Mirmiran P, Azizi F, Ghasemi A. 2018. Nitrate-rich dietary supplementation during pregnancy: The pros and cons. Pregnancy Hypertension 11:44–6.

27. Manassaram DM, Backer LC, Moll DM. 2006. A review of nitrates in drinking water: maternal exposure and adverse reproductive and developmental outcomes. Environmental Health Perspectives 114(3):320–7.

28. Cornblath M, Hartmann AF. 1948. Methemoglobinemia in young infants. The Journal of Pediatrics 33(4):421–5.

29. Kortboyer, Olling M, Zeilmaker MJ, Slob W, Boink A, Schothorst RC, Sips A, Meulenbelt J. 1997. The oral bioavailability of sodium nitrite investigated in healthy adult volunteers. National Institute of Public Health and the Environment, Bilthoven, The Netherlands, Report nr. 235892 007. Pp.94

30. L'hirondel JL, Avery AA, Addiscott T. 2006. Dietary nitrate: Where is the risk?. Environmental Health Perspectives 114(8):A458–A459.

31. Powlson DS, Addiscott TM, Benjamin N, Cassman KG, de Kok TM, van Grinsven H, L'hirondel J, Avery AA, van Kessel C. 2008. When does nitrate become a risk for humans?. Journal of Environmental Quality 37(2):291–5.

32. Sindelar JJ, Milkowski AL. 2012. Human safety controversies surrounding nitrate and nitrite in the diet. Nitric Oxide – Biology and Chemistry 26(4):259–66.

33. Berends JE, van den Berg, Lauri M. M, Guggeis MA, Henckens NFT, Hossein IJ, de Joode, Minke E. J. R, Zamani H, van Pelt, Kirsten A. A. J, Beelen NA, Kuhnle GG,

Theo MCM, vanBreda SGJ. 2010. Consumption of nitrate-rich beetroot juice with or without Vitamin C supplementation increases the excretion of urinary nitrate, nitrite, and N-nitroso compounds in humans 2019. International Journal of Molecular Sciences 20(9):2277. https://doi.org/10.3390/ijms20092277.

34. Etemadi A, Sinha R, Ward MH, Graubard BI, Inoue-Choi M, Dawsey SM, Abnet CC. 2017. Mortality from different causes associated with meat, heme iron, nitrates, and nitrites in the NIH-AARP Diet and Health Study: Population based cohort study. BMJ 2017;357:j1957. *doi: https://doi.org/10.1136/bmj.j1957*

35. Mensinga TT. 2003. Health implications of exposure to environmental nitrogenous compounds. Toxicological Reviews 22(1):41–51.

36. Spiegelhalder B, Eisenbrand G, Preussmann R. 1976. Influence of dietary nitrate on nitrite content of human saliva: Possible relevance to in vivo formation of N-nitroso compounds. Food and Cosmetics Toxicology 14(6):545–8.

37. Kapil V, Khambata RS, Jones DA, Rathod K, Primus C, Massimo G, Fukuto JM, Ahluwalia A. 1994. The noncanonical pathway for in vivo nitric oxide generation: The nitrate-nitrite-nitric oxide pathway. Pharmacological Reviews 72(3):692–766.

38. Benjamin N, O'Driscoll, Dougall H, Duncan C, Smith L, Golden M, McKenzie H. 1994. Stomach NO synthesis. Nature 368:502. https://doi.org/10.1038/368502a0.

39. Lundberg JO, Weitzberg E, Lundberg JM, Alving K. 1994. Intragastric nitric oxide production in humans: measurements in expelled air. Gut 35(11):1543–6.

40. Sukhovershin R, Cooke J. How May Proton pump inhibitors impair cardiovascular health? *American Journal of Cardiovascular Drugs.* 16:153–61.

41. Bondonno CP, Croft KD, Ward N, Considine MJ, Hodgson JM. 2015. Dietary flavonoids and nitrate: effects on nitric oxide and vascular function. Nutrition Revies 73(4):216–35.

42. Dykhuizen RS, R Frazer, C Duncan, C C Smith, M Golden, N Benjamin, C Leifert. 1996. Antimicrobial effect of acidified nitrite on gut pathogens: importance of dietary nitrate in host defense. Antimicrobial Agents and Chemotherapy 40(6):1422–5.

43. Xia DS, Liu Y, Zhang CM, Yang SH, Wang SL. 2006. Antimicrobial effect of acidified nitrate and nitrite on six common oral pathogens in vitro. Chinese Medical Journal 119(22):1904–9.

44. Modin A, Björne H, Herulf M, Alving K, Weitzberg E, Lundberg JON. 2001. Nitrite-derived nitric oxide: a possible mediator of 'acidic–metabolic' vasodilation. Acta Physiologica Scandinavica 171(1):9–16.

45. Castello PR, David PS, McClure T, Crook Z, Poyton RO. 2006. Mitochondrial cytochrome oxidase produces nitric oxide under hypoxic conditions: Implications for oxygen sensing and hypoxic signaling in eukaryotes. Cell 3(4):277–87.

46. Ghosh S, Kapil V, Fuentes-Calvo I, Bubb K, Pearl V, Milsom A, Khambata R, Maleki-Toyserkani S, Yousuf M, Benjamin N, Webb AJ, Caulfield MJ, Hobbs AJ, and Ahluwala A. 2013. Enhanced vasodilator activity of nitrite in hypertension: Critical role for erythrocytic xanthine oxidoreductase and translational potential. Hypertension 61:1091–1102.

47. Zollbrecht C, Persson AEG, Lundberg JO, Weitzberg E, Carlström M. 2016. Nitrite-mediated reduction of macrophage NADPH oxidase activity is dependent on xanthine oxidoreductase-derived nitric oxide but independent of S-nitrosation. Redox 10:119–27.

48. Zweier JL, Samouilov A, Kuppusamy P. 1999. Non-enzymatic nitric oxide synthesis in biological systems. Biochimica et Biophysica Acta – Bioeneretics 1411(2–3):250–62.

49. Mirmiran P, Bahadoran Z, Golzarand M, Asghari G, Azizi F. 2016. Consumption of nitrate containing vegetables and the risk of chronic kidney disease: Tehran Lipid and Glucose Study. Renal Failure 38(6):937–44.

50.  Cosola C, Sabatino A, di Bari I, Fiaccadori E, Gesualdo L. 2018. Nutrients, Nutraceuticals, and xenobiotics Affecting renal health. Nutrients 10(7):808. https://doi.org/10.3390/nu10070808.

51.  Bondonno CP, Croft KD, Hodgson JM. 2016. Dietary nitrate, nitric oxide, and cardiovascular health. Critical Reviews in Food Science and Nutrition 56(12):2036–52.

52.  Bondonno CP, Blekkenhorst LC, Liu AH, Bondonno NP, Ward NC, Croft KD, Hodgson JM. 2018. Vegetable-derived bioactive nitrate and cardiovascular health. Molecular Aspects of Medicine 61:83–91.

53.  Bedale W, Sindelar JJ, Milkowski AL. 2016. Dietary nitrate and nitrite: Benefits, risks, and evolving perceptions. Meat Science 120:85–92.

54.  van der Avoort, C. M. T., Loon LJv, Hopman MTE, Verdijk LB. 2018. Increasing vegetable intake to obtain the health promoting and ergogenic effects of dietary nitrate. 72(11):1485–9.

55.  McNally B, Griffin JL, Roberts LD. 2016. Dietary inorganic nitrate: From villain to hero in metabolic disease?. Molecular Nutrition and Food Research 60(1):67–78.

56.  Ma L, Hu L, Feng X, Wang S. 2018. Nitrate and nitrite in health and disease Aging and Disease 9(5): 938–45.

57.  Levine AB, Punihaole D, Levine TB. 2012. Characterization of the role of nitric oxide and its clinical applications. Cardiology (Switzerland) 122(1):55–68.

58.  Doel JJ, Benjamin N, Hector MP, Rogers M, Allaker RP. 2005. Evaluation of bacterial nitrate reduction in the human oral cavity. European Journal of Oral Sciences 113(1):14–19.

59.  Liddle L, Burleigh MC, Monaghan C, Muggeridge DJ, Sculthorpe N, Pedlar CR, Butcher J, Henriquez FL, Easton C. 2019. Variability in nitrate-reducing oral bacteria and nitric oxide metabolites in biological fluids following dietary nitrate administration: An assessment of the critical difference. Nitric Oxide – Biology and Chemistry 83:1–10.

60.  Mirvish SS. 1977. Role of N-nitroso compounds (NOC): Their chemical and in vivo formation and possible importance as environmental carcinogens. Journal of Toxicology and Environmental Health. 2(6):1267–77.

61.  Tannenbaum SR, Sinskey AJ, Weisman M, Bishop W. 1974. Nitrite in human saliva. Its possible relationship to nitrosamine formation. Journal of the National Cancer Institute 53(1):79–84.

62.  Mirvish SS. 1995. Role of N-nitroso compounds (NOC) and N-nitrosation in etiology of gastric, esophageal, nasopharyngeal and bladder cancer and contribution to cancer of known exposures to NOC. Cancer Letters 93(1):17–48.

63.  Weitzberg E, Lundberg JO. 2013. Novel aspects of dietary nitrate and human health. Annual Review of Nutrition 33:129–59.

64.  Huang Z, Shiva S, Kim-Sparito DB, Patel RP, Ringwood LA, Irby CE, Huang KT, Ho C, Hogg N, Schechter AN, and Gladwin MT. 2005. Enzymatic function of hemoglobin as a nitrite reductase that produces NO under allosteric control. Journal of Clinical Investigation 115(8):2099–2107.

65.  Li H, Liu X, Cui H, Chen Y, Cardounel AJ, Zweier JL. 2006. Characterization of the cytochrome P450 reductase-cytochrome P450-mediated nitric oxide and nitrosothiol generation from organic nitrates. Journal of Biological Chemistry 28(18):12546–54.

66.  Ignarro LJ, Lippton H, Edwards JC, Baricos WH, Hyman AL, Kadowitz PJ, Gruetter CA. 1981. Mechanism of vascular smooth muscle relaxation by organic nitrates, nitrites, nitroprusside and nitric oxide: evidence for the involvement of S-nitrosothiols as active intermediates. Journal of Pharmacology and Experimental Therapeutics 218(3):739–49.

67.  Ignarro LJ, Gruetter CS. 1980. Requirement of thiols for activation of coronary arterial guanylate cyclase by glyceryl trinitrate and sodium nitrite possible involvement of S-nitrosothiols. 631(2):221–31.
68.  McDonagh STJ, Wylie LJ, Thompson C, Vanhatalo A, Jones AM. 2019. Potential benefits of dietary nitrate ingestion in healthy and clinical populations: A brief review. European Journal of Sport Science 19(1):15–29.
69.  Zhao Y, Vanhoutte PM, Leung SWS. 2015. Vascular nitric oxide: Beyond eNOS. Journal of Pharmaceutical Sciences 129(2):83–94.
70.  Santolini J. 2019. What does "NO-Synthase" stand for?. Frontiers in Bioscience – Landmark 24(1):133–71.
71.  Stuehr DJ, Haque MM. 2018. Nitric oxide synthase enzymology in the 20 years after the Nobel Prize. British Journal of Pharmacology 176(2):177–88.
72.  Alderton WK, Cooper CE, Knowles RG. 2001. Nitric oxide synthases: structure, function and inhibition. Biochemical Journal. 357(3):593–615.
73.  Carvajal JA, Germain AM, Huidobro-Toro JP, Weiner CP. 2000. Molecular mechanism of cGMP-mediated smooth muscle relaxation. Journal of Cellular Physiology 184(3):409–20.
74.  Kehrer JP, Klotz L. 2015. Free radicals and related reactive species as mediators of tissue injury and disease: implications for health. Critical Reviews in Toxicology 45(9):765–98.
75.  Karwowska M, Kononiuk A. 2020. Nitrates/Nitrites in food-risk for nitrosative stress and benefits. Antioxidants 9(3):241. https://doi.org/10.3390/antiox9030241.
76.  Kietadisorn R, Juni RP, Moens AL. 2012. Tackling endothelial dysfunction by modulating NOS uncoupling: New insights into its pathogenesis and therapeutic possibilities. American Journal of Physiology – Endocrinology and Metabolism 302(5):E481–E495.
77.  Forstermann U, Sessa WC. 2012. Nitric oxide synthases: regulation and function. European Herat Journal 33(7):829–37.
78.  Szabó C, Ischiropoulos H, Radi R. 2007. Peroxynitrite: biochemistry, pathophysiology and development of therapeutics. Nature Reviews Drug Discovery 6(8):662–80.
79.  Anavi S, Tirosh O. 2020. iNOS as a metabolic enzyme under stress conditions. Free Radical Biology and Medicine 146:16–35.
80.  Kim-Shapiro DB, Gladwin MT. 2014. Mechanisms of nitrite bioactivation. Nitric Oxide – Biology and Chemistry 38(1):58–68.
81.  Zweier JL, Wang P, Samouilov A, Kuppusamy. 1995. Enzyme-independent formation of nitric oxide in biological tissues. Nature Medicine 1:804–9.
82.  McMahon TJ. 2003. Hemoglobin and nitric oxide. New England Journal of Medicine 349(4):402–5.
83.  Sarti P, Forte E, Giuffrè A, Mastronicola D, Magnifico MC, Arese M. 2012. The chemical interplay between nitric oxide and mitochondrial cytochromecoxidase: Reactions, effectors and pathophysiology. International Journal of Cell Biology 2012:571067. doi: 10.1155/2012/571067.
84.  Allen BW, Piantadosi CA. 2006. How do red blood cells cause hypoxic vasodilation? The SNO-hemoglobin paradigm Am. J. Physiol. Heart Circ. Physiol. 291(4):1507–12.
85.  Hess DT, Matsumoto A, Kim S, Marshall HE, Stamler JS. 2005. Protein S-nitrosylation: purview and parameters. Nat Rev Mol Cell Biol. 6(2):150–166.
86.  Fernando V, Zheng X, Walia Y, Sharma V, Letson J, Furuta S. 2019. S-Nitrosylation: An emerging paradigm of redox signaling. Antioxidants 8(9):404 doi: 10.3390/antiox8090404.
87.  Iammarino M, Di Taranto A, Cristino M. 2013. Monitoring of nitrites and nitrates levels in leafy vegetables (spinach and lettuce): a contribution to risk assessment. Journal of the Sceince of Food and Agriculture 94:773–8.

88. Salehzadeh H, Maleki A, Rezaee R, Shahmoradi B, Ponnet K. 2020. The nitrate content of fresh and cooked vegetables and their health-related risks PLOS ONE. 15(1):e0227551. https://doi.org/10.1371/journal.pone.0227551

89. Kalaycıoğlu Z, Erim FB. 2019. Nitrate and nitrites in foods: Worldwide regional distribution in view of their risks and benefits. J. Agric. Food Chem. 67(26):7205–22. doi: 10.1021/acs.jafc.9b01194.

90. Jackson JK, Patterson AJ, MacDonald-Wicks LK, Bondonno CP, Blekkenhorst LC, Ward NC, Hodgson JM, Byles JE, McEvoy MA. 2018. Dietary nitrate and diet quality: An examination of changing dietary intakes within a representative sample of Australian women Nutrients. 10(8):1005. doi: 10.3390/nu10081005.

91. Colla G, Kim H, Kyriacou MC, Rouphael Y. 2018. Nitrate in fruits and vegetables. Scientia Horticulturae 237:221–38.

92. Dahle HK. 1979. Nitrite as a food additive 2(2):17–24.

93. Lee HS. 2018. Exposure estimates of nitrite and nitrate from consumption of cured meat products by the U.S. population. Food Addit Contam Part A Chem Anal Control Expo Risk Assess. 35(1):29–39. doi:10.1080/19440049.2017.1400696.

94. Archer DL. 2002. Evidence that ingested nitrate and nitrite are beneficial to health J Food Prot. 2002 65(5):872–875. doi: 10.4315/0362-028x-65.5.872.95.

95. Knight TM, Forman D, Al-Dabbagh SA, Doll R. 1987. Estimation of dietary intake of nitrate and nitrite in Great Britain Food Chem Toxicol. 1987 25(4):277–85. doi: 10.1016/0278-6915(87)90123-2.

96. Hunault CC, van Velzen AG, Sips AJAM, Schothorst RC, Meulenbelt J. 2009. Bioavailability of sodium nitrite from an aqueous solution in healthy adults Toxicol Lett. 190(1):48–53. doi: 10.1016/j.toxlet.2009.06.865.97. Cassens. Use of sodium nitrite in cured meats today 1995. 49:72.

98. Rivera N, Bunning M, Martin J. 2019. Uncured-labeled meat products produced using plant-derived nitrates and nitrites: Chemistry, safety, and regulatory considerations J Agric Food Chem. 67(29):8074–84. doi: 10.1021/acs.jafc.9b01826.99.

99. Gassara F, Kouassi AP, Brar SK, Belkacemi K. 2015. Green alternatives to nitrates and nitrites in meat-based products–A review Crit Rev Food Sci Nutr. 56(13):2133–48. doi: 10.1080/10408398.2013.812610100.

100. Bloomer RJ, Butawan M, Pigg B, Martin KR. 2020. Acute ingestion of a novel nitrate-rich dietary supplement significantly increases plasma nitrate/nitrite in physically active men and women Nutrients 12(4):1176. doi: 10.3390/nu12041176.

101. Bloomer RJ, Farney TM, Trepanowski JF, McCarthy CG, Canale RE. 2011. Effect of betaine supplementation on plasma nitrate/nitrite in exercise-trained men. J Int Soc Sports Nutrition 8, 5 https://doi.org/10.1186/1550-2783-8-5.

102. Detopoulou P, Panagiotakos DB, Antonopoulou S, Pitsavos C, Stefanadis C. 2008. Dietary choline and betaine intakes in relation to concentrations of inflammatory markers in healthy adults: the ATTICA study Am J Clin Nutr. 87(2):424–30. doi: 10.1093/ajcn/87.2.424.

103. Trepanowski J, Farney T, McCarthy C, Schilling B, Craig S, Bloomer R. 2011. The effects of chronic betaine supplementation on exercise performance, skeletal muscle oxygen saturation and associated biochemical parameters in resistance trained men J Strength Cond Res. 2011 25(12):3461–71. doi: 10.1519/JSC.0b013e318217d48d.

104. Kapil V, Khambata R, Robertson A, Caulfield M, Ahluwalia A. 2015. Dietary nitrate provides sustained blood pressure lowering in hypertensive patients: A randomized, phase 2, double-blind, placebo-controlled study Hypertension. 65(2):320–7. doi: 10.1161/HYPERTENSIONAHA.114.04675. 105.

105. Miller GD, Marsh AP, Dove RW, Beavers D, Presley T, Helms C, Bechtold E, King SB, Kim-Shapiro D. 2012. Plasma nitrate and nitrite are increased by a high-nitrate supplement but not by high-nitrate foods in older adults Nutr Res. 32(3):160–8. doi: 10.1016/j.nutres.2012.02.002.

106. Vanhatalo A, Stephen J. Bailey, Jamie R. Blackwell, Fred J. DiMenna, Toby G. Pavey, Daryl P. Wilkerson, Nigel Benjamin, Paul G. Winyard, Andrew M. Jones. 2010. Acute and chronic effects of dietary nitrate supplementation on blood pressure and the physiological responses to moderate-intensity and incremental exercise Am J Physiol Regul Integr Comp Physiol. 299(4):R1121–31. doi: 10.1152/ajpregu.00206.2010.107.

107. Gilligan DM, Panza JA, Kilcoyne CM, Waclawiw MA, Casino PR, Quyyumi AA. 1994. Contribution of endothelium-derived nitric oxide to exercise-induced vasodilation Circulation. 90(6):2853–8. doi: 10.1161/01.cir.90.6.2853.

108. Hoon MW, Johnson NA, Chapman PG, Burke LM. 2013. The effect of nitrate supplementation on exercise performance in healthy individuals: A systematic review and meta-analysis Int J Sport Nutr Exerc Metab. 23(5):522–32. doi: 10.1123/ijsnem.23.5.522.

109. Jones A. 2014. Dietary nitrate supplementation and exercise performance. Sports Med. 44 Suppl 1 (Suppl 1):S35–45. doi: 10.1007/s40279-014-0149-y.110.

110. Jones AM, Thompson C, Wylie LJ, Vanhatalo A. 2018. Dietary nitrate and physical performance Annu Rev Nutr. 38:303–28. doi: 10.1146/annurev-nutr-082117-051622. 111.

111. Valenzuela P, Morales J, Emanuele E, Pareja-Galeano H, Lucia A. 2019. Supplements with purported effects on muscle mass and strength. Eur J Nutr. 58(8):2983–3008. doi: 10.1007/s00394-018-1882-z.

112. Clements W, Lee S, Bloomer R. 2014. Nitrate ingestion: A review of the health and physical performance effects Nutrients. 6(11):5224–64. doi: 10.3390/nu6115224.

113. Arrigo F.G. Cicero, Colletti A. 2017. Nutraceuticals and dietary supplements to improve quality of life and outcomes in heart failure patients Curr Pharm 23(8):1265–72. doi: 10.2174/1381612823666170124120518.

114. Hopper I, Connell C, Briffa T, De Pasquale CG, Driscoll A, Kistler PM, Macdonald PS, Sindone A, Thomas L, Atherton JJ. 2020. Nutraceuticals in patients with heart failure: A systematic review 26(2):166–79. DOI: 10.1016/j.cardfail.2019.10.014

115. Cicero A, Colletti A. 2017. Nutraceuticals and dieta.ry supplements to improve quality of life and outcomes in heart failure patients Current Pharmaceutical Design 23(999). DOI: 10.2174/1381612823666170124120518

116. Bryan N. 2018. Functional nitric oxide nutrition to combat cardiovascular disease. Curr Atheroscler Rep. 20(5):21. doi: 10.1007/s11883-018-0723-0.117.

117. Vitale K, Getzin A. 2019. Nutrition and supplement update for the endurance athlete: Review and recommendations Nutrients. 1(6):1289. doi: 10.3390/nu11061289.

118. Mortensen A, Aguilar F, Crebelli R, Di Domenico A, Dusemund B, Frutos MJ, Galtier P, Gott D, Gundert-Remy U, Lambre C, Leblanc JL, Lindtner O, Moldeus P, Mosesso P, Oskarsson A, Parent-Massin D, Stankovic I, Waalkens-Berendsen I, Woutersen RA, Wright M, van den Brandt P, Fortes C, Merino L, Toldrà F, Arcella D, Christodoulidou A, Cortinas Abrahantes J, Barrucci F, Garcia A, Pizzo F, Battacchi D, Younes M. 2017 Re-evaluation of potassium nitrite (E 249) and sodium nitrite (E 250) as food additives European Food Safety Authority, 15(6):e04786 DOI: 10.2903/j.efsa.2017.4786

119. Lundberg JO, Carlström M, Larsen FJ, Weitzberg E. 2011. Roles of dietary inorganic nitrate in cardiovascular health and disease Cardiovasc Res. 89(3):525–32. doi: 10.1093/cvr/cvq325. 120.

120. Vanhatalo A, Blackwell JR, L'Heureux JE, Williams DW, Smith A, van der Giezen M, Winyard PG, Kelly J, Jones AM. 2018. Nitrate-responsive oral microbiome modulates

nitric oxide homeostasis and blood pressure in humans Free Radic Biol Med. 24:21–30. doi: 0.1016/j.freeradbiomed.2018.05.078.121.

121. Gallardo EJ, Coggan AR. 2019. What is in your beet juice? Nitrate and nitrite content of beet juice products marketed to athletes International Journal of Sport Nutrition and Exercise 29(4):345–9. DOI: https://doi.org/10.1123/ijsnem.2018-0223.

122. Wruss J, Waldenberger G, Huemer S, Uygun P, Lanzerstorfer P, Müller U, Höglinger O, Weghuber J. 2015. Compositional characteristics of commercial beetroot products and beetroot juice prepared from seven beetroot varieties grown in Upper Austria. Journal of Food Composition and Analysis 42:46–55.

123. Zamani H, de Joode, MEJR, Hossein IJ, Henckens NFT, Guggeis MA, Berends JE, de Kok, TMCM, van Breda, SGJ. 2021. The benefits and risks of beetroot juice consumption: A systematic review Crit Rev Food Sci Nutr. 2021;61(5):788–804. doi: 10.1080/10408398.2020.1746629.

124. Domínguez R, Maté-Muñoz JL, Cuenca E, García-Fernández P, Mata-Ordoñez F, Lozano-Estevan MC, Veiga-Herreros P, da Silva SF, Garnacho-Castaño MV. 2018. Effects of beetroot juice supplementation on intermittent high-intensity exercise efforts J Int Soc Sports Nutr. 15:2. doi: 10.1186/s12970-017-0204-9.

125. Rothschild JA, Bishop DJ. 2019. Effects of dietary supplements on adaptations to endurance training Sports Med. 50(1):25–53. doi: 10.1007/s40279-019-01185-8.

126. Braakhuis AJ, Hopkins WG. 2015. Impact of dietary antioxidants on sport performance: A review. Sports Med. 45(7):939–55. doi: 10.1007/s40279-015-0323-x.

127. Peeling P, Castell LM, Derave W, de Hon O, Burke LM. 2019. Sports foods and dietary supplements for optimal function and performance enhancement in track-and-field athletes Int J Sport Nutr Exerc Metab. 29(2):198–209. doi: 10.1123/ijsnem.2018-0271.

128. Haun, CT, Kephart, WC, Holland AM, Mobley CB, McCloskey AE, Shake JJ, Pascoe DD, Roberts MD and colleagues. 2016. Differential vascular reactivity responses acutely following ingestion of a nitrate rich red spinach extract Eur J Appl Physiol. 16(11–12):2267–79. doi: 10.1007/s00421-016-3478-8.

129. Moore AN, Haun CT, Kephart WC, Holland AM, Mobley CB, Pascoe DD, Roberts MD, Martin JS. 2017. Red spinach extract increases ventilatory threshold during graded exercise testing. Sports (Basel) 5(4):80. doi: 10.3390/sports5040080

130. Gonzalez A, Accetta M, Spitz R, Mangine G, Ghigiarelli J, Sell K. 2019. Red spinach extract supplementation improves cycle time trial performance in recreationally active men and women J Strength Cond Res. doi: 10.1519/JSC.0000000000003173.

131. Wickham KA, Spriet LL. 2019. No longer beating around the bush: A review of potential sex differences with dietary nitrate supplementation Appl Physiol Nutr Metab. 44(9):915–24. doi: 10.1139/apnm-2019-0063.

132. Bryan NS, Ivy JL. 2015. Inorganic nitrite and nitrate: evidence to support consideration as dietary nutrients Nutr Res. 35(8):643–54. doi: 10.1016/j.nutres.2015.06.001.

133. Buchan GJ, Bonacci G, Fazzari M, Salvatore SR, Gelhaus Wendell S. 2018. Nitro-fatty acid formation and metabolism Nitric Oxide. 79:38–44. doi: 10.1016/j.niox.2018.07.003.

134. Rocha BS, Gago B, Barbosa RM, Cavaleiro C, Laranjinha J. 2015. Ethyl nitrite is produced in the human stomach from dietary nitrate and ethanol, releasing nitric oxide at physiological pH: potential impact on gastric motility Free Radic Biol Med. 82:160–6. doi: 10.1016/j.freeradbiomed.2015.01.021.

135. Habermeyer M, Roth A, Guth S, Diel P, Engel K, Epe B, Fuerst P, Heinz V, Humpf H, Joost H, Knorr D, DeKk T, Kulling S, Lampen A, Marko D, Rechkmmer G, Rietjens I, Stadler RH, Vieths S, Vogel R, Steinberg P, and Eisenbrand G. 2015. Nitrate and nitrite

in the diet: How to assess their benefit and risk for human health Mol Nutr Food Res. 59(1):106–28. doi: 10.1002/mnfr.201400286.

136. Hord NG, Tang Y, Bryan NS. 2009. Food sources of nitrates and nitrites: the physiologic context for potential health benefits Am J Clin Nutr. 90(1):1–10. doi: 10.3945/ajcn.2008.27131.

137. Siervo M, Scialò F, Shannon OM, Stephan BCM, Ashor AW. 2018. Does dietary nitrate say NO to cardiovascular ageing? Current evidence and implications for research. Proceedings of the Nutrition Society 2018. 77:112–23.

138. Jackson J, Patterson AJ, MacDonald-Wicks L, McEvoy M. 2017. The role of inorganic nitrate and nitrite in CVD. Nutr Res Rev. 30(2):247–64. doi: 10.1017/S0954422417000105.

139. Woessner MN, McIlvenna LC, Ortiz de Zevallos J, Neil CJ, Allen JD. 2018. Dietary nitrate supplementation in cardiovascular health: an ergogenic aid or exercise therapeutic? Am J Physiol Heart Circ Physiol. 314(2):H195–H212. doi: 10.1152/ajpheart.00414.2017.

140. Raubenheimer K, Bondonno C, Blekkenhorst L, Wagner K, Peake JM, Neubauer O. 2019. Effects of dietary nitrate on inflammation and immune function, and implications for cardiovascular health. Nutrition Reviews. 77(8)nuz25:584–99. doi: 10.1093/nutrit/nuz025.

141. Mayra ST, Johnston CS, Sweazea KL. 2019. High-nitrate salad increased plasma nitrates/nitrites and brachial artery flow-mediated dilation in postmenopausal women: A pilot study Nutr Res. 65:99–104. doi: 10.1016/j.nutres.2019.02.001.

142. Ormesher L, Myers JE, Chmiel C, Wareing M, Greenwood SL, Tropea T, Lundberg JO, Weitzberg E, Nihlen C, Sibley CP, Johnstone ED, and Cottrell EC. 2018. Effects of dietary nitrate supplementation, from beetroot juice, on blood pressure in hypertensive pregnant women: A randomised, double-blind, placebo-controlled feasibility trial Nitric Oxide. 80:37–44. doi: 10.1016/j.niox.2018.08.004.

143. Petersson J, Carlström M, Schreiber O, Phillipson M, Christoffersson G, Jägare A, Roos S, Jansson EA, Persson AEG, Lundberg JO, and Holm L. 2009. Gastroprotective and blood pressure lowering effects of dietary nitrate are abolished by antiseptic mouthwash. Free Radic Biol Med. 46(8):1068–75. doi: 10.1016/j.freeradbiomed.2009.01.011.

144. Govoni M, Jansson EÅ, Weitzberg E, Lundberg JO. 2008. The increase in plasma nitrite after a dietary nitrate load is markedly attenuated by an antibacterial mouthwash Nitric Oxide. 19(4):333–7. doi: 10.1016/j.niox.2008.08.003.

145. Briskey D, Tucker PS, Johnson DW, Coombes JS. 2016. Microbiota and the nitrogen cycle: Implications in the development and progression of CVD and CKD Nitric Oxide. 57:64–70. doi: 10.1016/j.niox.2016.05.002.

146. Tribble GD, Angelov N, Weltman R, Wang B, Eswaran SV, Gay IC, Parthasarathy K, Dao DV, Richardson KN, Ismail NM, Sharina IG, Hyde ER, Ajami NJ, Petrosino JF, and Bryan NS. 2019. Frequency of tongue cleaning impacts the human tongue microbiome composition and enterosalivary circulation of nitrate Frontiers in Cellular and Infection Microbiology. 9:39. DOI: 10.3389/fcimb.2019.00039.

147. Burleigh M, Liddle L, Muggeridge DJ, Monaghan C, Sculthorpe N, Butcher J, Henriquez F, Easton C. 2019. Dietary nitrate supplementation alters the oral microbiome but does not improve the vascular responses to an acute nitrate dose Nitric Oxide. 89:54–63. doi: 10.1016/j.niox.2019.04.010.

148. Tune J. 2004. Managing coronary blood flow to myocardial oxygen consumption J Appl Physiol (1985). 97(1):404–15. doi: 10.1152/japplphysiol.01345.2003.

149. Engan HK, Jones AM, Ehrenberg F, Schagaty E. 2012. Acute dietary nitrate supplementation improves dry static apnea performance Respir Physiol Neurobiol. 182(2–3):53–9. doi: 10.1016/j.resp.2012.05.007.

150. Kapil V, Milsom AB, Okorie M, Maleki-Toyserkani S, Akram F, Rehman F, Arghandawi S, Pearl V, Benjamin N, Loukogeorgakis S, Macallister R, Hobbs AJ, Webb Aj, and Ahluwalia A. 2010. Inorganic nitrate supplementation lowers blood pressure in humans: Role for nitrite-derived NO. Hypertension. 56(2):274–81. doi: 10.1161/HYPERTENSIONAHA.110.153536.

151. Asgary S, Afshani MR, Sahebkar A, Keshvari M, Taheri M, Jahanian E, Rafieian-Kopaei M, Malekian F, Sarrafzadegan N. 2016. Improvement of hypertension, endothelial function and systemic inflammation following short-term supplementation with red beet (Beta vulgaris L.) juice: a randomized crossover pilot study J Hum Hypertens. 30(10):627–32. doi: 10.1038/jhh.2016.34.

152. Velmurugan S, Gan JM, Rathod KS, Khambata RS, Ghosh SM, Hartley A, Van Eijl S, Sagi-Kiss V, Chowdhury TA, Curtis M, Kuhnle GG, Wade WG, and Ahluwalia A. 2016. Dietary nitrate improves vascular function in patients with hypercholesterolemia: a randomized, double-blind, placebo-controlled study Am J Clin Nutr. 3(1):25–38. doi: 10.3945/ajcn.115.116244.

153. Webb AJ, Patel N, Loukogeorgakis S, Okorie M, Aboud Z, Misra S, Rashid R, Miall P, Deanfield J, Benjamin N, MacAllister R, Hobbs AJ, and Ahluwalia A. 2008. Acute blood pressure lowering, vasoprotective, and antiplatelet properties of dietary nitrate via bioconversion to nitrite Hypertension. 51(3):784–90. doi: 10.1161/HYPERTENSIONAHA.107.103523.

154. Dejam A, Hunter CJ, Gladwin MT. 2007. Effects of dietary nitrate on blood pressure New England Journal of Medicine 356(15):1590. DOI: 10.1056/NEJMc070163.

155. d'El-Rei J, Cunha AR, Trindade M, Neves MF. 2016. Beneficial effects of dietary nitrate on endothelial function and blood pressure levels. Int J Hypertens. 2016:6791519. doi: 10.1155/2016/6791519.

156. Gee LC, Ahluwalia A. 2016. Dietary nitrate lowers blood pressure: Epidemiological, pre-clinical experimental and clinical trial evidence Curr Hypertens Rep. 8(2):17. doi: 10.1007/s11906-015-0623-4.

157. Siervo M, Lara J, Ogbonmwan I, Mathers JC. 2013. Inorganic nitrate and beetroot juice supplementation reduces blood pressure in adults: A systematic review and meta-analysis J Nutr. 43(6):818–26. doi: 10.3945/jn.112.170233.

158. Oggioni C, Jakovljevic DG, Klonizakis M, Ashor AW, Ruddock A, Ranchordas M, Williams E, Siervo M. 2018. Dietary nitrate does not modify blood pressure and cardiac output at rest and during exercise in older adults: A randomised cross-over study Int J Food Sci Nutr. 69(1):74–83. doi: 10.1080/09637486.2017.1328666.

159. Husmann F, Bruhn S, Mittlmeier T, Zschorlich V, Behrens M. 2019. Dietary nitrate supplementation improves exercise tolerance by reducing muscle fatigue and perceptual responses Frontiers in Physiology 10(APR):404 https://doi.org/10.3389/fphys.2019.00404

160. Hobbs DA, Goulding MG, Nguyen A, Malaver T, Walker CF, George TW, Methven L, Lovegrove JA. 2013. Acute ingestion of beetroot bread increases endothelium-independent vasodilation and lowers diastolic blood pressure in healthy men: A randomized controlled trial. J Nutr. 2013 43(9):1399–405. doi: 10.3945/jn.113.175778.

161. Hobbs DA, Kaffa N, George TW, Methven L, Lovegrove JA. 2012. Blood pressure-lowering effects of beetroot juice and novel beetroot-enriched bread products in normotensive male subjects Br J Nutr. 08(11):2066–74. doi: 10.1017/S0007114512000190.

162. Ahluwalia A, Gladwin M, Coleman GD, Hord N, Howard G, Kim-Shapiro DB, Lajous M, Larsen FJ, Lefer DJ, McClure LA, Nolan BT, Pluta R, Schechter A, Wang CY,

Ward MH, Harman JL. 2016. Dietary nitrate and the epidemiology of cardiovascular disease: Report from a national heart, lung, and blood institute workshop J Am Heart Assoc. 5(7):e003402. doi: 10.1161/JAHA.116.003402.

163. Dejam A, Hunter CJ, Schechter AN, Gladwin MT. 2004. Emerging role of nitrite in human biology. Blood Cells Mol Dis. 32(3):423–9. doi: 10.1016/j.bcmd.2004.02.002.

164. Keen JT, Levitt EL, Hodges GJ, Wong BJ. 2015. Short-term dietary nitrate supplementation augments cutaneous vasodilatation and reduces mean arterial pressure in healthy humans. Microvasc Res. 98:48–53. doi: 10.1016/j.mvr.2014.12.002.

165. Bahadoran Z, Mirmiran P, Kabir A, Azizi F, Ghasemi A. 2018. The nitrate-independent blood pressure–lowering effect of beetroot juice: A systematic review and meta-analysis Adv Nutr. 2017 Nov 15;8(6):830–8. doi: 10.3945/an.117.016717.

166. Larsen FJ, Ekblom B, Sahlin K, Lundberg JO, Weitzberg E. 2006. Effects of dietary nitrate on blood pressure in healthy volunteers. N Engl J Med. 355(26):2792–3. doi: 10.1056/NEJMc062800.

167. Remington J, Winters K. 2019. Effectiveness of dietary inorganic nitrate for lowering blood pressure in hypertensive adults: a systematic review JBI Database System Rev Implement Rep. 17(3):365–89. doi: 10.11124/JBISRIR-2017-003842.

168. Broxterman RM, La Salle DT, Zhao J, Reese VR, Richardson RS, Trinity JD. 2019. Influence of dietary inorganic nitrate on blood pressure and vascular function in hypertension: prospective implications for adjunctive treatment J Appl Physiol. 127(4):1085–94. doi: 10.1152/japplphysiol.00371.2019.

169. Bonilla Ocampo DA, Paipilla AF, Marín E, Vargas-Molina S, Petro JL, Pérez-Idárraga A. 2018. Dietary nitrate from beetroot juice for hypertension: A systematic review. Biomolecules. 8(4):134. doi: 10.3390/biom8040134.

170. Ashworth A, Cutler C, Farnham G, Liddle L, Burleigh M, Rodiles A, Sillitti C, Kiernan M, Moore M, Hickson M, Easton C, and Bescos R. 2019. Dietary intake of inorganic nitrate in vegetarians and omnivores and its impact on blood pressure, resting metabolic rate and the oral microbiome Free Radic Biol Med. 38:63–72. doi: 10.1016/j.freeradbiomed.2019.05.010.

171. Stanaway L, Rutherfurd-Markwick K, Page R, Ali A. 2017. Performance and health benefits of dietary nitrate supplementation in older adults: A systematic review. Nutrients. 9(11):1171. doi: 10.3390/nu9111171.

172. Nadtochiy SM, Redman EK. 2011. Mediterranean diet and cardioprotection: The role of nitrite, polyunsaturated fatty acids, and polyphenols. Nutrition. 27(7–8):733–44. doi: 10.1016/j.nut.2010.12.006.

173. Salloum FN, Sturz GR, Yin C, Rehman S, Hoke NN, Kukreja RC, Xi L. 2015. Beetroot juice reduces infarct size and improves cardiac function following ischemia–reperfusion injury: Possible involvement of endogenous H2S. Exp Biol Med (Maywood) 240(5):669–81. doi: 10.1177/1535370214558024

174. Wong BJ, Keen JT, Levitt EL. 2018. Cutaneous reactive hyperaemia is unaltered by dietary nitrate supplementation in healthy humans Clin Physiol Funct Imaging. 38(5):772–8. doi: 10.1111/cpf.12478.

175. Levitt EL, Keen JT, Wong BJ. 2015. Augmented reflex cutaneous vasodilatation following short-term dietary nitrate supplementation in humans Exp Physiol. 00(6):708–18. doi: 10.1113/EP085061.

176. Walker MA, Bailey TG, McIlvenna L, Allen JD, Green DJ, Askew CD. 2019. Acute dietary nitrate supplementation improves flow mediated dilatation of the superficial femoral artery in healthy older males Nutrients. 11(5):954. doi: 10.3390/nu11050954.

177. Stratton MA. 1984. Use of nitrates in patients with acute myocardial infarction Clin Pharm. 3(1):32–9.

178. Baker JE, Su J, Fu X, Hsu A, Gross GJ, Tweddell JS, Hogg N. 2007. Nitrite confers protection against myocardial infarction: Role of xanthine oxidoreductase, NADPH oxidase and KATP channels. J Mol Cell Cardiol. 43(4):437–44. doi: 10.1016/j.yjmcc.2007.07.057.

179. Jackson JK, Zong G, MacDonald-Wicks LK, Patterson AJ, Willett WC, Rimm EB, Manson JE, McEvoy MA. 2019. Dietary nitrate consumption and risk of CHD in women from the Nurses' Health Study. Br J Nutr. 121(7):831–8. doi: 10.1017/S0007114519000096.

180. Lara J, Ashor AW, Oggioni C, Ahluwalia A, Mathers JC, Siervo M. 2015. Effects of inorganic nitrate and beetroot supplementation on endothelial function: a systematic review and meta-analysis. Eur J Nutr. 55(2):451–9. doi: 10.1007/s00394-015-0872-7.

181. Holloway CJ, Montgomery HE, Murray AJ, Cochlin LE, Coeanu I, Hopwood N, Johnson AW, Rider OJ, Levett DZH, Tyler DJ, Francis JM, Neubauer S, Grocott MP, Clarke K; Caudwell Xtreme Everest Research Group. 2011. Cardiac response to hypobaric hypoxia: persistent changes in cardiac mass, function, and energy metabolism after a trek to Mt. Everest Base Camp. FASEB J. 25(2):792–6. doi: 10.1096/fj.10-172999.

182. Heather L, Cole M, Tan J, Ambrose L, Pope S, Abd-Jamil A, Carter E, Dodd M, Yeoh K, Schofield C, and Clarke K. 2012. Metabolic adaptation to chronic hypoxia in cardiac mitochondria. *Basic Res Cardiol.* 107:1–12.

183. Ashmore T, Fernandez BO, Branco-Price C, West JA, Cowburn AS, Heather LC, Griffin JL, Johnson RS, Feelisch M, Murray AJ. 2014. Dietary nitrate increases arginine availability and protects mitochondrial complex I and energetics in the hypoxic rat heart J Physiol. 592(21):4715–31. doi: 10.1113/jphysiol.2014.275263.

184. Pernow J, Jung C. 2013. Arginase as a potential target in the treatment of cardiovascular disease: reversal of arginine steal? Cardiovascular Research 98(3):334–43. doi: 10.1093/cvr/cvt036.

185. Ogoh S. 2017. Relationship between cognitive function and regulation of cerebral blood flow. *J Physiol Sci.* 67(3):345–51. doi: 10.1007/s12576-017-0525-0.

186. Peterson EC, Wang Z, Britz G. 2011. Regulation of cerebral blood flow. Int J Vasc Med. 2011:823525. doi: 10.1155/2011/823525.

187. Presley TD, Morgan AR, Bechtold E, Clodfelter W, Dove RW, Jennings JM, Kraft RA, Bruce King S, Laurienti PJ, Jack Rejeski W, Burdette JH, Kim-Shapiro DB, and Miller GD. 2011. Acute effect of a high nitrate diet on brain perfusion in older adults. Nitric Oxide. 24(1):34–42. doi: 10.1016/j.niox.2010.10.002.

188. Bescós R, Ferrer-Roca V, Galilea PA, Roig A, Drobnic F, Sureda A, Martorell M, Cordova A, Tur JA, Pons A. 2012. Sodium nitrate supplementation does not enhance performance of endurance athletes. Med Sci Sports Exerc. 44(12):2400–9. doi: 10.1249/MSS.0b013e3182687e5c.

189. Fan J, Bourdillon N, Meyer P, Kayser B. 2018. Oral nitrate supplementation differentially modulates cerebral artery blood velocity and prefrontal tissue oxygenation during 15 km time-trial cycling in Normoxia but not in hypoxia. Frontiers in Physiology 9(JUL):869.

190. Kelly J, Fulford J, Vanhatalo A, Blackwell JR, French O, Bailey SJ, Gilchrist M, Paul G. Winyard, Jones AM. 2013. Effects of short-term dietary nitrate supplementation on blood pressure, O2 uptake kinetics, and muscle and cognitive function in older adults. Am J Physiol Regul Integr Comp Physiol. 304(2):R73–83. doi: 10.1152/ajpregu.00406.2012.

191. Wightman EL, Haskell-Ramsay CF, Thompson KG, Blackwell JR, Winyard PG, Forster J, Jones AM, Kennedy DO. 2015. Dietary nitrate modulates cerebral blood flow parameters and cognitive performance in humans: A double-blind, placebo-controlled, crossover investigation Physiol Behav. 149:149–58. doi: 10.1016/j.physbeh.2015.05.035.

192. Masschelein E, Ruud Van Thienen, Xu Wang, Ann Van Schepdael, Martine Thomis, Peter Hespel. 2012. Dietary nitrate improves muscle but not cerebral oxygenation status during exercise in hypoxia J Appl Physiol (1985). 113(5):736–45. doi: 10.1152/japplphysiol.01253.2011.

193. Diaconu CC, Manea M, Marcu DR, Socea B, Spinu AD, Bratu OG. 2020. The erectile dysfunction as a marker of cardiovascular disease: a review 2019. Acta Cardiol. 75(4):286–92. doi: 10.1080/00015385.2019.1590498.

194. Burnett AL. 1997. Nitric oxide in the penis: Physiology and pathology 1997. J Urol. 157(1):320–4.

195. Andersson K. 2018. PDE5 inhibitors – pharmacology and clinical applications 20 years after sildenafil discovery. Br J Pharmacol. 175(13):2554–65. doi: 10.1111/bph.14205.

196. Kavoussi PK, Smith RP, Oliver JL, Costabile RA, Steers WD, Brown-Steinke K, de Ronde K, Lysiak JJ, Palmer LA. 2019. S-nitrosylation of endothelial nitric oxide synthase impacts erectile function. Int J Impot Res. 31(1):31–8. doi: 10.1038/s41443-018-0056-0.

197. Burnett AL. 2004. Novel nitric oxide signaling mechanisms regulate the erectile response. Int J Impot Res. 16:S15–S9. https://doi.org/10.1038/sj.ijir.3901209.

198. Bailey SJ, Jonathan Fulford, Anni Vanhatalo, Paul G. Winyard, Jamie R. Blackwell, Fred J. DiMenna, Daryl P. Wilkerson, Nigel Benjamin, Andrew M. Jone1. 2010. Dietary nitrate supplementation enhances muscle contractile efficiency during knee-extensor exercise in humans. J Appl Physiol (1985). 109(1):135–48. doi: 10.1152/japplphysiol.00046.2010.

199. Vanhatalo A, Fulford J, Bailey SJ, Blackwell JR, Winyard PG, Jones AM. 2011. Dietary nitrate reduces muscle metabolic perturbation and improves exercise tolerance in hypoxia. J Physiol. 589(Pt 22):5517–28. doi: 10.1113/jphysiol.2011.216341.

200. Lansley KE, Winyard PG, Bailey SJ, Vanhatalo A, Wilkerson DP, Blackwell JR, Gilchrist M, Benjamin N, Jones AM. 2011. Acute dietary nitrate supplementation improves cycling time trial performance. Med Sci Sports Exerc. 43(6):1125–31. doi: 10.1249/MSS.0b013e31821597b4.

201. Lansley KE, Winyard PG, Fulford J, Anni Vanhatalo, Bailey SJ, Blackwell JR, DiMenna FJ, Gilchrist M, Benjamin N, Jones AM. 2011. Dietary nitrate supplementation reduces the O2 cost of walking and running: a placebo-controlled study. J Appl Physiol (1985). 10(3):591–600. doi: 10.1152/japplphysiol.01070.2010.

202. Thompson KG, Turner L, Prichard J, Dodd F, Kennedy DO, Haskell C, Blackwell JR, Jones AM. 2014. Influence of dietary nitrate supplementation on physiological and cognitive responses to incremental cycle exercise. Respir Physiol Neurobiol. 93:11–20. doi: 10.1016/j.resp.2013.12.015. Epub 2013 Dec 31.

203. Bailey SJ, Winyard P, Vanhatalo A, Blackwell JR, DiMenna FJ, Wilkerson DP, Tarr J, Nigel Benjamin, Jones AM. 2000. Dietary nitrate supplementation reduces the O2 cost of low-intensity exercise and enhances tolerance to high-intensity exercise in humans. J Appl Physiol (1985). 07(4):1144–55. doi: 10.1152/japplphysiol.00722.2009.

204. Larsen FJ, Weitzberg E, Lundberg JO, Ekblom B. 2010. Dietary nitrate reduces maximal oxygen consumption while maintaining work performance in maximal exercise. Free Radic Biol Med. 48(2):342–7. doi: 10.1016/j.freeradbiomed.2009.11.006.

205. Larsen FJ, Schiffer TA, Borniquel S, Sahlin K, Ekblom B, Lundberg JO, Weitzberg E. 2011. Dietary inorganic nitrate improves mitochondrial efficiency in humans. Cell Metab. 13(2):149–59. doi: 10.1016/j.cmet.2011.01.004.

206. Ferguson SK, Hirai DM, Copp SW, Holdsworth CT, Allen JD, Jones AM, Musch TI, Poole DC. 2013. Effects of nitrate supplementation via beetroot juice on contracting rat skeletal muscle microvascular oxygen pressure dynamics. 187(3):250–55.

207. Affourtit C, Bailey SJ, Jones AM, Smallwood MJ, Winyard PG. 2015. On the mechanism by which dietary nitrate improves human skeletal muscle function. Front Physiol. 6:211. doi: 10.3389/fphys.2015.00211.

208. Wylie LJ, Park JW, Vanhatalo A, Kadach S, Black MI, Stoyanov Z, Schechter AN, Jones AM, Piknova B. 2019. Human skeletal muscle nitrate store: influence of dietary nitrate supplementation and exercise. J Physiol. 597(23):5565–76. doi: 10.1113/JP278076.

209. Kelly J, Vanhatalo A, Wilkerson D. 2013. Effects of nitrate on the power–duration relationship for severe-intensity exercise. Med Sci Sports Exerc. 45(9):1798–806. doi: 10.1249/MSS.0b013e31828e885c.

210. Thompson C, Vanhatalo A, Jell H, Fulford J, Carter J, Nyman L, Bailey SJ, Jones AM. 2016. Dietary nitrate supplementation improves sprint and high-intensity intermittent running performance. Nitric Oxide. 61:55–61. doi: 10.1016/j.niox.2016.10.006.

211. Thompson C, Wylie L, Fulford J, Kelly J, Black M, McDonagh S, Jeukendrup A, Vanhatalo A, Jones A. 2015. Dietary nitrate improves sprint performance and cognitive function during prolonged intermittent exercise. Eur J Appl Physiol. 15(9):1825–34. doi: 10.1007/s00421-015-3166-0.

212. Margaret M, Eliot K, Heuertz RM, Weiss E. 2012. Whole beetroot consumption acutely improves running performance . J Acad Nutr Diet. 112(4):548–52. doi: 10.1016/j.jand.2011.12.002.

213. Bond H, Morton L, Braakhuis AJ. 2012. Dietary nitrate supplementation improves rowing performance in well-trained rowers. Int J Sport Nutr Exerc Metab. 2012 Aug;22(4):251–6. doi: 10.1123/ijsnem.22.4.251.

214. Muggeridge DJ, Howe CCF, Spendiff O, Pedlar C, James PE, Easton C. 2013. The effects of a single dose of concentrated beetroot juice on performance in trained flatwater kayakers. Int J Sport Nutr Exerc Metab. 2013 23(5):498–506. doi: 10.1123/ijsnem.23.5.498.

215. Cermak NM, Res P, Stinkens RE, Lundberg JO, Gibala MJ, van Loon, L. J. C. 2012. No improvement in endurance performance after a single dose of beetroot juice. Int J Sport Nutr Exerc Metab. 22(6):470–8. doi: 10.1123/ijsnem.22.6.470.

216. Bescós R, Rodríguez FA, Iglesias X, Ferrer MD, Pons A. 2011. Acute administration of inorganic nitrates reduces VO2 peak in endurance athletes. Med Sci Sports Exerc. 43(10):1979–86. doi: 10.1249/MSS.0b013e318217d439.

217. Wilkerson D, Hayward G, Bailey S, Vanhatalo A, Blackwell J, Jones A. 2012. Influence of acute dietary nitrate supplementation on 50 mile time trial performance in well-trained cyclists. Eur J Appl Physiol. 112(12):4127–34. doi: 10.1007/s00421-012-2397-6.

218. Garnacho-Castaño MV, Palau-Salvà G, Cuenca E, Muñoz-González A, García-Fernández P, del Carmen Lozano-Estevan M, Veiga-Herreros P, Maté-Muñoz JL, Domínguez R. 2018. Effects of a single dose of beetroot juice on cycling time trial performance at ventilatory thresholds intensity in male triathletes. J Int Soc Sports Nutr. 15(1):49. doi: 10.1186/s12970-018-0255-6.

219. Wylie L, Mohr M, Krustrup P, Jackman S, Ermιdis G, Kelly J, Black M, Bailey S, Vanhatalo A, Jones A. 2013. Dietary nitrate supplementation improves team sport-specific intense

intermittent exercise performance. Eur J Appl Physiol. 13(7):1673–84. doi: 10.1007/s00421-013-2589-8.

220. Kramer SJ, Baur DA, Spicer MT, Vukovich MD, Ormsbee MJ. 2016. The effect of six days of dietary nitrate supplementation on performance in trained CrossFit athletes. Journal of the International Society of Sports Nutrition. 13(1):39.

221. Lee J. Wylie, James Kelly, Stephen J. Bailey, Jamie R. Blackwell, Philip F. Skiba, Paul G. Winyard, Asker E. Jeukendrup, Anni Vanhatalo, Andrew M. Jones. 2013. Beetroot juice and exercise: pharmacodynamic and dose-response relationships. J Appl Physiol (1985). 115(3):325–36. doi: 10.1152/japplphysiol.00372.2013.

222. Kenjale AA, Katherine L. Ham, Thomas Stabler, Jennifer L. Robbins, Johanna L. Johnson, Mitch VanBruggen, Grayson Privette, Eunji Yim, William E. Kraus, Jason D. Allen. 2011. Dietary nitrate supplementation enhances exercise performance in peripheral arterial disease. J Appl Physiol (1985). 110(6):1582–91. doi: 10.1152/japplphysiol.00071.2011.

223. Kerley C. 2017. Dietary nitrate as modulator of physical performance and cardiovascular health. Curr Opin Clin Nutr Metab Care. 20(6):440–6. doi: 10.1097/MCO.0000000000000414.

224. Kerley CP, Dolan E, James PE, Cormican L. 2018. Dietary nitrate lowers ambulatory blood pressure in treated, uncontrolled hypertension: a 7-d, double-blind, randomised, placebo-controlled, cross-over trial. Br J Nutr. 119(6):658–63. doi: 10.1017/S0007114518000144.

225. Breese BC. 2013. Beetroot juice supplementation speeds o2 uptake kinetics and improves exercise tolerance during severe-intensity exercise initiated from an elevated metabolic rate. Am J Physiol Regul Integr Comp Physiol. 305(12):R1441–50. doi: 10.1152/ajpregu.00295.2013.

226. Van De Walle G, Vukovich M. 2018. The effect of nitrate supplementation on exercise tolerance and performance: A systematic review and meta-analysis J Strength Cond Res. 2018 Jun;32(6):1796–1808. doi: 10.1519/JSC.0000000000002046.

227. McMahon NF, Leveritt MD, Pavey TG. 2017. The effect of dietary nitrate supplementation on endurance exercise performance in healthy adults: A systematic review and meta-analysis. Sports Med. 47(4):735–56. doi: 10.1007/s40279-016-0617-7.

228. Campos HO, Drummond LR, Rodrigues QT, Machado FSM, Pires W, Wanner SP, Coimbra CC. 2018. Nitrate supplementation improves physical performance specifically in non-athletes during prolonged open-ended tests: a systematic review and meta-analysis. Br J Nutr. 119(6):636–57. doi: 10.1017/S0007114518000132.

229. Larsen FJ, Weitzberg E, Lundberg JO, Ekblom B. 2007. Effects of dietary nitrate on oxygen cost during exercise. Acta Physiol (Oxf). 191(1):59–66. doi: 10.1111/j.1748-1716.2007.01713.x.

230. Bahadoran Z, Ghasemi A, Mirmiran P, Azizi F, Hadaegh F. 2015. Beneficial effects of inorganic nitrate/nitrite in type 2 diabetes and its complications. Nutr Metab (Lond). 12(1):16. doi: 10.1186/s12986-015-0013-6.

231. Mirmiran P, Houshialsadat Z, Gaeini Z, Bahadoran Z, Azizi F. 2020. Functional properties of beetroot (Beta vulgaris) in management of cardio-metabolic diseases. Nutr Metab (Lond) 17:3. doi: 10.1186/s12986-019-0421-0. eCollection 2020.

232. Wootton-Beard PC, Brandt K, Fell D, Warner S, Ryan L. 2014. Effects of a beetroot juice with high neobetanin content on the early-phase insulin response in healthy volunteers. J Nutr Sci 3e9. doi: 10.1017/jns.2014.7. eCollection 2014.

233. Omar SA, Webb AJ, Lundberg JO, Weitzberg E. 2016. Therapeutic effects of inorganic nitrate and nitrite in cardiovascular and metabolic diseases. J Intern Med 279(4):315–30. doi: 10.1111/joim.12441.

234. Olumese FE, Oboh H. 2016. Effects of daily intake of beetroot juice on blood glucose and hormones in young healthy subjects. Nigerian Quarterly Journal of Hospital Medicne. 26(2):455–62.

235. Beals JW, Binns SE, Davis JL, Giordano GR, Klochak AL, Paris HL, Schweder MM, Peltonen GL, Scalzo RL, Bell C. 2017. Concurrent beet juice and carbohydrate ingestion: Influence on glucose tolerance in obese and nonobese adults. J Nutr Metab 6436783. doi: 10.1155/2017/6436783.

# 15 Nitrates and Methemoglobinemia

*Sarah Fossen Johnson*

## CONTENTS

15.1  Introduction..................................................................................................347
15.2  Nitrate Toxicokinetics...................................................................................348
15.3  Congenital Methemoglobinemia....................................................................349
15.4  Acquired Methemoglobinemia......................................................................350
15.5  Examples of Methemoglobinemia Caused by Nitrate or Nitrite...............351
15.6  Conclusion.....................................................................................................352
References.................................................................................................................353

## 15.1  INTRODUCTION

Red blood cells contain a tetramer protein called hemoglobin. Hemoglobin binds iron and delivers oxygen to the tissues of the body. Methemoglobin is created when oxidizing chemicals or pharmaceuticals oxidize the iron in hemoglobin, leading to a conformational change. The hemoglobin can no longer bind and deliver oxygen easily, so the tissues of the body do not receive enough oxygen. If methemoglobin concentration in the blood reaches above 1–2 percent, it is considered to be methemoglobinemia, although most people with methemoglobin of 3 percent or below are asymptomatic (Fossen Johnson 2019).

Because there are also endogenous sources of oxidizing chemicals such as oxygen free radicals (Kuiper-Prins et al. 2016), the body has developed several pathways to reduce methemoglobin back to hemoglobin. The dominant pathway utilizes the NAD cytochrome b5 reductase enzyme to reduce methemoglobin to hemoglobin. In addition to its role in methemoglobinemia, NAD cytochrome b5 reductase plays a role in maintaining CoQ10 and ascorbic acid in their reduced states, and the prevention of lipid peroxidation in membranes, at both the cellular and mitochondrial levels (Fusco et al. 2011).

There is a lesser pathway that utilizes the enzyme NADPH methemoglobin reductase. This enzyme is likely only responsible for about 5 percent of all reductions (Mansouri and Lurie 1993). This path is minor; the enzyme is more of a non-specific reductase that can work on dyes that are usually found exogenously. Methylene blue is an example of the type of dye that NADPH methemoglobin reductase can work on. Interestingly, NADPH methemoglobin reductase reduces the dye which, in

DOI: 10.1201/9780429326806-20

turn, reduces methemoglobin, making it a good treatment for methemoglobinemia (Kuiper-Prins et al. 2016). This is a very important pathway for people who have a deficiency in NAD cytochrome b5 reductase because it may be the only path functioning to reduce methemoglobin. These two pathways work to maintain homeostasis (Ash-Bernal, Wise, and Wright 2004).

When more than 1 percent of a person's blood is methemoglobin, they have methemoglobinemia (Ash-Bernal, Wise, and Wright 2004). In babies, it may present as nonspecific symptoms such as fussiness and lethargy. In adults, it may present as shortness of breath, weakness, headache, hypoxia that cannot be improved with additional oxygen, a blue color to their skin – also described as a gray or dusky color, and chocolate-colored blood (Bayat and Kosinski 2011, Taleb et al. 2013). When methemoglobin levels reach 70 percent and higher, it can be fatal (Taleb et al. 2013).

Often, mild cases of methemoglobinemia will spontaneously resolve without treatment. In more serious cases, methylene blue will be administered intravenously, or ascorbic acid, N-adenylcystein (NAC), and/or Q10 will be administered. Methylene blue works by NADPH methemoglobin reductase reducing it to leukomethylene blue, which then reduces methemoglobin to hemoglobin. Interestingly, methylene blue at high concentrations can be an oxidizing agent that can create methemoglobin. Ascorbic acid can also be used to treat methemoglobinemia. It works directly on the methemoglobin to reduce it to hemoglobin (Kang et al. 2018). N-adenylcysteine acts as a cofactor in reducing the methemoglobin, or by intracellular glutathione cysteine, both of which are products of NAC metabolism (Tanen, LoVecchio, and Curry 2000).

Compared to adults, infants have different physiology that makes them more susceptible to methemoglobinemia than older children and adults. Adults reduce methemoglobinemia almost twice as fast as infants can and have higher levels of NAD cytochrome b5 reductase, the enzyme associated with a reduction of methemoglobin (Yip and Spyker 2018). Infants also have a higher percentage of fetal hemoglobin, as compared to adults. It is easier to oxidize fetal hemoglobin than adult hemoglobin (Kuiper-Prins et al. 2016). When infants are exposed to nitrate in water, those under six months drink more water for their weight as compared to adults, and their exposure will be higher (U.S. Environmental Protection Agency – Office of Research and Development 2011).

The study of methemoglobinemia and nitrate can be traced back more than seventy years in the United States. It started as a hunch about the well water from the father of an infant that had cyanosis and associated methemoglobinemia. It led to the first discovery that well water contaminated with nitrate could cause methemoglobinemia. Over the years there has been debate about the role of diarrheal illness in methemoglobinemia development in infants exposed to nitrate in well water. Regardless of the source, methemoglobinemia has been documented in many age groups including infants, older children, adults, and older adults.

## 15.2   NITRATE TOXICOKINETICS

When an adult ingests nitrate, it is rapidly absorbed by the small intestine and widely distributed into all body fluids (Dusdieker et al. 1996, Hord et al. 2011). Once it has been absorbed, some of the nitrate in the blood concentrates into the salivary glands

(Dusdieker et al. 1996, Qu et al. 2016). Within ten minutes, it is secreted into the mouth, where it interacts with commensal facultative anaerobes and is reduced to nitrite (Kanady et al. 2012, Dusdieker et al. 1996). Excretion of nitrate occurs mainly through the urinary route; however, there is some excretion through feces (Health Canada 2017). Interestingly, nitrate concentrations in serum are not a good indicator of the concentration of nitrate secreted in breastmilk (Dusdieker et al. 1996).

There are differences between infants and older children and adults. Infants have a very low level of the enzyme responsible for reducing nitrate to nitrite in their mouths (Kanady et al. 2012). In adults, the mouth is the major site for nitrate reduction to nitrite. This reduction happens in infants' stomachs, where the pH is high enough to allow for the growth of bacteria that converts nitrate to nitrite.

Infants also differ in that they have a lower expression of the enzyme that converts methemoglobin to hemoglobin (Yip and Spyker 2018, Mensinga, Speijers, and Meulenbelt 2003), so the half-life of nitrite in the body is likely longer in infants than in adults and children. This can cause an elevated level of methemoglobin. Finally, nitrate is excreted by infants only through the urinary route (Health Canada 2017).

There are two types of methemoglobinemia: Congenital and acquired. Congenital methemoglobinemia is due to a deficiency in the enzyme CYB5R3 (Kedar et al. 2018) or a mutation in the CYB5R gene (Lorenzo et al. 2011). Congenital methemoglobinemia can be a very severe disorder that can lead to death (DomBourian et al. 2015). It can also be a less severe, more treatable disease. The type of methemoglobinemia associated with nitrate exposure is acquired. Acquired methemoglobinemia can be caused by a wide variety of agents, including the topical anesthetics Lidocaine and Prilocaine (Bloom 2001), gastrointestinal infections, and nitrate/nitrite (Ash-Bernal, Wise, and Wright 2004). Unlike congenital methemoglobinemia, which is due to genetic factors, acquired methemoglobinemia is caused by chemicals that have an oxidizing effect on the iron center in the hemoglobin protein. For example, nitrate must first be reduced to nitrite by bacteria in the mouth, or stomach in the case of infants, and the nitrite can be absorbed into the bloodstream where it can oxidize hemoglobin directly.

## 15.3 CONGENITAL METHEMOGLOBINEMIA

Methemoglobinemia is an illness that can be acquired from an array of different things – for example, oxidizing chemicals, food with naturally occurring nitrate or nitrite, food that has been intentionally preserved with nitrite, pharmaceuticals, and water from nitrate-contaminated water wells. It is also a congenital disease. There are three types of congenital methemoglobinemia: Type I, Type II, and Hemoglobin M Disease (HbM).

Acquired methemoglobinemia occurs more frequently than congenital methemoglobinemia. Mannino and coworkers (2017) published the first case study in the United States of Type II methemoglobinemia that had been genetically confirmed. In contrast, Ash-Bernal (2004) noted 138 cases of acquired methemoglobinemia in just 28 months in two hospitals. There has been no actual incidence or prevalence of Type II congenital methemoglobinemia calculated due to its rarity (Da-Silva, Sajan, and Underwood 2003).

HbM is a congenital methemoglobinemia characterized by a mutation in the gene that codes for the globin chain of the hemoglobin protein. This results in hemoglobin that is resistant to reduction. This makes treatment for methemoglobinemia more difficult, but treatment is rarely needed, as most people with the mutation live a normal life (Bayat and Kosinski 2011, DomBourian et al. 2015). Treatment for methemoglobinemia for these patients is usually only cosmetic, as there are no other symptoms beyond slightly blue skin. There are variant types of HbM: Boston, Iwate, Hyde Park, Saskatoon, Milwaukee, and Osaka. Each variant has a different amino acid substitution (Mansouri and Lurie 1993).

Type I is the most common type of congenital methemoglobinemia and is characterized by a deficiency in the CYB5R3 gene that causes the loss of the erythrocyte form of the enzyme CYB5R. This loss has autosomal recessive inheritance and causes only minor symptoms, such as cyanosis, similar to HbM (DomBourian et al. 2015, Nicolas-Jilwan 2019, Fusco et al. 2011).

Unlike HbM and Type I, which are mostly mild, Type II is associated with serious illness in infancy, with many affected infants dying before their first birthday (DomBourian et al. 2015). It is a more rare form of methemoglobinemia than Type I or HbM. Type II is caused by full stops or deletions in the CYB5R3 gene that results in an inactive Cytochrome B5 Reductase enzyme, or loss of expression altogether. This has severe consequences, especially for the neurological system. It has been hypothesized that this is due to altered fatty acid desaturation which can affect demyelination in the brain (Fusco et al. 2011, Mansouri and Lurie 1993). Infants present with a variety of different symptoms: mild cyanosis, progressive microcephaly, severe encephalopathy, generalized dystonia, seizures, strabismus, severe hypotonia, developmental delay, and white matter changes on brain imaging (Nicolas-Jilwan 2019, Fusco et al. 2011, Mannino et al. 2018). To help address symptoms, infants can be treated with ascorbic acid and Q10, or methylene blue if the methemoglobin level is above 20 percent (Nicolas-Jilwan 2019). There is no cure for Methemoglobinemia II. Any treatment is palliative in nature.

## 15.4 ACQUIRED METHEMOGLOBINEMIA

Acquired methemoglobinemia occurs when a person comes into contact with chemicals, pharmaceuticals, some vegetables, prepackaged food, and/or preserved meats that can oxidize hemoglobin into methemoglobin. There are two main ways in which this happens – with direct action to the hemoglobin protein, or by indirect means, where the compound itself is not able to oxidize hemoglobin, but can through other pathways create an oxidizing molecule. Nitrate is of concern for infants eating vegetables such as spinach and beets. Older children and adults can come into contact with preserved meat and prepackaged foods, which can be a significant source of nitrite.

Studies have been done looking at factors that cause infants to develop methemoglobinemia after eating vegetables. In one study, one- and two-year-old infants had developed constipation. The treating doctor for both infants recommended to the parents that they make zucchini soup, and use that to reconstitute formula. In both cases, the infant developed methemoglobinemia. The researcher found that in addition to zucchini being high in nitrate, both soups had been allowed to sit for a period of time, which increased the conversion from nitrate to nitrite in the soup. Both infants were treated with 1 percent methylene blue and recovered (Savino et al. 2006).

Another study centered around seven infants who had developed methemoglobinemia after eating green vegetables. The infants' ages ranged from seven to thirteen months. The researchers analyzed data surrounding the methemoglobinemia and found that all of the cases had eaten baby food with beets. In order of highest nitrate content to lowest nitrate content, beets are the highest, followed by spinach, lettuces, leeks, cabbages, pumpkins, and green beans. Carrots show no, or very little, nitrate. The researchers note that these cases are unusual due to the infants being over 6 months of age, and so not having the usual physiological challenges associated with methemoglobinemia and infants (Sanchez-Echaniz, Benito-Fernandez, and Mintegui-Raso 2001). The amount of nitrogen in particular foods can vary due to different practices of farmers using nitrate fertilizer. Baby food prepared in the home imparts the largest chance for methemoglobinemia, as the nitrate cannot be controlled like with commercial baby foods.

An example of a pharmaceutical that can cause acquired methemoglobinemia is silver nitrate. It can be used to prevent secondary infections in people with severe burns. It is a topical agent that can easily pass into the blood stream where it becomes nitrite. Methemoglobinemia can be difficult to treat in these patients as the skin may not be intact, silver nitrate causes changes in skin color, there may be extensive bandaging, and burn patients already exhibit cardiovascular changes as a result of the burns. The best diagnostic test is looking for chocolate-colored blood (Geffner, Powars, and Choctaw 1981).

Although there is an emphasis on infants and their unique physiology in much of the methemoglobinemia literature, older children and adults can also develop it. The section below gives several examples of the types of acquired methemoglobinemia that have been published. Many of the situations described show classic symptoms of methemoglobinemia.

## 15.5 EXAMPLES OF METHEMOGLOBINEMIA CAUSED BY NITRATE OR NITRITE

In 1953, a story called "Eleven Blue Men" appeared in *New Yorker* magazine (Roueché 1953). It chronicled the appearance of eleven blue men one morning in New York City, and the subsequent investigation of what was causing the blue color. In this case, the men had used salt cellars that had accidentally been partially filled with sodium nitrite instead of sodium chloride at a diner. They received high doses of sodium nitrite. The nitrite acted quickly on their hemoglobin to convert it into methemoglobin in their bodies, and they developed very serious cases of methemoglobinemia. One of the men died.

Just prior to "Eleven Blue Men" being published, Dr. Hunter Comely published a case report on nitrate in well water causing methemoglobinemia in infants, also known as Blue Baby Syndrome (Comley 1945). He described two cases of methemoglobinemia in infants younger than 6 months old. The infants had been drinking formula that had been reconstituted with well water. Once the nitrate-containing well water was replaced with water from another source, the methemoglobinemia resolved.

In 1948 and 1950 two epidemiology studies were published that described more than a hundred methemoglobinemia cases in southwestern Minnesota, and cases all over the nation (Rosenfield and Huston 1950, Cornblath and Hartmann 1948). The same pattern was repeated as Comely described; the cases were infants under the age of 6 months, and they had been drinking well water with high concentrations of nitrate in it. Many of the wells were shallow and were close to animal enclosures, two attributes that make it easy for nitrate to migrate into well water.

There have also been cases of nitrate-induced methemoglobinemia documented more recently in adults, children, and infants (Knobeloch and Proctor 2001, Knobeloch et al. 2000, Centers for Disease Control 1997, Funke et al. 2018). The infant cases follow the same pattern as described above. The adult and child cases have various toxic exposures associated with them. For example, in New Jersey in the 1990s there were two incidents involving nitrite being introduced into hot tap water by a failing backflow prevention valve on boilers. In the first case, 29 students were sickened after eating soup that had been reconstituted with the nitrite-containing boiler water. The majority of the students exhibited symptoms such as cyanosis, nausea, abdominal pain, vomiting, and dizziness. Fourteen of the children were hospitalized and treated with methylene blue; the methemoglobinemia resolved in the remaining 15 students within 36 hours after ingestion (Centers for Disease Control 1997).

The second case occurred in adults who had also been subjected to boiler water entering their hot water tap. In this case, the people who were sickened all drank coffee prepared with water from the hot water tap. Physicians determined they had methemoglobinemia due to the presence of elevated levels of methemoglobin in their blood. For those who were most seriously affected, oxygen and intravenous methylene blue was administered. All of the affected people recovered within 24 hours (Centers for Disease Control 1997).

Another example of a person developing methemoglobinemia induced by nitrate they had ingested is the case of a 71-year-old-man who opened and drank the fluid in a lava lamp. He had a history of alcohol abuse and had wrongly assumed the fluid inside the lamp was alcohol. It actually contained 76 percent calcium nitrate. When he was taken to the emergency room, he at first appeared normal, but after six hours in the emergency room his skin turned a grayish color; his methemoglobin was 45.6 percent, his oxygen saturation decreased, and additional oxygen did not bring his saturation level up. Based on these and other symptoms he was diagnosed with methemoglobinemia. He was treated with methylene blue and survived (Funke et al. 2018).

## 15.6  CONCLUSION

Although methemoglobinemia is not very frequently seen outside medical situations, it is still an illness that is worth understanding. Whether it is Type II congenital methemoglobinemia or acquired methemoglobinemia, it can pose significant problems for people who are affected. In congenital methemoglobinemia, the Cy5b3r enzyme is either only partially functional or not functional at all and that causes the illness. Acquired methemoglobinemia is characterized by the action of an oxidizing agent on the tetrameric hemoglobin protein. Both acquired and congenital methemoglobinemia share some symptoms, such a blueish color to the person's skin, shortness of breath, chocolate-colored blood, and a low dissolved oxygen concentration that cannot be improved with the addition of accessory oxygen.

Some forms of congenital methemoglobinemia are so rare their incidence and prevalence are unknown. Acquired methemoglobin, in contrast, happens in medical facilities with some frequency. It is a known side effect of many topical anesthetic medications although, when used properly, this will rarely be a side-effect. Additionally, there have been well-documented cases of accidental poisonings that have occurred in adults and children that resulted in acquired methemoglobinemia.

Infants have an increased risk of methemoglobinemia due to physiological factors. They have a lower concentration of enzyme necessary to convert methemoglobin to hemoglobin than adults, and they have a higher concentration of fetal hemoglobin, which is easier to convert to methemoglobin. Infants are often harder to diagnose as they present with nonspecific symptoms such as fussiness and lethargy. Because the speed of converting from methemoglobin to hemoglobin is slower in infants, this also results in a longer half-life of nitrate in the body.

In closing, nitrate is an important chemical when it comes to methemoglobinemia. It is an exposure factor that can come both from food and contaminated water, and it is tasteless, so people are unaware when they have been exposed. It is well-known that in the world, nitrate in drinking water is a major risk factor for methemoglobinemia. Although this is usually associated with an environmental exposure that pertains to babies, it is also true that older children and adults can develop methemoglobinemia. In most acquired cases, the methemoglobinemia is reversible, but in a small percentage of cases, methemoglobin can build up in the bloodstream to high concentrations that can result in death.

## REFERENCES

Ash-Bernal, R., R. Wise, and S.M. Wright. 2004. "Acquired methemoglobinemia: A retrospective series of 138 cases at 2 teaching hospitals." *Medicine (Baltimore)* 83 (5):265–273.

Bayat, A., and R.W. Kosinski. 2011. "Methemoglobinemia in a newborn: a case report." *Pediatr Dent* 33 (3):252–254.

Bloom, John C., and Brandt, John T. 2001. *Casarett and Doull's Toxicology: The Basic Science of Poisons*. Edited by Curtis D. Klaassen, 395. New York: McGraw-Hill.

Centers for Disease Control. 1997. "Methemoglobinemia attributable to nitrate contamination of potable water through boiler fluid additives – New Jersey 1992 and 1996." *MMWR Weekly* 46 (9):202–204.

Comley, Hunter H. 1945. "Cyanosis in infants caused by nitrates in well water." *Journal of the American Medical Association* 129:112–116.

Cornblath, M., and A.F. Hartmann. 1948. "Methemoglobinemia in young infants." *J Pediatr* 33 (4):421–425.

Da-Silva, S.S., I.S. Sajan, and J.P. Underwood, 3rd. 2003. "Congenital methemoglobinemia: a rare cause of cyanosis in the newborn – a case report." *Pediatrics* 112 (2):e158–161.

DomBourian, M., A. Ezhuthachan, A. Kohn, B. Berman, and E. Sykes. 2015. "A 5-hour-old male neonate with cyanosis." *Lab Med* 46 (1):60–63; quiz e14. doi:10.1309/LMBQIZ5DRS4K9IMD.

Dusdieker, L.B., P.J. Stumbo, B.C. Kross, and C.I. Dungy. 1996. "Does increased nitrate inges-tion elevate nitrate levels in human milk?" *Arch Pediatr Adolesc Med* 150 (3):311–314.

Fossen Johnson, S. 2019. "Methemoglobinemia: Infants at risk." *Curr Probl Pediatr Adolesc Health Care* 49 (3):57–67. doi:10.1016/j.cppeds.2019.03.002.

Funke, M.E., C.E. Fischetti, A.M. Rodino, and S.P. Shaheen. 2018. "Methemoglobinemia in-duced by ingesting lava lamp contents." *Clin Pract Cases Emerg Med* 2 (3):207–210. doi:10.5811/cpcem.2018.5.38261.

Fusco, C., G. Soncini, D. Frattini, E. Della Giustina, C. Vercellati, E. Fermo, and P. Bianchi. 2011. "Cerebellar atrophy in a child with hereditary methemoglobinemia type II." *Brain Dev* 33 (4):357–360. doi:10.1016/j.braindev.2010.06.015.

Geffner, M.E., D.R. Powars, and W.T. Choctaw. 1981. "Acquired methemoglobinemia." *West J Med* 134 (1):7–10.

Health Canada. 2017. *Nitrate and Nitrite in Drinking Water*. Edited by the Federal-Provincial-Territorial Committee on Drinking Water.

Hord, N.G., J.S. Ghannam, H.K. Garg, P.D. Berens, and N.S. Bryan. 2011. "Nitrate and nitrite content of human, formula, bovine, and soy milks: implications for dietary nitrite and nitrate recommendations." *Breastfeed Med* 6 (6):393–399. doi:10.1089/bfm.2010.0070.

Kanady, J.A., A.W. Aruni, J.R. Ninnis, A.O. Hopper, J.D. Blood, B.L. Byrd, L.R. Holley, M.R. Staker, S. Hutson, H.M. Fletcher, G.G. Power, and A.B. Blood. 2012. "Nitrate reductase activity of bacteria in saliva of term and preterm infants." *Nitric Oxide* 27 (4):193–200. doi:10.1016/j.niox.2012.07.004.

Kang, C., D.H. Kim, T. Kim, S.H. Lee, J.H. Jeong, S.B. Lee, J.H. Kim, M.H. Jung, K.W. Lee, and I.S. Park. 2018. "Therapeutic effect of ascorbic acid on dapsone-induced methemo-globinemia in rats." *Clin Exp Emerg Med* 5 (3):192–198. doi:10.15441/ceem.17.253.

Kedar, P.S., V. Gupta, P. Warang, A. Chiddarwar, and M. Madkaikar. 2018. "Novel mutation (R192C) in CYB5R3 gene causing NADH-cytochrome b5 reductase deficiency in eight Indian patients associated with autosomal recessive congenital methemoglobinemia type-I." *Hematology* 23 (8):567–573. doi:10.1080/10245332.2018.1444920.

Knobeloch, L., B. Salna, A. Hogan, J. Postle, and H. Anderson. 2000. "Blue babies and nitrate-contaminated well water." *Environ Health Perspect* 108 (7):675–678. doi:10.1289/ehp.00108675.

Knobeloch, L., and M. Proctor. 2001. "Eight blue babies." *WMJ* 100 (8):43–47.

Kuiper-Prins, E., G.F. Kerkhof, C.G. Reijnen, and P.J. van Dijken. 2016. "A 12-day-old boy with methemoglobinemia after circumcision with local anesthesia (Lidocaine/Prilocaine)." *Drug Saf Case Rep* 3 (1):12. doi:10.1007/s40800-016-0033-9.

Lorenzo, F.R., J.D. Phillips, R. Nussenzveig, B. Lingam, P.A. Koul, S.L. Schrier, and J.T. Prchal. 2011. "Molecular basis of two novel mutations found in type I methemoglobin-emia." *Blood Cells Mol Dis* 46 (4):277–281. doi:10.1016/j.bcmd.2011.01.005.

Mannino, E.A., T. Pluim, J. Wessler, M.T. Cho, J. Juusola, and S.A. Schrier Vergano. 2018. "Congenital methemoglobinemia type II in a 5-year-old boy." *Clin Case Rep* 6 (1):170–178. doi:10.1002/ccr3.1310.

Mansouri, A., and A.A. Lurie. 1993. "Concise review: Methemoglobinemia." *Am J Hematol* 42 (1):7–12. doi:10.1002/ajh.2830420104.

Mensinga, T.T., G.J. Speijers, and J. Meulenbelt. 2003. "Health implications of exposure to environmental nitrogenous compounds." *Toxicol Rev* 22 (1):41–51.

Nicolas-Jilwan, M. 2019. "Recessive congenital methemoglobinemia type II: Hypoplastic basal ganglia in two siblings with a novel mutation of the cytochrome b5 reductase gene." *Neuroradiol J* 32 (2):143–147. doi:10.1177/1971400918822153.

Qu, X.M., Z.F. Wu, B.X. Pang, L.Y. Jin, L.Z. Qin, and S.L. Wang. 2016. "From nitrate to nitric oxide: The role of salivary glands and oral bacteria." *J Dent Res* 95 (13):1452–1456. doi:10.1177/0022034516673019.

Rosenfield, A.B., and R. Huston. 1950. "Infant methemoglobinemia in Minnesota due to nitrates in well water." *Minn Med* 33 (8):789–796.

Roueché, Berton. 1953. *Eleven Blue Men, and Other Narratives of Medical Detection.* 1st ed. Boston: Little Brown.

Sanchez-Echaniz, J., J. Benito-Fernandez, and S. Mintegui-Raso. 2001. "Methemoglobinemia and consumption of vegetables in infants." *Pediatrics* 107 (5):1024–1028. doi:10.1542/peds.107.5.1024.

Savino, F., S. Maccario, C. Guidi, E. Castagno, D. Farinasso, F. Cresi, L. Silvestro, and G.C. Mussa. 2006. "Methemoglobinemia caused by the ingestion of courgette soup given in order to resolve constipation in two formula-fed infants." *Ann Nutr Metab* 50 (4):368–71. doi:10.1159/000094301.

Taleb, M., Z. Ashraf, S. Valavoor, and J. Tinkel. 2013. "Evaluation and management of acquired methemoglobinemia associated with topical benzocaine use." *Am J Cardiovasc Drugs* 13 (5):325–330. doi:10.1007/s40256-013-0027-2.

Tanen, D.A., F. LoVecchio, and S.C. Curry. 2000. "Failure of intravenous N-acetylcysteine to reduce methemoglobin produced by sodium nitrite in human volunteers: A randomized controlled trial." *Ann Emerg Med* 35 (4):369–373.

U.S. Environmental Protection Agency – Office of Research and Development. 2011. *Exposure Factors Handbook.*

Yip, L., and D.A. Spyker. 2018. "NADH-methemoglobin reductase activity: adult versus child." *Clin Toxicol (Phila)* 56 (9):866–868. doi:10.1080/15563650.2018.1444768.

# 16 Inorganic Nitrate and Nitrite

## Dietary Nutrients or Poisons?

*Nathan S. Bryan*

## CONTENTS

16.1   Introduction.................................................................................................357
16.2   Contributions of Nitrate and Nitrite from Fertilizer Use .........................358
16.3   Potential Risks of Over-exposure to Nitrite and Nitrate..........................359
16.4   Nitrate and Nitrite Metabolism in Humans..............................................360
16.5   Nitric Oxide Production from Nitrate and Nitrite.....................................362
16.6   Health Benefits of Nitrate and Nitrite .....................................................363
16.7   Nitrite and Nitrate Effects on Host Defense ............................................366
16.8   Risk-Benefit Analysis...............................................................................367
16.9   Conclusion ................................................................................................368
References.............................................................................................................369

## 16.1 INTRODUCTION

All life on Earth requires nitrogen. Nitrogen is one of the most abundant elements on Earth. The air we breathe is 78 percent nitrogen, and that nitrogen is fixed or oxidized by the environment, primarily through lightning, to form nitrite and nitrate in the soil. This is referred to as the Atmospheric Nitrogen Cycle. Nitrite and nitrate are the most usable forms of nitrogen for plant growth and nutrient assimilation. Nitrogen-based fertilizers are used worldwide to enhance plant growth and vitality and are necessary to feed the world's population. Nitrogen is considered the most important nutrient, and plants absorb more nitrogen than any other element. Nitrogen is essential to making sure plants are healthy as they develop and nutritious to eat after they are harvested, primarily through the formation of protein, and protein makes up much of the tissues of most living things.

In order to improve plant growth and crop yields to feed a growing population, nitrogen-based fertilizers are commonly used in farming. Nitrate salts are among the key components within inorganic fertilizers, and the increased dependency of farming practices on such fertilizers over several decades has led to increasing levels of human

DOI: 10.1201/9780429326806-21

exposure to nitrate. The plants or vegetables humans consume and, in many cases, the well water consumed in rural areas, provides a source of nitrate to humans. For years, people have viewed dietary sources of nitrate, including drinking water, as harmful to humans, causing methemoglobinemia and increasing the risks of certain cancers. However, methemoglobinemia is rare, especially in adults, and evidence suggests that it is primarily due to infective enteritis from drinking bacterial-contaminated water rather than with exposure to nitrate alone. Also, epidemiological evidence for an association between cancers of the digestive tract and nitrate intake is inconclusive in terms of increased risks of cancer [1, 2]. Historically, nitrite and nitrate have been used as medicinal agents [3]. Today evidence reveals they are necessary for normal vascular and immune function. However, public perception is that they are harmful substances in our food and water supply [4]. Ironically, most people would not argue that a diet rich in fruits and vegetables is healthy. In fact, the mechanism of action for the protective effects of vegetables can be explained in part by their nitrate and nitrite content [5]. The discovery of the nitric oxide pathway in the early 1980s revealed that nitrate is actually produced endogenously from the oxidation of nitric oxide in the body, changing our perception of nitrate safety [6]. Recently, benefits of dietary sources of nitrate for cardiovascular health and protection against infections have been unveiled, calling for an assessment of the risk and benefits associated with nitrate in our food and water supply. The scope of this chapter is to review the current state of the science on nitrite and nitrate in human health and disease and put this in the context of current regulations and safety.

## 16.2   CONTRIBUTIONS OF NITRATE AND NITRITE FROM FERTILIZER USE

Since 1950 global fertilizer use has increased by a factor of five [7] which has resulted in a doubling of the intake of inorganic nitrate through food consumption. The increased use of fertilizer and nitrate consumption has been most pronounced for high-value vegetable crops such as lettuce and spinach, increasing nitrate levels over the past century up to fourfold. However, there is a highly variable nitrate and nitrite content of vegetables across vegetable categories and also with geographic location of where the vegetables are grown [8]. Vegetables grown and sold in Dallas and Los Angeles contain higher amounts of nitrate than the same vegetables grown in New York [8]. Furthermore, organically grown vegetables have on average about ten times less nitrate than conventionally grown vegetables [8]. This is likely due to a lack of nitrogen-based fertilizers added to organic vegetables. There are also vast differences in the amount of nitrate consumed based on ethnic dietary norms. For example, the typical Japanese diet contributes 18.8 mg/kg nitrate or over 1200 mg per day for a 65 kg adult and has been shown to maintain normal blood pressure [9]. On the contrary, the standard American diet only contributes about 2.3 mg/kg or only about 150 mg nitrate per day [5]. In this case, the American diet is deficient in nitrate to the extent that this deficiency may be contributing to the epidemic of hypertension and many other chronic diseases in many Americans [10].

Vegetables contribute over 85 percent of the daily dietary intake of nitrate. Endogenous production of nitrite and nitrate from the oxidation of nitric oxide is also an important contributor to humans' overall exposure of nitrate [11]. Hord and colleagues [5] estimated that approximately 80 percent of dietary nitrate are derived from vegetable consumption. However, most people associate nitrite and nitrate exposure to their presence in cured and processed meats. Recent reports have shown that less than 5 percent of the ingested nitrite and nitrate are derived from cured meat sources, with the remainder coming from vegetables and saliva [12–14]. Relative intake of nitrate from drinking water typically ranges from 10–20 percent. However, those living in rural agricultural areas using shallow wells may have much higher exposure to nitrate from their drinking water. The primary sources of nitrate that contaminate groundwater and surface water are fertilizers, manure, or organic compost applied to the soil and also from local human waste from septic tanks and wastewater treatment systems [15]. In the Southwestern United States and other agricultural areas, inorganic fertilizer and animal manure are the most common nitrate source, while urban areas without proper sewer containment contribute to the nitrate levels in groundwater. Further, ammonia will volatilize from manures and fertilizers to be transported through the atmosphere to be deposited and nitrified by microbial action and to become an additional source of nitrate pollution of waters.

## 16.3  POTENTIAL RISKS OF OVER-EXPOSURE TO NITRITE AND NITRATE

As Paracelsus famously proclaimed, "the dose makes the poison." There are known toxicities of nitrite and nitrate at extremely high levels. The major concerns of acute toxicity are methemoglobinemia, hypotension, and chronically the potential for harmful nitrosation reactions that can occur in the stomach, forming low molecular weight nitrosamines, some of which have been shown to cause a wide range of tumors in more than forty animal species [16]. If nitrate is at extremely high levels, acute toxicity can arise, such as methemoglobinemia [17]. The LD50 for sodium nitrate, the dose that would cause death in 50 percent of laboratory animals, is greater than 2000mg/kg. However, emerging evidence indicates that, in moderation, nitrate can play an important and beneficial role in human physiology [18]. As discussed earlier, the doses needed to normalize blood pressure that can be consumed through foods and diets is less than 20m/kg or one hundred times less than any cause for toxicity. In this regard, nitrate is extremely safe. Sodium nitrite is weakly toxic. The LD50 in rats is 180 mg/kg and its human LDLo is 71 mg/kg, meaning a 65 kg person would likely have to consume at least 4.6 g to result in death. Published clinical studies indicate that 1-2 mg/kg sodium nitrite in humans is safe and effective in some human diseases [10]. Excess nitrate and nitrite exposure is, in specific contexts, associated with an increased risk of negative health outcomes, as mentioned above.

Dietary Reference Intake (DRI) categories are set by the Food and Nutrition Board of the National Academy of Sciences for essential nutrients in order to clearly define, where possible, the contexts in which intakes are deficient, safe, or potentially excessive. These DRI categories include the Recommended Dietary Allowance (RDA), Adequate Intake (AI), Tolerable Upper Level Intake (TUL) and Estimated Average Intake (EAR) [19]. There are a number of considerations when establishing DRIs for nutrients, including nutritional status and potential toxicities. It is important to note that nitrate exposure limits are set on the basis of nitrate in drinking water, the route of exposure associated with nitrate toxicities, although accidental toxic exposures of nitrate and nitrites have occurred through other means [20].

The US Environmental Protection Agency (EPA) has established a maximum contaminant level (MCL) for nitrate in drinking water of 10 mg/L nitrate-nitrogen (nitrate-N) (equivalent to 45 mg/L as nitrate). The World Health Organization (WHO) guideline and European Union (EU) Council Directive 98/83/EC established 50 mg/L as nitrate (equivalent to 11 mg/L as nitrate-N) as higher limits to protect against methemoglobinemia or "blue baby syndrome," to which infants are especially susceptible [21]. The US and the EU maximum contaminant level for nitrate were promulgated (US: 1962; EU: 1980) without scientific explanation. During the 1970s, the WHO began to consider the increased risk of nitrate exposure for some cancers. In 1993, the WHO recognized that the epidemiological evidence for an association between dietary nitrate and cancer was not sufficient for action, and that the guideline value should be established solely to prevent methemoglobinemia. As a result, the US EPA limits human exposure to inorganic nitrate to >10 mg/L or 10 ppm nitrate nitrogen, and nitrites to 1 ppm nitrite nitrogen [22]. The Joint Food and Agricultural Organization/World Health Organization has set the Acceptable Daily Intake (ADI) for the nitrate ion at 3.7 mg/kg body weight and, for the nitrite ion, at 0.06 mg/kg body weight [23]. This translates into an exposure limit of 222 mg nitrate and 3.6 mg nitrite for a 60 kg adult. Likewise, EPA has set a Reference Dose (RfD) for nitrate of 1.6 mg nitrate nitrogen/kg body weight per day (equivalent to about 7.0 mg nitrate ion/kg body weight per day). The regulatory level is usually met for public water supplies, which are routinely monitored. Much less is known about private wells which, in the United States, are usually required to be tested only when the well is constructed or when the property is sold. Due to their very high solubility, nitrite and nitrate can enter groundwater. Tentative estimates for Western Europe and the United States are that 2 to 3 percent of the population is potentially exposed to drinking water exceeding 50 mg/L [24], which is still orders of magnitude less than amounts that would cause any toxicity.

## 16.4   NITRATE AND NITRITE METABOLISM IN HUMANS

The risk-benefit spectrum from nitrite and nitrate may very well depend upon the specific metabolism and the presence of other components that may be concomitantly ingested. The stepwise reduction of nitrate to nitrite and NO may account for the benefits while pathways leading to nitrosation of low molecular weight amines or amides may account for the long-term health risks of nitrate and nitrite exposure, since acute toxicity from methemoglobinemia and hypotension are extremely rare at best and almost nonexistent. Understanding and affecting those pathways will certainly help in mitigating the risks. The discussion below describes these two pathways.

Human exposure to nitrate is dependent upon an individual's intake of vegetables and the local concentration of nitrate in drinking water as well as the total amount of nitric oxide produced in the body. Nitrate is rapidly absorbed in the small intestine and readily distributed throughout the body [25]. Approximately 25 percent of nitrate is concentrated in our salivary glands [26], causing salivary nitrate concentration to reach approximately ten times higher than plasma nitrate levels. Approximately 20 percent of that salivary nitrate is reduced to nitrite in the mouth by facultative anaerobic bacteria, which are found on the surface of the tongue [27–29], resulting in about a 5 percent reduction of total ingested nitrate to nitrite. As a result, nitrate ingestion is the main source of nitrite exposure. This biochemistry has been demonstrated stoichiometrically in mice fed twenty times higher nitrate than nitrite [30]. Humans do not have a functional nitrate reductase enzyme, so they are dependent upon oral nitrate reducing bacteria to perform this first metabolic step of nitrate reduction to nitrite and nitric oxide. Nitrate, when consumed orally, reaches a peak plasma concentration in about one hour [31]. The half-life of plasma nitrate is approximately five hours [31]. Since nitrate is a relatively small anion and is not protein-bound, it is reabsorbed in the renal tubules. Nitrate is also excreted in the urine directly or after conversion to urea [32]. Clearance of nitrate from blood to urine approximates 20 ml/min in adults [33] indicating considerable renal tubular reabsorption of the ion. It is estimated that 96 percent of the filtered nitrite and nitrate is reabsorbed in the renal tubules [34]. Other studies in dogs suggest that approximately 80 percent of filtered nitrate is reabsorbed [35]. The high concentration of nitrate in saliva, continuous production from nitric oxide, and the re-absorption from renal tubules strongly suggest that nitrate and nitrite have an important role in normal human physiology. The human body is designed to excrete toxins, not reabsorb them and produce them endogenously.

Nitrate itself is inert in humans. Nitrite, on the other hand, is biologically active and can be reactive, especially in the acid environment of the stomach, where it can nitrosate other molecules, including proteins, amines, and amides. Nitrite administered orally is 98 percent bioavailable [36]. Nitrite is occasionally found in the environment, but most human exposure occurs through ingested nitrate that is reduced to nitrite by oral bacteria [28]. The risk of the formation of N-nitrosamines formed in the stomach from ingesting foods enriched in nitrite and nitrate has been a concern for decades. This is because some low molecular weight amines can be converted (nitrosated) to their carcinogenic N-nitroso derivatives by reaction with nitrite (see reaction below) [26]. Nitrosamines are a class of chemical compounds that were first described in the chemical literature over a hundred years ago, but not until 1956 did they receive much attention. In that year two British scientists, John Barnes and Peter Magee, reported that dimethylnitrosamine produced liver tumors in rats [37]. This discovery was made during a routine screening of chemicals that were being proposed for use as solvents in the dry-cleaning industry. Approximately three hundred low molecular weight nitrosamines have been tested, and 90 percent of them are carcinogenic in a wide variety of experimental animals. Most nitrosamines are mutagens, and a number are transplacental carcinogens. Most are organ-specific. For instance, dimethylnitrosamine causes liver cancer in experimental animals, whereas some of the tobacco-specific nitrosamines cause lung cancer. Since nitrosamines are metabolized the same in human and animal tissues, it seems highly likely that humans are susceptible to the carcinogenic properties of nitrosamines. Amines can occur commonly, and sodium nitrite is derived from nitrate. Nitrosamines can occur because their chemical precursors – amines and nitrosating agents – occur commonly, and the chemical reaction for nitrosamine formation is quite facile. The reactions below illustrate the nitrosation reactions.

$R_2NH$ (amines) + $NaNO_2$ (sodium nitrite) $\rightarrow$ $R_2N\text{-}N=O$ (nitrosamine)

In the presence of acid (such as in the stomach) or heat (such as via cooking), nitrosamines are converted to diazonium ions.

$R_2N\text{-}N=O$ (nitrosamine) + (acid or heat) $\rightarrow$ $R\text{-}N^+\text{-}N=O$ (diazonium ion)

Certain nitrosamines, such as dimethylnitrosamine and N-nitrosopyrrolidine, form carbocations that react with biological nucleophiles (such as DNA or an enzyme) in the cell.

$R\text{-}N^+\text{-}N=O$ (diazonium ion) $\rightarrow$ $R^+$ (carbocation) + $N_2$ (leaving group)
+: Nu (biological nucleophiles) $\rightarrow$ R-Nu

If this reaction occurs at a crucial site in a biomolecule, it can disrupt normal cell function, leading to cancer or cell death.

About 1970 it was discovered that ascorbic acid inhibits nitrosamine formation [38]. Another antioxidant, alpha-tocopherol (vitamin E), has also been shown to inhibit nitrosamine formation [39]. Ascorbic acid, erythorbic acid, and alpha-tocopherol inhibit nitrosamine formation due to their oxidation-reduction properties. For example, when ascorbic acid is oxidized to dehydroascorbic acid, nitrous anhydride, a potent nitrosating agent formed from sodium nitrite, is reduced to nitric oxide, which is not a nitrosating agent. Most vegetables that are enriched in nitrate are also rich in antioxidants such as vitamins C and E that can act to prevent the unwanted nitrosation chemistry. Controlling the metabolic fate of nitrate and nitrite away from nitrosation and toward reduction to NO, may provide a strategy to promote health benefits while mitigating the health risks. Adverse health effects may be the result of a complex interaction of the amount of nitrate ingested, the concomitant ingestion of nitrosation co-factors and precursors, and specific medical conditions that increase nitrosation such as chronic inflammation.

## 16.5 NITRIC OXIDE PRODUCTION FROM NITRATE AND NITRITE

Now that we have established the potential risks of nitrite and nitrate ingestion and exposure as being context- and dose-dependent, we can focus on the potential benefits of these molecules. The protective and beneficial effects of nitrate and nitrite can be explained by the production of nitric oxide (NO), which has a very large number of effects that are thought to be beneficial in mammals. NO is the most important molecule in regulating blood pressure and maintaining vascular homeostasis. Nitric oxide rapidly reacts with oxyhemoproteins, such as oxyhemoglobin to form nitrate and methemoglobin [40]. As a result, nitrate is formed within the body as a result of NO production. In 1994 two groups independently presented evidence for the generation of NO in the stomach resulting from the acidic reduction of inorganic nitrite

[41, 42], demonstrating a recycling pathway whereby nitrate could be reduced back to NO. Benjamin and colleagues demonstrated that the antibacterial effects of acid alone was markedly enhanced by the addition of nitrite, which is present in saliva. High levels of NO is produced in exhaled air from the stomach in humans after a nitrate-rich meal. The NO production is abolished after pretreatment with a proton pump inhibitor (PPI) and markedly increased after ingestion of nitrate, showing the importance of both luminal pH and the conversion of nitrate to nitrite for stomach NO generation. These were the first reports of NO synthase-independent formation of NO *in vivo*. In the classical NO synthase pathway, NO is formed by oxidation of the guanidino nitrogen of L-arginine with molecular oxygen as the electron acceptor [43]. This complex reaction is catalyzed by specific heme-containing enzymes; the NO synthases, and the reaction requires several co-factors. The nitrate-nitrite-nitric oxide pathway is fundamentally different; instead of L-arginine it used the simple inorganic anions nitrate ($NO_3^-$) and nitrite ($NO_2^-$) as substrates in a stepwise reduction process that did not require NO synthase or multiple co-factors. Oral commensal bacteria are essential for the first step in the nitrate-nitrite-NO pathway, since humans lack a nitrate reductase gene. It was known from the literature that the salivary glands extract nitrate from plasma, but the reason for this active process was not explained. Oral facultative anaerobic bacteria residing mainly in the crypts of the tongue then reduce nitrate to nitrite by the action of nitrate reductase enzymes [44, 45]. These bacteria use nitrate as an alternative electron acceptor to gain Adenosine Triphosphate (ATP) in the absence of oxygen. This highly effective bacterial nitrate reduction results in salivary levels of nitrite that are a thousand-fold higher than those found in plasma [46]. When nitrite-rich saliva meets the acidic gastric juice, nitrite is protonated to form nitrous acid ($HNO_2$), which then decomposes to NO and a variety of other nitrogen oxides [41, 42]. It is now established that oral commensal bacteria are pivotal in gastric NO formation, and gastric NO levels are consistently low in animals reared under completely germ-free conditions [47]. If the oral bacteria are selectively removed with an antiseptic mouthwash, the gastric NO levels decrease drastically [48]. This NO, generated as a result of nitrate, promotes a number of beneficial health effects. Nitric oxide is known to modulate a number of mechanisms involved from increasing gastric blood flow and mucus production to the regulation of the epithelial barrier. Increased dietary nitrate and the application of nitrite have both been shown to increase gastric blood flow and gastric mucosal thickness in a nitric oxide dependent manner [29, 49, 50]. This has been shown to confer protection against non-steroidal anti-inflammatory drugs (NSAID) induced ulceration [51]. Thus, when endogenous production of one of the major determinants of gastric protection, prostaglandins, is depressed by cyclo-oxygenase inhibition by NSAID's, the integrity of the gastric mucosa can be preserved by nitrate.

## 16.6 HEALTH BENEFITS OF NITRATE AND NITRITE

It is known that a diet rich in fruits and vegetables has a positive influence on blood pressure [52]. It has been speculated that the blood-pressure-lowering effects of certain diets may be due to the nitrate content [53]. In fact, a recent food sample survey indicated that a DASH (Dietary Approach to Stop Hypertension) diet exceeds the ADI for nitrate by greater than 500 percent and may account for the modest blood pressure effects [5]. Webb and colleagues demonstrated a significant decrease in systolic and diastolic blood pressure within three hours of ingestion of 0.5 liters of beetroot juice (a rich source of nitrate) [54]. These data were consistent with earlier results from Lundberg's group showing a significant blood-pressure-lowering effect from three days supplementation with sodium nitrate [55]. The maximal effect in Webb's study coincided with peak plasma levels of nitrite and was abolished if the study participants did not swallow their saliva, suggesting that bacterial reduction of nitrate to nitrite and nitrite entering the acid stomach to produce NO is necessary in order for it to have biological activity. This concept was confirmed in a study showing treating with antimicrobial mouthwashes is sufficient to abolish any beneficial effect in animals given supplementary nitrate [49]. The stepwise reduction of nitrate to nitrite to nitric oxide is, by necessity, an inefficient process by which each step yields a 3-log lower concentration of product than substrate [56]. Although the reduction efficacy from nitrate to NO is very inefficient, it is clear that a diet rich in nitrate can provide a source of bioactive NO due to this human nitrogen cycle. This pathway is illustrated in Figure 16.1.

Recent data are also emerging implicating nitrate and nitrite on endothelial function. Nitric oxide synthesized by endothelial NOS plays a key role in vascular homeostasis by maintaining vessels in their relaxed state. Experimental and clinical studies provide evidence that defects of endothelial NO function, referred to as endothelial dysfunction, is not only associated with all major cardiovascular risk factors such as hyperlipidemia, diabetes, hypertension, smoking, and severity of atherosclerosis, but also has a profound predictive value for future atherosclerotic disease progression [57–60]. Increasing numbers of risk factors results in a stepwise reduction in plasma nitrite, a surrogate for eNOS activity [61]. In conventional and eNOS knockout mice dietary nitrate and nitrite supplementation restores NO homeostasis and protects against ischemia reperfusion injury [30, 62]. Nitrite supplementation ameliorates the microvascular inflammation and endothelial dysfunction in mice subject to a high cholesterol diet [63]. Research has also shown a biological activity of nitrate on platelet function [54]. The role of platelet adhesion, activation, and aggregation in atherosclerosis and thrombosis is well known [64]. Platelet adhesion and aggregation are inhibited by nitric oxide, either from within the platelet itself or produced by the vascular endothelium [65, 66]. Dietary supplementation with either potassium nitrate in the range 0.5-2mmol [67] or using beetroot juice [54] results in a significant inhibition of platelet aggregation.

Exercise performance is a direct reflection of our body's ability to accommodate increased blood flow to working muscles. Endothelial production of NO provides vessel dilatation in response to exercise. Plasma nitrite levels increase in response to exercise in healthy individuals, whereby in aged patients with endothelial dysfunction there is no increase in nitrite from exercise [68]. Nitrite has also been shown to predict exercise capacity in humans [69]. Short-term (three days) dietary supplementation with sodium nitrate results in improved muscular efficiency and a reduction in oxygen consumption during sub-maximal exercise in healthy subjects and enhance tolerance to high-intensity exercise in humans [70, 71]. The amount of nitrate used in these studies was 6-8mg/kg, exceeding the ADI for nitrate in both studies.

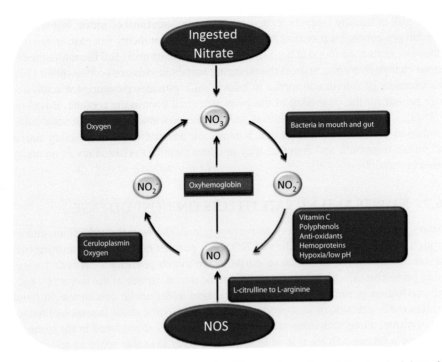

**FIGURE 16.1** The human nitrogen cycle: Dietary nitrate is rapidly absorbed into the bloodstream through the proximal gut, where it mixes with endogenous nitrate from the NOS/NO pathway. A large portion of nitrate is taken up by the salivary glands, secreted with saliva, and reduced to nitrite by symbiotic bacteria in the oral cavity. Salivary-derived nitrite is further reduced to NO and other biologically active nitrogen oxides in the acidic stomach. The remaining nitrite is rapidly absorbed and accumulates in tissues, where it serves to regulate cellular functions via reduction to NO or possibly by direct reactions with protein and lipids. NO and nitrite are ultimately oxidized to nitrate, which again enters the enterosalivary circulation or metabolized and excreted in urine, sweat, and feces.

Nitrate is also found naturally in a mother's breast milk. Human breast milk is recommended to serve as the exclusive food for the first six months of life and continue, along with safe, nutritious complementary foods, up to two years [72, 73]. Breast milk is nature's most perfect food. In fact, the US Centers for Disease Controls in 2010 stated, "Breast milk is widely acknowledged as the most complete form of nutrition for infants, with a range of benefits for infants' health, growth, immunity, and development." Breast milk is a unique nutritional source for infants, one that cannot adequately be replaced by any other food, including infant formula. Human milk is known to confer significant nutritional and immunological benefits for the infant [74–76]. Several research studies have previously demonstrated relatively high concentrations of nitrite and nitrate in human breast milk. Ohta and colleagues found high concentrations of nitrite and nitrate (166-1246µM or 1-7 mg/100ml) in the breast milk of Japanese mothers from days 1-8 [77]. Cekmen and colleagues found, in the breast milk of healthy mothers, extremely high concentrations of nitrite that was reduced in pre-eclampsia patients [78]. Hord and colleagues recently demonstrated that nitrite and nitrate are found at high levels in human breast milk, and the ratio of these anions change over time to meet the changing metabolic demands of the infant [79]. The presence of nitrate and nitrite in human milk provides evidence for a physiologic benefit for the protection of the gastrointestinal tract in the neonate. Based on exposure rates of infants from breast milk nitrite and nitrate, the most reasonable conclusion that can be made is that humans are adapted to receive dietary nitrite and nitrate from birth and therefore may not pose significant risks at levels naturally found in certain foods [79].

## 16.7 NITRITE AND NITRATE EFFECTS ON HOST DEFENSE

Dietary nitrate also plays a very important role in the host defense of mammals through the conversion to nitrite and NO, both precursors for potent antimicrobial substances. Reduction of nitrate to nitrite in the mouth provides a first site for exerting protective effects. Nitrite produced on the dorsal surface of the tongue by bacterial reduction is further reduced to nitric oxide under acidic conditions [80] and result in the destruction of acid-producing organisms. Nitric oxide from acidic reduction of nitrite, along with other reactive nitrogen species, is produced in the stomach following a nitrate-rich meal in quantities far greater than that required to produce vasodilatation, suggesting a role other than modulation of gastric blood flow [31]. A number of important human pathogens (*C albicans*, *E.coli* 0157, *salmonella*, *shigella*) are resistant to a low pH found in the stomach that enables them to survive passage through the human stomach [81]. The addition of nitrite in concentrations found in human saliva results in a much more effective killing of these pathogens than in acid alone [41, 82]. Indeed, even *H pylori*, a bacteria well adapted to colonize the human stomach, is susceptible to acidified nitrite [83]. The acidification of nitrite derived from dietary nitrate appears to play a pivotal role in protection against ingested pathogens. Further support for this can be found in the reduced levels of post-prandial salivary nitrite found in subjects treated with the broad-spectrum antibiotic amoxicillin [84]. Patients on broad-spectrum antibiotics are at increased risk of opportunistic gastrointestinal tract infections from species such as *C albicans* and *Clostridium difficile*. It may be that the killing of commensal bacteria by the overuse of antibiotics is responsible for the disruption of salivary nitrate reduction and subsequent suppression of salivary nitrite that allows these organisms to establish themselves in the gastrointestinal tract.

It appears that nature has also designed an effective strategy for wound healing. Both man and animals will lick their wounds upon sustaining an injury, an action that promotes healing [85]. This effect is likely to be due in part to saliva's antimicrobicidal properties, which can substantially reduce bacterial contamination of the wound [86]. Licking of human skin results in the production of nitric oxide as salivary nitrite is reduced on the acidic skin surface [87]. A number of common skin pathogens are effectively killed by acidified nitrite [88]. It seems likely that the beneficial effect of saliva on wound healing is due in part to the acidification of nitrite that results from dietary sources of nitrate.

There are disruptions in this pathway that now clearly put people at risk for a number of chronic diseases, including cardiovascular disease, the number one killer of men and women worldwide. These are listed below:

1. Insufficient dietary intake of nitrate/nitrite rich foods (green leafy vegetables, beets, etc.)
2. Problems with nitrate uptake in duodenum (sialin (SLC17A5) transporter mutations – Salla Disease)
3. Insufficient saliva production (Sjogrens syndrome)
4. Lack of oral commensal bacteria to reduce nitrate to nitrite (use of antibiotics (over 200 million users)/antiseptic mouthwash (over 200 million users), poor oral hygiene)
5. Insufficient stomach acid production – Achlorhydria (use of PPI's (over 200 million users), H. Pylori infection, iron overload)
6. Increased oxidative stress that scavenges NO

If we increase our nitrate consumption and stop using mouthwash and antacids, the human body will reward itself with the production of NO.

## 16.8 RISK-BENEFIT ANALYSIS

There are inherent risks of almost every human activity. We must make decisions that minimize risks while maximizing potential benefits. If we consider the risk-benefit of nitrate and nitrite exposure based on basic science but also on human clinical data, the benefits far outweigh any risks, especially at doses and levels that are achieved through diet. Insufficient NO production is known to cause many, if not all, age-related chronic diseases [89]. Loss of NO production causes hypertension, erectile dysfunction, vascular dementia – Alzheimers, diabetes, peripheral artery disease (PAD), small vessel disease, atherosclerosis, blood clotting disorders, acute respiratory distress syndrome (ARDS), heart failure, acute kidney injury – renal failure, immune dysfunction, and many more serious and currently unmanaged diseases. Clinical evidence also reveals that restoration of NO production through the repletion of nitrite and nitrate in the diet can positively affect most, if not all, of the above medical conditions [10, 90]. The risks of nitrite and nitrate from our diet is minimal with regards to methemoglobinemia and low blood pressure, and the benefits are profound with regards to combatting most of the chronic diseases that face the world today. One can only conclude that nitrite and nitrate should be considered indispensable nutrients that must be accounted for and standardized in our diets and foods. The fact that typical consumption patterns of vegetables and fruit exceed regulatory limits for dietary nitrate calls into question these regulatory standards. These data call into question the rationale for regulating the concentration of nitrate from food sources. The long-standing and recognized health benefits of diets rich in fruits and vegetables, and which are enriched in nitrate, suggest that at levels found in our foods, nitrate exposure through vegetables or fruits should be considered safe [91]. Part of the health benefits of vegetable sources of nitrate may be in context with the high antioxidant capacity of vegetables. Vitamin C and other antioxidants are very potent inhibitors of nitrosation reactions [38, 92] and many vegetables are rich in antioxidants. Nature itself has provided an inherent protective mechanism that prevents any and all unwanted and unintended consequences of nitrate and nitrite ingestion while promoting the mechanism and benefits of NO production.

Randomized, placebo-controlled clinical trials using standardized nitrite supplements have shown remarkable benefits. A nitric oxide lozenge containing sodium nitrite has been shown to normalize blood pressure in hypertensive patients, restore endothelial function and increase blood flow by 34 percent [93]. In another study, the nitrite lozenge normalized blood pressure in pre-hypertensive patients and increased the functional capacity of the heart while improving quality-of-life measures [94]. The nitrite lozenge lowers triglycerides [95], reverses carotid plaque [96], and corrects conditions of high blood pressure, heart disease, kidney disease, and cognition in a patient with Argininosuccinic aciduria (ASA) [97]. All of the above studies and dosing were without any adverse side effects.

## 16.9   CONCLUSION

The known and scientifically proven benefits of nitrite and nitrate in our food supply and diet far outweigh any potential risks from their exposure. In this regard, we must now move forward with standardization of nitrate in farming practices so we can then make dietary recommendations and establish dietary guidelines for a recommended daily allowance for nitrite and nitrate that can improve the health of a growing population. Results from a report from the National Research Council (*The Health Effects of Nitrate, Nitrite, and N-Nitroso Compounds*, NRC 1981) show that the amount of nitrite and nitrate consumed through our diet matches what our body makes from NO if we assume most of the endogenous NO goes to stepwise oxidation to nitrite and nitrate. Therefore, our steady-state levels of NOx, which are routinely used as clinical biomarkers of NO activity, come almost 50 percent from the diet. In fact, swallowing saliva alone is a primary source of exposure to nitrite and nitrate. The enterosalivary concentration and circulation of nitrate and, ultimately, nitrite provides an essential pathway for health and host defense [50, 98]. In the last ten years, there has been an explosion of nitrite- and nitrate-based dietary supplements and nutritional products due to the fact that the science is very clear. Too little nitrite and nitrate in your diet puts you at risk for a number of chronic diseases. Repletion of nitrite and nitrate from your diet can protect you from many diseases. There will very likely be considerable debate about the need for reviews by regulatory agencies and potential action on nitrite and nitrate in our food and water supply. It is hard to conceive that the ingestion of nitrate from fruits and vegetables could have any impact on potential adverse toxic outcomes.

Due to the intense social and governmental desire to identify causes of cardiovascular disease and cancer, we as a society have come to a point where some mechanistic/physiological observations and weak epidemiological associations can mislead us, even when the totality of evidence shows such findings to be biologically implausible. The risk/benefit balance should be a strong consideration before there are any suggestions for new regulatory or public health guidelines for nitrite and nitrate exposures. The preponderance of data reveals that nitrite and nitrate at levels consumed through diet are not only safe but absolutely essential and beneficial for human health.

# REFERENCES

[1]   Bryan, N.S., et al., Ingested nitrate and nitrite and stomach cancer risk: An updated review. *Food Chem Toxicol*, 2012. 50(10): p. 3646–3665.

[2]   Milkowski, A., et al., Nutritional epidemiology in the context of nitric oxide biology: a risk-benefit evaluation for dietary nitrite and nitrate. *Nitric Oxide*, 2010. 22(2): p. 110–119.

[3]   Butler, A.R. and M. Feelisch, Therapeutic uses of inorganic nitrite and nitrate: From the past to the future. *Circulation*, 2008. 117(16): p. 2151–2159.

[4]   L'Hirondel, J.L., *Nitrate and Man: Toxic, Harmless or Beneficial?* 2001, Wallingford, UK: CABI Publishing.

[5]   Hord, N.G., Y. Tang, and N.S. Bryan, Food sources of nitrates and nitrites: the physiologic context for potential health benefits, *Am J Clin Nutr*, 2009. 90(1): p. 1–10.

[6]   Tannenbaum, S.R., et al., Nitrite and nitrate are formed by endogenous synthesis in the human intestine. *Science*, 1978. 200(4349): p. 1487–1489.

[7]   Erisman, J.W., et al., How a century of ammonia synthesis changed the world. *Nature Geoscience*, 2008. 1: p. 636–639.

[8]   Nunez de Gonzalez, M.T., et al., A survey of nitrate and nitrite concentrations in conventional and organic-labeled raw vegetables at retail. *J Food Sci*, 2015. 80(5): p. C942–949.

[9]   Sobko, T., et al., Dietary nitrate in Japanese traditional foods lowers diastolic blood pressure in healthy volunteers. *Nitric Oxide*, 2010. 22(2): p. 136–140.

[10]  Bryan, N.S., Functional nitric oxide nutrition to combat cardiovascular disease. *Curr Atheroscler Rep*, 2018. 20(5): p. 21.

[11]  Gangolli, S.D., et al., Nitrate, nitrite and N-nitroso compounds. *Eur J Pharmacol*, 1994. 292(1): p. 1–38.

[12]  Cassens, R.G., Residual nitrite in cured meat. *Food Technol*, 1997. 51: p. 53–55.

[13]  Archer, D.L., Evidence that ingested nitrate and nitrite are beneficial to health. *J Food Prot*, 2002, 65(5): p. 872–875.

[14]  Milkowski, A., et al., Nutritional epidemiology in the context of nitric oxide biology: a risk-benefit evaluation for dietary nitrite and nitrate. *Nitric Oxide*, 22(2): p. 110–119.

[15]  Harter, T., Agricultural impacts on groundwater nitrate. *Hydrology*, 2009. 8(4): p. 22–23.

[16]  Tricker, A.R. and R. Preussmann, Carcinogenic N-nitrosamines in the diet: occurrence, formation, mechanisms and carcinogenic potential. *Mutat Res*, 1991. 259(3-4): p. 277–289.

[17]  Marcus, H. and J.R. Joffe, Nitrate methemoglobinemia. *N Engl J Med*, 1949. 240(15): p. 599–602.

[18]  Lundberg, J.O., E. Weitzberg, and M.T. Gladwin, The nitrate-nitrite-nitric oxide pathway in physiology and therapeutics. *Nat Rev Drug Discov*, 2008. 7(8): p. 156–167

[19]  Otten JJ, H.J., Meyers, LD, Eds. *Dietary Reference Intakes: The Essential Guide to Nutrient Requirements*. National Academy Press, Food and Nutrition Board, Institute of Medicine, National Academy of Sciences, 2006.

[20]  Methemoglobinemia following unintentional ingestion of sodium nitrite – New York, 2002. *MMWR Morb Mortal Wkly Rep*, 2002. 51(29): p. 639–642.

[21]  WHO, *Recommendations: Nitrate and Nitrite. Guidelines for Drinking Water Quality*, 3rd ed. 2004, Geneva: World Health Organization, 417–420.

[22]  Agency, U.E.P., National Primary Drinking Water Regulations: Final Rule, 40. *Fed Regist.*, 1991. *CFR Parts 141–143*(56 (20)): p. 3526–3597.

[23]  Authority, E.F.S., Nitrate in vegetables: scientific opinion of the panel on contaminants in the food chain. *The EFSA Journal*, 2008. 689: p. 1–79.

[24] Van Grinsven, H.J., et al., Does the evidence about health risks associated with nitrate ingestion warrant an increase of the nitrate standard for drinking water? *Environ Health*, 2006. 5: p. 26.

[25] Walker, R., The metabolism of dietary nitrites and nitrates. *Biochem Soc Trans*, 1996. 24(3): p. 780–785.

[26] Spiegelhalder, B., G. Eisenbrand, and R. Preussmann, Influence of dietary nitrate on nitrite content of human saliva: Possible relevance to in vivo formation of N-nitroso compounds. *Food Cosmet Toxicol*, 1976. 14(6): p. 545–548.

[27] Doel, J.J., et al., Evaluation of bacterial nitrate reduction in the human oral cavity. *Eur J Oral Sci*, 2005. 113(1): p. 14–19.

[28] Lundberg, J.O., et al., Nitrate, bacteria and human health. *Nat Rev Microbiol*, 2004. 2(7): p. 593–602.

[29] Bjorne, H.H., et al., Nitrite in saliva increases gastric mucosal blood flow and mucus thickness. *J Clin Invest*, 2004. 113(1): p. 106–114.

[30] Bryan, N.S., et al., Dietary nitrite supplementation protects against myocardial ischemia-reperfusion injury. *Proc Natl Acad Sci U S A*, 2007. 104(48): p. 19144–19149.

[31] McKnight, G.M., et al., Chemical synthesis of nitric oxide in the stomach from dietary nitrate in humans. *Gut*, 1997. 40(2): p. 211–214.

[32] Green, L.C., et al., Nitrate biosynthesis in man. *Proc Natl Acad Sci U S A*, 1981. 78(12): p. 7764–7768.

[33] Wennmalm, A., et al., Metabolism and excretion of nitric oxide in humans. An experimental and clinical study. *Circ Res*, 1993. 73(6): p. 1121–1127.

[34] Rahma, M., et al., Effects of furosemide on the tubular reabsorption of nitrates in anesthetized dogs. *Eur J Pharmacol*, 2001. 428(1): p. 113–119.

[35] Godfrey, M. and D.S. Majid, Renal handling of circulating nitrates in anesthetized dogs. *Am J Physiol*, 1998. 275(1 Pt 2): p. F68–73.

[36] Hunault, C.C., et al., Bioavailability of sodium nitrite from an aqueous solution in healthy adults. *Toxicol Lett*, 2009. 190(1): p. 48–53.

[37] Magee, P.N. and J.M. Barnes, The production of malignant primary hepatic tumours in the rat by feeding dimethylnitrosamine. *Br J Cancer*, 1956. 10(1): p. 114–122.

[38] Mirvish, S.S., Blocking the formation of N-nitroso compounds with ascorbic acid in vitro and in vivo. *Ann N Y Acad Sci*, 1975. 258: p. 175–180.

[39] Mirvish, S.S., Inhibition by vitamins C and E of in vivo nitrosation and vitamin C occurrence in the stomach. *Eur J Cancer Prev*, 1996. 5 Suppl 1: p. 131–136.

[40] Yoshida, K. and K. Kasama, Biotransformation of nitric oxide. *Environ Health Perspect*, 1987. 73: p. 201–205.

[41] Benjamin, N., et al., Stomach NO synthesis. *Nature*, 1994. 368(6471): p. 502.

[42] Lundberg, J.O., et al., Intragastric nitric oxide production in humans: measurements in expelled air. *Gut*, 1994. 35(11): p. 1543–1546.

[43] Moncada, S. and A. Higgs, The L-arginine-nitric oxide pathway. *N Engl J Med*, 1993. 329(27): p. 2002–2012.

[44] Spiegelhalder, B., G. Eisenbrand, and R. Preussman, Influence of dietary nitrate on nitrite content of human saliva: possible relevance to in vivo formation of N-nitroso compounds. *Food Cosmet Toxicol*, 1976. 14: p. 545–548.

[45] Duncan, C., et al., Chemical generation of nitric oxide in the mouth from the enterosalivary circulation of dietary nitrate. *Nat Med*, 1995. 1(6): p. 546–551.

[46] Lundberg, J.O. and M. Govoni, Inorganic nitrate is a possible source for systemic generation of nitric oxide. *Free Radic Biol Med*, 2004. 37(3): p. 395–400.

[47] Sobko, T., et al., Gastrointestinal nitric oxide generation in germ-free and conventional rats. *Am J Physiol Gastrointest Liver Physiol*, 2004. 287(5): p. G993–997.

[48]   Petersson, J., Nitrate, nitrite and nitric oxide in gastric mucosal defense, in *Uppsala University*. 2008, Uppsala: Uppsala.

[49]   Petersson, J., et al., Gastroprotective and blood pressure lowering effects of dietary nitrate are abolished by an antiseptic mouthwash. *Free Radic Biol Med*, 2009. 46(8): p. 1068–1075.

[50]   Petersson, J., et al., Dietary nitrate increases gastric mucosal blood flow and mucosal defense. *Am J Physiol Gastrointest Liver Physiol*, 2007. 292(3): p. G718–724.

[51]   Jansson, E.A., et al., Protection from nonsteroidal anti-inflammatory drug (NSAID)-induced gastric ulcers by dietary nitrate. *Free Radic Biol Med*, 2007. 42(4): p. 510–518.

[52]   Appel, L.J., et al., A clinical trial of the effects of dietary patterns on blood pressure. DASH Collaborative Research Group. *N Engl J Med*, 1997. 336(16): p. 1117–1124.

[53]   Lundberg, J.O., et al., Cardioprotective effects of vegetables: Is nitrate the answer? *Nitric Oxide*, 2006. 15(4): p. 359–362.

[54]   Webb, A.J., et al., Acute blood pressure lowering, vasoprotective, and antiplatelet properties of dietary nitrate via bioconversion to nitrite. *Hypertension*, 2008. 51(3): p. 784–790.

[55]   Larsen, F.J., et al., Effects of dietary nitrate on blood pressure in healthy volunteers. N *Engl J Med*, 2006. 355(26): p. 2792–2793.

[56]   Jansson, E.A., et al., A mammalian functional nitrate reductase that regulates nitrite and nitric oxide homeostasis. *Nat Chem Biol*, 2008. 4(7): p. 411–417.

[57]   Schachinger, V., M.B. Britten, and A.M. Zeiher, Prognostic impact of coronary vaso-dilator dysfunction on adverse long-term outcome of coronary heart disease. *Circulation*, 2000. 101(16): p. 1899–1906.

[58]   Halcox, J.P., et al., *Prognostic value of coronary vascular endothelial dysfunction. Circulation*, 2002. 106(6): p. 653–658.

[59]   Bugiardini, R., et al., Endothelial function predicts future development of coronary artery disease: A study of women with chest pain and normal coronary angiograms. *Circulation*, 2004. 109(21): p. 2518–2523.

[60]   Lerman, A. and A.M. Zeiher, Endothelial function: Cardiac events. *Circulation*, 2005. 111(3): p. 363–368.

[61]   Kleinbongard, P., et al., Plasma nitrite concentrations reflect the degree of endothelial dysfunction in humans. *Free Radic Biol Med*, 2006. 40(2): p. 295–302.

[62]   Bryan, N.S., et al., Dietary nitrite restores NO homeostasis and is cardioprotective in endothelial nitric oxide synthase-deficient mice. *Free Radic Biol Med*, 2008. 45(4): p. 468–474.

[63]   Stokes, K.Y., et al., Dietary nitrite prevents hypercholesterolemic microvascular inflam-mation and reverses endothelial dysfunction. *Am J Physiol Heart Circ Physiol*, 2009. 296(5): p. H1281–1288.

[64]   Davi, G. and C. Patrono, Platelet activation and atherothrombosis. *N Engl J Med*, 2007. 357(24): p. 2482–2494.

[65]   Radomski, M.W., R.M. Palmer, and S. Moncada, The anti-aggregating properties of vas-cular endothelium: interactions between prostacyclin and nitric oxide. *Br J Pharmacol*, 1987. 92(3): p. 639–646.

[66]   Radomski, M.W., R.M. Palmer, and S. Moncada, An L-arginine/nitric oxide pathway present in human platelets regulates aggregation. *Proc Natl Acad Sci U S A*, 1990. 87(13): p. 5193–5197.

[67]   Richardson, G., et al., The ingestion of inorganic nitrate increases gastric S-nitrosothiol levels and inhibits platelet function in humans. *Nitric Oxide*, 2002. 7(1): p. 24–29.

[68]   Lauer, T., et al., Age-dependent endothelial dysfunction is associated with failure to increase plasma nitrite in response to exercise. *Basic Res Cardiol*, 2008. 103(3): p. 291–297.

[69]   Rassaf, T., et al., Nitric oxide synthase-derived plasma nitrite predicts exercise capacity. *Br J Sports Med*, 2007. 41(10): p. 669–73; discussion 673.

[70]   Larsen, F.J., et al., Effects of dietary nitrate on oxygen cost during exercise. *Acta Physiol (Oxf)*, 2007. 191(1): p. 59–66.

[71]   Bailey, S.J., et al., Dietary nitrate supplementation reduces the O2 cost of low-intensity exercise and enhances tolerance to high-intensity exercise in humans. *J Appl Physiol*, 2009. 106(6): pp. 1875–1887.

[72]   Heinig, M.J., The American Academy of Pediatrics recommendations on breastfeeding and the use of human milk. *J Hum Lact*, 1998. 14(1): p. 2–3.

[73]   Gartner, L.M., et al., Breastfeeding and the use of human milk. *Pediatrics*, 2005. 115(2): p. 496–506.

[74]   Hoddinott, P., D. Tappin, and C. Wright, Breast feeding. *Bmj*, 2008. 336(7649): p. 881–887.

[75]   James, D.C. and R. Lessen, Position of the American Dietetic Association: promoting and supporting breastfeeding. *J Am Diet Assoc*, 2009. 109(11): p. 1926–1942.

[76]   Ip, S., et al., *Breastfeeding and Maternal and Infant Health Outcomes in Developed Countries. Evidence Report/Technology Assessment, A. Publication*, 2007, Tufts-New England Medical Center Evidence-based Practice Center. Rockville, MD.

[77]   Ohta, N., et al., Nitric oxide metabolites and adrenomedullin in human breast milk. *Early Hum Dev*, 2004. 78(1): p. 61–65.

[78]   Cekmen, M.B., et al., Decreased adrenomedullin and total nitrite levels in breast milk of preeclamptic women. *Clin Biochem*, 2004. 37(2): p. 146–148.

[79]   Hord, N.G., et al., Nitrate and nitrite content of human, formula, bovine and soy milks: implications for dietary nitrite and nitrate recommendations. *Breastfeeding Medicine, 2010* (In press).

[80]   Duncan, C., et al., Chemical generation of nitric oxide in the mouth from the enterosalivary circulation of dietary nitrate. *Nat Med*, 1995. 1(6): p. 546–551.

[81]   Duncan, C., et al., Protection against oral and gastrointestinal diseases: importance of dietary nitrate intake, oral nitrate reduction and enterosalivary nitrate circulation. *Comp Biochem Physiol A Physiol*, 1997. 118(4): p. 939–948.

[82]   Dykhuizen, R.S., et al., Antimicrobial effect of acidified nitrite on gut pathogens: importance of dietary nitrate in host defense. *Antimicrob Agents Chemother*, 1996. 40(6): p. 1422–1425.

[83]   Dykhuizen, R.S., et al., Helicobacter pylori is killed by nitrite under acidic conditions. *Gut*, 1998. 42(3): p. 334–337.

[84]   Dougall, H.T., et al., The effect of amoxycillin on salivary nitrite concentrations: an important mechanism of adverse reactions? *Br J Clin Pharmacol*, 1995. 39(4): p. 460–462.

[85]   Bodner, L., Effect of parotid submandibular and sublingual saliva on wound healing in rats. *Comp Biochem Physiol A Comp Physiol*, 1991. 100(4): p. 887–890.

[86]   Hart, B.L. and K.L. Powell, Antibacterial properties of saliva: role in maternal periparturient grooming and in licking wounds. *Physiol Behav*, 1990. 48(3): p. 383–386.

[87]   Benjamin, N., et al., Wound licking and nitric oxide. *Lancet*, 1997. 349(9067): p. 1776.

[88]   Weller, R., et al., Antimicrobial effect of acidified nitrite on dermatophyte fungi, Candida and bacterial skin pathogens. *J Appl Microbiol*, 2001. 90(4): p. 648–652.

[89]   Torregrossa, A.C., M. Aranke, and N.S. Bryan, Nitric oxide and geriatrics: Implications in diagnostics and treatment of the elderly. *Journal of Geriatric Cardiology*, 2011. 8: p. 230–242.

[90]  Lundberg, J.O., E. Weitzberg, and M.T. Gladwin, The nitrate-nitrite-nitric oxide pathway in physiology and therapeutics. *Nat Rev Drug Discov*, 2008. 7(8): p. 156–167.

[91]  Authority, E.F.S., Nitrate in vegetables: scientific opinion of the panel on contaminants in the food chain. *The EFSA Journal*, 2008. 289: p. 1–79.

[92]  Wilms, L.C., et al., Protection by quercetin and quercetin-rich fruit juice against induction of oxidative DNA damage and formation of BPDE-DNA adducts in human lymphocytes. *Mutat Res*, 2005. 582 (1–2): p. 155–162.

[93]  Houston, M. and L. Hays, Acute effects of an oral nitric oxide supplement on blood pressure, endothelial function, and vascular compliance in hypertensive patients. *J Clin Hypertens (Greenwich)*, 2014. 16(7): p. 524–529.

[94]  Biswas, O.S., V.R. Gonzalez, and E.R. Schwarz, Effects of an Oral Nitric Oxide Supplement on Functional Capacity and Blood Pressure in Adults with Prehypertension. *J Cardiovasc Pharmacol Ther*, 2014. 20(1): p. 52–59.

[95]  Zand, J., et al., All-natural nitrite and nitrate containing dietary supplement promotes nitric oxide production and reduces triglycerides in humans. *Nutr Res*, 2011. 31(4): p. 262–269.

[96]  Lee, E., Effects of nitric oxide on carotid intima media thickness: A pilot study. *Alternative Therapies in Health and Medicine*, 2016. 22(S2): p. 32–34.

[97]  Nagamani, S.C., et al., Nitric-oxide supplementation for treatment of long-term complications in argininosuccinic aciduria. *Am J Hum Genet*, 2012. 90(5): p. 836–846.

[98]  Erzurum, S.C., et al., Higher blood flow and circulating NO products offset high-altitude hypoxia among Tibetans. *Proc Natl Acad Sci U S A*, 2007. 104(45): p. 17593–17598.

# 17 Nitrate and Cardiovascular Systems

*Andrew Xanthopoulos, Apostolos Dimos,*
*Alexandros Zagouras, Nikolaos Iakovis,*
*John Skoularigis, and Filippos Triposkiadis*

## CONTENTS

17.1 Introduction .................................................................................................376
17.2 Nitrogen Balance in Mammals .....................................................................376
17.3 L-Arginine Nitric Oxide Synthase Pathway ................................................376
17.4 Enterosalivary Circulation of Nitrate/Nitrite/NO.......................................377
17.5 Nitrosative Chemistry of Nitrite and Nitrate...............................................377
17.6 Nitrovasodilators...........................................................................................378
17.7 Dietary Sources of NO..................................................................................379
17.8 Metabolism of Dietary Nitrate .....................................................................380
17.9 Beneficial Effects of Dietary Nitrate on Cardiovascular Health................381
    17.9.1 Blood Pressure.................................................................................381
    17.9.2 Endothelial Function .......................................................................382
    17.9.3 Ischemia Reperfusion Injury ..........................................................382
    17.9.4 Arterial Stiffness..............................................................................383
    17.9.5 Platelet Function..............................................................................383
    17.9.6 Improvement of Exercise Performance ..........................................383
17.10 Amount of Dietary Nitrate Required to Promote Cardiovascular
    Health.............................................................................................................384
17.11 NO in Chronic Cardiovascular Diseases......................................................384
17.12 NO in Acute Cardiovascular Events ............................................................386
17.13 Nitrate–Nitrite–Nitric Oxide Pathway and the Metabolic
    Syndrome ......................................................................................................386
17.14 Tolerance to Circulatory Effects of Nitrate.................................................388
17.15 Water Nitrate in Human Health ...................................................................388
    17.15.1 Methemoglobinemia.......................................................................390
    17.15.2 Adverse Pregnancy Outcomes.......................................................390
    17.15.3 Cancer..............................................................................................390
    17.15.4 Thyroid Disease...............................................................................390
    17.15.5 Other Health Effects .......................................................................391
17.16 Conclusions...................................................................................................391
References..................................................................................................................391

DOI: 10.1201/9780429326806-22

## 17.1  INTRODUCTION

The discovery of the nitric oxide (NO) pathway in the 1980s represented a critical advance in the understanding of cell signaling and especially into major new advancements in many clinical areas, including, but not limited to cardiovascular medicine (1). NO, which is derived from the nitrite and nitrate storage pools complementing the NO synthase (NOS) dependent pathway is one of the simplest biological molecules in nature, and it is very important in almost every phase of biology and medicine. It plays a significant role as an endogenous regulator of blood flow and thrombosis; it is also a principal neurotransmitter-mediating penile erectile function and a major pathophysiological mediator of inflammation and host defense. Continuous generation of NO is essential for the integrity of the cardiovascular system, and decreased production and/ or bioavailability of NO leads to the development of many cardiovascular disorders. Nevertheless, NO is just as important in other organ systems (2). NO is important for communication in the nervous system and is a critical molecule used by the immune system to kill invading pathogens, including bacteria and cancer cells. The production of NO from L-arginine is a complex and complicated biochemical process of many steps and factors that may be altered and affect ultimate NO production (3).

Regular intake of nitrate- and nitrite-containing foods may ensure that blood and tissue levels of nitrite and NO pools are maintained at a level sufficient to compensate for any disturbances in endogenous NO synthesis (4). Since low levels of supplemental nitrite and nitrate have been shown to enhance blood flow, dietary sources of NO metabolites can, therefore, improve blood flow and oxygen delivery and protect against various cardiovascular diseases or other conditions associated with NO insufficiency. The consumption of nitrite- and nitrate-rich foods, such as fruits, leafy vegetables, and some meats along with antioxidants, such as vitamin C and polyphenols, can increase NO production (1, 4, 5).

In this chapter, the main sources of dietary nitrate will be discussed, and its metabolism will be briefly reviewed, highlighting its effect on human health and especially on the cardiovascular system.

## 17.2  NITROGEN BALANCE IN MAMMALS

In humans, nitrogen is principally ingested as protein. Only a tiny proportion of this nitrogen is converted to NO via the nitric oxide synthase pathway. In healthy humans, about 1 mmol of nitrate is generated from L-arginine and then NO oxidation, which represents about one-thousandth of the amount of nitrogen ingested. During illness, such as gastroenteritis, this can increase by as much as eightfold (6). Ruminants, and other mammals that rely on symbiotic bacteria to metabolize cellulose, can also make use of non-protein nitrogen sources for protein synthesis. Rumen bacteria can convert urea to ammonia, which is used to produce amino acids that can be incorporated into mammalian proteins (7).

## 17.3  L-ARGININE NITRIC OXIDE SYNTHASE PATHWAY

Three distinct NO synthase isoforms exist in mammals: inducible, neuronal, and endothelial. The loci of each indicate the pluripotent effects of NO in mammals, with roles in host defense, neuronal and other cellular signaling pathways, and vascular control. L-arginine provides organic nitrogen as a substrate which, along with $O_2$ and NADPH, is converted to NO and L-citrulline. The NO produced has a very short half-life, being rapidly oxidized in the presence of superoxide or oxyhemoglobin to nitrate, which enters the mammalian nitrate cycle (3, 5).

## 17.4  ENTEROSALIVARY CIRCULATION OF NITRATE/NITRITE/NO

The symbiosis between legumes and nitrogen-fixing rhizobia finds its counterpoint in the relationship between nitrate-reducing bacteria hidden in crypts in mammalian tongues. Nitrate from the diet is rapidly and completely absorbed from the upper gastro-intestinal tract. This, along with nitrate derived from the oxidation of NO synthesized by the L-arginine NO synthase pathway, is actively taken up in the salivary glands. The resulting salivary nitrate concentration may be ten times greater than the plasma nitrate concentration. In crypts on the dorsum of the tongue, facultative utilizes nitrate as an alternative electron acceptor (8). The nitrite released elevates salivary nitrite to levels a thousand times that of plasma in the resting state. In the presence of acid-generating plaque bacteria, some nitrite is chemically reduced to NO (9). The remaining salivary nitrite is then swallowed. In the acidic environment of the stomach, some of this nitrite is further reduced to NO (10), which has an important role in both protecting against enteric pathogens and regulating gastric blood flow and mucous production (11, 12). Some of the nitrite is absorbed from the stomach with important consequences for mammalian vascular physiology. NO is continually released from the surface of normal human skin (13). It is excreted in human sweat and reduced to nitrite by skin bacteria. As normal skin is slightly acidic (pH around 5.5), nitrite is reduced to NO. It has been proposed that NO inhibits skin pathogens and facilitates wound healing (14, 15).

Also, breast milk has been shown to contain variable amounts of nitrite and nitrate, and it has been suggested that conversion of these anions to more reactive nitrogen oxides may contribute to the protective action of breastfeeding against infant gastro-enteritis (16).

## 17.5  NITROSATIVE CHEMISTRY OF NITRITE AND NITRATE

Regardless of the source of exposure, the primary concern of exposure to nitrite, nitrate, or other nitrogen oxides is the nitrosative chemistry that can occur (17). The risk–benefit spectrum from nitrite and nitrate may depend upon the specific metabolism and the presence of other components that may be concomitantly ingested. The stepwise reduction to nitrite and NO may account for the benefits while pathways leading to nitrosation of low molecular weight amines or amides may account for the health risks of nitrate exposure. Understanding and affecting those pathways will help in mitigating those risks (1).

Nitrate itself is generally considered to be harmless at low concentrations. Nitrite, on the other hand, is reactive, especially in the acid environment of the stomach, where it can nitrosate other molecules, including proteins, amines, and amides. Nitrite is occasionally found in the environment, but most human exposure occurs through ingested nitrate that can be chemically reduced to nitrite by commensal bacteria often found in saliva (18). Nitrite and nitrate have been associated with an increased risk of cancer. This risk may result from the exposure to preformed N-nitrosamines or N-nitrosamines formed in the stomach from ingesting foods enriched in nitrite and nitrate. Some low molecular weight amines can be converted to their carcinogenic N-nitroso derivatives by reaction with nitrite (19). Most nitrosamines are mutagens, and a number are transplacental carcinogens. Most are organspecific. For instance, dimethylnitrosamine causes liver cancer in experimental animals, whereas some of the tobacco-specific nitrosamines cause lung cancer (20). Nitrosamines can occur because their chemical precursors – amines and nitrosating agents – occur commonly.

It has been reported that ascorbic acid inhibits nitrosamine formation (21). Another antioxidant, a-tocopherol (vitamin E), has also been shown to inhibit nitrosamine formation (22). Ascorbic acid, erythorbic acid and a-tocopherol inhibit nitrosamine formation due to their oxidation–reduction properties. Most vegetables that are enriched in nitrate are also rich in antioxidants such as vitamins C and E that can act to prevent the unwanted nitrosation chemistry. These compounds are also now almost universally added to cured processed meats. Controlling the metabolic fate of nitrate and nitrite away from nitrosation and toward reduction of NO may provide a strategy to promote health benefits while mitigating the health risks. Adverse health effects may be the result of a complex interaction of the amount of nitrite and nitrate ingested, the concomitant ingestion of nitrosation cofactors and precursors, and specific medical conditions such as chronic inflammation that increase nitrosation. Controlling for such factors is essential for defining the safety of nitrite and nitrate.

## 17.6   NITROVASODILATORS

The pharmacological effect of nitrovasodilators is mediated by their vasoactive metabolite NO. The target of NO is the soluble guanylyl cyclase (sGC), which exists as a heterodimer containing a heme moiety (23). NO interacts with the ferrous ion, forming a ferrous–nitrosyl–heme complex resulting in a conformational change in the sGC, resulting in activation of its catalytic activity. With divalent magnesium as a cofactor, sGC catalyzes guanosine-5¢-triphosphate to cyclic guanosine monophosphate (cGMP) which, in turn, activates various protein kinases such as protein kinase G. The signaling cascade results in the alteration of several key mediators of the vasotone. Intracellular calcium is sequestered in the sarcoplasmic reticulum via increased uptake by activation of $Ca^{2+}Mg^{2+}ATPase$ and decreased efflux from calcium channels. In addition, dephosphorylation of the myosin light chain results in decreased sensitivity to calcium (24). The messenger effect of cGMP is terminated by its hydrolysis by phosphodiesterases (PDEs). The inhibition of PDE in the penis by drugs like sildenafil for erectile dysfunction results in potentiation of organic nitrate action, thus leading to possible severe hypotension (25). Organic nitrate possess dose-dependent vasoselectivity which can be exploited to best manage various cardiovascular pathologies. The venous vasculature is significantly more sensitive to the effects of organic nitrate than the arterial side and is preferentially dilated upon administration of low doses. This results in an increased capacitance in the veins and decreased venous return to the atria, thus reducing cardiac preload and left ventricle end-diastolic pressure (26). Doses that have substantial effects on venodilation have little effect on arterial vascular resistance and systemic blood pressure. Part of this selectivity is likely due to an increased amount of bioactivating enzymes in the venous smooth muscles cells compared to those on the arterial side (27). Further venodilation and arterial relaxation occur with higher doses which decrease cardiac afterload and systemic blood pressure.

Organic nitrates have no intrinsic effect on cardiac contractility or heart rate, but reflex tachycardia may develop in response to the reduction in blood pressure (26). Profound hypotension may occur in patients concurrently receiving agents such as beta blockers or alpha-adrenergic receptor blockers which block the sympathetic tone. Sodium nitroprusside as a more spontaneous donor of NO, has an indiscriminate effect on venous and arterial relaxation (28). Organic nitrates have a complex effect on coronary blood flow; their beneficial effects may not be due to an indiscriminate coronary artery vasodilatation but rather a redistribution of cardiac perfusion (29). When nitroglycerin and isosorbine dinitrate are administered during periods of increased oxygen demand and ischemia, an unequal reduction in coronary blood flow is experienced, flow to well-perfused areas is decreased to a greater extent than those that are occluded and experiencing ischemia (30). While total myocardial flow is reduced due to decreased cardiac output resulting from decreased preload, ischemia is relieved via reduction of myocardial oxygen demand and balancing of the flow based on metabolic demand. In addition, coronary flow to regions supplied by severely occluded arteries is improved by nitrate due to an increase in perfusion through collateral vascular beds which circumvent the occlusion (31). The culmination of these effects is a reduction of myocardial oxygen demand while improving flow to the ischemic regions of the heart. Due to a decrease in preload, ventricular wall tension is reduced which results in a reduction in oxygen demand (26, 32). Reduction in wall tension also decreases subendocardial resistance allowing for improved perfusion through the cardiac wall (29). Higher doses of organic nitrate reduce systemic vascular resistance, which in turn decreases afterload (26). In severe ventricular dysfunction, the combination of an organic nitrate with hydralazine (an arterial vasodilator) results in decreased preload and afterload which decreases oxygen demand while improving cardiac output (29). If hypotension develops, a reduction in coronary perfusion pressure may lead to an increased oxygen debt in the myocardial tissue worsening ischemia.

## 17.7   DIETARY SOURCES OF NO

Certain foods, such as green leafy vegetables and beetroot, are particularly rich in nitrate. Consumption of a typical Western diet results in the ingestion of approximately 1–2 mmol nitrate per day. Nitrite, because of its anti-botulism effect, has been used as preservative and colorant for centuries. In addition, humans are also exposed to biologically active nitrogen oxides from the combustion of fossil fuels or inhalation of tobacco smoke (5, 33).

The two main sources of nitrate in the diet are vegetables and drinking water. Since nitrate is absorbed effectively with a bioavailability of 100 percent, total nitrate consumption is largely dependent on the nitrate content of vegetables and quantity of vegetable intake, as well as the level of nitrate in local drinking water (34, 35). Certain vegetables, especially beetroot, lettuce, rocket, and spinach are naturally rich in nitrate, while other vegetables, such as peas, potato, and tomato, contain nitrate at lower concentrations (36). The level of nitrate in vegetables also varies according to soil conditions, time of year, nitrate content of fertilizers, growing conditions, the storage, and transport environment as well as cooking procedures (35, 37–39). The nitrate content of drinking water, which is regulated in most countries due to health concerns, fluctuates and is dependent on bacterial nitrogen fixation, the decay of organic matter in the soils, manure from large-scale livestock production, and fertilizer use (40). Nitrate intake varies greatly between individuals and populations. Mean daily intakes are estimated to be between 0.4 to 2.6 mg/kg or 31 to 185 mg (41), and actual individual daily nitrate intakes ranging from less than 20 mg to greater than 400 mg (37, 38, 42). Indeed, individuals who follow the Dietary Approaches to Stop Hypertension (DASH) diet may consume as much as 1000 mg/d. The European Food Safety Authority has set the Acceptable Daily Intake (ADI) for nitrate at 3.7 mg/kg (approximately 260 mg for a 70 kg adult). The consumption of a diet rich in high-nitrate vegetables could markedly exceed the ADI. It seems unlikely that this would be related to detrimental health effects. Although the consumption of green leafy vegetables is widely promoted, the diets of many populations are low in these vegetables and therefore low in nitrate with intakes often less than 100 mg/d. Different patterns in the consumption of vegetables can yield differences in nitrate intake of over 700 percent and can be significantly greater than the ADI (43, 44). It has also been suggested that the Mediterranean diet, well known for its cardioprotective properties (see below), might provide substantial amounts of nitrate because of its high content of nitrate-rich vegetables. In addition, prospective cohort studies suggest that green leafy vegetables have protective effects against cardiovascular diseases but have not assessed nitrate intake (45).

There is, however, concern among some researchers about the toxicity associated with intake of nitrate and nitrite, particularly in relation to methemoglobinemia, colorectal cancer, and cardiovascular disease. Nevertheless, due to the health benefits observed and the uncertainty surrounding the toxic effects, a number of researchers are calling for a re-evaluation of the guidelines for the acceptable levels of nitrate in water and intake from food (45).

## 17.8   METABOLISM OF DIETARY NITRATE

Inorganic nitrate was associated with an increased risk of gastrointestinal cancer and methaemoglobinaemia in infants and it is considered as a contaminant in vegetables and water (46). It was reported that dietary nitrate consumption could lead to the formation of endogenous N-nitrosamines, most of which were shown to be carcinogens in rodents. However, epidemiological studies in vegetarians have shown the opposite effect, as vegetarian diets are associated with low mortality rate and are protective against cardiovascular diseases and different types of cancer (47, 48). Furthermore, a meta-analysis on this issue found that a high nitrate intake was associated with a weak but statistically significant reduced risk of cancer in humans (49). Also, it has been indicated that dietary nitrate could be viewed as a bioactive food component.

Dietary nitrate is rapidly absorbed from the gastrointestinal tract into the circulation with plasma levels remaining high for 5–6h after ingestion (50). Subsequently, part of this nitrate is actively taken up by the salivary glands, concentrated up to twenty times higher than plasma levels, and secreted into the oral cavity in the saliva (50). Oral bacteria reduce salivary nitrate to nitrite which is swallowed and absorbed across the upper gastrointestinal tract and into the circulation. It is not known how nitrite crosses the gut wall into the circulation, but there have been suggestions that anion exchanger-1 in erythrocytes may be involved (51). When swallowed, the acidic environment of the stomach facilitates the reduction of nitrite to NO and other N forms (52). This reaction is enhanced by the presence of vitamin C and polyphenols in the stomach, both of which can be obtained from vegetables (53). Circulatory nitrite can also be reduced to NO through the activity of a number of enzymes with nitrite reductase capability present in several cells such as hemoglobin, myoglobin, and mitochondria (50). The key role of oral bacteria in the conversion of dietary nitrate to nitrite has been demonstrated by many studies. There is a correlation between oral bacteria, dietary nitrate ingestion, increased circulatory levels of nitrite, and reduced blood pressure (54).

Patients with hypertension have impaired NO bioavailability, caused by increased levels of reactive oxygen species which scavenge NO (55). Reduced NO bioavailability is a major risk factor for both hypertension and cardiovascular diseases (55). Consequently, many studies in the last two decades have investigated the effects of L-arginine supplementation and NO bioavailability in patients with hypertension (56). However large amounts of oral L-arginine increase the risk of mortality in both animals with septic shock and patients with acute myocardial infarction (57). Thus, current research is now focusing on the effect of dietary nitrate on NO metabolism and vascular function.

## 17.9 BENEFICIAL EFFECTS OF DIETARY NITRATE ON CARDIOVASCULAR HEALTH

People who follow the Mediterranean Diet have a low incidence of cardiovascular disease (58). This type of diet includes high intake of fruit and vegetables, low levels of red meat, a large amount of fish and white and a moderate intake of red wine. The cardioprotective effect of the Mediterranean Diet is probably partially due to the high nitrate/nitrite content because Mediterranean Diet contains as much as 20 times the nitrate/nitrite present in a typical Western Diet (4, 43, 59). Nitrate intake is associated with vegetable intake and with other nutrients such as fiber, carotenoids, and folate (60).

The beneficial effects of dietary nitrate on cardiovascular health are the following:

### 17.9.1 BLOOD PRESSURE

A diet rich in fruit and vegetables lowers blood pressure with the greatest protective effect from consumption of green leafy and cruciferous vegetables with high nitrate content (61–63). While the exact compounds responsible for this blood pressure-lowering effect are not known, the possibility that nitrate is a key molecule was highlighted with the discovery of the enterosalivary pathway of, and the evidence of decreased NO production, in hypertension (10, 64–66). Many studies have shown a reduction in blood pressure after an acute or chronic dose of nitrate, whether in the form of beetroot juice, a high green leafy vegetable diet, or nitrate salts (67–69). The enterosalivary nitrate–nitrite–NO pathway has an important role in maintaining plasma nitrite levels. Even in the absence of dietary nitrate intake, endogenous nitrate derived from the L-arginine–NOS pathway contributes to the physiological modulation of blood flow. Finally, the blood pressure-lowering effect of sunlight has been attributed to the conversion of nitrate in sweat to nitrite by commensal bacteria on the surface of the skin (70, 71).

### 17.9.2 ENDOTHELIAL FUNCTION

Endothelial dysfunction with attenuated NO production is central to the pathogenesis of cardiovascular disorders. Dietary nitrate could improve endothelial function by serving as an alternate source of vasoactive NO through the nitrate–nitrite– NO pathway. Flow-mediated vasodilation (FMD) of the brachial artery is the gold standard for noninvasive measurement of conduit artery endothelial function (72). The size of FMD response is prognostic not only for the occurrence of a cardiovascular event in high- and low-risk individuals, but also the severity of the disease (73–75). Also, FMD provides a measure of in vivo endothelium-derived NO bioavailability (76). A small but significant improvement in FMD has been observed in healthy volunteers following ingestion of 200 mg spinach (182 mg nitrate) and other foods that contain nitrate (77).

### 17.9.3 ISCHEMIA REPERFUSION INJURY

Ischemia reperfusion injury is the tissue damage that occurs on the restoration of circulation after a period of ischemia or lack of oxygen. The consequences of ischemia reperfusion injury can be mild, resulting in reversible cell dysfunction, or severe, with multiple organ failure and death. Particularly in the heart (see below) and brain, ischemia reperfusion injury is a major cause of death and morbidity (78). Decreased NO bioavailability is a central event in ischemia reperfusion injury, and importantly contributes to the vascular dysfunction (79). Nitrite through its conversion to NO could prevent or limit the damage. This is enhanced in hypoxia when oxygen tension and pH falls, rendering the L-arginine–NOS pathway inactive. The administration of nitrite protects against ischemia-reperfusion injury in liver, heart, brain, kidney, and chronic hind-limb ischemia (80–85). Nevertheless, this type of protection was not observed in renal ischemia-reperfusion injury indicating that nitrite protection may be tissue specific (86). Increasing plasma and tissue nitrite levels, therefore, would enlarge the storage pool of NO and could prevent or limit the damage caused by ischemia–reperfusion injury. This could be of therapeutic value in human diseases such as myocardial infarction, stroke, organ transplantation, and cardiopulmonary arrest. Nitrite is equally effective when it is delivered by oral or systemic administration before, during, or after ischemia (45). Nitrate, through the nitrate–nitrite–NO pathway, would have a similar effect (81).

### 17.9.4 ARTERIAL STIFFNESS

Arterial stiffness is a decreased capacity of expansion and contraction in response to change in pressure (87). It is increasingly recognized as an important assessment of cardiovascular risk. NO influences vascular tone, and therefore arterial stiffness, through effects on smooth muscle relaxation. A number of noninvasive methods for measuring arterial stiffness have been developed which are generally derived from pulse wave velocity (PWV), peripheral arterial waveforms or arterial distensibility measures (88). PWV is considered the gold-standard measurement of arterial stiffness and is independently predictive of cardiovascular events (89). PWV is derived from the time taken for a pulse pressure to travel between two points (generally the carotid and femoral artery) a measured distance apart. Inhibition of NO alters PWV through effects on mean arterial pressure (90). Improving NO status via the enterosalivary nitrate– nitrite–NO pathway could potentially improve arterial stiffness (91).

### 17.9.5 PLATELET FUNCTION

Platelet aggregation is a key event in the pathogenesis of atherosclerosis and the development of acute thrombotic events, including myocardial infarction and unstable angina (92). Anti-platelet therapy highlights the importance of platelet aggregation in these disorders by reducing cardiovascular risk (93). The aggregation and adherence of platelets to the endothelium is normally prevented by the production of endogenous antiplatelet agents, including NO (94, 95). This NO is produced by the vascular endothelium or within the platelet itself (96). An effect of nitrate, through the nitrate–nitrite–NO pathway, to reduce platelet activity could have a large impact on cardiovascular disease. Indeed, this has been demonstrated after dietary supplementation with beetroot juice and potassium nitrate (97).

### 17.9.6 IMPROVEMENT OF EXERCISE PERFORMANCE

Exercise is cardioprotective with a number of cardiovascular disease risk factors reduced with regular exercise (98). In addition, exercise confers protection following a heart attack in humans (99). While the exact mechanisms behind the cardioprotective effects of exercise are not clear, enhanced endothelial NO synthases (eNOS) expression and elevated NO in response to shear stress during exercise could play a role (71, 100, 101). NO plays an important role in the physiological adaptation to exercise by enhancing muscle blood flow, altering glucose uptake and metabolism, calcium homeostasis, and modulating muscle contraction (102, 103). It also controls cellular respiration through its interactions with mitochondrial respiratory chain enzymes (104). Because NO production is increased in the vascular endothelium by exercise, it has been suggested that plasma nitrite levels postexercise are a good predictor of exercise capacity (105). Whether the increase in systemic nitrite after administration of dietary or inorganic nitrite would have an effect on oxygen consumption during exercise has been investigated in humans (106, 107). It was found that four to six days of dietary nitrate supplementation in the form of beetroot juice reduced blood pressure and O2 cost, in low and moderate-intensity exercise, while the time to exhaustion during high-intensity exercise was significantly increased. It was found that three days of supplementation with sodium nitrate resulted in a reduction in oxygen consumption during submaximal and maximal exercise (69, 108). The exact mechanisms still need to be determined but there is some evidence to suggest that it involves the mitochondrion and/or an increase in blood flow to the muscles with an improved balance between blood flow and oxygen uptake (109–111).

## 17.10  AMOUNT OF DIETARY NITRATE REQUIRED TO PROMOTE CARDIOVASCULAR HEALTH

Inorganic nitrate and beetroot juice supplements are associated with significant reductions in systolic blood pressure (112). A range of doses has been reported. The average amount of nitrate used in trials can be calculated as about 500mg/d (range 139–1042mg/d). Environmental factors such as the season of the year, the light, the irrigation of the field, the use of fertilizers, and the storage conditions (temperature, humidity, and light), could affect the quantity of nitrate in the beetroot juice (113).

Using pharmacological salts (potassium nitrate), Kapil and colleagues investigated the effect of two different doses providing 248 and 744mg of nitrate, respectively (67). While both doses were similarly effective in reducing diastolic blood pressure, the lower dose did not show a significant effect in reducing systolic blood pressure. This suggests that a minimum amount of nitrate might be needed to induce changes in systolic blood pressure. In accord with current evidence, this threshold could be up to about 500mg of nitrate. This amount clearly exceeds the ADI of 3-7mg/kg per day (about 260mg/d for a 70kg adult) as well as exceeding the average intake of vegetarians (53, 114).

Daily consumption of at least one portion of high-nitrate vegetables could increase nitrate intake in order to lower blood pressure. However, the intake of excessive amounts of nitrate could be associated with an increased risk of unfavorable health outcomes (113).

## 17.11  NO IN CHRONIC CARDIOVASCULAR DISEASES

Endothelial dysfunction precedes chronic cardiovascular disease and is caused by several risk factors, including aging, hypercholesterolemia, hypertension, diabetes mellitus, smoking, obesity, chronic systemic inflammation, hyperhomocysteinemia, and a family history of premature atherosclerosis. An intact endothelium, however, is required for the maintenance of vascular tone and architecture, blood fluidity, and antithrombotic protection. NO signaling plays a crucial role in maintaining these functions by regulation of vascular tone and smooth muscle cell proliferation, white blood cell adhesion, and platelet function (5, 79). eNOS produce NO from the amino acid L-arginine and several co-factors (115). Disturbance in the NOS-dependent NO pathway is a key element in the development of endothelial dysfunction, which leads to the onset and progression of atherosclerosis. In endothelial dysfunction, reduced NO bioavailability is caused by an inhibition of eNOS function rather than by lower eNOS protein expression (116). One of the events leading to altered eNOS activity has been related to decreased levels of the cofactor, (6R)-5,6,7,8-tetrahydro-l-biopterin, and has been termed "eNOS uncoupling" (117). Under these pathological conditions, eNOS produces potentially harmful reactive oxygen species (ROS, e.g., superoxide) rather than NO. The extent of endothelial dysfunction relates to the risk of patients for developing chronic cardiovascular diseases and acute ischemic events, e.g., myocardial infarction and stroke. Another key element in the maintenance of endothelial function is endothelial progenitor cells (EPCs, also referred to as circulating angiogenic cells). EPCs participate in the repair of vessel injury and neovascularization (118–120). NO modulates the recruitment of EPCs, and balanced NO homeostasis is, therefore, required for appropriate interaction between resident endothelial cells and circulating EPCs. Imbalanced NO levels may thus contribute to impaired vascular regenerative processes, for example, in patients with chronic ischemic heart disease (121, 122).

NO produced from eNOS can react with hemoglobin in the circulation to undergo oxidation to nitrate. Owing to the immense concentrations of hemoglobin in blood, signaling by NO has therefore been considered to occur only at the site of NO production. Indeed, evidence suggests that free hemoglobin is an effective NO scavenger that can attenuate NO bioavailability and NO-related cardiovascular functions; in patients with end-stage renal disease, impaired vascular function occurs, in part, due to hemodialysis-related release of free hemoglobin into the circulation (123). Thus, the maintenance of NO bioavailability throughout the cardiovascular system is important for vascular homeostasis.

In vivo, several biochemical mechanisms exist, allowing NO to be transported as NO itself, to react with plasma compounds to form nitrosospecies, and to be oxidized to bioactive nitrite. These mechanisms ensure NO bioavailability and signaling throughout the body (124–127). The complexity of the circulating NO opens new, considerable pharmacological and therapeutical possibilities in the diagnosis and therapy of cardiovascular diseases. In order to detect endothelial function and to evaluate the impact of dietary interventions on NO-mediated vascular functions, several invasive and noninvasive techniques have been used. The measurement of the flow-mediated dilation (FMD) has been used in numerous interventional trials to assess endothelial function. In FMD the relative diameter change of the brachial artery from pre-ischemia to reactive hyperemia is measured via high-resolution ultrasound. Ischemia is induced via inflation of a blood pressure cuff around the forearm, distal to the part of the artery assessed by ultrasound. During the ischemic period, the forearm vasculature vasodilates, and the final pressure release from the cuff consequently leads to increased flow in the conduit brachial artery. This is accompanied by an enhanced vessel wall shear stress, which stimulates eNOS to produce NO, which in turn causes vascular smooth muscle relaxation and accompanying arterial dilation. FMD is, therefore, generally regarded to be the gold standard in the characterization of endothelial function in vivo. Notably, the extent of endothelial function detected via FMD also correlates with the function in the coronary conduit arteries (3, 5, 128).

## 17.12   NO IN ACUTE CARDIOVASCULAR EVENTS

Acute myocardial infarction caused by rupture of an atherosclerotic plaque and subsequent occlusion of a large coronary artery by thrombus is one of the leading causes of death worldwide. During an acute myocardial infarction, the immediate and successful recanalization of the occluded coronary artery is the optimal therapy which leads to an infarct size reduction and improves the prognosis of the patient. The process of restoring blood flow to the ischemic myocardium, however, can induce injury itself. This phenomenon, termed ischemia/reperfusion injury reduces the beneficial effects of myocardial reperfusion (129). Apart from acute impairment of left ventricular (LV) function, large myocardial infarctions lead to a pathologic remodeling of the LV, limiting prognosis. Four phenomena, namely, myocardial stunning, the no-reflow phenomenon, arrhythmias, and lethal reperfusion injury, contribute to the extent and impact of the ultimate myocardial ischemia/reperfusion injury (130–133). The key mediators of these patho-biological mechanisms are ROS, deranged calcium levels ($Ca^{2+}$), pH, the mitochondrial permeability transition pore (mPTP), metabolic changes, and the immune responses (130, 134, 135). As demonstrated in numerous experimental studies, myocardial ischemia/reperfusion injury can be modulated. Two of the most powerful protective mechanisms in this context are ischemic preconditioning (IPC) and ischemic post conditioning (PostC), which are capable of reducing final infarct size by ~30–60 percent of the myocardium at risk (135, 136).

Several animal experimental studies point to a protective role of the NO pool during myocardial ischemia/reperfusion (80, 137–139). Administration of nitrite in an attempt to modulate ischemia/reperfusion injury leads to an elevation of the circulating NO pool and exerts protective tissue effects of ischemia/reperfusion injury in vivo (80, 138, 140). The protective effects derive from a hemeprotein-dependent reduction of nitrite to NO (138). NO derived from nitrite reduction via cardiac myoglobin reversibly modulates mitochondrial electron transport, thus decreasing reperfusion-derived oxidative stress and inhibiting cellular apoptosis leading to a smaller final infarct size. A recent study demonstrated that while dietary nitrate reduced blood pressure, calcium-handling protein content, SERCA enzymatic properties, and left ventricular function was not altered in the left ventricle of healthy rats (141).

In summary, disturbed NO homeostasis is a hallmark of the development of chronic cardiovascular disease and plays a key role in acute ischemic events, such as an acute myocardial infarction (Table 17.1). However, the involvement of NO in all of the mechanisms described above also points to the possibility of therapeutic interventions that modulate NO bioavailability (3).

## 17.13   NITRATE–NITRITE–NITRIC OXIDE PATHWAY AND THE METABOLIC SYNDROME

Obesity and hypertension are increasing problems worldwide resulting in an enormous economic burden to society (142, 143). The metabolic syndrome is a combination of medical abnormalities, including central obesity, dyslipidemia,

## TABLE 17.1
## Nitrate's Effects on the Cardiovascular System

| Pros | Cons |
|---|---|
| 1. Decrease in preload | 1. Chronic and long-term therapy with organic nitrates are not associated with long term survival-mortality when used in patients with coronary artery disease |
| 2. Antianginal effect | |
| 3. Anti-aggregatory effects on platelets | |
| 4. Reduction of myocardial oxygen demand and improvement of flow to the ischemic regions of the heart | 2. Organic nitrate therapy leads to increased vascular oxidative stress which in turn can produce endothelial dysfunction |
| 5. Higher doses of organic nitrates reduce systemic vascular resistance, which in turn decreases afterload | 3. Rapid development of tolerance to the vasodilatory effects of nitrate |
| 6. Nitrates possess dose-dependent vasoselectivity which can be exploited to best manage various cardiovascular pathologies | 4. Reflex tachycardia may develop in response to the reduction in blood pressure |
| 7. Use of nitrate (nitroglycerin) in the management of myocardial infarction | |
| 8. Intravenous nitroglycerin and sodium nitroprusside is indicated for the management of hypertensive crises | |
| 9. Use of drugs that promote the synthesis of nitric oxide in the management of the pulmonary hypertension | |

hyperglycemia, and hypertension, and it is estimated that around 25 percent of the world's adult population has the metabolic syndrome (143). This clustering of metabolic abnormalities that occur in the same individual appears to present a considerably higher cardiovascular risk compared with the sum of the risk associated with each abnormality.

The lack of the gene for eNOS results not only in the development of hypertension, but also displays key features of the metabolic syndrome including dyslipidemia, obesity, and insulin resistance (144–146). In addition, polymorphism in the eNOS gene is associated with metabolic syndrome in humans (147). So, a reduced NO bioavailability is a central event in the pathogenesis of metabolic syndrome. The stimulation of the nitrate–nitrite–NO pathway can partly compensate for disturbances in endogenous NO generation from eNOS. These findings may have implications for novel nutrition-based preventive and therapeutic strategies against cardiovascular disease and type 2 diabetes (5). The endogenous nitrate levels are already sufficient to affect cellular processes. Thus, in addition to the second-by-second regulation of vascular tone by eNOS-derived NO, its oxidative end product nitrate may serve as a long-lived reservoir for NO-like bioactivity in tissues (5). These bioactive compounds may be mediating the observed effects, but the exact mechanism for this and the signaling pathways involved remains to be elucidated.

## 17.14    TOLERANCE TO CIRCULATORY EFFECTS OF NITRATE

The efficacy of oral nitrate in the therapy for angina pectoris has been questioned. It was demonstrated in animals that the nitrate underwent extensive first-pass hepatic metabolism, and that systemic drug levels were low (148). However, demonstrations in humans that oral isosorbide dinitrate prolonged hemodynamic effects and improved exercise tolerance in angina pectoris have been followed by the increasing use of the oral nitrate (149). However, uncertainty persists regarding long-term use, as tolerance has been shown to develop quickly to the circulatory effects of the nitrate (150). It is generally assumed that such tolerance is not clinically relevant, but as the beneficial effects of the nitrate in angina pectoris are, at least in part, related to their systemic vascular effects, circulatory tolerance might be expected to be associated with changes in anti-anginal effects (151).

The mechanism of nitrate tolerance is unknown. It was proposed that circulatory tolerance to organic nitrate involved the oxidation of a critical sulfhydryl group in the glyceryl trinitrate receptor (152). This hypothesis was supported by the reversal of in vivo and in vitro tolerance induced by the disulfide reducing agent "dithiothreitol." Recent observations that higher isosorbide dinitrate plasma levels seen during sustained therapy are associated with high levels of the metabolites may indicate displacement of the parent compound from receptor sites, and this could account for the diminished sensitivity (153). The effect of nitrate in angina is at least in part due to the effects of these agents on the peripheral arterioles and capacitance veins (151).

## 17.15    WATER NITRATE IN HUMAN HEALTH

Nitrate levels in water resources have increased in many areas of the world largely due to applications of inorganic fertilizer and animal manure in agricultural areas (17). The regulatory limit for nitrate in public drinking water supplies was set to protect against infant methemoglobinemia. The risk of specific cancers and birth defects may be increased when nitrate is ingested under conditions that increase the formation of N-nitroso compounds. Water nitrates can be associated with colorectal cancer, bladder, and breast cancer, and thyroid disease (154). Also, many studies observed increased risk with ingestion of water nitrate levels that were below regulatory limits (155, 156).

In the last century the natural rate at which nitrogen is deposited onto land – through the production and application of nitrogen fertilizers, the combustion of fossil fuels, and replacement of natural vegetation with nitrogen-fixing crops such as soybeans – has been doubled (157, 158). The major anthropogenic source of nitrogen in the environment is nitrogen fertilizer, the application of which increased exponentially after the development of the Haber–Bosch process in the 1920s. Most synthetic fertilizer applications to agricultural land occurred after 1980 (159). Since approximately half of all applied nitrogen drains from agricultural fields to contaminate surface and groundwater, nitrate concentrations in our water resources have also increased (157). The maximum contaminant level (MCL) for nitrate in public drinking water supplies in the United States is 10 mg/Lasnitrate-nitrogen (NO3-N). This concentration is approximately equivalent to the World Health Organization

(WHO) guideline of 50 mg/L as NO3 or 11.3 mg/L NO3-N (multiply NO3 mg/L by 0.2258). The MCL was set to protect against infant methemoglobinemia, but other health effects, including cancer and adverse reproductive outcomes, were not considered (160). Through endogenous nitrosation, nitrate is a precursor in the formation of N-nitroso compounds (NOC); most NOC are carcinogens and teratogens. Thus, exposure to NOC formed after ingestion of nitrate from drinking water and dietary sources may result in cancer, birth defects, or other adverse health effects. Nitrate is found in many foods, with the highest levels occurring in some green leafy and root vegetables (161, 162). Average daily intakes from food are in the range of 30–130 mg/day as $NO_3$ (7–29 mg/day NO3-N) (161). Because NOC formation is inhibited by ascorbic acid, polyphenols, and other compounds present at high levels in most vegetables, dietary nitrate intake may not result in substantial endogenous NOC formation (161, 163).

In the United States nitrate levels in groundwater under agricultural land were about three times the national background level (164). In Europe, the Nitrates Directive was set in 1991 (165, 166) to reduce or prevent nitrate pollution from agriculture. Areas most affected by nitrate pollution are designated as "nitrate vulnerable zones" and are subject to mandatory Codes of Good Agricultural Practice (166). Average nitrate levels in groundwater in most other European countries have been stable at around 17.5 mg/L NO3 (4 mg/L $NO_3$-N) across Europe over a 20-year period (1992–2012), with some differences between countries both in trends and concentrations (167). Average concentrations are lowest in Finland (around 1 mg/L NO3 in 1992–2012) and highest in Malta (58.1 mg/L in 2000–2012). Average annual nitrate concentrations at river monitoring stations in Europe showed a steady decline from 2.7 NO3-N in 1992 to 2.1 mg/L in 2012, with the lowest average levels in Norway and highest in Greece (167).

Drinking-water nitrate is readily absorbed in the upper gastrointestinal tract and distributed in the human body. When it reaches the salivary glands, it is actively transported from the blood into saliva, and levels may be up to 20 times higher than in the plasma (168, 169). In the oral cavity, 6–7 percent of the total nitrate can be reduced to nitrite, predominantly by nitrate-reducing bacteria (19, 170, 171). The secreted nitrate, as well as the nitrite generated in the oral cavity, re-enter the gastrointestinal tract when swallowed. Under acidic conditions in the stomach, nitrite can be protonated to nitrous acid ($HNO_2$) and subsequently yield dinitrogen trioxide ($N_2O_3$), nitric oxide (NO), and nitrogen dioxide ($NO_2$). Since the discovery of endogenous NO formation, it has become clear that NO is involved in a wide range of NO-mediated physiological effects. These comprise the regulation of blood pressure and blood flow by mediating vasodilation (172–174), the maintenance of blood vessel tonus (175), the inhibition of platelet adhesion and aggregation (176), modulation of mitochondrial function (110), and several other processes (177–180).

On the contrary, various nitrate and nitrite-derived metabolites such as nitrous acid (HNO2) are powerful nitrosating agents and known to drive the formation of N-nitroso compounds (NOC), which are suggested to be the causal agents in many of the nitrate-associated adverse health outcomes. NOC comprises N-nitrosamines and N-nitrosamides, and may be formed when nitrosating agents encounter N-nitrosatable amino acids, which are also from dietary origins. The nitrosation process depends on the reaction mechanisms involved, on the concentration of the compounds involved, the pH of the reaction environment, and further modifying factors, including the presence of catalysts or inhibitors of N-nitrosation (180–182). Endogenous nitrosation can also be inhibited by, for instance, dietary compounds like vitamin C, which has the capacity to reduce HNO2 to NO; and alpha-tocopherol or polyphenols, which can reduce nitrite to NO (170, 183–185). Inhibitory effects on nitrosation have also been described for dietary flavonoids such as quercetin, ferulic and caffeic acid, betel nut extracts, garlic, coffee, and green tea polyphenols (186). The effect of these fruits and vegetables is unlikely to be due solely to ascorbic acid (187). The protective potential of such dietary inhibitors depends not only on the reaction rates of N-nitrosatable precursors and nitrosation inhibitors but also on their biokinetics, since an effective inhibitor needs to follow gastrointestinal circulation kinetics similar to nitrate (188).

### 17.15.1 Methemoglobinemia

Drinking-water nitrate can lead to methemoglobinemia in infants under six months of age (189, 190). Ingested nitrate is reduced to nitrite by bacteria in the mouth and in the infant's stomach, which is less acidic than those of adults. Nitrite binds to hemoglobin to form methemoglobin, which interferes with the oxygen-carrying capacity of the blood. Methemoglobinemia is a life-threatening condition that occurs when methemoglobin levels exceed about 10 percent (189, 190). Risk factors for infant methemoglobinemia include formula made with water containing high nitrate levels, foods, and medications that have high nitrate levels and enteric infections (190–192).

### 17.15.2 Adverse Pregnancy Outcomes

Drinking-water nitrate intake during pregnancy has been investigated as a risk factor for a range of pregnancy outcomes, including spontaneous abortion, fetal deaths, prematurity, intrauterine growth retardation, low birth weight, congenital malformations, and neonatal deaths (155).

### 17.15.3 Cancer

In many studies, historical nitrate levels have been evaluated in relation to several cancers. Most of these studies evaluated potential confounders and factors affecting nitrosation (155). It was reported that ovarian and bladder cancers were positively associated with the long-term average nitrate levels in public water supplies (193). There were observed inverse associations for uterine and rectal cancer, but no association with cancers of the breast, colon, rectum, pancreas, kidney, lung, melanoma, non-Hodgkin lymphoma, or leukemia. In addition, breast cancer risk was not associated with water nitrate levels (155).

### 17.15.4 Thyroid Disease

In animals, it was demonstrated that ingestion of nitrate at high doses can competitively inhibit iodine uptake and induce hypertrophy of the thyroid gland (194). The consumption of water with nitrate levels at or above the MCL was associated with an increased prevalence of thyroid hypertrophy (195). In addition, an increased prevalence of subclinical hypothyroidism was found among children in an area with high nitrate levels (51–274 mg/L NO3) in water supplies. Also, dietary nitrate was associated with an increased prevalence of hypothyroidism but not hyperthyroidism (155).

### 17.15.5 OTHER HEALTH EFFECTS

Associations between nitrate in drinking water and other non-cancer health effects, including Type 1 childhood diabetes, blood pressure, and acute respiratory tract infections in children were evaluated. It was indicated that NOC may play a role in the pathology of Type 1 diabetes mellitus through damage to pancreatic beta cells (189, 196). Experimental studies in animals and controlled feeding studies in humans have demonstrated mixed evidence of these effects and on other cardiovascular endpoints, such as vascular hypertrophy, heart failure, and myocardial infarction (197–199). Recent findings have suggested that plasma nitrate was associated with an increased overall risk of death, but no association was observed for incident cardiovascular disease (200).

On the contrary, diet nitrite has been associated with favorable health outcomes not only in the cardiovascular system (see above) but also in other parts of the human body such as increased blood flow in certain parts of the brain (201). Another potential beneficial effect of nitrate is protection against bacterial infections via reduction of nitrite by enteric bacteria. Nitrite in drinking water was associated with both preventive and therapeutic effects in colonic inflammation (202).

## 17.16 CONCLUSIONS

Although NO is one of the simplest biological molecules in nature, it is essential for the integrity of the cardiovascular system, and decreased production and/or bioavailability of NO leads to the development of many cardiovascular disorders. The two main sources of nitrate in the diet are vegetables and drinking water. The beneficial effects of dietary nitrate on cardiovascular health include blood pressure reduction, improved endothelial function, protection against ischemia–reperfusion injury, mitigation of arterial stiffness, reduction of platelet activity, and enhanced exercise performance. On the contrary, various nitrate- and nitrite-derived metabolites have been associated with adverse outcomes such as methemoglobinemia, adverse pregnancy outcomes, cancer, thyroid disease, and acute respiratory tract infections. Daily consumption of at least one portion of high-nitrate vegetables may be beneficial. However, excessive amounts of nitrate intake could be associated with an increased risk of unfavorable health outcomes.

## REFERENCES

1. Ma L, Hu L, Feng X, Wang S. Nitrate and nitrite in health and disease. *Aging Dis.* 2018: 9(5):938–45.
2. Levine AB, Punihaole D, Levine TB. Characterization of the role of nitric oxide and its clinical applications. *Cardiology.* 2012: 122(1):55–68.

3.  Lee PM, Gerriets V. *Nitrates.* StatPearls: Treasure Island, FL. 2020.

4.  Bloomer RJ, Butawan M, Pigg B, Martin KR. Acute ingestion of a novel nitrate-rich dietary supplement significantly increases plasma nitrate/nitrite in physically active men and women. *Nutrients.* 2020: 12(4).

5.  Bryan NS, Loscalz J. *Nitrite and nitrate in human health and disease.* New York: Humana Press. 2011.

6.  Forte P, Dykhuizen RS, Milne E, McKenzie A, Smith CC, Benjamin N. Nitric oxide synthesis in patients with infective gastroenteritis. *Gut.* 1999: 45(3):355–61.

7.  Huntington GB, Archibeque SL. Practical aspects of urea and ammonia metabolism in ruminants. *J Anim Sci.* 2000: 77(E-Suppl):1–11.

8.  Doel JJ, Benjamin N, Hector MP, Rogers M, Allaker RP. Evaluation of bacterial nitrate reduction in the human oral cavity. *Eur J Oral Sci.* 2005: 113(1):14–9.

9.  Duncan C, Dougall H, Johnston P, Green S, Brogan R, Leifert C, et al. Chemical generation of nitric oxide in the mouth from the enterosalivary circulation of dietary nitrate. *Nat Med.* 1995: 1(6):546–51.

10. Benjamin N, O'Driscoll F, Dougall H, Duncan C, Smith L, Golden M, et al. Stomach NO synthesis. *Nature.* 1994: 368(6471):502.

11. Wallace JL, Miller MJ. Nitric oxide in mucosal defense: A little goes a long way. *Gastroenterology.* 2000: 119(2):512–20.

12. Dykhuizen RS, Frazer R, Duncan C, Smith CC, Golden M, Benjamin N, et al. Antimicrobial effect of acidified nitrite on gut pathogens: Importance of dietary nitrate in host defense. *Antimicrob Agents Chemother.* 1996: 40(6):1422–5.

13. Weller R, Pattullo S, Smith L, Golden M, Ormerod A, Benjamin N. Nitric oxide is generated on the skin surface by reduction of sweat nitrate. *J Invest Dermatol.* 1996: 107(3):327–31.

14. Weller R, Price RJ, Ormerod AD, Benjamin N, Leifert C. Antimicrobial effect of acidified nitrite on dermatophyte fungi, Candida and bacterial skin pathogens. *J Appl Microbiol.* 2001: 90(4):648–52.

15. Benjamin N, Pattullo S, Weller R, Smith L, Ormerod A. Wound licking and nitric oxide. *Lancet.* 1997: 349(9067):1776.

16. Hord NG, Ghannam JS, Garg HK, Berens PD, Bryan NS. Nitrate and nitrite content of human, formula, bovine, and soy milks: Implications for dietary nitrite and nitrate recommendations. *Breastfeed Med.* 2011: 6(6):393–9.

17. Sato Y, Ishihara M, Fukuda K, Nakamura S, Murakami K, Fujita M, et al. Behavior of nitrate-nitrogen and nitrite-nitrogen in drinking water. *Biocontrol Sci.* 2018: 23(3):139–43.

18. Lundberg JO, Weitzberg E, Cole JA, Benjamin N. Nitrate, bacteria and human health. *Nat Rev Microbiol.* 2004: 2(7):593–602.

19. Spiegelhalder B, Eisenbrand G, Preussmann R. Influence of dietary nitrate on nitrite content of human saliva: possible relevance to in vivo formation of N-nitroso compounds. *Food Cosmet Toxicol.* 1976: 14(6):545–8.

20. Magee PN, Barnes JM. The production of malignant primary hepatic tumours in the rat by feeding dimethylnitrosamine. *Br J Cancer.* 1956: 10(1):114–22.

21. Mirvish SS. Blocking the formation of N-nitroso compounds with ascorbic acid in vitro and in vivo. *Ann N Y Acad Sci.* 1975: 258:175–80.

22. Mirvish SS. Inhibition by vitamins C and E of in vivo nitrosation and vitamin C occurrence in the stomach. *Eur J Cancer Prev.* 1996: 5 Suppl 1:131–6.

23. Garthwaite J. New insight into the functioning of nitric oxide-receptive guanylyl cyclase: Physiological and pharmacological implications. *Mol Cell Biochem.* 2010: 334(1–2):221–32.

24. Lucas KA, Pitari GM, Kazerounian S, Ruiz-Stewart I, Park J, Schulz S, et al. Guanylyl cyclases and signaling by cyclic GMP. *Pharmacol Rev.* 2000: 52(3):375–414.

25. Simonsen U. Interactions between drugs for erectile dysfunction and drugs for cardiovascular disease. *Int J Impot Res.* 2002: 14(3):178–88.

26. Bassenge E, Zanzinger J. Nitrates in different vascular beds, nitrate tolerance, and interactions with endothelial function. *Am J Cardiol.* 1992: 70(8):23B–9B.

27. Bauer JA, Fung HL. Arterial versus venous metabolism of nitroglycerin to nitric oxide: A possible explanation of organic nitrate venoselectivity. *J Cardiovasc Pharmacol.* 1996: 28(3):371–4.

28. Friederich JA, Butterworth JF. Sodium nitroprusside: Twenty years and counting. *Anesth Analg.* 1995: 81(1):152–62.

29. Winbury WM. Pharmacology of nitrates in relation to antianginal action. In: Lichtlen PR, Engel H-J, Schrey A, Swan HJC, editors. *Nitrates III: Cardiovascular Effects.* New York: Springer. 1981. p. 4–11.

30. Engel H-J, Wolf R, Pretschner P, Hundeshagen H, Lichtlen PR. Effects of nitrates on myocardial blood flow during angina: Comparison of results obtained by inert gas clearance and 201 thallium imaging. In: Lichtlen PR, Engel H-J, Schrey A, Swan HJC, editors. *Nitrates III: Cardiovascular Effects.* New York: Springer. 1981. p. 184–91.

31. Mason DT, Klein RC, Awan NA. Effects of systemic nitroglycerin on perfusion of ischemic myocardium in clinical coronary artery disease. In: Lichtlen PR, Engel H-J, Schrey A, Swan HJC, editors. *Nitrates III: Cardiovascular Effects.* New York: Springer. 1981. p. 193–201.

32. Rodgers EM, De Boeck G. Nitrite-induced reductions in heat tolerance are independent of aerobic scope in a freshwater teleost. *J Exp Biol.* 2019: 222(Pt 23).

33. Karwowska M, Kononiuk A, Wojciak KM. Impact of sodium nitrite reduction on lipid oxidation and antioxidant properties of cooked meat products. *Antioxidants* (Basel). 2019: 9(1).

34. Van Velzen AG, Sips AJ, Schothorst RC, Lambers AC, Meulenbelt J. The oral bioavailability of nitrate from nitrate-rich vegetables in humans. *Toxicol Lett.* 2008: 181(3):177–81.

35. Pennington JAT. Dietary exposure models for nitrates and nitrites. *Food Control.* 1998: 9:385–95.

36. Santamaria P. Nitrate in vegetables: Toxicity, content, intake and EC regulation. *J. Sci. Food Agr.* 2006: 86:10–7.

37. Petersen A, Stoltze S. Nitrate and nitrite in vegetables on the Danish market: Content and intake. *Food Addit Contam.* 1999: 16(7):291–9.

38. Ysart G, Miller P, Barrett G, Farrington D, Lawrance P, Harrison N. Dietary exposures to nitrate in the UK. *Food Addit Contam.* 1999: 16(12):521–32.

39. Anjana US, Iqbal M, Abrol YP. Are nitrate concentrations in leafy vegetables within safe limits? *Curr. Sci.* 2007: 92:355–60.

40. Addiscott TM. *Nitrate, Agriculture and the Environment.* CABI Publishing, Oxford. 2005.

41. Gangolli SD, van den Brandt PA, Feron VJ, Janzowsky C, Koeman JH, Speijers GJ, et al. Nitrate, nitrite and N-nitroso compounds. *Eur J Pharmacol.* 1994: 292(1):1–38.

42. Bahadoran Z, Ghasemi A, Mirmiran P, Mehrabi Y, Azizi F, Hadaegh F. Estimation and validation of dietary nitrate and nitrite intake in Iranian population. *Iran J Public Health.* 2019: 48(1):162–70.

43. Hord NG, Tang Y, Bryan NS. Food sources of nitrates and nitrites: The physiologic context for potential health benefits. *Am J Clin Nutr.* 2009: 90(1):1–10.

44. Kapil V, Webb AJ, Ahluwalia A. Inorganic nitrate and the cardiovascular system. *Heart.* 2010: 96(21):1703–9.

45. Bondonno CP, Croft KD, Hodgson JM. Dietary nitrate, nitric oxide, and cardiovascular health. *Crit Rev Food Sci Nutr.* 2016: 56(12):2036–52.

46. Bryan NS, Alexander DD, Coughlin JR, Milkowski AL, Boffetta P. Ingested nitrate and nitrite and stomach cancer risk: An updated review. *Food Chem Toxicol.* 2012: 50(10):3646–65.

47. Key TJ, Appleby PN, Spencer EA, Travis RC, Roddam AW, Allen NE. Mortality in British vegetarians: Results from the European prospective investigation into cancer and nutrition (EPIC-Oxford). *Am J Clin Nutr.* 2009: 89(5):1613S-9S.

48. Huang T, Yang B, Zheng J, Li G, Wahlqvist ML, Li D. Cardiovascular disease mortality and cancer incidence in vegetarians: A meta-analysis and systematic review. *Ann Nutr Metab.* 2012: 60(4):233–40.

49. Song P, Wu L, Guan W. Dietary nitrates, nitrites, and nitrosamines intake and the risk of gastric cancer: A meta-analysis. *Nutrients.* 2015: 7(12):9872–95.

50. Lundberg JO, Weitzberg E, Gladwin MT. The nitrate-nitrite-nitric oxide pathway in physiology and therapeutics. *Nat Rev Drug Discov.* 2008: 7(2):156–67.

51. Vitturi DA, Teng X, Toledo JC, Matalon S, Lancaster JR, Jr., Patel RP. Regulation of nitrite transport in red blood cells by hemoglobin oxygen fractional saturation. *Am J Physiol Heart Circ Physiol.* 2009: 296(5):H1398–407.

52. Lundberg JO, Govoni M. Inorganic nitrate is a possible source for systemic generation of nitric oxide. *Free Radic Biol Med.* 2004: 37(3):395–400.

53. Lidder S, Webb AJ. Vascular effects of dietary nitrate (as found in green leafy vegetables and beetroot) via the nitrate-nitrite-nitric oxide pathway. *Br J Clin Pharmacol.* 2013: 75(3):677–96.

54. Kapil V, Haydar SM, Pearl V, Lundberg JO, Weitzberg E, Ahluwalia A. Physiological role for nitrate-reducing oral bacteria in blood pressure control. *Free Radic Biol Med.* 2013: 55:93–100.

55. Hermann M, Flammer A, Luscher TF. Nitric oxide in hypertension. *J Clin Hypertens* (Greenwich). 2006: 8(12 Suppl 4):17–29.

56. Dong JY, Qin LQ, Zhang Z, Zhao Y, Wang J, Arigoni F, et al. Effect of oral L–arginine supplementation on blood pressure: a meta-analysis of randomized, double-blind, placebo-controlled trials. *Am Heart J.* 2011: 162(6):959–65.

57. Venho B, Voutilainen S, Valkonen VP, Virtanen J, Lakka TA, Rissanen TH, et al. Arginine intake, blood pressure, and the incidence of acute coronary events in men: The Kuopio Ischaemic heart disease risk factor study. *Am J Clin Nutr.* 2002: 76(2):359–64.

58. Keys A. Mediterranean diet and public health: personal reflections. *Am J Clin Nutr.* 1995: 61(6 Suppl):1321S-3S.

59. Lundberg JO, Feelisch M, Bjorne H, Jansson EA, Weitzberg E. Cardioprotective effects of vegetables: is nitrate the answer? *Nitric Oxide.* 2006: 15(4):359–62.

60. Bryan NS, Hord NG. Dietary nitrates and nitrites: The physiological context for potential health benefits. *Food, Nutrition and the Nitric Oxide Pathway: Biochemistry and Bioactivity.* Lancaster, PA: DESTech Publications. 2010.

61. Rouse IL, Beilin LJ, Armstrong BK, Vandongen R. Blood-pressure-lowering effect of a vegetarian diet: Controlled trial in normotensive subjects. *Lancet.* 1983: 1(8314–5):5–10.

62. Joshipura KJ, Ascherio A, Manson JE, Stampfer MJ, Rimm EB, Speizer FE, et al. Fruit and vegetable intake in relation to risk of ischemic stroke. *JAMA.* 1999: 282(13):1233–9.

63. Joshipura KJ, Hu FB, Manson JE, Stampfer MJ, Rimm EB, Speizer FE, et al. The effect of fruit and vegetable intake on risk for coronary heart disease. *Ann Intern Med.* 2001: 134(12):1106–14.

64. Lundberg JO, Weitzberg E, Lundberg JM, Alving K. Intragastric nitric oxide production in humans: measurements in expelled air. *Gut.* 1994: 35(11):1543–6.

65. Zweier JL, Wang P, Samouilov A, Kuppusamy P. Enzyme-independent formation of nitric oxide in biological tissues. *Nat Med.* 1995: 1(8):804–9.

66. Giansante C, Fiotti N. Insights into human hypertension: the role of endothelial dysfunction. *J Hum Hypertens.* 2006: 20(10):725–6.

67. Kapil V, Milsom AB, Okorie M, Maleki-Toyserkani S, Akram F, Rehman F, et al. Inorganic nitrate supplementation lowers blood pressure in humans: role for nitrite-derived NO. *Hypertension.* 2010: 56(2):274–81.

68. Larsen FJ, Ekblom B, Sahlin K, Lundberg JO, Weitzberg E. Effects of dietary nitrate on blood pressure in healthy volunteers. *N Engl J Med.* 2006: 355(26):2792–3.

69. Larsen FJ, Weitzberg E, Lundberg JO, Ekblom B. Effects of dietary nitrate on oxygen cost during exercise. *Acta Physiol (Oxf).* 2007: 191(1):59–66.

70. Gilchrist M, Shore AC, Benjamin N. Inorganic nitrate and nitrite and control of blood pressure. *Cardiovasc Res.* 2011: 89(3):492–8.

71. Da Silva RF, Trape AA, Reia TA, Lacchini R, Oliveira–Paula GH, Pinheiro LC, et al. Association of endothelial nitric oxide synthase (eNOS) gene polymorphisms and physical fitness levels with plasma nitrite concentrations and arterial blood pressure values in older adults. *PLoS One.* 2018: 13(10):e0206254.

72. Celermajer DS, Sorensen KE, Gooch VM, Spiegelhalter DJ, Miller OI, Sullivan ID, et al. Non–invasive detection of endothelial dysfunction in children and adults at risk of atherosclerosis. *Lancet.* 1992: 340(8828):1111–5.

73. Modena MG, Bonetti L, Coppi F, Bursi F, Rossi R. Prognostic role of reversible endothelial dysfunction in hypertensive postmenopausal women. *J Am Coll Cardiol.* 2002: 40(3):505–10.

74. Witte DR, Westerink J, de Koning EJ, van der Graaf Y, Grobbee DE, Bots ML. Is the association between flow-mediated dilation and cardiovascular risk limited to low-risk populations? *J Am Coll Cardiol.* 2005: 45(12):1987–93.

75. Neunteufl T, Katzenschlager R, Hassan A, Klaar U, Schwarzacher S, Glogar D, et al. Systemic endothelial dysfunction is related to the extent and severity of coronary artery disease. *Atherosclerosis.* 1997: 129(1):111–8.

76. Green D. Point: Flow-mediated dilation does reflect nitric oxide-mediated endothelial function. *J Appl Physiol (1985).* 2005: 99(3):1233–4: discussion 7–8.

77. Bondonno CP, Yang X, Croft KD, Considine MJ, Ward NC, Rich L, et al. Flavonoid-rich apples and nitrate-rich spinach augment nitric oxide status and improve endothelial function in healthy men and women: A randomized controlled trial. Free *Radic Biol Med.* 2012: 52(1):95–102.

78. Murray CJ, Lopez AD. Alternative projections of mortality and disability by cause 1990–2020: Global Burden of Disease Study. *Lancet.* 1997: 349(9064):1498–504.

79. Munzel T, Daiber A. Inorganic nitrite and nitrate in cardiovascular therapy: A better alternative to organic nitrates as nitric oxide donors? *Vascul Pharmacol.* 2018: 102:1–10.

80. Duranski MR, Greer JJ, Dejam A, Jaganmohan S, Hogg N, Langston W, et al. Cytoprotective effects of nitrite during in vivo ischemia-reperfusion of the heart and liver. *J Clin Invest.* 2005: 115(5):1232–40.

81. Webb A, Bond R, McLean P, Uppal R, Benjamin N, Ahluwalia A. Reduction of nitrite to nitric oxide during ischemia protects against myocardial ischemia-reperfusion damage. *Proc Natl Acad Sci U S A.* 2004: 101(37):13683–8.

82. Baker JE, Su J, Fu X, Hsu A, Gross GJ, Tweddell JS, et al. Nitrite confers protection against myocardial infarction: role of xanthine oxidoreductase, NADPH oxidase and K(ATP) channels. *J Mol Cell Cardiol.* 2007: 43(4):437–44.

83. Jung KH, Chu K, Ko SY, Lee ST, Sinn DI, Park DK, et al. Early intravenous infusion of sodium nitrite protects brain against in vivo ischemia-reperfusion injury. *Stroke*. 2006: 37(11):2744–50.

84. Tripatara P, Patel NS, Webb A, Rathod K, Lecomte FM, Mazzon E, et al. Nitrite-derived nitric oxide protects the rat kidney against ischemia/reperfusion injury in vivo: Role for xanthine oxidoreductase. *J Am Soc Nephrol*. 2007: 18(2):570–80.

85. Kumar D, Branch BG, Pattillo CB, Hood J, Thoma S, Simpson S, et al. Chronic sodium nitrite therapy augments ischemia-induced angiogenesis and arteriogenesis. *Proc Natl Acad Sci U S A*. 2008: 105(21):7540–5.

86. Basireddy M, Isbell TS, Teng X, Patel RP, Agarwal A. Effects of sodium nitrite on ischemia-reperfusion injury in the rat kidney. *Am J Physiol Renal Physiol*. 2006: 290(4):F779–86.

87. Avolio A. Arterial stiffness. *Pulse* (Basel). 2013: 1(1):14–28.

88. Boudoulas KD, Vlachopoulos C, Raman SV, Sparks EA, Triposciadis F, Stefanadis C, et al. Aortic function: From the research laboratory to the clinic. *Cardiology*. 2012: 121(1):31–42.

89. Willum-Hansen T, Staessen JA, Torp-Pedersen C, Rasmussen S, Thijs L, Ibsen H, et al. Prognostic value of aortic pulse wave velocity as index of arterial stiffness in the general population. *Circulation*. 2006: 113(5):664–70.

90. Stewart AD, Millasseau SC, Kearney MT, Ritter JM, Chowienczyk PJ. Effects of inhibition of basal nitric oxide synthesis on carotid-femoral pulse wave velocity and augmentation index in humans. *Hypertension*. 2003: 42(5):915–8.

91. Bahra M, Kapil V, Pearl V, Ghosh S, Ahluwalia A. Inorganic nitrate ingestion improves vascular compliance but does not alter flow-mediated dilatation in healthy volunteers. *Nitric Oxide*. 2012: 26(4):197–202.

92. Davi G, Patrono C. Platelet activation and atherothrombosis. *N Engl J Med*. 2007: 357(24):2482–94.

93. Awtry EH, Loscalzo J. Aspirin. *Circulation*. 2000: 101(10):1206–18.

94. Alheid U, Frolich JC, Forstermann U. Endothelium-derived relaxing factor from cultured human endothelial cells inhibits aggregation of human platelets. *Thromb Res*. 1987: 47(5):561–71.

95. Radomski MW, Palmer RM, Moncada S. Endogenous nitric oxide inhibits human platelet adhesion to vascular endothelium. *Lancet*. 1987: 2(8567):1057–8.

96. Radomski MW, Palmer RM, Moncada S. An L-arginine/nitric oxide pathway present in human platelets regulates aggregation. *Proc Natl Acad Sci U S A*. 1990: 87(13): 5193–7.

97. Webb AJ, Patel N, Loukogeorgakis S, Okorie M, Aboud Z, Misra S, et al. Acute blood pressure lowering, vasoprotective, and antiplatelet properties of dietary nitrate via bioconversion to nitrite. *Hypertension*. 2008: 51(3):784–90.

98. Cornelissen VA, Fagard RH. Effects of endurance training on blood pressure, blood pressure-regulating mechanisms, and cardiovascular risk factors. *Hypertension*. 2005: 46(4):667–75.

99. Hull SS, Jr., Vanoli E, Adamson PB, Verrier RL, Foreman RD, Schwartz PJ. Exercise training confers anticipatory protection from sudden death during acute myocardial ischemia. *Circulation*. 1994: 89(2):548–52.

100. Sessa WC, Pritchard K, Seyedi N, Wang J, Hintze TH. Chronic exercise in dogs increases coronary vascular nitric oxide production and endothelial cell nitric oxide synthase gene expression. *Circ Res*. 1994: 74(2):349–53.

101. Hambrecht R, Adams V, Erbs S, Linke A, Krankel N, Shu Y, et al. Regular physical activity improves endothelial function in patients with coronary artery

disease by increasing phosphorylation of endothelial nitric oxide synthase. *Circulation.* 2003: 107(25): 3152–8.

102. Stamler JS, Meissner G. Physiology of nitric oxide in skeletal muscle. *Physiol Rev.* 2001: 81(1):209–37.

103. Axton ER, Beaver LM, St Mary L, Truong L, Logan CR, Spagnoli S, et al. Treatment with nitrate, but not nitrite, lowers the oxygen cost of exercise and decreases glycolytic intermediates while increasing fatty acid metabolites in exercised zebrafish. *J Nutr.* 2019: 149(12):2120–32.

104. Moncada S, Erusalimsky JD. Does nitric oxide modulate mitochondrial energy generation and apoptosis? *Nat Rev Mol Cell Biol.* 2002: 3(3):214–20.

105. Rassaf T, Lauer T, Heiss C, Balzer J, Mangold S, Leyendecker T, et al. Nitric oxide synthase-derived plasma nitrite predicts exercise capacity. *Br J Sports Med.* 2007: 41(10): 669–73; discussion 73.

106. Bailey SJ, Fulford J, Vanhatalo A, Winyard PG, Blackwell JR, DiMenna FJ, et al. Dietary nitrate supplementation enhances muscle contractile efficiency during knee-extensor exercise in humans. *J Appl Physiol* (1985). 2010: 109(1):135–48.

107. Bailey SJ, Winyard P, Vanhatalo A, Blackwell JR, Dimenna FJ, Wilkerson DP, et al. Dietary nitrate supplementation reduces the O2 cost of low-intensity exercise and enhances tolerance to high-intensity exercise in humans. *J Appl Physiol* (1985). 2009: 107(4):1144–55.

108. Larsen FJ, Weitzberg E, Lundberg JO, Ekblom B. Dietary nitrate reduces maximal oxygen consumption while maintaining work performance in maximal exercise. *Free Radic Biol Med.* 2010: 48(2):342–7.

109. Clerc P, Rigoulet M, Leverve X, Fontaine E. Nitric oxide increases oxidative phosphorylation efficiency. *J Bioenerg Biomembr.* 2007: 39(2):158–66.

110. Larsen FJ, Schiffer TA, Borniquel S, Sahlin K, Ekblom B, Lundberg JO, et al. Dietary inorganic nitrate improves mitochondrial efficiency in humans. *Cell Metab.* 2011: 13(2):149–59.

111. Lundberg JO, Carlstrom M, Larsen FJ, Weitzberg E. Roles of dietary inorganic nitrate in cardiovascular health and disease. *Cardiovasc Res.* 2011: 89(3):525–32.

112. Siervo M, Lara J, Ogbonmwan I, Mathers JC. Inorganic nitrate and beetroot juice supplementation reduces blood pressure in adults: A systematic review and meta-analysis. *J Nutr.* 2013: 143(6):818–26.

113. Ashworth A, Bescos R. Dietary nitrate and blood pressure: Evolution of a new nutrient? *Nutr Res Rev.* 2017: 30(2):208–19.

114. Mitek M, Anyzewska A, Wawrzyniak A. Estimated dietary intakes of nitrates in vegetarians compared to a traditional diet in Poland and acceptable daily intakes: Is there a risk? *Rocz Panstw Zakl Hig.* 2013: 64(2):105–9.

115. Moncada S, Higgs A. The L-arginine-nitric oxide pathway. *N Engl J Med.* 1993: 329(27):2002–12.

116. Li H, Wallerath T, Munzel T, Forstermann U. Regulation of endothelial-type NO synthase expression in pathophysiology and in response to drugs. *Nitric Oxide.* 2002: 7(3):149–64.

117. Thum T, Fraccarollo D, Schultheiss M, Froese S, Galuppo P, Widder JD, et al. Endothelial nitric oxide synthase uncoupling impairs endothelial progenitor cell mobilization and function in diabetes. *Diabetes.* 2007: 56(3):666–74.

118. Dimmeler S, Zeiher AM. Nitric oxide and apoptosis: Another paradigm for the double-edged role of nitric oxide. *Nitric Oxide.* 1997: 1(4):275–81.

119. Werner N, Kosiol S, Schiegl T, Ahlers P, Walenta K, Link A, et al. Circulating endothelial progenitor cells and cardiovascular outcomes. *N Engl J Med.* 2005: 353(10):999–1007.

120. Shantsila E, Watson T, Lip GY. Endothelial progenitor cells in cardiovascular disorders. *J Am Coll Cardiol.* 2007: 49(7):741–52.

121. Heiss C, Jahn S, Taylor M, Real WM, Angeli FS, Wong ML, et al. Improvement of endothelial function with dietary flavanols is associated with mobilization of circulating angiogenic cells in patients with coronary artery disease. *J Am Coll Cardiol.* 2010: 56(3):218–24.

122. Dimmeler S. Regulation of bone marrow-derived vascular progenitor cell mobilization and maintenance. *Arterioscler Thromb Vasc Biol.* 2010: 30(6):1088–93.

123. Meyer C, Heiss C, Drexhage C, Kehmeier ES, Balzer J, Muhlfeld A, et al. Hemodialysis-induced release of hemoglobin limits nitric oxide bioavailability and impairs vascular function. *J Am Coll Cardiol.* 2010: 55(5):454–9.

124. Rassaf T, Bryan NS, Kelm M, Feelisch M. Concomitant presence of N-nitroso and S-nitroso proteins in human plasma. *Free Radic Biol Med.* 2002: 33(11):1590–6.

125. Rassaf T, Bryan NS, Maloney RE, Specian V, Kelm M, Kalyanaraman B, et al. NO adducts in mammalian red blood cells: Too much or too little? *Nat Med.* 2003: 9(5):481–2; author reply 2–3.

126. Rassaf T, Kleinbongard P, Preik M, Dejam A, Gharini P, Lauer T, et al. Plasma nitrosothiols contribute to the systemic vasodilator effects of intravenously applied NO: experimental and clinical study on the fate of NO in human blood. *Circ Res.* 2002: 91(6):470–7.

127. Elrod JW, Calvert JW, Gundewar S, Bryan NS, Lefer DJ. Nitric oxide promotes distant organ protection: evidence for an endocrine role of nitric oxide. *Proc Natl Acad Sci U S A.* 2008: 105(32):11430–5.

128. Tsujimoto T, Kajio H. Use of nitrates and risk of cardiovascular events in patients with heart failure with preserved ejection fraction. *Mayo Clin Proc.* 2019: 94(7):1210–20.

129. Piper HM, Garcia-Dorado D, Ovize M. A fresh look at reperfusion injury. *Cardiovasc Res.* 1998: 38(2):291–300.

130. Yellon DM, Hausenloy DJ. Myocardial reperfusion injury. *N Engl J Med.* 2007: 357(11):1121–35.

131. Bolli R, Marban E. Molecular and cellular mechanisms of myocardial stunning. *Physiol Rev.* 1999: 79(2):609–34.

132. Ito H. No-reflow phenomenon and prognosis in patients with acute myocardial infarction. *Nat Clin Pract Cardiovasc Med.* 2006: 3(9):499–506.

133. Heusch G. Stunning–great paradigmatic, but little clinical importance. *Basic Res Cardiol.* 1998: 93(3):164–6.

134. Murphy E, Steenbergen C. Mechanisms underlying acute protection from cardiac ischemia-reperfusion injury. *Physiol Rev.* 2008: 88(2):581–609.

135. Skyschally A, Schulz R, Heusch G. Pathophysiology of myocardial infarction: protection by ischemic pre- and postconditioning. *Herz.* 2008: 33(2):88–100.

136. Zhao ZQ, Corvera JS, Halkos ME, Kerendi F, Wang NP, Guyton RA, et al. Inhibition of myocardial injury by ischemic postconditioning during reperfusion: comparison with ischemic preconditioning. *Am J Physiol Heart Circ Physiol.* 2003: 285(2):H579–88.

137. Hamid SA, Totzeck M, Drexhage C, Thompson I, Fowkes RC, Rassaf T, et al. Nitric oxide/cGMP signalling mediates the cardioprotective action of adrenomedullin in reperfused myocardium. *Basic Res Cardiol.* 2010: 105(2):257–66.

138. Hendgen-Cotta UB, Merx MW, Shiva S, Schmitz J, Becher S, Klare JP, et al. Nitrite reductase activity of myoglobin regulates respiration and cellular viability in myocardial ischemia-reperfusion injury. *Proc Natl Acad Sci U S A.* 2008: 105(29):10256–61.

139. Rassaf T, Flogel U, Drexhage C, Hendgen-Cotta U, Kelm M, Schrader J. Nitrite reductase function of deoxymyoglobin: Oxygen sensor and regulator of cardiac energetics and function. *Circ Res.* 2007: 100(12):1749–54.

140. Bryan NS, Calvert JW, Elrod JW, Gundewar S, Ji SY, Lefer DJ. Dietary nitrite supplementation protects against myocardial ischemia-reperfusion injury. *Proc Natl Acad Sci U S A.* 2007: 104(48):19144–9.

141. Kirsh AJ, Juracic ES, Petrick HL, Monaco CMF, Barbeau PA, Tupling AR, et al. Dietary nitrate does not alter cardiac function, calcium handling proteins, or SERCA activity in the left ventricle of healthy rats. *Appl Physiol Nutr Metab.* 2020.

142. Eckel RH, Grundy SM, Zimmet PZ. The metabolic syndrome. *Lancet.* 2005: 365(9468):1415–28.

143. Despres JP, Lemieux I. Abdominal obesity and metabolic syndrome. *Nature.* 2006: 444(7121):881–7.

144. Huang PL. eNOS, metabolic syndrome and cardiovascular disease. *Trends Endocrinol Metab.* 2009: 20(6):295–302.

145. Cook S, Hugli O, Egli M, Vollenweider P, Burcelin R, Nicod P, et al. Clustering of cardiovascular risk factors mimicking the human metabolic syndrome X in eNOS null mice. *Swiss Med Wkly.* 2003: 133(25–26):360–3.

146. Nisoli E, Clementi E, Paolucci C, Cozzi V, Tonello C, Sciorati C, et al. Mitochondrial biogenesis in mammals: The role of endogenous nitric oxide. *Science.* 2003: 299(5608):896–9.

147. Monti LD, Barlassina C, Citterio L, Galluccio E, Berzuini C, Setola E, et al. Endothelial nitric oxide synthase polymorphisms are associated with type 2 diabetes and the insulin resistance syndrome. *Diabetes.* 2003: 52(5):1270–5.

148. Needleman P, Lang S, Johnson EM, Jr. Organic nitrates: Relationship between biotransformation and rational angina pectoris therapy. *J Pharmacol Exp Ther.* 1972: 181(3):489–97.

149. Lee G, Mason DT, De Maria AN. Effects of long-term oral administration of isosorbide dinitrate on the antianginal response to nitroglycerin. Absence of nitrate cross-tolerance and self-tolerance shown by exercise testing. *Am J Cardiol.* 1978: 41(1):82–7.

150. Needleman P. Tolerance to the vascular effects of glyceryl trinitrate. *J Pharmacol Exp Ther.* 1970: 171(1):98–102.

151. Smolen V.F. (1981) Pharmacokinetics – Pharmacodynamics and the bioavailability of organic nitrate drug products. In: Lichtlen PR, Engel HJ, Schrey A, Swan HJC, editors. *Nitrates III.* Berlin, Heidelberg: Springer.

152. Needleman P, Johnson EM, Jr. Mechanism of tolerance development to organic nitrates. *J Pharmacol Exp Ther.* 1973: 184(3):709–15.

153. Schelling JL, Lasagna L. A study of cross-tolerance to circulatory effects of organic nitrates. *Clin Pharmacol Ther.* 1967: 8(2):256–60.

154. Donoso MR, Cortes AS. Exposure to nitrates in drinking water and its association with thyroid gland dysfunction. *Rev Med Chil.* 2018: 146(2):223–31.

155. Ward MH, Jones RR, Brender JD, de Kok TM, Weyer PJ, Nolan BT, et al. Drinking water nitrate and human health: An updated review. *Int J Environ Res Public Health.* 2018: 15(7).

156. Valiyeva GG, Bavasso I, Di Palma L, Hajiyeva SR, Ramazanov MA, Hajiyeva FV. Synthesis of Fe/Ni Bimetallic nanoparticles and application to the catalytic removal of nitrates from water. *Nanomaterials* (Basel). 2019: 9(8).

157. Davidson EA, David MB, Galloway JN, Goodale CL, Haeuber R, Harrison JA, Howarth RW, Jaynes DB, Lowrance RR, Nolan BT. Excess nitrogen in the

U.S. environment: Trends, risks, and solutions. In *Issues in Ecology*. Washington, DC: Ecological Society of America. 2012.

158. Vitousek PM, Aber JD, Howarth RW, Likens GE, Matson PA, Schindler DW. Human alteration of the global nitrogen cycle: Sources and consequences. *Ecological Applications*. 1997: 7(3):737–50.

159. Howarth RW. Coastal nitrogen pollution: A review of sources and trends globally and regionally. *Harmful Algae*. 2008: 8(1):14–20.

160. USEPA. Regulated Drinking Water Contaminants: Inorganic Chemicals. Available online: www. epa.gov/ground-water-and-drinking-water/table-regulated-drinking- water-contaminants (accessed on 23 September 2017).

161. International Agency for Research on Cancer (IARC). *IARC Monographs on the Evaluation of Carcionogenic Risks to Humans: Ingested Nitrate and Nitrite and Cyanobacterial Peptide Toxins*. IARC: Lyon, France. 2010.

162. National Research Council (NRC). *The Health Effects of Nitrate, Nitrite, and N-Nitroso Compounds*. Washington, DC: NRC. 1981.

163. Mirvish SS. Role of N-nitroso compounds (NOC) and N-nitrosation in etiology of gastric, esophageal, nasopharyngeal and bladder cancer and contribution to cancer of known exposures to NOC. *Cancer Lett*. 1995: 93(1):17–48.

164. Dubrovsky NM, Burow KR, Clark GM, Gronberg JM, Hamilton PA, Hitt KJ, Mueller DK, Mun, MD, Nolan BT, Puckett LJ. *The Quality of Our Nation's Waters – Nutrients in the Nation's Streams and Groundwater*. U.S. Geological Survey: Reston, VA. 2010.

165. European Commission. The Nitrates Directive. Available online: http://ec.europa.eu/ environment/water/ water-nitrates/index_en.html (accessed on 10 May 2018).

166. European Union (EU). Council Directive 91/676/EEC of 12 December 1991 Concerning the Protection of Waters against Pollution Caused by Nitrates from Agricultural Sources. European Union (EU): Brussels, 1991.

167. European Environment Agency (EEA). *Ground water Nitrate*.

168. Lv J, Neal B, Ehteshami P, Ninomiya T, Woodward M, Rodgers A, et al. Effects of intensive blood pressure lowering on cardiovascular and renal outcomes: A systematic review and meta-analysis. *PLoS Med*. 2012: 9(8):e1001293.

169. Tricker AR, Kalble T, Preussmann R. Increased urinary nitrosamine excretion in patients with urinary diversions. *Carcinogenesis*. 1989: 10(12):2379–82.

170. Eisenbrand G, Spiegelhalder B, Preussmann R. Nitrate and nitrite in saliva. *Oncology*. 1980: 37(4):227–31.

171. Eisenbrand G. The significance of N-Nitrosation of drugs. Nicolai HV, Eisenbrand G, Bozler G, editors. Gustav Fischer Verlag, Stuttgart and New York: 1990: 47–69.

172. Ceccatelli S, Lundberg JM, Fahrenkrug J, Bredt DS, Snyder SH, Hokfelt T. Evidence for involvement of nitric oxide in the regulation of hypothalamic portal blood flow. *Neuroscience*. 1992: 51(4):769–72.

173. Moncada S, Palmer RM, Higgs EA. Nitric oxide: Physiology, pathophysiology, and pharmacology. *Pharmacol Rev*. 1991: 43(2):109–42.

174. Rees DD, Palmer RM, Moncada S. Role of endothelium-derived nitric oxide in the regulation of blood pressure. *Proc Natl Acad Sci U S A*. 1989: 86(9):3375–8.

175. Palmer RM, Ferrige AG, Moncada S. Nitric oxide release accounts for the biological activity of endothelium-derived relaxing factor. *Nature*. 1987: 327(6122):524–6.

176. Radomski MW, Palmer RM, Moncada S. The anti-aggregating properties of vascular endothelium: Interactions between prostacyclin and nitric oxide. *Br J Pharmacol*. 1987: 92(3):639–46.

177. Ceccatelli S, Hulting AL, Zhang X, Gustafsson L, Villar M, Hokfelt T. Nitric oxide synthase in the rat anterior pituitary gland and the role of nitric oxide in regulation of luteinizing hormone secretion. *Proc Natl Acad Sci U S A*. 1993: 90(23):11292–6.

178. Green SJ, Scheller LF, Marletta MA, Seguin MC, Klotz FW, Slayter M, et al. Nitric oxide: Cytokine-regulation of nitric oxide in host resistance to intracellular pathogens. *Immunol Lett.* 1994: 43(1–2):87–94.

179. Langrehr JM, Hoffman RA, Lancaster JR, Jr., Simmons RL. Nitric oxide–a new endogenous immunomodulator. *Transplantation.* 1993: 55(6):1205–12.

180. Wei XQ, Charles IG, Smith A, Ure J, Feng GJ, Huang FP, et al. Altered immune responses in mice lacking inducible nitric oxide synthase. *Nature.* 1995: 375(6530):408–11.

181. d'Ischia M, Napolitano A, Manini P, Panzella L. Secondary targets of nitrite-derived reactive nitrogen species: Nitrosation/nitration pathways, antioxidant defense mechanisms and toxicological implications. *Chem Res Toxicol.* 2011: 24(12):2071–92.

182. Ridd, JH. Nitrosation, diazotisation, and deamination. *Q. Rev.* 1961. 15:418–41.

183. Akuta T, Zaki MH, Yoshitake J, Okamoto T, Akaike T. Nitrative stress through formation of 8-nitroguanosine: insights into microbial pathogenesis. *Nitric Oxide.* 2006: 14(2):101–8.

184. Loeppky RN, Bao YT, Bae JY, Yu L, Shevlin G. Blocking nitrosamine formation: Understanding the chemistry of rapid nitrosation. In Loeppky RN, Michejda CJ, editors. *Nitrosamines and Related N- Nitroso Compounds: Chemistry and Biochemistry.* Washington, DC: American Chemical Society. 1994.

185. Qin L, Liu X, Sun Q, Fan Z, Xia D, Ding G, et al. Sialin (SLC17A5) functions as a nitrate transporter in the plasma membrane. *Proc Natl Acad Sci U S A.* 2012: 109(33):13434–9.

186. Vermeer IT, Moonen EJ, Dallinga JW, Kleinjans JC, van Maanen JM. Effect of ascorbic acid and green tea on endogenous formation of N-nitrosodimethylamine and N-nitrosopiperidine in humans. *Mutat Res.* 1999: 428(1–2):353–61.

187. Helser MA, Hotchkiss JH, Roe DA. Influence of fruit and vegetable juices on the endogenous formation of N-nitrosoproline and N-nitrosothiazolidine-4-carboxylic acid in humans on controlled diets. *Carcinogenesis.* 1992: 13(12):2277–80.

188. Zeilmaker MJ, Bakker MI, Schothorst R, Slob W. Risk assessment of N–nitrosodimethylamine formed endogenously after fish-with-vegetable meals. *Toxicol Sci.* 2010: 116(1):323–35.

189. Ward MH, deKok TM, Levallois P, Brender, J., Gulis G, Nolan BT, VanDerslice J, *Workgroup report: Drinking-water nitrate and health – recent findings and research needs.* International Society for Environmental Epidemiology (2005). *Environ Health Perspect.* 2005: 113(11):1607–14.

190. Greer FR, Shannon M. Infant methemoglobinemia: The role of dietary nitrate in food and water. *Pediatrics.* American Academy of Pediatrics Committee on Environmental Health. 2005: 116(3):784–6.

191. Sanchez-Echaniz J, Benito-Fernandez J, Mintegui-Raso S. Methemoglobinemia and consumption of vegetables in infants. *Pediatrics.* 2001: 107(5):1024–8.

192. Charmandari E, Meadows N, Patel M, Johnston A, Benjamin N. Plasma nitrate Concentrations in children with infectious and noninfectious diarrhea. *J Pediatr Gastroenterol Nutr.* 2001: 32(4):423–7.

193. Weyer PJ, Cerhan JR, Kross BC, Hallberg GR, Kantamneni J, Breuer G, et al. Municipal drinking water nitrate level and cancer risk in older women: The Iowa Women's Health Study. *Epidemiology.* 2001: 12(3):327–38.

194. De Groef B, Decallonne BR, Van der Geyten S, Darras VM, Bouillon R. Perchlorate versus other environmental sodium/iodide symporter inhibitors: Potential thyroid-related health effects. *Eur J Endocrinol.* 2006: 155(1):17–25.

195. van Maanen JM, Welle IJ, Hageman G, Dallinga JW, Mertens PL, Kleinjans JC. Nitrate contamination of drinking water: Relationship with HPRT variant frequency in lymphocyte DNA and urinary excretion of N-nitrosamines. *Environ Health Perspect.* 1996: 104(5):522–8.

196. Longnecker MP, Daniels JL. Environmental contaminants as etiologic factors for diabetes. *Environ Health Perspect.* 2001: 109 Suppl 6: 871–6.

197. Ahluwalia A, Gladwin M, Coleman GD, Hord N, Howard G, Kim-Shapiro DB, et al. *Dietary nitrate and the epidemiology of cardiovascular disease.* Report from a national heart, lung, and blood institute workshop. *J Am Heart Assoc.* 2016: 5(7).

198. Kapil V, Khambata RS, Robertson A, Caulfield MJ, Ahluwalia A. Dietary nitrate provides sustained blood pressure lowering in hypertensive patients: A randomized, phase 2, double-blind, placebo-controlled study. *Hypertension.* 2015: 65(2):320–7.

199. Omar SA, Webb AJ, Lundberg JO, Weitzberg E. Therapeutic effects of inorganic nitrate and nitrite in cardiovascular and metabolic diseases. *J Intern Med.* 2016: 279(4):315–36.

200. Maas R, Schwedhelm E, Kahl L, Li H, Benndorf R, Luneburg N, et al. Simultaneous assessment of endothelial function, nitric oxide synthase activity, nitric oxide-mediated signaling, and oxidative stress in individuals with and without hypercholesterolemia. *Clin Chem.* 2008: 54(2):292–300.

201. Presley TD, Morgan AR, Bechtold E, Clodfelter W, Dove RW, Jennings JM, et al. Acute effect of a high nitrate diet on brain perfusion in older adults. *Nitric Oxide.* 2011: 24(1): 34–42.

202. Jadert C, Phillipson M, Holm L, Lundberg JO, Borniquel S. Preventive and therapeutic effects of nitrite supplementation in experimental inflammatory bowel disease. *Redox Biol.* 2014: 2:73–81.

# Part 6

## Regulations

# 18 Regulations on Nitrate Use and Management

*Fayçal Bouraoui, Panos Panagos, Anna Malagó,*
*Alberto Pistocchi, and Christophe Didion*

## CONTENTS

18.1    Introduction ................................................................................................405
18.2    The EU Nitrate Pollution Prevention Regulations and Policies ................406
     18.2.1    The Water Framework Directive .................................................406
     18.2.2    Nitrates Directive .......................................................................407
     18.2.3    Urban Wastewater Treatment Directive .....................................409
     18.2.4    The Marine Strategy Framework Directive .................................415
     18.2.5    The CAP ......................................................................................417
     18.2.6    The Soil Thematic Strategy ........................................................418
     18.2.7    Green Deal/Farm 2 Fork/Zero Pollution ....................................420
References ..........................................................................................................421

## 18.1  INTRODUCTION

Nitrogen, and in particular reactive nitrogen, is an essential element for life. Reactive nitrogen was originally produced by a few natural sources and its availability was a limiting factor to crop production (ENA 2011). The development of synthetic nitrogen fertilizer through the Haber–Bosch process (Smil 2001) led to the development of industrial agriculture. The agricultural intensification and associated excessive use of nitrogen fertilizer led to a large nitrogen surplus, with the amount of applied fertilizers, both synthetic and manure forms, largely exceeding crop requirements, and often resulting in a large diffuse release of nitrogen in the environment. In times of excess, dissolved nutrients lead to severe water-quality degradation, causing eutrophication and putting at risk human health and ecosystems. Diffuse pollution, mostly from agriculture, is the second-largest pressure on European surface waters (EEA 2018a), and nitrate is the most common pollutant causing groundwater bodies to have poor chemical status (EEA 2018a). While only 1.8 percent of agricultural surface water exceeds 50 mg $NO_3$ L-1, about 19 percent of agricultural surface water is classified as eutrophic or hypertrophic (EC 2018). Eutrophication also affects coastal and marine waters. Soil pollution due to excessive nitrogen fertilization has been shown to severely damage soil functioning, reducing productivity.

DOI: 10.1201/9780429326806-24

Nitrate being found in different media, affecting crop production and human and environmental health, is controlled in Europe by several pieces of legislation. Originally the European legislation to control nitrate was very sector-specific. Indeed the Nitrates Directive (EC 1991b) controls nitrate from agricultural sources, while emission of nitrate and nitrogen in general from point sources was controlled by the Urban Waste Water Treatment Directive (EC 1991a). Limits of nitrate concentration are also given in both the Groundwater, a daughter of the Water Framework Directive, and the Drinking Water Directive. The Common Agriculture Policy has shifted to promote a more friendly, agriculture-promoting agri-environment and climate measures and investments. Marine waters are protected through the Marine Strategy Framework Directive and by international conventions by controlling the amount of land-based nitrate and other nitrogen compounds discharged into European Seas. The European Commission lately published the European Green Deal (EC 2019a), setting policy objectives for the coming years with climate and environment as key pillars of the growth strategy. Key building blocks of the European Green Deal include the Farm to Fork Strategy that lays out the direction for the future food production system in Europe and the Zero Pollution Action Plan that aims at securing a healthy planet for healthy people.

This chapter will detail the European policies on nitrate control and management. It will also give some facts on the efficiency of the various policies in controlling the amount of nitrate present in the environment.

## 18.2  THE EU NITRATE POLLUTION PREVENTION REGULATIONS AND POLICIES

### 18.2.1  THE WATER FRAMEWORK DIRECTIVE

The Water Framework Directive WFD (EC 2000) was introduced to rationalize and streamline existing water legislation. As such two Directives were repealed in 2007 (Drinking Water Directive and Sampling Drinking Water Directive) and four more in 2013. The Directive requires Member States to prevent the deterioration and enhance the status of all water bodies with the aim of achieving good ecological and chemical status for all waters. The deadline for achieving this was originally set for 2015, but the deadline could be postponed if certain exceptions are invoked, until the maximum limit of 2027.

The WFD was a breakthrough in the field of water management by putting forward many novel aspects calling for a "sustainable approach to manage an essential resource," "holistic ecosystem protection," "integration of planning," the "polluter pays principle," and the "right geographical scale" (EC 2007). Indeed, the Directive requires the management to be performed at a river-basin level. By going beyond the administrative boundaries, the WFD led to a new reorganization of water management, where river basins became the central geographical unit of analysis.

The WFD also called for integration of other policies and addressing the management of water from a holistic point of view (Wiering et al. 2020). A leading principle of the Directive is that environmental problems should be tackled at the source, and to do so it relies on other specific Directives such as the Nitrates Directive for controlling nitrate from agricultural sources.

In the WFD, ecological status is evaluated using the abundance of aquatic life, nutrient availability, pollution by chemicals including nutrient conditions, and morphological characteristics. The ecological status is broken down into five categories, ranging from high status with low or no human impact, to bad status. The WFD does not provide nutrient concentration targets but requires the Member States to derive nutrient concentrations that can sustain sound and healthy ecosystems (Poikane et al. 2019). Analyzing the approaches used by the Member States, Poikane and colleagues (2019) reported that 20 countries used nitrate concentrations as one of the physical-parameters in the assessment of the ecological status, with values ranging from 5.6 mg N L-1 (25 mg L-1 NO3) and 11.3 mg N L-1 (50 mg L-1 NO3). Only six countries included nitrate concentration as criteria for ecological assessment in lakes. According to the European Environment Agency, in 2015 around 40 percent of surface water bodies were in "good ecological status" (EC 2019b). Groundwaters are in a better situation, with 74 percent in good chemical status and 89 percent in good quantitative status.

The assessment of the trend in ecological status is difficult, since the condition is measured according to several parameters that do not evolve uniformly. It has been shown that the main factors that have a negative influence on the status of water bodies include diffuse pollution from agriculture (including pesticides and nitrate), structural changes to water bodies (dams, impoundments, dredging, changes to the water level, etc.), and over-abstraction of water for irrigation or other uses (EEA 2018a). The contribution of nitrate from point sources is rather limited (See Malagó et al. 2019) for the river basins draining into the Mediterranean Sea).

In December 2019, the European Commission published its evaluation of the WFD, the so-called fitness check (EC Fitness Check 2019), concluding that the implementation of the directive had a fundamental role in protecting water resources, but some issues in the implementation and new challenges have emerged that need to be addressed. It appeared in particular that diffuse pollution remained a critical issue (Wiering et al. 2020).

## 18.2.2 Nitrates Directive

The Nitrates Directive (EC 1991b) complements the Water Framework Directive and is one of the key instruments in the protection of waters against agricultural pressures, nitrate in particular. The EU Nitrates Directive (ND) aims to protect water quality across Europe by preventing nitrate from agricultural sources polluting ground and surface waters (including rivers, lakes, coastal and transitional waters). To achieve these objectives Member States (MS) are required to identify waters affected by pollution or potentially affected if no action is taken – pollution refers to waters where nitrate concentration exceeds 50 mg NO3 L-1 or could be so if no measures are taken, and water bodies affected by eutrophication or that could be if no measures are taken. MS are required to designate Nitrate Vulnerable Zones (NVZ) which are the areas draining polluted waters where Action Programmes, including specific measures for minimizing pollution from nitrate, have to be implemented. Member States can also choose to apply measures to their whole territory rather than designating specific zones (whole territory approach). The specific measures focus on reducing the pollution at the source, and comprise:

- period of prohibition of fertilizer application to prevent nutrient losses to waters;
- restrictions for application on sloped soils, and on soaked, frozen, or snow-covered soils to prevent nitrate losses from leaching and runoff;
- restrictions for application near watercourses (buffer strips);
- requirements for effluent storage works and capacity of manure storage;
- rational fertilization (e.g. balanced fertilization corresponding to crops needs, splitting fertilization, limitations);
- crop rotation, permanent crop enhancement, and vegetation cover in rainy periods;
- fertilization plans, spreading records;
- limits on the total amount of livestock manure that may be applied to land.

The Directive requires MS to assess the effectiveness of the measures and report on their implementation every four years based on an extensive monitoring network. The report must include a detailed description of: (i) nitrate concentrations in groundwaters and surface waters; (ii) eutrophication of surface waters; (iii) summary of the Action Programme(s) on water quality and agricultural practices; (iv) revision of NVZs and Action Programme(s); and (v) an estimation of future trends in water quality.

The designated nitrate vulnerable zones (period 2012–2015) extended over an area of 2.17 km$^2$, representing nearly 61 percent of the agricultural area. The designated zones have expanded by 11 percent from the reporting period 2008–2011.

For the 2012–2015 period, the European Commission reported that 13.2 percent of the groundwater monitoring points exceed the 50 mg NO3 L-1 threshold (Figure 18.1). Concerning surface waters, 1.8 percent of the surface monitoring points exceed the 50 mg NO3 L-1 threshold (Figure 18.1). A higher trophic status (eutrophic or hypertrophic) was detected for 19 percent and 21 percent of the rivers and lakes monitoring points, respectively. Data for saline waters were reported for a limited number of countries. The data indicated that more than 50 percent of the saline monitoring stations were classified as eutrophic. It is important to stress that there is a large variability of data among and within countries. We aggregated the data based on the locations of the river basins outlet. The data by region is shown in Figure 18.2.

The highest nitrate concentrations are found for the monitoring stations located in river basins draining into the North Sea, both for surface and groundwaters. The highest percentage of freshwater monitoring stations in the eutrophic classes are again located in river basins draining into the North Sea. For saline water, the large majority of the stations are classified as eutrophic in the Baltic Sea, while all monitoring points are eutrophic in the North and Black Seas. This clearly highlights the special case of the Baltic Sea where the nutrient loads discharged into the Sea were significantly reduced, however without solving the trophic problem of the Baltic.

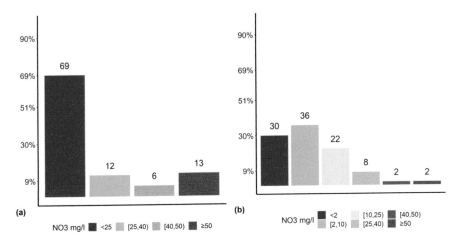

**FIGURE 18.1** Surface waters (b) in the period 2012–2015.

Figure 18.3 and Figure 18.4 display the spatial distribution of average annual $NO_3$ concentration by administrative region (NUTS3) in groundwater and surface water, respectively.

### 18.2.3 URBAN WASTEWATER TREATMENT DIRECTIVE

The Urban Waste Water Treatment Directive (EC 1991b) was adopted to protect the environment from discharges of wastewater from households and certain industries (e.g. food processing) with characteristics similar to those of domestic waters. The Directive requires that all agglomerations with a 2,000 population equivalent (p.e.) and higher are equipped with collecting systems and that all wastewater discharged be subject to at least secondary treatment. The 2019 REFIT Evaluation (REFIT 2019, Pistocchi et al. 2019) found that the Directive is overall effective and efficient, but there is room for improvement in order to address those sources of pollution not properly covered so far. These include combined sewer overflows, urban runoff, individual systems, and small settlements. In addition, there is a need to modernize the Directive to ensure that it appropriately deals with problems such as contaminants of emerging concern (e.g. pharmaceuticals and microplastics) and climate change, and is adjusted to the latest technological progress (EC 2020d).

To further protect endangered waters, the Directive requires MS to designate Sensitive Areas (SA), which include (i) freshwater bodies, estuaries, and eutrophic coastal waters, or which may become eutrophic if protective action is not taken; (ii) surface freshwaters used for drinking that contain or are likely to contain more than mg NO3 L-1; (iii) and areas where further treatment is necessary to comply with other Council Directives.

Sensitive areas may therefore be designated due to a need to control pollution from Nitrogen (N), Phosphorus (P), or for other reasons. The Directive requires that wastewater discharged into these sensitive areas undergo a more stringent treatment

**a) Groundwater**

| Region | Percentage of stations by classes of $NO_3$ concentration | | | |
|---|---|---|---|---|
| | <25mg/l | [25,40] mg/l | [40,50] mg/l | ≥50mg/l |
| Arctic Ocean | 100 | 0 | 0 | 0 |
| Baltic Sea | 84.2 | 6.5 | 3.1 | 6.2 |
| Black Sea | 75 | 10.1 | 4.6 | 10.3 |
| Caribbean Sea | 91.7 | 8.3 | 0 | 0 |
| Indian Ocean | 86.9 | 10.5 | 2.6 | 0 |
| Mediterranean Sea | 66.4 | 13.3 | 5.7 | 15.5 |
| North Atlantic Sea | 59.7 | 18.5 | 7.7 | 14.1 |
| North Sea | 62.1 | 14.1 | 7.8 | 16 |

**b) Surface water**

| Region | Percentage of stations by classes of $NO_3$ concentration | | | | | |
|---|---|---|---|---|---|---|
| | <2mg/l | [2,10) mg/l | [10,25) mg/l | [25,40) mg/l | [40,50) mg/l | ≥50mg/l |
| Arctic Ocean | 100 | 0 | 0 | 0 | 0 | 0 |
| Baltic Sea | 84.2 | 6.5 | 0 | 3.1 | 0 | 6.2 |
| Black Sea | 75 | 10.1 | 0 | 4.6 | 0 | 10.3 |
| Caribbean Sea | 91.7 | 8.3 | 0 | 0 | 0 | 0 |
| Indian Ocean | 86.9 | 10.5 | 0 | 2.6 | 0 | 0 |
| Mediterranean Sea | 66.4 | 13.3 | 0 | 5.7 | 0 | 15.5 |
| North Atlantic Sea | 59.7 | 18.5 | 0 | 7.7 | 0 | 14.1 |
| North Sea | 62.1 | 14.1 | 0 | 7.8 | 0 | 16 |

**c) Freshwater**

| Region | Percentage of stations by trophic status | | |
|---|---|---|---|
| | Eutrophic | Could become eutrophic | Non eutrophic |
| Arctic Ocean | 0 | 0 | 100 |
| Baltic Sea | 24.5 | 27.9 | 47.6 |
| Mediterranean Sea | 20.2 | 35.1 | 44.7 |
| North Atlantic Sea | 12.2 | 24.3 | 63.5 |
| North Sea | 18.9 | 27.4 | 53.7 |

**d) Saline waters**

| Region | Percentage of stations by trophic status | | |
|---|---|---|---|
| | Eutrophic | Could become eutrophic | Non eutrophic |
| Arctic Ocean | 69.1 | 28.7 | 2.2 |
| Baltic Sea | 100 | 0 | 0 |
| Mediterranean Sea | 19.8 | 17 | 63.2 |
| North Atlantic Sea | 13 | 21.9 | 65.1 |
| North Sea | 100 | 0 | 0 |

**FIGURE 18.2**   Distribution of the monitoring points by classes of concentration/trophic status and association to a sea.

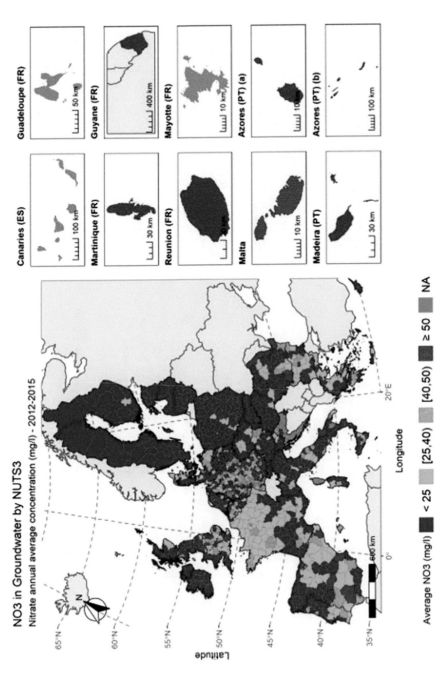

**FIGURE 18.3** Map of average NO3 annual concentration in groundwater (period 2012–2015) by NUTS3 (NUTS-2013 version).

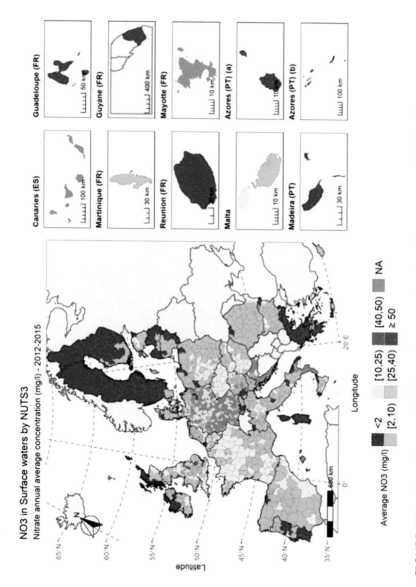

**FIGURE 18.4**  Map of average NO3 annual concentration in surface waters (period 2012–2015) by NUTS3 (NUTS-2013 version).

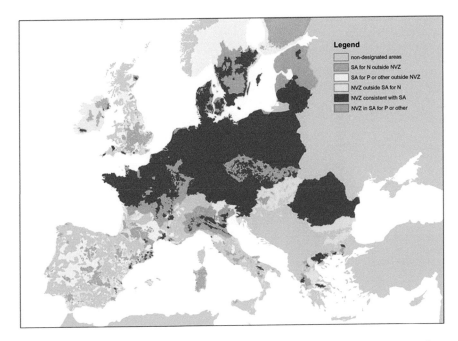

**Legend**
- non-designated areas
- SA for N outside NVZ
- SA for P or other outside NVZ
- NVZ outside SA for N
- NVZ consistent with SA
- NVZ in SA for P or other

**FIGURE 18.5** Comparison of sensitive areas SA (EEA 2018b) and the most recent nitrate vulnerable zones (NVZ) for the period 2016–2019, including drafted zones in Spain. SA for Croatia and Bulgaria were not available and are therefore ignored in this representation.

than the secondary level. It is still unclear whether the designation of sensitive areas is efficient and effective in controlling nutrient loads from urban wastewater. At present, the designated sensitive areas (SA) are not fully consistent with the vulnerable zones identified under the Nitrates Directive (NVZ).

Figure 18.5 shows a comparison between the two zones, highlighting regions where NVZ are designated outside SA for N. These NVZ should be potentially designated also as SA for N. Likewise, there are regions where SA for N are designated outside of NVZ, representing areas requiring a potential designation as NVZ.

It should be stressed that both the Urban Waste Water Treatment Directive (UWWTD) and the Nitrogen Directive (ND) leave EU countries the freedom to avoid designating SA or NVZ if they decide to apply the provisions for these areas to the whole of their territory. While this choice causes heavier overall requirements for the treatment of wastewater and the management of fertilizers, it may also enable a simpler and more homogeneous administrative approach at country level and a better level playing field for the economic operators within the country.

We quantified the implications of different ways of designating SA with respect to NVZ. To this end, we consider the current fleet of wastewater treatment plants in Europe (EEA 2020). Table 18.1 shows the number of plants, and their cumulative capacity in population equivalents (p.e.), in the different types of designated areas shown in Figure 18.5. The same information is also reported for plants above a

**TABLE 18.1**
**Analysis of the Number of Plants and their Cumulative Capacity in Population Equivalents (p.e.) in the Different Combinations of SA and NVZ**

| Combination | NVZ | SA | label | Area farmland (CLC18=12, 13, 18 or 19) (ha) | Number of plants | Capacity (p.e.) | Number of plants > 10,000 (p.e.) | Capacity (p.e.) only plants > 10,000 (p.e.) |
|---|---|---|---|---|---|---|---|---|
| 1 | no | no | non-designated areas | 45220948 | 5294 | 1.62E+08 | 1697 | 150505072 |
| 2 | no | yes, for N or N&P | SA for N not in NVZ | 11955086 | 3009 | 7.16E+07 | 835 | 64040538 |
| 3 | no | yes, for other (including P) | SA For P or other not NVZ | 8073039 | 2008 | 6.25E+07 | 672 | 57895156 |
| 4 | yes | no | NVZ outside SA | 16254949 | 2091 | 7.00E+07 | 813 | 64974254 |
| 5 | yes | yes, for N or N&P | SA coinciding with NVZ | 81654414 | 11725 | 3.83E+08 | 5024 | 352406835 |
| 6 | yes | yes, for other (including P) | NVZ on SA for P or other | 12873669 | 1565 | 5.20E+07 | 666 | 47824244 |

capacity of 10,000 PE, i.e. those subject to a requirement of nutrient removal in SA according to the UWWTD.

We explore the following scenarios:

- SA for N are fully aligned with NVZ (i.e. if a SA for N is not an NVZ, it becomes such, and the other way around).
- A "whole territory" approach is adopted for both the UWWTD and the ND in all countries.
- All plants are upgraded to N removal in all SA (irrespective of whether these are designated for N, P, or another).
- NVZ are fully aligned with SA for N as well as SA for P and others (i.e., if an area is a SA for P or other, it also becomes an NVZ).

The results of this assessment are reported in Table 18.2. Strategies pursuing an alignment of NVZ and SA would require upgrading to N removal for about 3,700 plants including 1,500 plants exceeding 10,000 p.e. with a cumulative capacity of slightly more than 120 million p.e. (Scenario 1). The number of plants would be less than a half if we limited the upgrade to plants of capacity > 10,000 PE, but the cumulative capacity to upgrade would not change much (about 113 additional p.e.). Similar efforts correspond to the requirement of N removal also in SA currently designated for P or other.

This corresponds to an increase of about 25 percent on the current fleet of WWTPs providing N removal. Much heavier would be the effort to adopt a "whole territory" approach everywhere, an upgrade being required for more than 10,000 plants (or close to 4,000 if limited to capacity > 10,000 p.e.) with a cumulative capacity of more than 300 million PE. This corresponds to an increase of about 70 percent on the current fleet.

It is also useful to evaluate the additional extent of the agricultural area that would be subject to the provisions of the ND (constraints on the management of fertilizers) under the different scenarios. Starting from the baseline distribution of the total extent of arable land (Corine land cover classes 12, 13, 18, and 19 in 2018; ETC/ULS 2019) we can estimate that aligning all the NVZ to the SA for N would constrain about 120,000 additional km2 of arable land (about 10% of the currently constrained farmland). If we aligned NVZ to all SA for N, P or another, this extent would almost double. A "whole territory" approach scenario would constrain more than 650,000 additional km2 (about 60% of the currently constrained farmland). The costs entailed by these additional requirements should be considered in relation to the benefits they could bring in terms of reducing N pollution. While some countries are already adopting a "whole country" approach, others might face a challenge due to the need for investments for the upgrading of plants. The costs associated to constraining the whole territory might be lower in the case of the ND, as requirements to follow good agricultural practice may already exist in many countries, also outside the NVZ.

### 18.2.4 THE MARINE STRATEGY FRAMEWORK DIRECTIVE

The Marine Strategy Framework Directive (EC 2008) aims to protect the marine environment, prevent its deterioration or, where practicable, restore marine environments

**TABLE 18.2**
**Number of Plants Requiring Nitrogen Removal and their Cumulative Capacity, under Different Scenarios of Designation of NVZ/SA**

| scenario | description | plants to upgrade | plants to upgrade >10,000 PE | % change from baseline | % change from baseline >10,000 PE | cumulative capacity (PE) | cumulative capacity (PE) 10,000 PE | % change from baseline | % change from baseline >10,000 PE | new farmland to constrain (ha) | % change from baseline |
|---|---|---|---|---|---|---|---|---|---|---|---|
| #1 | SA due to N fully aligned with NVZ | 3656 | 1479 | 24.8% | 25.2% | 1.22E+08 | 1.13E+08 | 26.8% | 27.1% | 11955086 | 10.8% |
| #2 | "whole territory" approach | 10958 | 3848 | 74.4% | 65.7% | 3.46E+08 | 3.21E+08 | 76.2% | 77.1% | 65249073 | 58.9% |
| #3 | N, P removal in all SA | 3573 | 1338 | 24.3% | 22.8% | 1.15E+08 | 1.06E+08 | 25.2% | 25.4% | 0 | 0.0% |
| #4 | NVZ fully aligned with SA due to N, P, other | 0 | 0 | 0.0% | 0.0% | 0 | 0 | 0.0% | 0.0% | 20028125 | 18.1% |

in areas where they have been adversely affected. The implementation is ecosystem-based with four marine regions to be considered: the Baltic, the Northeast Atlantic Ocean, the Mediterranean Sea, and the Black Sea. Each Member State shall develop a "Marine Strategy" to achieve or maintain Good Environmental Status (GES) of their waters within that region by 2020. The implementation. similar to the WFD, includes the analysis of the "essential characteristics and current environmental status," and an analysis of the predominant pressures and impacts, which include, among others, nutrient enrichment coming from direct discharges from point sources and/or losses from diffuse sources, including agriculture and atmospheric deposition.

The Marine Directive complements also existing marine conventions, including OSPAR, HELCOM, Black Sea, and Barcelona. OSPAR is the convention for the protection of the North Sea and has been active since 1998. It has five domains of activity, including eutrophication. The strategy of OSPAR is to maintain a healthy marine environment with no eutrophication. This strategy relies on the commitment of contracting parties to reduce the load of nitrogen and phosphorus by 50 percent, compared to 1985, in sensitive areas where pollution exists or could occur. A significant decrease of total nitrogen loads was observed for rivers discharging into the Greater North Sea and Celtic Seas since 1997 (OSPAR 2018). The Helsinki Commission (HELCOM) aims, since 1974, at protecting the Baltic Sea from inland, airborne, and sea pollution. The Baltic Sea is still suffering from a severe eutrophication problem due to nutrient enrichment (HELCOM 2018). About 97 percent of the sea area is affected by eutrophication, mainly due to excessive nutrient input from land-based sources. Despite the decrease of riverine inputs of nutrients, the impact on the sea environment is yet to be seen (HELCOM 2018). The Mediterranean Action Programme (MAP) was established in 1975 as the institutional framework for cooperation in addressing marine environmental degradation in the Mediterranean Sea. It was followed by the adoption of the Convention for the Protection of the Marine Environment and the Coastal Region of the Mediterranean (Barcelona Convention). Contracting parties committed to combat pollution from land-based sources, including nutrients.

### 18.2.5 The CAP

Launched in 1962, the EU's common agricultural policy (CAP) is a partnership between agriculture and society, and between Europe and its farmers. The CAP is a common policy for all EU countries, and it aims to support farmers and improve agricultural productivity, ensuring a stable supply of affordable food; to safeguard European Union farmers to make a reasonable living; to help tackle climate change and the sustainable management of natural resources; to maintain rural areas and landscapes across the EU; to keep the rural economy alive by promoting jobs in farming, agri-foods industries, and associated sectors.

Under the 2014–2020 CAP there are 13 statutory management requirements (SMRs) and 7 good agricultural environmental conditions (GAECs). The GAECs include the purpose of maintaining agricultural activities, avoiding the abandonment of agricultural land, and sustaining the environment – in particular regarding the soils (Angileri et al. 2011). More specifically, the GAEC represent a piece of legislation that farmers shall be compliant within order to receive full CAP payments.

On 1 June 2018, the European Commission presented the legislative proposals on the future of the CAP for the period after 2020. This new CAP will introduce a new way of working to modernize and simplify the EU's policy on agriculture. It is based on a more-flexible, performance, and results-based approach that takes into account local conditions and needs while increasing the European Union's ambitions in terms of sustainability. It is proposed that all EU farmers will have to implement an obligatory nutrient management tool – "Farm Sustainability Tool for Nutrients" (FaST) – that would contribute to reducing ammonia and N2O emissions from the agriculture sector and improve water quality in Europe. Such a tool would compile information from satellite data, soil sampling, and land-parcel information and would be directly accessible to farmers in order to help them make informed decisions on nutrient requirements.

Member States will implement the future CAP with the so-called "CAP strategic plan" at the national level. These plans will combine a wide range of targeted interventions addressing the Member States' specific needs and delivering tangible results in relation to EU-level objectives while contributing to the environment ambition laid down in the Green Deal (see paragraph 18.2.7). Member States thus produce a thorough assessment of their needs based on a Strengths, Weaknesses, Opportunities and Threats analysis (SWOT) of their territory and agri-food sector.

### 18.2.6 The Soil Thematic Strategy

The Soil Thematic Strategy (EC 2006) is the main EU policy strategy directed at soil protection. The European Union and most EU Member States do not have specific legislation targeting soils, but instead, aspects of soil protection are determined by other sectoral policies such as agriculture, forestry, water, waste, and land-use planning. The Thematic Strategy for Soil Protection was adopted by the European Commission in September 2006 and consists of a Communication from the Commission to the other European Institutions, a proposal for a European framework Directive, and an Impact Assessment. Even if an adequate EU legislation for soil protection was missing, the Soil Framework Directive was not approved, blocked by a minority of five countries, and the proposed directive was withdrawn in 2014 (Panagos and Montanarella 2018). Therefore, only the Soil Thematic Strategy remains in place.

The Soil Thematic Strategy sought to change this by establishing actions at EU level: integration of soil protection aspects in other sectoral policies, development of the knowledge base through studies and research projects, and raising public awareness about the role that soil plays in the economy and the ecosystem (Ronchi et al. 2019). Soil diffuse pollution is among the threats described in the Soil Thematic Strategy. The main strength of the Soil Thematic Strategy is the call for stronger integrated policy and additional research related to enhanced soil protection in the EU. The most important weakness is the non-binding nature of the proposed measures to reduce the negative impact of soil threats and enhance soil functions. In this policy framework, the EU monitors the nutrients in soils with the soil survey LUCAS. Soil Nitrogen (N), Phosphorus (P), and Potassium (K) are assessed for the first time in the EU using measured data for about 22,000 sampled locations with the LUCAS 2009/12 and LUCAS 2015 campaign (Orgiazzi et al. 2018). Table 18.3 provides the

**TABLE 18.3**
**Mean Nitrogen (N) Concentration for All Lands, Arable Land and Grassland based on LUCAS Topsoil Survey**

| Country | N in soils g/kg | | |
|---|---|---|---|
| | *All land* | Arable land | Grassland |
| Austria | 3.11 | 1.96 | 3.62 |
| Belgium | 2.92 | 1.36 | 3.53 |
| Bulgaria | 1.47 | 1.30 | 1.59 |
| Cyprus | 1.31 | 1.05 | 1.38 |
| Czech Republic | 2.14 | 1.79 | 2.13 |
| Denmark | 2.23 | 1.95 | 3.29 |
| Estonia | 5.72 | 2.10 | 4.73 |
| Finland | 5.10 | 3.64 | 4.00 |
| France | 2.34 | 1.80 | 3.11 |
| Germany | 2.46 | 1.69 | 3.84 |
| Greece | 1.59 | 1.17 | 1.61 |
| Hungary | 1.94 | 1.75 | 2.55 |
| Ireland | 7.13 | 2.64 | 6.17 |
| Italy | 1.90 | 1.60 | 2.18 |
| Latvia | 3.76 | 2.26 | 2.78 |
| Lithuania | 2.96 | 1.84 | 2.52 |
| Luxembourg | 2.67 | 2.80 | |
| Malta | 2.02 | 2.10 | 2.03 |
| Netherlands | 2.29 | 1.69 | 3.19 |
| Poland | 1.93 | 1.29 | 3.33 |
| Portugal | 1.66 | 1.39 | 1.74 |
| Romania | 1.95 | 1.60 | 2.28 |
| Slovakia | 2.33 | 1.82 | 2.49 |
| Slovenia | 4.00 | 2.01 | 3.90 |
| Spain | 1.56 | 1.26 | 2.22 |
| Sweden | 5.02 | 2.52 | 3.95 |
| UK | 3.80 | 2.43 | 4.48 |
| **Grand Total** | **2.78** | **1.70** | **3.12** |

mean nitrogen values for all lands, arable lands, and grasslands. After the quality check, the LUCAS topsoil database included 21,682 samples; the soil samples in arable lands are about 8,675 (40%) and the ones in grassland are 4,661 (21.5%). It interesting to see that the grasslands have almost double values in nitrogen compared to arable lands.

The distribution of topsoil nitrogen is highly correlated with soil organic carbon, given that nitrogen is a main component of soil organic matter (Ballabio et al. 2019). While the Carbon/Nitrogen (C/N) ratio can vary, some carbon-rich soils are also nitrogen-rich, at least in terms of absolute quantities. On top of this, the vegetation cover and climate also influence nitrogen distribution. Therefore, forests and grasslands areas tend to have higher nitrogen content compared to arable lands. Forests of

Scandinavia, or those of the mountain areas, are clearly outlined by the map (Ballabio et al. 2019). Climate also acts as a main driving force influencing nitrogen content along the Atlantic area; in particular, the United Kingdom and Ireland show higher N concentrations due to a fresh and humid climate, which favors organic matter accumulation. Soil texture also plays a role in stabilizing organic matter and thus nitrogen. Areas with coarser soils, such as most of Poland, tend to have less nitrogen even if other conditions are favorable (e.g. vegetation, climate).

While the nitrogen concentration is relevant for assessing stocks and potential $N_2O$ emissions, the ratio between carbon and nitrogen (C/N) can better represent the differences in the organic matter composition. Where higher C/N rates correspond to more oligotrophic soils typical of coniferous forests, or to peatland soils, lower C/N rates are typical of more balanced nutrient-rich soils.

### 18.2.7 GREEN DEAL/FARM 2 FORK/ZERO POLLUTION

The European Green Deal launches a new growth strategy for the EU that responds to the challenges posed by climate change and environmental degradation. It sets policy initiatives aiming at climate neutrality for Europe in 2050 and to protect human life, animals, and plants, by reducing pollution, notably through the Biodiversity and Farm to Fork strategies. The European Green Deal has the ambition to make the European Union the first climate-neutral continent by 2050, responding to the escalating climate crisis (Haines and Scheelbeek 2020). The Biodiversity Strategy 2030 (EC 2020b), the Farm to Fork Strategy (EC 2020c), the European Climate law, and the upcoming zero pollution action plan are parts of the EU Green Deal.

The Biodiversity Strategy for 2030 is a comprehensive, ambitious, and long-term plan to protect nature and reverse the degradation of ecosystems, while the Farm to Fork strategy specifically addresses the challenges of sustainable food systems. They both set an ambitious objective of reducing nutrient pollution by 50 percent in 2030, while ensuring that there is no deterioration in soil fertility. This will reduce the use of fertilizers by at least 20 percent and will be achieved by implementing and enforcing the relevant environmental and climate legislation in full, by identifying with the Member States the nutrient load reductions needed to achieve these goals, applying balanced fertilization and sustainable nutrient management, and by managing nitrogen and phosphorus better throughout their life cycle. The Biodiversity Strategy 2030 has envisaged an update of the EU Soil Thematic Strategy in 2021. Soil health will benefit from ambitious objectives to be reached by 2030 within the EU Green Deal: 50 percent reduction of pesticides, 50 percent decrease of excess nutrients, 20 percent fertilizer reduction, organic farming at 25 percent of agricultural lands, 10 percent increase of landscape features, increase of protected lands at 30 percent, wetlands restoration, and halting land degradation (Montanarella and Panagos 2021).

To do so, the Commission will develop with the Member States an integrated nutrient management action plan, which was already announced in the Circular Economy Action Plan (EC 2020a). The aim is to ensure a more sustainable application of nutrients, address nutrient pollution at the source, stimulate the markets for recovery of nutrients, and increase the sustainability of the livestock sector. This could be achieved by a more systematic application of precise fertilization techniques and sustainable agricultural practices, notably in hotspot areas of intensive livestock farming. This will be done by means of measures that Member States will include in their CAP Strategic Plans, such as the Farm Sustainability Tool for nutrient management, investments, advisory services, and EU space technologies (Copernicus, Galileo). Stimulating the market for recovered nutrients could also boost the recycling of organic waste into renewable fertilizers, in line with the circular economy principles.

## REFERENCES

Angileri, V., Loudjani, P. and Serafini, F. 2011. GAEC implementation in the European Union: Situation and perspectives. *Ital. J. Agron.* 6 6–9 Online: www.agronomy.it/index.php/agro/article/view/ija.2011.6.s1.e2.

Ballabio, C., Lugato, E., Fernández-Ugalde, O., Orgiazzi, A., Jones, A., Borrelli, P., Montanarella, L. and Panagos, P. 2019. Mapping LUCAS topsoil chemical properties at European scale using Gaussian process regression. *Geoderma.* 355 113912.

EC 1991a Council Directive 91/271/EEC of 21 May 1991 concerning urban waste-water treatment Online: https://eur-lex.europa.eu/legal-content/EN/TXT/?uri=CELEX:31991L0271.

EC 1991b Council of the European Communities CEC, 1991. Council Directive 91/676/EEC concerning the protection of waters against pollution caused by nitrates from agricultural sources Off. J. L 375. Online: https://eur-lex.europa.eu/legal-content/EN/TXT/PDF/?uri=CELEX:31991L0676&from=EN.

EC 2000 Establishing a framework for community action in the field of water policy. Directive 2000/60/EC of the European Parliament and of the Council of 23 October 2000 Off. *J. Eur. Communities*, Brussels Online: https://eur-lex.europa.eu/resource.html?uri=cellar:5c835afb-2ec6-4577-bdf8-756d3d694eeb.0004.02/DOC_1&format=PDF.

EC 2006 Communication from the Commission (COM/2006/231). The Soil Thematic Strategy Online: https://eur-lex.europa.eu/legal-content/EN/TXT/PDF/?uri=CELEX:52006DC0231&from=EN.

EC 2007 *Towards Sustainable Water Management in the European Union.* First stage in the implementation of the Water Framework Directive 2000/60/EC-[SEC(2007) 362] [SEC(2007) 363].

EC 2008 Directive 2008/56/EC of the European Parliament and of the Council of 17 June 2008 establishing a framework for community action in the field of marine environmental policy (Marine Strategy Framework Directive) Online: https://eur-lex.europa.eu/legal-content/EN/TXT/?uri=CELEX%3A32008L0056.

EC 2018 Implementation of Council Directive 91/676/EEC concerning the protection of waters against pollution caused by nitrates from agricultural sources based on Member State reports for the period 2012-2015. Communication and Staff Working Documents Online: https://ec.europa.eu/environment/water/water-nitrates/index_en.html.

EC Fitness Check 2019. Fitness Check of the Water Framework Directive, Groundwater Directive, Environmental Quality Standards Directive and Floods Directive 2000/60/EC.

EC 2019a Communication from the Commission (COM/2019/640 final) The European Green Deal Online: https://eur-lex.europa.eu/legal-content/EN/TXT/?qid=1588580774040&uri=CELEX%3A52019DC0640.

EC 2019b Implementation report: *Water Framework Directive and Floods Directive Online*: https://ec.europa.eu/info/news/implementation-report-water-framework-directive-and-floods-directive-questions-and-answers-2019-feb-26_en.

EC 2020a Communication from the Commission (COM/2020/98 final) A new Circular Economy Action Plan for a Cleaner and More Competitive Europe Online: https://eur-lex.europa.eu/legal-content/EN/TXT/?qid=1583933814386&uri=COM:2020:98:FIN.

EC 2020b Communication from the Commission (COM/2020/380 final) EU Biodiversity Strategy for 2030 – Bringing nature back into our lives Online: https://eur-lex.europa.eu/legal-content/EN/TXT/?qid=1590574123338&uri=CELEX%3A52020DC0380.

EC 2020c Communication from the Commission (COM/2020/381 final). A Farm to Fork Strategy for a fair, healthy and environmentally friendly food system Online: https://eur-lex.europa.eu/legal-content/EN/TXT/?uri=CELEX:52020DC0381.

EC 2020d Urban Waste Water: 10th Report on Implementation. Online: https://ec.europa.eu/commission/presscorner/detail/en/QANDA_20_1562

EEA 2018a *European Waters – Assessment of Status and Pressures 2018*. European Environment Agency Online: www.eea.europa.eu/publications/state-of-water.

EEA 2018b Waterbase – UWWTD: Urban Waste Water Treatment Directive – v8, 2018 Online: www.eea.europa.eu/data-and-maps/data/waterbase-uwwtd-urban-waste-water-treatment-directive-7.

EEA 2020 *Urban Waste Water Treatment in Europe*. European Environment Agency Online: www.eea.europa.eu/data-and-maps/indicators/urban-waste-water-treatment/urban-waste-water-treatment-assessment-5.

ENA 2011 *European Nitrogen Assessment* (ENA) | Nitrogen in Europe Online: www.nine-esf.org/node/360/ENA-Book.html.

ETC/ULS 2019 *Updated CLC Illustrated Nomenclature Guidelines* Online: https://land.copernicus.eu/user-corner/technical-library/corine-land-cover-nomenclature-guidelines/docs/pdf/CLC2018_Nomenclature_illustrated_guide_20190510.pdf.

Haines, A. and Scheelbeek, P. 2020. European Green Deal: a major opportunity for health improvement *Lancet* 395 1327–9 Online: https://pubmed.ncbi.nlm.nih.gov/32394894/.

HELCOM 2018 *State of the Baltic Sea*. Second HELCOM holistic assessment 2011–2016 (Baltic Sea Environment Proceedings 155) Online: https://helcom.fi/baltic-sea-trends/holistic-assessments/state-of-the-baltic-sea-2018/.

Malagó, A., Bouraoui, F., Grizzetti, B., and De Roo, A. 2019. *Modelling nutrient fluxes into the Mediterranean Sea. J. Hydrol. Reg. Stud.* 22: 100592.

Montanarella, L. and Panagos, P. 2021. The relevance of sustainable soil management within the European Green Deal Land Use Policy. 100: 104950.

Orgiazzi, A., Ballabio, C., Panagos, P., Jones, A., and Fernández-Ugalde, O. 2018. LUCAS Soil, the largest expandable soil dataset for Europe: A review *Eur. J. Soil Sci.* 69: 140–53 Online: http://doi.wiley.com/10.1111/ejss.12499.

OSPAR 2018 *Status Report on the OSPAR Network of Marine Protected Areas*.

Panagos, P., and Montanarella, L. 2018. Soil thematic strategy: An important contribution to policy support, research, data development and raising the awareness *Curr. Opin. Environ. Sci. Heal.* 5, 38–41.

Pistocchi, A., Dorati, C., Grizzetti, B., Udias Moinelo, A., Vigiak, O., and Zanni, M. 2019. *Water Quality in Europe: Effects of the Urban Wastewater Treatment Directive Online*: https://ec.europa.eu/jrc/en/publication/eur-scientific-and-technical-research-reports/water-quality-europe-effects-urban-wastewater-treatment-directive.

Poikane, S., Kelly, M.G., Salas Herrero, F., Pitt. J.A., Jarvie, H.P., Claussen, U., Leujak, W., Lyche Solheim, A., Teixeira, H. and Phillips, G. 2019. Nutrient criteria for surface waters under the European Water Framework Directive: Current state-of-the-art, challenges and future outlook. *Sci. Total Environ.* 695: 133888.

REFIT .2019. Commission Staff Working Document Evaluation of the Council Directive 91/ 271/EEC of 21 May 1991, concerning urban waste-water treatment Online: https:// ec.europa.eu/environment/water/water-urbanwaste/pdf/UWWTD   Evaluation   SWD 448-701 web.pdf.

Ronchi, S., Salata, S., Arcidiacono, A., Piroli, E., and Montanarella. L. 2019. Policy instruments for soil protection among the EU member states. *A Comparative Analysis Land Use Policy*. 82, 763–780.

Smil, V. 2001. *Cycles of Life: Civilization and the Biosphere* (Scientific American Library, New York) Online: http://faculty.washington.edu/stevehar/smil_ch.1.PDF.

Wiering, M., Boezeman, D. and Crabbé, A. 2020. *The Water Framework Directive and Agricultural Diffuse Pollution: Fighting a Running Battle?* Water 12 1447 Online: www. mdpi.com/2073-4441/12/5/1447.

# Index

**A**

Ammonium
  emissions 108
  in fertilizers 29
  mineralization 12
  in soils 27
  transformation 7, 108

**D**

Diabetes 328
Eutrophication 112, 223

**G**

Glycemia 328

**H**

Hemoglobin 47, 86, 188
Hypoxia 37, 113, 197, 200

**I**

Insulin 310, 328

**M**

Methemoglobinemia 347
  acquired 350
  congenital 349

**N**

Nitrate
  antimicrobial activity 292, 294
  assimilation 14, 16, 77
  biogeochemistry 4
  circulatory effects 388
  definition 304
  determination in waters 241, 242, 244
  dietary exposure 266
  dietary metabolism
  dietary supplement 314
  estimation in real time 141
  guidelines 187
  human health
    arterial stiffness 383
    blood pressure. 283, 309, 310, 314, 316, 317,
      318, 319, 320, 321, 322, 324, 329, 359,
      362, 363, 364, 368
    cardioprotection 314, 322, 323,
      328, 329
    cardiovascular health 381
    diabetes 328, 384, 387, 391
    endothelian function 382

    erectile dysfunction 325
    exercise performance 383
    glycemia 328, 387
    insulin resistance 328, 397
    ischemia 382
    methemoglobinemia
    platelet function 383
  in foods
    animal based foods 267
    dietary supplements 314
    food additives 313
    plant foods
      beets 316
      red spinach 317
      vegetables 267
  in human body 303
  in plants 74
  in soils 11
  in waters 185
    in St. Hellas, Greece 209
    canal waters 225
    groundwater 228
    lake waters 224
    river waters 219
  management 137
  metabolism in humans 360
  pollution 85, 186, 187, 188, 190
  potential harm 306
    cancer 390
    methemoglobinemia 390
    pregnancy outcomes 390
    thyroid disease 390
    toxicity from ingestion 319
  potential health benefits 309
    aerobic exercise performance 325
    cardioprotection 322
    diabetes 328
    erectile dysfunction 325
    glycemia 328
    improved blood flow 320
    improved cognition 324
    increased NO 319
    insulin resistance 328
    reduction in blood pressure 321
  potential risks 359
  properties of molecule 23
  reduction
    assimilatory 34
    dissimilatory 36
  regulations

green deal/farm 2 fork/zero pollution 420
common agricultural policy 417
marine strategy framework directive 409
nitrate directive 407
soil thematic strategy 418
water framework directive 406
urban water treatment directive 409
remobilization 97
sources of nitrate in groundwater 86
storage in plants organs 82
threshold value 407
toxicity to aquatic life 188
transport in agricultural systems 45
transport in plants 92
transport processes in groundwaters 58
transporters 81
uptake mechanisms 75
Nitric oxide
emissions 110
nitrate link 83
Nitrification 13, 27, 75, 92
Nitrite
formation of nitrosamines 307
function 310
in food chain 308, 309, 313
in fresh water 186, 196
in human body 308
in soils 107, 305
in plants 315
in waters 219, 224, 225, 228
quality standards 219
Nitrobacter 13, 14, 27, 29, 186
Nitrogen
assimilation 79
balance in mammals 376
biological fixation 92
budget 31
compounds 14
concentration in topsoils 419
cycle in soils 11
depolymerization mineralization 12
nitrification 13
denitrification 18
ecological impact 120
estimation

biosensors 141, 244
in field systems 166
in plants 167, 168, 169, 171
fertilizers 103
fixation 11, 80
human cycle 365
inhibitors 124
in plants 85
in soils 419
use efficiency 84
Nitrogen oxide
cardiovasular diseases 384
pathway 386
Nitrosamines 307, 310, 313, 319, 359, 362,
378, 380

**S**
Sensors
biosensors 244, 245, 254
electromagnetic sensors 248
plant sensors 166, 167, 168, 169
soil sensors 139, 141
Soil
acidification 14, 115–17, 121
degradation 32, 36, 117
erosion 34, 104, 116–18, 125
management 36
properties' variability 138, 139, 142
salinization 117
Syndrome
blue baby 235, 306, 351, 360

**W**
Water
analysis 235
canal waters 216, 225, 227, 228
groundwater 210, 211, 214, 215, 216, 217, 228,
229, 230, 231, 232
lake waters 224, 225, 241, 244, 251
river waters 211, 216, 219, 242, 247
quality 46, 47, 53, 59, 60, 61, 62, 86, 187,
210, 211, 217, 219, 253, 254, 405, 407,
408, 418
pollution 36, 117, 210, 211
salinity 228